Arid Lands Management

Arid Lands Management

Toward Ecological Sustainability

Thomas W. Hoekstra and Moshe Shachak

Technical Editors

Published by the

University of Illinois Press • Urbana and Chicago

in cooperation with the International Arid Lands Consortium

Publication of this book was supported by a grant from the International Arid Lands Consortium.

Manufactured in the United States of America
P 5 4 3 2 1

This book is printed on acid-free paper.

Library of Congress Cataloging-in-Publication Data
Arid lands management: toward ecological sustainability / technical editors, Thomas W. Hoekstra and Moshe Shachak.
p. cm.
Includes bibliographical references and index.
ISBN 0-252-06717-7 (pbk.: acid-free paper)
1. Arid regions ecology.
2. Ecosystem management.
3. Arid regions—Management.
I. Hoekstra, T. W.
II. Shachak, Moshe.
QH541.5.A74A75 1999
577.54—ddc21 98-8908
CIP

Contents

Foreword

David Nahmias and Kenneth E. Foster

If I can see any national problem that I shall tackle, it doesn't have a national answer but a regional answer. Water, markets, economy, deserts: we have to deal with nature, not with politics.
—Shimon Peres, Prime Minister, State of Israel

Prime Minister Peres's observation, made in his remarks to those attending the conference in Jerusalem from which this volume takes its name, is as true of the Sonoran Desert, which is shared by the United States and Mexico, as it is of the deserts of the Middle East, which are shared by more than half a dozen nations. It is our role in nature, ultimately, that determines whether we prosper or perish.

As the contemporary philosopher K. M. Meyer-Abich phrased it: "This is the starting point for enlightenment: that in common with millions of other species of plants and animals, the species *Homo sapiens* is part of natural history" (Meyer-Abich 1993).

We do well to keep that point in mind when we reconsider the now-famous definition of sustainable development offered in the Brundtland Report (World Commission on Environment and Development 1987), the document that got so many of us thinking and talking about sustainability (ecological, economic, and otherwise) in the first place: "Sustainable development is development that meets the needs of the present without compromising the ability of future generations to meet their own needs."

We do well, too, to consider the caveat to that definition added by David Brooks of Canada's International Development Research Center: "Because misconceptions surround the term [sustainable development], a few clarifiers should be kept in mind. The adjective is *sustainable,* not *sustained.* The noun is *development,* not *growth.* And the word *economic* does not appear. . . . Although these distinctions may appear simplistic, they are important. Growth . . . means to increase in size by adding material. Development, on the other hand, is the realization of potential" (Brooks 1993).

As we strive toward ecological sustainability in the world's drylands, is it not really "the realization of potential" that we seek? Our role is the management of ecological potential, including enhancement, maintenance, and restoration of arid ecological systems. Al Gore, the vice president of the United States, recognized the importance of ecological sustainability when he stated that "desertification, which knows no borders, can foster genuine cooperation solutions."

It is *if* we are realists. After all, the sum of water on the planet, which is the bottom line in our equation, is not likely to increase. Political control of available water shifts all too frequently, it is true. However, as Prime Minister Peres reminds us, it is nature, not politics, that finally holds the key to the success or failure of human endeavors. And it is to a better (which is to say, less imperfect) understanding of nature that the International Arid Lands Consortium and the authors, editors, and publisher of this volume invite you now.

References

Brooks, D. 1993. Beyond catch phrases: What does sustainable development really mean? Arid Lands Newsletter 33:4.

Meyer-Abich, K. M. 1993. Revolution for nature: From the environment to the connatural world. Trans. M. Armstrong. Denton: University of North Texas Press (copublisher with White Horse Press, U.K.), 35.

World Commission on Environment and Development. 1987. Our common future. The Brundtland Report. Oxford: Oxford University Press.

Preface

Thomas W. Hoekstra and Menachem Sachs

The purpose of the workshop on Arid Lands Management: Toward Ecological Sustainability, which culminated in this volume, was to establish the state of knowledge on arid lands ecological systems and their management for two uses. The first use for the information is for the International Arid Lands Consortium and other granting institutions that support research as it is important for them to know what the state of knowledge is to determine where to best expend limited resources to add to the state of knowledge in arid lands ecology. The second use is for arid lands managers who wish to implement the state of knowledge in their management programs.

The technical editors of this book have tried to assist readers who are specifically interested in the management and research implications of the chapters in two ways. First, as you read through these, you will find a portion of text identified with a single dagger (†) to indicate the previous sentence(s) have management implications and with a double dagger (††) to indicate research implications of the preceding material. Second, the editors have synthesized these management and research implications in the last chapter to provide the reader with easier access to the information.

The chapters that comprise this book came out of a workshop held in Israel in June 1994. That workshop, like this book, was entitled "Arid Lands Management: Toward Ecological Sustainability." The workshop and book were made possible through the generous support of the members and sponsors of the International Arid Lands Consortium (IALC). The IALC provided funds for travel expenses to and from Israel and considerable administrative assistance in assembling participants from around the world. The JNF-Keren Kayemeth Leisrael graciously hosted the invited participants in Israel while they were exchanging their ideas and working together on initial drafts of multiauthored papers that are now incorporated into this book.

Special acknowledgment is due to the IALC Research and Development Advisory Committee (RADAC), who conceived the idea for this workshop and book, identified the best scientists around the world that could address the subjects published in this book, and secured their participation in the workshop and the subsequent text. The RADAC provides technical support to the IALC Board of Directors and is composed of members of each member institution.

Special thanks are also due to the lead organizers of the discussions and chapters on populations (Moshe Shachak and David Saltz), communities (Steve Archer and No'am Seligman), and ecosystems (Linda Joyce and Joe Landsberg). These individuals provided discussion papers at the workshop and wrote the summarizing drafts and final papers from the contributions of many individuals at the workshop. In addition to the planned invited papers listed above, several single- and multiauthored chapters originated from the discussions that occurred during the workshop. These authors also are due special recognition for their significant contributions to the success of the workshop and this book. Finally, we would like to recognize the authors who offered papers on the current research on arid lands ecology.

We would like to provide a special thank you to the JNF-Keren Kayemeth Leisrael and the people of Israel for being outstanding hosts for this workshop.

PART I *The Ecological Framework for Sustainability*

Introduction

This section includes chapters that provide the framework for understanding ecological sustainability. It should come as no surprise that there are as many opinions about such a framework as there are ecologists. However, we were fortunate in having three authors develop a consistent logic for the framework. Paul G. Risser begins with a selected history of the development of ecological sciences, which provides a perspective of how we arrived at our present ecological understanding. He describes the current status of ecological sciences and arid lands ecology with a description of three major problem areas recognized by the Ecological Society of America: global change, biological diversity, and sustainable ecological systems. His chapter concludes with a description of the importance of sustainability and challenges for the future.

The second chapter, by Steward T. A. Pickett, Moshe Shachak, Bertrand Boeken, and Juan J. Armesto, takes the next step by defining an approach to management that requires the accounting of interactions between ecological systems. Management is defined to include three objectives: restoration, conservation (maintenance), and development (intentional alteration) of existing ecological systems. These objectives are then put in the context of large filters, such as productivity from ecosystems, biological diversity from communities, and patchiness from landscapes. These large filters are described in terms of flows for individual ecological systems and for the integrated metaecological system. Interactions among the ecological systems are considered for metaecological systems, which include the states and processes of two or more ecological systems such as communities, ecosystems, and landscapes. The final section of their chapter describes one case study in Israel where trees are planted in a desertified landscape (a process known as savannization) in terms of the objectives, metrics, and interactions of metaecological systems.

The final chapter in the framework series, by Thomas W. Hoekstra and Linda A. Joyce, describes ecological systems and scale as a basis for management of arid lands systems. They use the term "ecological system" as defined by the large body of literature attendant to the subdisciplines of ecology (namely, biome, population, organism, community, ecosystem, and landscape). They stress the point that these ecological systems do not represent ecological scales in and of themselves but are, in fact, each multiscaled criteria. By definition, ecological systems inherently include humans as a biophysical attribute, not as a separate explanation for ecological system existence or use. Management of ecological systems is contrasted to the traditional approach of management for ecological system products such as water, timber, or livestock. Therefore, the concept of management has significant ties to ecological system structure and function, with the traditional products being a result of system function and management directed at manipulating system function to sustain the system and to increase the product output. Hoekstra and Joyce conclude with a description of adaptive management.

1 Arid Lands Research and Biosphere Sustainability

Paul G. Risser

Brief History of the Ecological Sciences

The ecological sciences have evolved over the past century, beginning with the early interest in the late 1800s in the general topic of how plants and animals are adapted to their environment (Walter 1973). It then became obvious that plants and animals in different environmental conditions shared some common adaptive characteristics and, moreover, that there was a repetitive pattern to groups of organisms with similar apparent adaptations (Goodland 1975). This led to considerations and definitions of communities, largely based on similarity of vegetation. In the 1940s, more attention was drawn to practical applications of ecological principles, such as soil conservation and range management (Heady 1975). By the 1950s, there was an early emphasis on the mathematical representation of both plant and animal species. This was also the time in which the concept of succession, which formed the basis of range management and many conservation recommendations, was reevaluated in the context of a view that individual plants acted independently rather than as a cohesive, integrated collection of species (Whittaker 1962).

Beginning in the 1970s, there was increased attention to ecosystem research. Initial questions focused on patterns of primary production, the relationships between plant and animal communities, and biogeochemical cycling (Likens and Bormann 1975). The 1980s brought the extension of ecosystem science to multiple spatial scales and a focus on landscapes. For the first time since the management emphasis in the 1940s, ecology began to treat the human dimension much more seriously. This combination was enhanced by the availability of remotely sensed data and computerized geographic information systems. Finally, in the late 1980s and early 1990s, ecology had become an important component in global change research, including an explicit recognition of the human component (Risser 1992). As this global perspective is applied to arid systems, there is a greater appreciation that drought contributes to desertification and, in some instances, may be the cause, but also that human activities and direct intervention play significant roles in expanding desert ecosystems (Graetz 1991). In addition, there has been increased emphasis on the interplay between desert ecosystems, especially soil processes, and major atmospheric processes at the global scale (Schlesinger et al. 1990).

Current Status of Ecological Sciences and Arid Lands Ecology

Much ecological research continues to focus on populations and communities, particularly with respect to questions about adaptive strategies, interrelations among plant and animal populations, and the organisms' responses to stresses such as pollution, desertification, and habitat fragmentation. For example, in some arid lands, there are surprisingly complex food interactions between the foraging activity of granivorous birds and rodents (Thompson and Brown 1991). Although rodents, ants, and birds compete for seed, there are longer-term ways in which rodents affect the habitat, causing an increase in seed-bearing annual plants. These annuals provide more seeds for birds who, because of their mobility, can focus on localized areas where seed densities are high. Whether or not seedlings actually become established may depend on the presence of other plants

(Silvertown and Wilson 1994), as well as on the prevailing weather and climate conditions.

In addition to these more conventional lines of investigation, much more attention is currently paid to paleoecology and the reconstruction of past biogeographic patterns and to research across ecosystem, landscape, and global spatial scales (Gillis 1992). Under some scenarios of climate change, the area occupied by deserts worldwide is predicted to increase more than 15%. In addition, deserts export chemicals and dust to the atmosphere, affecting global atmospheric processes. As an illustration of the importance of these arid lands processes, it is now recognized that the limited nutrient supply in many desert soils is not the result of slow nutrient cycling. Decomposition is unexpectedly rapid in many arid systems, and in alkaline soils considerable ammonia is produced. This is important because ammonia is the only significant generator of alkalinity in rainfall (Schlesinger et al. 1990). Thus, our knowledge of ecological processes in arid systems continues to expand in many directions. As in virtually all areas of ecology, there are innumerable directions in which arid lands research could expand. However, since there are limited resources to support research and many questions of particular interest to society, there is the question of whether or not these research programs should have more focus and be directed to a set of priorities.††

In 1991, the Ecological Society of America recognized that the United States could not support an unlimited array of ecological research topics and that some priorities should be set (Lubchenco et al. 1991, Risser et al. 1991). In setting these priorities, two criteria were used:

- the potential to contribute to fundamental ecological knowledge, and
- the potential to respond to major human concerns about the sustainability of the biosphere.

From these two criteria, three research priorities were established:

- global change, including the ecological causes and consequences of changes in climate, in atmospheric soil and water chemistry (including pollutants), and in land-use and water-use patterns;
- biological diversity, including natural and anthropogenic changes in patterns of genetic, species, and habitat diversity, ecological determinants and consequences of diversity, the conservation of rare and declining species, and the effects of global and regional change on biological diversity; and
- sustainable ecological systems, including the definition and detection of stress in natural and managed ecological systems, the restoration of damaged systems, the role of pests, pathogens, and disease, and the interface between ecological processes and human social systems.

Although each of these priorities is distinctive, many of the fundamental processes of underlying mechanisms and the implications of the research are common on two or three of the priorities. Thus, a research program directed toward these three priorities not only results in a focused effort but also represents some economy through shared investigations. Moreover, addressing these priorities requires a multidisciplinary approach, including the biological, physical, and social sciences. Since these three priorities are global in scale, the ecological sciences of today and tomorrow must be international in scope and organization. By and large, these three priorities are not only applicable to arid lands research but also follow the directions set by the research community. Research on biodiversity is fragmented throughout the world, but there are coordinated efforts addressing global processes in arid lands. The concept of sustainability in deserts and arid lands is becoming more appreciated, especially as these lands become more conspicuous in social unrest.††

Sustainability: Its Importance and Application

On a worldwide basis, each of the three ecological research priorities is at a different stage of development. Global change research is well organized internationally through the International Geosphere-Biosphere Programs and the spin-off projects, especially those under the terrestrial systems component. However, too little attention has been devoted to questions dealing with the ecological component. For example, what regulates the large-scale dynamics of plant and animal populations? What regulates the fluxes of energy and materials (including nutrients and pollutants) within and among ecosystems? For instance, we know that long-term grazing of arid rangelands increases the spatial and temporal heterogeneity of water, nitrogen, and other resources. This heterogeneity of soil resources promotes invasion by shrubs, and soil fertility is reduced by soil erosion and gaseous emissions in the areas between shrubs (Schlesinger et al. 1990). We also know that the patterns of El Niño and La Nina episodes in arid lands affect the establishment of species by seedlings and the relative abundance of perennials, depending upon life history characteristics. From these examples, it is clear that arid lands research should continue to focus on these ecosystems' role in global processes.

Research on biodiversity has focused on the ecology of specific species, on some principles for making decisions about the optimum management strategies for protecting and conserving biodiversity, and on inventories of biological diversity. Insufficient attention has been directed toward understanding how ecological processes produce and maintain biological diversity or how biodiversity affects ecological processes. For example, how would the loss of one or more species affect ecological processes such as primary production or nutrient cycling? In most warm deserts, annual species account for most of the species diversity and annual biomass production (Inouye 1991). Thus, El Niño effects that alter precipitation patterns have the potential to change species diversity and productivity. Or, in another example, disruption of the microphytic crusts on desert soils can have significant effects on both herbaceous and woody species (West 1990).

Developing and managing sustainable ecological systems are goals espoused by many. However, current research efforts are fragmented, and there is no consistent agreement on the underlying goals or concepts. Some of this confusion arises because of the need to connect different areas of sciences, such as ecology and economics. Another challenge is the incorporation of different space and time scales. For example, it may be possible to develop sustainable ecosystems when these systems are large, but smaller systems are not sustainable. Or there may be quite different time scales. For example, international markets may make a certain cropping system economically profitable nearly overnight, but the ecological consequences of changing cropping systems may be obvious only after years or even decades.††

Another challenge to developing sustainable ecological systems is the need to decide upon specific indicators of sustainability. These indicators must be robust enough to detect even long-term, subtle deterioration of the ecosystems, yet they must be simple measurements that can be made in the field. Indicators of arid ecosystems might include measures of primary production, biodiversity, soil fertility and stability, and perhaps some index of ecosystem processes such as decomposition or nutrient cycling rates (Whitford 1988). We should, however, be alert to more subtle conditions that change the ecosystem in fundamental ways, such as the effects of grazing, which also affects the albedo and, in turn, regional climate patterns (Charney et al. 1977).††

The Challenge

Arid lands have always been important to the world's human populations, but their significance has increased over the past few decades because of population demographics and continued use of natural resources (Swift et al. 1989). No topic is ultimately more important than sustaining the portion of the biosphere and providing for long-term support of a high quality life for inhabitants of arid lands ecosystems. Thus, a significant portion of the world's research program should be directed toward developing sustainable arid lands. This will require several steps: using the results of basic research in making management decisions (Boeken and Shachak 1994), the integration of many traditional disciplines in new ways of organizing and supporting research (Lubchenco et al. 1991), and making this information available to decision makers and the public. By combining the state of knowledge with ways that knowledge can be accessed by practitioners (e.g., managers), we hope this book will be a step toward this goal.

References

Boeken, B., and M. Shachak. 1994. Desert plant communities in human-made patches: Implications for management. Ecological Applications 4:702–16.

Charney, J. G., W. G. Quirk, S. H. Chow, and J. Kornfield. 1977. A comparative study of effects of albedo change on drought in semiarid regions. Journal of Atmospheric Science 34:1366–85.

Gillis, A. M. 1992. Israeli researchers planning for global climate change on the local level. BioScience 42:587–89.

Goodland, R. J. 1975. The tropical origin of ecology: Eugene Warming's jubilee. Oikos 26:240–45.

Graetz, R. D. 1991. Desertification: A tale of two feedbacks. *In* E. Medina, D. W. Schindler, E. D. Schulze, B. H. Walker, and H. A. Mooney, eds. Ecosystem experiments. SCOPE 45. New York: John Wiley and Sons. 59–88.

Heady, H. F. 1975. Rangeland management. New York: McGraw-Hill.

Inouye, R. S. 1991. Population biology of desert annual plants. *In* G. A. Polis, ed. The ecology of desert communities. Tucson: University of Arizona Press. 27–52.

Likens, G. E., and F. G. Bormann. 1975. An experimental approach to New England landscapes. *In* A. D. Hasler, ed. Coupling of land and water systems. London: Chapman and Hall. 7–30.

Lubchenco, J., A. M. Olson, L. B. Brubaker, S. R. Carpenter, M. M. Holland, S. P. Hubbell, S. A. Levin, J. A. MacMahon, P. A. Matson, J. M. Mellilo, H. A. Mooney, C. H. Peterson, H. R. Pulliam, L. A. Real, P. G. Regal, and P. G. Risser. 1991. The sustainable biosphere initiative: An ecological research agenda. Ecology 72:371–412.

Risser, P. G., ed. 1992. Long-term ecological research: An international perspective. Chichester: John Wiley and Sons.

Risser, P. G., J. Lubchenco, and S. A. Levin. 1991. Biological research priorities: A sustainable biosphere. BioScience 41:625–27.

Schlesinger, W. H., J. F. Reynolds, C. L. Cunningham, L. F. Huenneke, W. M. Jarrell, R. A. Virginia, and W. G. Whitford. 1990. Biological feedback in global desertification. Science 247:1043–48.

Silvertown, J., and J. B. Wilson. 1994. Community structure in a desert perennial community. Ecology 75:409–17.

Swift, M. J., P. G. H. Frost, B. M. Campbell, J. C. Hatton, and K. B. Wilson. 1989. Nitrogen cycling in farming systems derived from savanna: Perspectives and challenges. *In* M. Clarholm and L. Bergstrom, eds. Ecology of arable land. Norwell, Mass.: Kluwer Academic Publishers. 63–76.

Thompson, D. B., and J. H. Brown. 1991. Indirect facilitation of granivorous birds by desert rodents: Experimental evidence from foraging patterns. Ecology 72:852–63.

Walter, H. 1973. Vegetation of the earth in relation to climate and the ecophysiological conditions. London: English Universities Press.

West, N. E. 1990. Structure and function of microphytic soil crusts in wildland ecosystems of arid to semi-arid regions. Advances in Ecological Research 20:179–223.

Whitford, W. G. 1988. Decomposition and nutrient cycling in disturbed arid ecosystems. *In* E. B. Allen, ed. Reconstruction of disturbed arid lands: An ecological approach. Boulder, Colo.: Westview Press. 135–61.

Whittaker, R. H. 1962. Classification of natural communities. Botanical Review 28:1–239.

2 The Management of Ecological Systems

Steward T. A. Pickett, Moshe Shachak, Bertrand Boeken, and Juan J. Armesto

This chapter explores the relationship between ecological science and management. We begin by discussing the important concepts of ecological system function and regulation and the background concepts relevant to the broad issue of management from an ecological perspective. We then explore a generalized strategy for management application to ecological systems, whatever their hierarchical level or spatial scale. Finally, we show how the contemporary ecological paradigm can be used to generate alternative management scenarios; we then work through a case study that illustrates the possible effects of incorporating contemporary ecology into management.

There is a growing concern among land managers and planners to accommodate functional ecological systems and processes within human-settled areas and spatially arrange modified and natural patches so that ecological functions can be maintained within the landscape. The explicit inclusion of ecological understanding in regional management programs holds the promise of both reducing the current rates of land degradation and achieving greater sustainability at the regional level (Jolly and Torrey 1993, Lee et al. 1992, Myers 1993, Slocombe 1993).

The Nature of Ecological Systems

What is the ecological system to be managed? What controls the working and persistence of an ecological system? These are key questions for conserving, restoring, or harvesting ecological systems. Ecological science offers several insights concerning these questions and identifies the relevant processes and features of ecological systems that should be accommodated in management planning and decision making (NRC 1986). We focus on measurable parameters of fundamental ecological significance and of practical interest.

Productivity is the generation of living matter from nonliving materials and solar energy or certain chemotropic environments measured over area and time. This basic process generates the resource base that humans and other organisms depend on and is the foundation for the long-term sustainability of the biosphere. Changes in a site's productivity can be used as an indicator of environmental stress or physical disruption of ecological systems.

Diversity refers to the variety in kinds of organisms and of the array of biological or physical structures they generate in the interaction with their environments. Diversity is a fundamental property of the biosphere. There is diversity of genetics and physiologies among individuals, diversity of species and higher taxa, diversity of communities and ecosystems, and diversity of landscapes (Falk 1990, Franklin 1993). These kinds of diversity are products of the unique evolutionary potential of living things and their interactions with the past and current opportunities and limits in the environment (Wilson 1992).

These two fundamental features of the living world—productivity and diversity—are expressed differentially over space. Such heterogeneity from place to place can often be discerned as discrete patches. Patchiness results from birth, growth, and death of organisms, from biotic interactions, from biological effects on physical conditions, and from the underlying heterogeneity of the physical environment. Patches, which differ in quality, are defined here as their productivity, diversity, and suitability to various human uses.

Finally, any array of patches is subject to alteration in its composition and spatial arrangement through time. Together, such changes are known as *patch dynamics* (Pickett and Thompson 1978). Patchiness can, therefore, be considered the potentially alterable template in which diversity and productivity exist and a key feature in successful management (Loehle 1991).

When we consider management in the context of regulating productivity, diversity, and patch dynamics, our model of ecological systems should incorporate the major interactions and transformations that affect these fundamental features. In any ecological system to be managed, we recognize at least four different transformation processes:

1. the flow of energy, which represents the transformation of solar energy into biomass (fig. 2.1a);
2. the flow of species, which represents the transformation of one assemblage of coexisting species into another (fig. 2.1b);
3. the flow of patches, which represents the transformation of one patch type into another (fig. 2.1c); and
4. the flow of chemical elements and materials, which enables the flow of energy, species, and patch dynamics (fig. 2.1d).

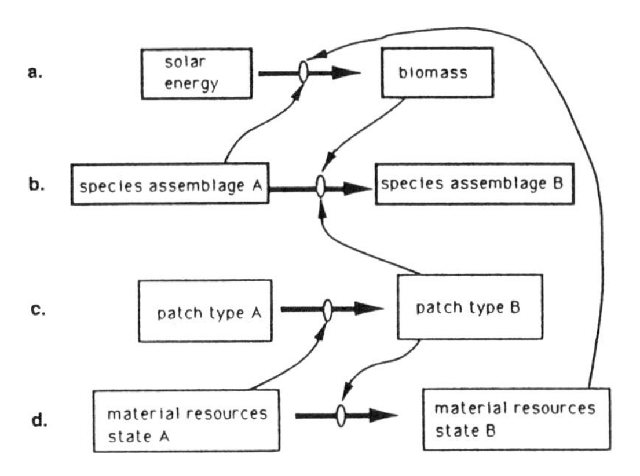

Figure 2.1. Multiflow ecological system. The system represents the relationships among four flow types: (a) energy, (b) species, (c) patches, and (d) materials, all of which are essential flows for assessing ecological management. Each flow is shown as an ecological flow chain (EFC) (i.e., the transformation of a given flow between states). Combining the interacting ecological flow chains results in an ecological system. Several critical connections between individual flow chains in which the donor chain controls the function of the target flow chain are shown by thin lines running from states of the controlling flow chain to the arrow indicating fluxes in the target flow chain. Such connections are called controllers.

We refer to transformations (or states) within a given flow type (1–4 above) as an *ecological flow chain* (Shachak and Jones 1995). An ecological flow chain represents a series of states in ecological systems and their transformations. Flow chains are two or more states that are directly linked through the flow of energy, species, or matter. It is unlikely, however, that ecological flow chains exist in isolation. Flow chains interact through the controlling effect of a state in one flow chain over a transformation process in the other. A network of ecological flow chains and their interactions represents a multiflow model of an ecological system (fig. 2.1).†

An ecological system connotes interactions among organisms and their products and interactions among organisms and the processes on which organisms depend. A definition of ecological systems, in terms of flow chains, requires that the units, fluxes, and interactions among those flow chains be identified. Management starts from understanding the flow chains that constitute the ecological system. The flow chains chosen to describe the ecological system in question constitute a project-specific model.

In ecological management aiming to control productivity and diversity, the core of the ecological system is a network of four ecological flow chains (fig. 2.1a–d). Additional flow chains and their interactions with the core flow chains can be integrated into the ecological system in order to produce a more comprehensive model for management.

The Nature and Goals of Management

Management, in general, refers to the "handling, direction, or control" of some entity or process, according to the *Random House Dictionary*. Management requires knowing how to intervene in a system to achieve a desired state or goal. A variety of ecological systems can be managed, including species, communities, ecosystems, and landscapes, among others (Risser 1985). However, management is often focused narrowly on maximizing the yield of particular system products or commodities, such as timber or crops (Schroeder and Keller 1990). In order to accommodate a broader view of management (Anderson 1994, Hunter 1990, NRC 1992), ecological systems should be managed for multiple uses, such as to yield commodities, amenities, or services, for economic, social, aesthetic, cultural, or other values. Specific activities considered within this view of management are defined below.†

Restoration (e.g., Barrett and Niering 1993, Jordan 1993) focuses on the reconstruction of ecological systems (populations, communities, or landscapes) that have been destroyed or seriously damaged by people or a natural disaster (e.g., climate change). The key is that some loss in the capacity of the ecological system to provide the expected products or services is perceived and attempts are made to recover the lost values. Restoration can be practiced on the site of the degraded system or can attempt to compensate the loss elsewhere. It can seek to reconstitute wetlands, to allow a moribund population to resume reproduction and regeneration, to reestablish a species that had been extirpated, and so on.†

Conservation (e.g., Fiedler and Jain 1992, NRC 1993) usually aims to maintain some ecological entity, state, or process (Holt and Talbot 1978, Niering 1987). Rare ecosystems, species, populations, or landscapes can be targeted. Nature reserves, inhabited areas, and production lands can all be involved in conservation. The most effective conservation, however, goes beyond land acquisition and protection to consider the biological and material exchanges between the protected unit and the entire landscape matrix (Hansson 1992).†

Another form of management refers to the creation or modification of landscapes to produce a land base for habitation, agriculture, forestry, recreation, or other societal pursuits.†

Management can involve some or all of the above activities. In addition to the planning and execution phases, management requires monitoring the ecological systems once the plan is in place, although this ongoing commitment is often neglected. Land-use changes associated with management have traditionally been viewed as conflicting with ecological concerns (Bean et al. 1991); however, the constructive and continuous involvement with the managed system can provide a common ground where ecologists and managers can meet.

The remainder of this chapter proposes a strategy for managing ecological systems. We believe that a scenario that integrates the controlling or limiting ecological factors with the social and economical concerns of managers can reduce the apparent conflict between ecology and development (Pickett 1993).

A Strategy for Ecological Systems Management

A generalized scheme for the process of ecological management may be applied by any discipline or social group concerned with conservation, restoration, or resource use. The scheme should accommodate multiple societal goals, be they economic, agricultural, cultural, or wildlife preservation, among others. However, traditional management has been conceived within the artificial boundaries of a professional discipline or approach. As a result, specific management practices reside within disciplinary paradigms. The background assumptions that a discipline makes are a fundamental part of its paradigm, and such assumptions constrain the parameters and goals of management (Pickett 1993). We explore below the intersections of disciplines concerned with management, especially attending to the insights gained by including ecological knowledge.

A management plan should be based on a model of the ecological system's initial and future states. The model must specify the constraints and characteristics of the starting system as well as the desired and unwanted features. Successful model building requires an understanding of how the ecological system works or is regulated. Following that, within the environmental regulations dictated by the law, management generates several development scenarios among which the one scenario to be applied in a given situation can be chosen (e.g., Houck 1994). Management plans should also accommodate eventual changes in such regulations and societal goals.

In order to explore the implications and connections between ecological science and management, a more detailed definition of the concept of a disciplinary paradigm is needed. Technically, a paradigm is the worldview of a discipline. It comprises the background assumptions, modes of problem solving, and belief system of the discipline's practitioners (Kuhn 1970). The paradigm constrains management decisions because it determines what components are recognized in the ecological system and the model of system regulation. A discipline that assumes biodiversity is an important value or property of ecological systems may generate a different set of management scenarios than a discipline that assumes that sustained yield of a certain commodity is the central purpose of management (Society of American Foresters 1993).

The paradigm also suggests the role of values as part of the belief system. Disciplinary paradigms often reflect social values. Policy makers may choose to consult only those disciplines that seem to focus on objects and phenomena that they consider important. Each discipline has a domain that specifies what objects, processes, and phenomena it will address. Economic value and production of commodities are often major guiding variables in the choice of disciplines to contribute to the management scenario. Such values are most likely to be understood by the public or policy makers.

Where does ecological science enter into the planning and execution of management scenarios? Currently, ecology is often used only to evaluate the environmental impact of scenarios that have already been designed by managers and engineers and to assess their compliance with certain environmental quality laws or mandates about endangered species, for example. In such a role, ecologists are asked to provide controls or mitigating measures on other disciplines and their products. Such an adversarial relationship is a poor way to use ecological knowledge. Moreover, it can lead to the public perception that ecologists and environmentalists alike are forces that obstruct progress. We propose here that a more constructive and less confrontational approach should bring ecological knowledge into the process of planning, executing, and monitoring management. In this way, we will use ecological science to evaluate the alternative development scenarios and will do so in the broadest context of environmental, social, and economical influences on the ecological systems to be managed.

Ecological Systems and Paradigms

A new ecological paradigm has emerged over the last several decades (Botkin 1990, Pickett et al. 1992). It emphasizes key features of ecological systems (see below) that can be critical for successful management. The insights of this paradigm need to be more readily available for managers if ecological science is to be incorporated earlier and more effectively into the planning and decision-making process.

In brief, the contemporary ecological paradigm emphasizes that ecological systems can be open to unexpected or episodic external influences (i.e., to fluxes other than the obvious solar energy input or climatic influences that are clearly external in origin). The new paradigm further recognizes that ecological systems can be subject to regulating factors external to the focal system. Classically, ecologists emphasized the internal regulation of their ecological systems (Kingsland 1985). The contemporary paradigm also recognizes that the trajectories of system dynamics can have a stochastic or nondeterministic component (Botkin and Sobel 1975). The remaining points of the contemporary paradigm are (1) natural disturbance may be an important influence on ecological systems (Pickett and White 1985), (2) systems may rarely be at equilibrium with respect to species composition and diversity (Davis 1986), and (3) humans influence ecological systems even in certain regions thinly populated currently or in the past (Turner et al. 1990).

We can simplify this list of characteristics of the contemporary paradigm in order to apply it to the management of ecological systems. The structure of ecological systems, as depicted in the model of interacting flow chains (fig. 2.1), is the basic determinant of how an ecological system will respond to its environment. The capacity for indeterminism must be built into the flows of the systems model. *Disturbance,* defined as a physical disruption of the system structure (Pickett et al. 1989), accommodates the nonequilibrium nature of some ecological systems and the catastrophic effects of humans, whereas openness indicates the potential for regulation of ecological systems by the interaction of internal and external flow chains (fig. 2.1). When such interactions control either the input or output of organisms, materials, energy, or other influences relative to the ecological system, a system is open. Therefore, we use susceptibility to disturbance and the degree of openness to outside influence to summarize the ecological system behavior in light of the contemporary paradigm.

Disturbance, for purposes of this discussion, can also result indirectly from a disruption of flows critical to maintain system structure, including elements of patchiness. Openness can change as a result of the interaction of internal flow chains with flow chains originating outside the reference ecological system. The degree of openness can be detected by a change in internal flux relative to external input or a change in output relative to change of input. The paradigm recognizes that humans can exert the sorts of influences just described.

The analysis of ecological systems (communities, ecosystems, or landscapes) as connected arrays of flow chains shows that they can respond in a wide variety of ways to changes in the processes driving them. Such an analysis is important because management can alter system structure, the degree of openness, or disturbance dynamics (fig. 2.2). Some or all of the flow chains (fig. 2.1) can be altered by management. For instance, change in system structure can be depicted by transforming patch type B to patch type C within the patch flow chain (fig. 2.2). Disturbance can be represented by eliminating species assemblage A, while changes in the degree of openness can be shown by adding an external element or material flow. One example of such changes in a concrete management situation is a clear-cut of a forest patch that will change forest patchiness, cause disturbance (elimination of the patch), and change system openness to water and nutrient flow due to increased runoff and decreased evapotranspiration. The new clear-cut patch can feed back further on the internal structure by generating, for example, a new species assemblage C. Furthermore, if the new species assemblage C is susceptible to disturbance, such as fire, the susceptibility of the whole system to disturbance increases. If species assemblage C is a consequence of the invasion of exotic species, then the new clear-cut patch changes the openness of the system to species flow.

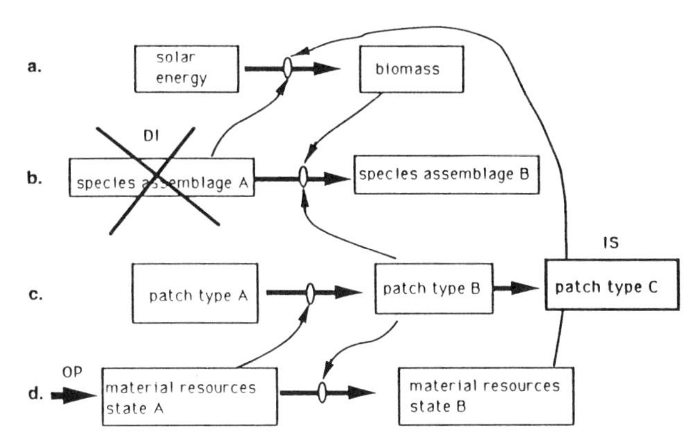

Figure 2.2. Management and changes in multiflow ecological system. The figure shows changes in internal structure (IS—transformation of patch type B to C), openness (OP—adding an external flow of material resources at state A), and disturbance (DI—eliminating species assemblage A). The system can respond to each change by further changes in structure, openness, and disturbance regime.

Generating Scenarios for Management

We will focus on how ecological parameters can generate improved scenarios for management. The generation of scenarios will exploit the contemporary ecological paradigm, combining it with the strategy of modeling ecological systems as connected arrays of flow chains (figs. 2.1 and 2.2). We open with a case history that supports the conceptual approach developed to this point.

Developing a Savanna in a Desertified Landscape

The open areas of the northern Negev Desert, Israel, have been under a grazing regime for several thousand years that has presumably led to a decrease in vegetation cover and increase in soil erosion (Evenari et al. 1983). Within the framework of restoring parts of the Negev and settling certain areas, the Jewish National Fund (JNF), which functions as the national development agency in wild and pastoral lands, decided to develop, as an ecological park, the desertified area between the cities of Beer Sheva and Ofakim. The aim was to convert the area into a human-made savanna (i.e., to add a tree stratum and to increase the productivity of the herb stratum). Because this management strategy produces a savanna-like physiognomy in the desert, the project is called *savannization.*

We use this project as a case study to apply the concept of management of ecological systems. In particular, the contemporary ecological paradigm is used to guide appropriate manipulation of the site and to evaluate the degree of openness, self-regulation versus external regulation, stochasticity versus determinism in dynamics, equilibrium versus nonequilibrium phenomena, and human impact in the ecological system to be managed. Although the case study states explicitly among its goals the maintenance and enhancement of ecological diversity and productivity, in principle the strategy of manipulating the patchiness and fluxes among patches can be used to enhance other ecological functions or values. Among the concerns for this case study are:

1. *Open system.* Any specific area in the Negev is open to flows of water (input via rainfall; output via runoff), soil (input and output via dust, output via sediments in runoff), organic matter, seeds, and animals. The openness is typical of many spatial scales from small plots or vegetation patches to a network of watersheds.
2. *Role of external regulating factors.* At the regional scale, the Negev's productivity and diversity are controlled by rainfall, heat waves, and wind storms. At the spatial scale of patches in the landscape, the fluxes noted above for open systems can become important contributors.
3. *Susceptibility to disturbance.* Major potential disturbance agents in the Negev are floods, drought, and animal burrowing.
4. *Human impacts.* Human activities, herding, and runoff farming have been a part of the ecological system throughout the last few thousand years (Evenari et al. 1983).

Generating the Ecological System Model

The system model of the savannization project was constructed in three stages. First, we selected the key flow chains to describe the premanagement system. Second, we identified the types of manipulations introduced by management. Third, we identified the flows or elements of the system that were added, removed, or altered as a result of the management.

Premanagement Model of the Ecological System Five ecological flow chains were selected to construct a model of the system. The first is the productivity flow chain (i.e., the process of transformation of solar energy into living biomass). The second is the species diversity flow chain that describes the relationships among the regional and local species pools. We included the productivity and species flow chains because these are the two key ecological targets of management. We employed three more flow chains describing fluxes of patches, water, and soil in order to explain the changes in the two target flow chains (fig. 2.3).

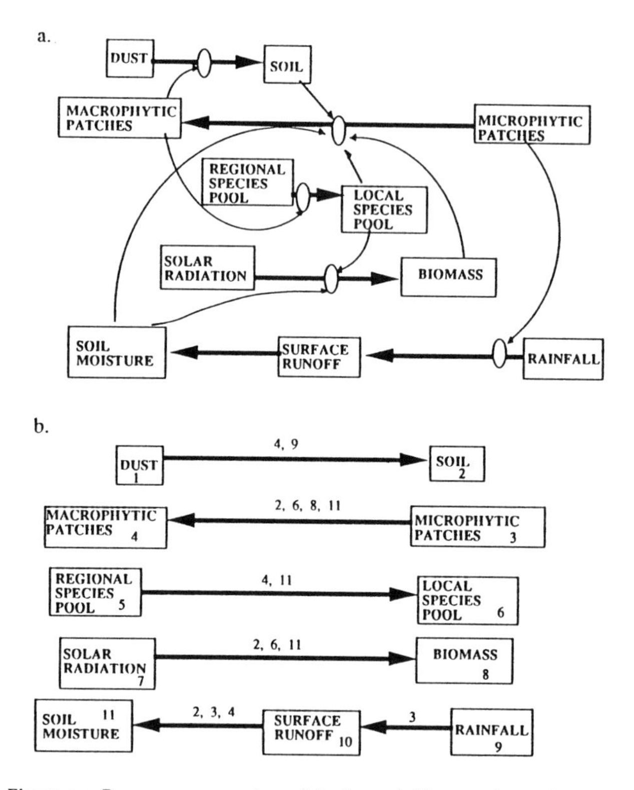

Figure 2.3. Premanagement model of a multiflow ecological system in the northern Negev. The model is a translation of the general concept of multiflow ecological system (fig. 2.1) into the specific situation of the northern Negev. The system represents the relationships among five ecological flow chains: soil, patches, species, energy, and water. In (a), several critical connections among the flow chains that control productivity (biomass) and diversity (local species pool) are shown by thin lines running from states of the controlling flow chains to the fluxes in the target flow chain. In (b), to simplify the diagram, the critical connections are shown by numbers of state variables on the flux that controls the target flow chain. For example, the numbers 4 and 11 on the species flow chain mean that the flow is influenced by controller 4 (macrophytic patches) and by controller 11 (soil moisture). The entire system is explained in the text.

In the northern Negev, productivity and diversity are dependent on the patchiness and patch dynamics. Two major kinds of patches can be recognized in the area: (1) microphytic patches that have soil covered by a crust of bacteria, cyanobacteria, algae, mosses, and lichens (Shachak et al. 1992, West 1990), and (2) macrophytic patches consisting of shrubs and understory herbs with a slightly raised soil mound at their bases (Allen 1991, Garner and Steinberger 1989, Schlesinger et al. 1990, Weinstein 1975, West 1989). Both patch types differ in (1) soil microtopography (either a mound or a flat surface), (2) soil cover (either loose soil covered by litter or a firm crust), and (3) microclimate (either a shrub canopy or exposed surface). As a result of these differences, the two patch types differ in their role in controlling water and soil flows and productivity and diversity in the landscape (Shachak et al. chap. 18, this volume).

Well-developed microphytic crusts are hydrophobic, which reduces rainfall infiltration into the soil and increases surface runoff water generation (West 1990). Since soil flow, or the transport of soil particles, is controlled by runoff and dust exchanges, the microphytic patches also control the soil flow in the system. Macrophytic patches, due to their structure, are able to absorb the flows generated by the microphytic patches.

The relationship among productivity, diversity, patchiness, water availability, and soil characteristics is embodied in a conceptual model (fig. 2.3) that describes the processes affecting productivity and diversity. The higher plants of the established macrophytic patch cause increased dust deposition and accumulation that results in the formation of the soil mound of the patch (fig. 2.3b, controller #4 of dust to soil). The macrophytic patches with their soil mounds open new sites for seed accumulation (fig. 2.3b, controller #4 of regional species pool to local species pool).

The relatively high soil moisture of the macrophytic patch is due to the higher infiltration of direct rainfall and additional runoff induced by the microphytic soil crust (fig. 2.3b, controller #3 of rainfall to runoff). When combined with high nutrient availability, this enables seed germination, seedling establishment, growth, and reproduction (fig. 2.3b, controller #11 of the flow from the regional species pool to the local species pool). These ecological processes increase species diversity of the heterogeneous landscape in relation to the area covered only by the microphytic soil crust.

Each species that colonizes the macrophytic patch will respond to the soil accumulation in the mound, soil moisture content, and nutrient availability by biomass accumulation. This will determine the total biomass of the patch (fig. 2.3b, controllers #2, 6, and 11 of solar radiation to biomass). This network of processes (fig. 2.3a) increases the productivity of the patchy landscape in relation to the area covered only by the microphytic soil crust (Boeken and Shachak 1994). The productivity and diversity of a landscape unit, which contains more than one macrophytic patch with its adjacent microphytic patch, will depend on the number of the microphytic-to-macrophytic pairs and their source-sink properties, size, and spatial arrangement.

Management Additions to the Ecological System If, due to slow natural succession or desertification, the macrophytic patches cannot absorb the runoff and nutrients generated by the microphytic soil crust, then human-made macrophytic patches that can store runoff water and nutrients can be added as a part of the managed landscape. One way to do this is by constructing pits that will increase the number of sinks in the landscape for runoff water and nutrients (Shachak et al. chap. 18, this volume). Therefore, in the human-made macrophytic patches, productivity and diversity will be enhanced. In essence, the novel feature added to the patch flow chain is the human-made macrophytic patch (fig. 2.4, box 12). By adding the human-made macrophytic patch, we are also adding an external controller to the human-managed ecological system (fig. 2.4, box 15). The mode of pit construction, pit size, and spatial arrangement in the landscape may vary according to the objectives of development. Thus, we can expect different specific scenarios in the responses of the ecological system to the introduction of human-made macrophytic patches.

Planting various species of trees in the human-made macrophytic patches is a second novel feature added to the eco-

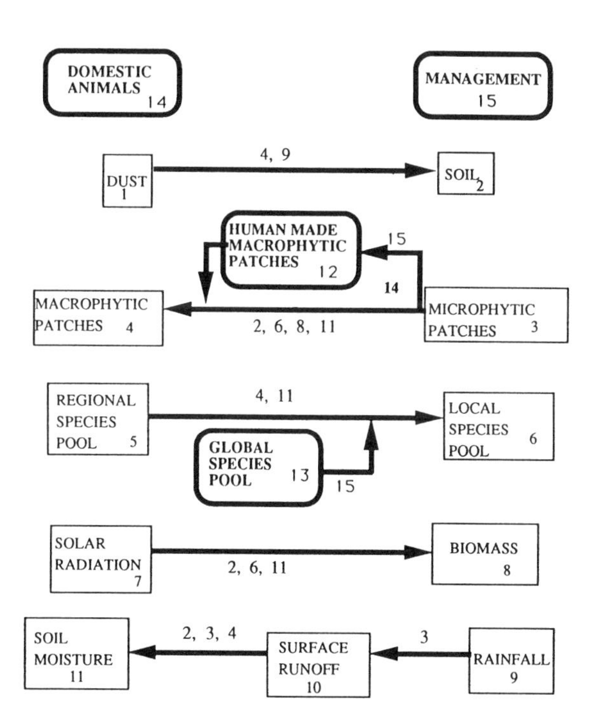

Figure 2.4. A system model of the Savannization Project. The bold-bordered variables (12–15) are the novel features added to the system (fig. 2.3b). Management adds two new variables to the patches and the species flow chains (12 and 13) and two new controllers (14 and 15). The effects of the new features on controlling flow chains are shown in bold numbers.

logical system. In the model, the effect of tree planting is represented by a new flow from "global species pool" to the species flow chain (fig. 2.4, box 13). The selection of species of trees may vary according to the objectives of the management project.

Management Alteration or Removal of System Elements or Flows In analyzing the effect of grazing on the system, it was concluded that uncontrolled grazing will reduce the number and size of the macrophytic patches and decrease the productivity and diversity of the managed area. On the other hand, if grazing is removed, then, due to natural succession, the number and size of the macrophytic patches will increase, resulting in a decrease of the source function provided by the microphytic patches. This will reduce the resources available to the human-made patches. Thus, the management decision was not to remove grazing but to modify the grazing regime. Management requires monitoring to adjust patchiness by grazing in such a way that the microphytic patches will supply water and nutrients to both natural and human-made macrophytic patches. In the model, the role of grazing is depicted as domestic animal control on the patch flow chain (fig. 2.4, box 14). This control can be established by permitting grazing by sheep at specified levels. Experimental assessment of the effects of various densities and seasons of grazing can be made to generate management recommendations.

Ecological System Model and Paradigm

The savannization project model (fig. 2.4) was constrained by the ecological paradigm that assumes that productivity and biodiversity are the most important values of the ecological system within the context of management. This leads to the selection of two flow chains, energy and species, as the core of the model. However, these two flow chains did not provide all the necessary patches, flows, and regulations to determine the system structure and its dynamics. The water, soil, and patch flow chains were also required.

In the savannization project, management alters internal system structure, including patchiness, and the drivers of the system (disturbance and openness). Changes in the internal structure of the model system due to management are described by the integration of two new states and flows (human-made macrophytic patches and species additions from a global pool) (fig. 2.4, boxes 12 and 13) and new and modified regulators (domestic animals and water flow) (fig. 2.4, boxes 14 and 15) into the original five ecological flow chains (fig. 2.3).

The addition of human-made macrophytic patches by constructing dikes (*shikim*) is a direct physical disruption of the existing patchiness (i.e., a disturbance to the preexisting system). The new patches absorb water, soil, and nutrients generated upslope by microphytic patches but not absorbed by the downslope macrophytic patches. The manipulations thus affect the openness of the system to external rainfall and internal runoff inputs. The new patch structure will regulate the flow of water from rainfall to runoff (fig. 2.4).

What are the possible responses of the managed ecological system? We analyze the effect of changes in system structure, degree of openness, and susceptibility to disturbance:

1. *Internal feedbacks.* (a) Colonization of plant species from the natural patches will affect the human-made macrophytic patches. In this way, the local species pool will affect the diversity and productivity of the human-made macrophytic patches. (b) Trees planted in macrophytic patches will affect the microclimate and productivity of the herb stratum (i.e., the introduced tree species will control the energy flow of local species). (c) Finally, tree litter, a new component, will cover a portion of the microphytic patches and change the controllers over the water and nutrients flow chain by decreasing runoff and increasing infiltration.

2. *Changes in the internal structure affect the openness.* (a) Trees will trap dust; human-made macrophytic patches will increase system openness to dust, which will result in soil accumulation. (b) The human-made macrophytic patches will reduce the energy of runoff water and, therefore, soil output. (c) Trees will change the openness of the system to birds, which can change the composition of the local species pool by altering the seed rain and bird predation on seeds and insects. (d) Finally, the human-made patches on the slopes will reduce water and nutrient leakage from the watershed.

3. *Changes in the internal structure affect susceptibility to disturbance.* (a) The mounds of the shikim are susceptible to soil disturbances by burrowing and grazing animals. (b) The trees are susceptible to disturbance by wind.

4. *Changes in openness affect the internal structure. Thymelea hirsuta* (Thymeleaceae) will colonize the shikim. This species, which is unpalatable to sheep and usually grows in the valley, has the potential to invade disturbed areas where soil moisture is increased due to the construction of the shikim. The human-made macrophytic patches increase the openness of the system to invasion by *Thymelea.* Large new patches of *Thymelea* on the slope change the internal structure by increasing the cover of macrophytic patches on the slope. This can reduce the water input and output of the shikim (Kosovsky 1994) by adding a new controller on the water flow chain.

5. *Changes in openness to one flow affect the openness of a second flow.* (a) A decrease in water leakage, due to the construction of the shikim, changes the openness of the water flow chain, which affects the openness of the soil flow chain by reducing soil erosion. (b) An increase in the openness of the system to birds may increase the openness of seed flow, resulting in greater rates of seed deposition or seed predation.

6. *Openness affects the disturbance regime.* A decrease in water leakage will increase the amount of dry herbaceous

phytomass in the dry season in the human-made macrophytic patches. This will increase system susceptibility to disturbance by fire.

7. *Disturbance affects the internal structure.* Soil disturbance of the mounds of the shikim by burrowing and trampling animals will increase soil erosion. Erosion of the loose soil produced may add soil to the macrophytic patches, change the sizes of mounds, and thus affect the patchiness.

8. *Disturbance affects openness.* (a) Fire will increase wind erosion of soil and remaining dead plant materials. (b) Erodible soil produced by the burrowing and trampling animals may be moved out of the system by wind.

9. *Disturbance affects susceptibility to disturbance.* The burrowing of animals in the mounds of the shikim may lead to the breakdown of the mound by high-energy, high-flow runoff events.

Generalizing the Approach

The savannization project shows nine possible types of responses to management, including changes in internal structure, degree of openness, and rate or magnitude of disturbance (figs. 2.2 and 2.4). These nine types of responses can be used to assess the effects of management on any ecological system regardless of structure, flow chains, and spatial scale. As a cause of changes in the ecological system, internal structure may act through density-dependent regulation of consumer populations. Other aspects of change in internal structure, such as succession, can change the degree of openness of the system. For example, forest succession affects both the retention of materials within a watershed (Bormann and Likens 1979) and the susceptibility of the ecological system to subsequent disturbance as older trees grow more slowly and become more prone to windthrow than younger ones.

System structure can also change through, for example, the amount of patch edge versus interior in the landscape, thus altering the degree of openness to influences from the matrix surrounding the patches. The effects of changes in openness on subsequent measures of the degree of openness of the system are indirect. Often, such changes will operate through changes in the internal structure of the system. For example, the edge to interior ratio of patches may determine how much water is lost from the entire landscape. Degree of openness affects disturbance by determining the entry of certain fluxes that have the potential to disrupt the system (e.g., predators, exotic species, toxification). If the degree of openness affects successional rates and, through them, the susceptibility to subsequent disturbance, it is an indirect effect of openness on disturbance.

Disturbance has direct effects on the internal structure of the system by removing or creating patches or other elements of the ecological flow chains of which the system is built. Disturbance indirectly affects the degree of openness through changes in internal structure of the system, including patchiness, flow chains, and succession rates.

Note that the variety of kinds of control on system response is quite large. Classically, ecological theory emphasized only internal feedbacks. The current approach identifies many more spatial and indirect effects that can be important in framing and informing management decisions. The nine types of responses to management can be used to identify positive or negative effects on different ecological flow chains, to assess both biotic and abiotic effects, to examine the alteration of the environmental context, and to assess differential sensitivity to different kinds of environmental impact.

Management of Ecological Systems: An Overview

We have shown how to link the contemporary ecological paradigm, the models of ecological systems as connected chains of interactions and fluxes, and the desire to incorporate ecological science early in the process of generating management scenarios. Constructing a model of the ecological system to be managed that can be experimental and perfectible over time is the first key point at which ecological ideas can be applied. The model should include the patches to be developed by management as well as the natural patches to remain in the landscape. The ecological, as well as social and cultural, fluxes and influences can be joined to generate a model for management of ecological systems (e.g., Slocombe 1993, Smith and Theberge 1986, Wagner and Kay 1993, Woodley et al. 1993). The unique questions suggested by this approach are: What ecological flow chains are critical for management? What is the nature of the connections between human-managed and "natural" components of the flow chains? How does the connection between managed and "natural" components of the flow chains affect the structure, openness, and disturbance regime of the system? How can the patchiness of the ecological system be arranged on the ground? How can the system structure, openness, and disturbance regime be managed?

After a model of the ecological system is constructed, it can help determine a preferred management strategy. Management for sustained ecological function integrating human and natural components requires knowing the dynamics of the system or ecological flow chains (Luken 1990, Pickett and Thompson 1978, Webb and Thomas 1994). Are the dynamics deterministic or probabilistic? Are there equilibrium states of the system? If so, on what time and space scales do they appear? What management tactics can produce the desired dynamics of nondeterministic, nonequilibrium landscapes over the long term? Are the dynamics of various patches in the system controlled by internal or external fluxes? What is the role of disturbance in the system?

Are either subtle or catastrophic human effects an embedded component in the system (McDonnell and Pickett 1993)?

Once these questions are answered, one or more specific management scenarios can be planned and evaluated in the context of the desired management goals. Each scenario will indicate what the patchiness of the system is to be, what ecological, geochemical, and anthropogenic fluxes will and should exist across patch boundaries, and what compensatory tactics are appropriate to maintain diversity, productivity, and ecological values for the society, including services and amenities. In general, managed landscapes can be expected to sustain their various economic and nonpecuniary values. A model of the network of flow chains and the physical patchiness of the ecological system facilitates decision making and permits managers to prioritize and accomplish their goals. Bringing ecological science into the management process early enough to target these concerns can reduce the confrontational stance of ecology and development.

Note

We are grateful to M. L. Cadenasso for helpful criticisms of an earlier version of this chapter. J. J. Armesto acknowledges the support of the Institute of Ecosystem Studies.

References

Allen, E. B. 1991. Temporal and spatial organization of desert plant communities. *In* J. Skujins, ed. Semiarid lands and desert: Soil resource and reclamation. New York: Dekker. 295–332.

Anderson, H. M. 1994. Reforming national forest policy. Issues in Science and Technology 10:40–47.

Barrett, N. E., and W. A. Niering. 1993. Tidal marsh restoration: Trends in vegetation change using a geographic information system. Restoration Ecology 1:18–28.

Bean, M. J., S. G. Fitzgerald, and M. A. O'Connell. 1991. Reconciling conflicts under the endangered species act: The habitat conservation experience. Washington, D.C.: World Wildlife Fund.

Boeken, B., and M. Shachak. 1994. Desert plant communities in human-made patches: Implications for management. Ecological Applications 4:702–16.

Bormann, F. H., and G. E. Likens. 1979. Catastrophic disturbance and the steady-state in northern hardwood forests. American Scientist 67:660–69.

Botkin, D. B. 1990. Discordant harmonies: A new ecology for the twenty-first century. New York: Oxford University Press.

Botkin, D. B., and M. J. Sobel. 1975. Stability in time-varying ecosystems. American Naturalist 109:625–46.

Davis, M. B. 1986. Climatic instability, time lags, and community disequilibrium. *In* J. Diamond and T. J. Case, eds. Community ecology. New York: Harper and Row. 269–84.

Evenari, M., L. Shanan, and N. Tadmor. 1983. The Negev: Challenge of a desert. London: Oxford University Press.

Falk, D. A. 1990. The theory of integrated conservation strategies for biological diversity. *In* R. S. Mitchell, C. J. Sheviak, and D. J. Leopold, eds. Ecosystem management: Rare species and significant habitats. Albany: New York State Museum. 5–10.

Fiedler, P. L., and S. Jain, eds. 1992. Conservation biology: The theory and practice of nature conservation, preservation and management. New York: Chapman and Hall.

Franklin, J. F. 1993. Preserving biodiversity: Species ecosystems, or landscapes. Ecological Applications 3:202–5.

Garner, W., and Y. Steinberger. 1989. A proposed mechanism for the formation of "fertile islands" in desert ecosystems. Journal of Arid Environments 16:257–62.

Hansson, L., ed. 1992. Ecological principles of nature conservation: Applications in temperate and boreal environments. New York: Elsevier Applied Science.

Holt, S. J., and L. M. Talbot. 1978. New principles for the conservation of wild living resources. Wildlife Monograph No. 59. Washington, D.C.: Wildlife Society.

Houck, O. A. 1994. Of bats, birds and B-A-T: The convergent evolution of environmental law. Mississippi Law Journal 63:403–71.

Hunter, M. L. 1990. Wildlife, forests, and forestry: Principles of managing forests for biological diversity. Engelwood Cliffs, N.J.: Regents/Prentice-Hall.

Jolly, C. L., and B. B. Torrey, eds. 1993. Population and land use in developing countries: Report of a workshop. Washington, D.C.: National Academy Press.

Jordan, W. R., III. 1993. Restoration as a technique for identifying and characterizing human influences on ecosystems. *In* McDonnell and Pickett, eds. Humans as components of ecosystems. 271–79.

Kingsland, S. E. 1985. Modeling nature: Episodes in the history of population ecology. Chicago: University of Chicago Press.

Kosovsky, Adar. 1994. Overland flow in a semiarid region: Lahav Hills, Israel. Master's thesis. Jerusalem: Institute of Earth Science, Hebrew University of Jerusalem. [Hebrew with English summary]

Kuhn, T. S. 1970. The structure of scientific revolutions. 2d ed. Chicago: University of Chicago Press.

Lee, R. G., R. Flamm, M. G. Turner, C. Bledsoe, P. Chandler, C. DeFarrari, R. Gottfried, R. J. Naiman, N. Schumaker, and D. Wear. 1992. Integrating sustainable development and environmental vitality: A landscape ecology approach. *In* R. J. Naiman, ed. Watershed management: Balancing sustainability and environmental change. New York: Springer-Verlag. 499–521.

Loehle, C. 1991. Managing and monitoring ecosystems in the face of heterogeneity. *In* J. Kolasa and S. T. A. Pickett, eds. Ecological heterogeneity. New York: Springer-Verlag. 144–59.

Luken, J. O. 1990. Directing ecological succession. New York: Chapman and Hall.

McDonnell, M. J., and S. T. A. Pickett, eds. 1993. Humans as components of ecosystems: The ecology of subtle human effects and populated areas. New York: Springer-Verlag.

Myers, N. 1993. The question of linkages in environment and development. BioScience 43:302–10.

National Research Council (NRC). 1986. Ecological knowledge and environmental problem-solving: Concepts and case studies. Washington, D.C.: National Academy Press.

———. 1992. Conserving biodiversity: A research agenda for development agencies. Washington, D.C.: National Academy Press.

———. 1993. Setting priorities for land conservation. Washington, D.C.: National Research Council.

Niering, W. A. 1987. Vegetation dynamics (succession and climax) in relation to plant community management. Conservation Biology 1:287–95.

Pickett, S. T. A. 1993. An ecological perspective on population change and land use. *In* Jolly and Torrey, eds. Population and land use in developing countries. 37–41.

Pickett, S. T. A., J. Kolasa, J. J. Armesto, and S. L. Collins. 1989. The ecological concept of disturbance and its expression at various hierarchical levels. Oikos 54:129–36.

Pickett, S. T. A., V. T. Parker, and P. Fiedler. 1992. The new paradigm in ecology: Implications for conservation biology above the species level. *In* Fiedler and Jain, eds. Conservation biology. 65–88.

Pickett, S. T. A., and J. N. Thompson. 1978. Patch dynamics and the design of nature reserves. Biological Conservation 13:27–37.

Pickett, S. T. A., and P. S. White, eds. 1985. The ecology of natural disturbance and patch dynamics. Orlando, Fla.: Academic Press.

Risser, P. G. 1985. Toward a holistic management perspective. BioScience 35:414–18.

Schlesinger, W. H., J. F. Reynolds, G. L. Cunningham, L. F. Huenneke, W. M. Jarrell, R. A. Virginia, and W. G. Whitford. 1990. Biological feedbacks in global desertification. Science 247:1043–48.

Schroeder, R. L., and M. E. Keller. 1990. Setting objectives: A prerequisite of ecosystem management. *In* R. S. Mitchell, C. J. Sheviak, and D. J. Leopold, eds. Ecosystem management: Rare species and significant habitats. Albany: New York State Museum. 1–4.

Shachak, M., B. Boeken, J. Cepeda-Pizarro, J. Gutièrrez-Camus, J. Wrann, S. Benedetti, W. Canto, and G. Soto. 1992. Savannization: An ecological answer to desertification. *In* J. Cepeda-Pizarro, ed. Proceedings of the savannization workshop, La Serena, Chile, November. La Serena, Chile: University of La Serena Press. 7–15.

Shachak, M., and C. G. Jones. 1995. Ecological flow chains and ecological systems: Concepts for linking species and ecosystem perspectives. *In* C. G. Jones and J. H. Lawton, eds. Linking species and ecosystems. New York: Chapman and Hall. 280–94.

Slocombe, D. S. 1993. Environmental planning, ecosystem science, and ecosystem approaches for integrating environment and development. Environmental Management 17:289–303.

Smith, P. G. R., and J. B. Theberge. 1986. A review of criteria for evaluating natural areas. Environmental Management 10:715–34.

Society of American Foresters. 1993. Task force report on sustaining long-term forest health and productivity. Bethesda, Md.: Society of American Foresters.

Turner, B. L., W. C. Clark, R. W. Kates, J. F. Richards, J. T. Matthews, and W. B. Meyer, eds. 1990. The earth as transformed by human action: Global and regional changes in the biosphere over the past 300 years. New York: Cambridge University Press.

Wagner, F. H., and C. E. Kay. 1993. "Natural" or "healthy" ecosystems: Are U.S. national parks providing them? *In* McDonnell and Pickett, eds. Humans as components of ecosystems. 257–70.

Webb, N. R., and J. A. Thomas. 1994. Conserving insect habitats in heathland biotopes: A question of scale. *In* P. J. Edwards, R. M. May, and N. R. Webb, eds. Large-scale ecology and conservation biology. Boston: Blackwell Scientific Publications. 129–51.

Weinstein, N. 1975. The effects of a desert shrub on its micro-environment and the herbaceous plants. Master's thesis. Jerusalem: Hebrew University.

West, N. E. 1989. Spatial pattern: Functional interactions in shrub dominated plant communities. *In* C. M. McKell, ed. The biology and utilization of shrubs. London: Academic Press. 283–305.

———. 1990. Structure and function of microphytic soil crusts in wildland ecosystems of arid and semi-arid regions. Advances in Ecological Research 20:180–223.

Wilson, E. O. 1992. The diversity of life. New York: Norton.

Woodley, S., J. Kay, and G. Francis, eds. 1993. Ecological integrity and the management of ecosystems. Waterloo: St. Lucie Press.

3 Management of Arid and Semi-Arid Ecological Systems

Thomas W. Hoekstra and Linda A. Joyce

This chapter continues the discussion on the ecological basis for sustainability in terms of kinds of ecological systems and their interactions. It also describes how research and management can work together in the process of learning and sharing conclusions about the responses of arid and semi-arid ecological systems to management. The discussion of ecological system management concepts presented in this chapter includes both the human and the biophysical dimensions. The human dimension defines our relationship to the ecological system as a user; the biophysical dimension defines each kind of ecological system in terms unique to that kind of system (i.e., community, ecosystem, etc.). We will present the human perspective of biophysical ecology, a description of biophysical ecology as ecological systems, the human relationships to ecological systems, and some guiding principles of ecological system management and the adaptive management process where research and management collaborate to learn about these systems.

Biophysical Ecology: The Human Perspective

Human views and impacts on ecological systems and their management are scale biased. Humans have a limited range of sensing capability—we live seventy years, more or less, we see a limited range of visible radiation, we hear a limited range of sounds, we have the capacity to travel at a limited range of speeds across the landscape. These ranges of human perception effectively influence our understanding of the entities with which we interact. Since we are innovative organisms, we have developed tools, such as microscopes, telescopes, radiation sensors, radiocarbon-dating techniques, and so on, to expand our temporal and spatial range of perception of systems.

In addition to the characteristics of our perception of systems, we have influenced system structures and processes through the application of administrative boundaries and length of land tenure. The scale of this human-imposed context adds complexity to the understanding and implementation of ecological system management.

We would like to further describe how an anthropocentric view of systems results in scale-biased interpretations. We have developed a generalized picture model to help explain this phenomenon (fig. 3.1) (Allen et al. 1987). In the upper left corner of this figure, the general model is depicted: the circles represent individual systems; the lines represent processes that occur between the systems; and the smaller circles and lines within the larger circles are, respectively, the component structures and processes of the system. In other words, the circle represents a system that receives and gives material and information to other systems. Furthermore, it is itself composed of systems and is part of other systems. An eye represents different positions of the human observer in the other views of the general model.

In the upper right portion of the figure, the same typology shows that humans see the product of the system not the internal structures and processes that produce those products. For foresters, this would be analogous to seeing tree diameter growth but not the anatomical structures or physiological processes that produced the diameter growth.

In the lower left portion of the figure, the model depicts humans looking out at the boundary of a system where they observe the average effect of the environmental influence affecting that system. The system boundary moderates the environmental affects, and the observer senses attenuated

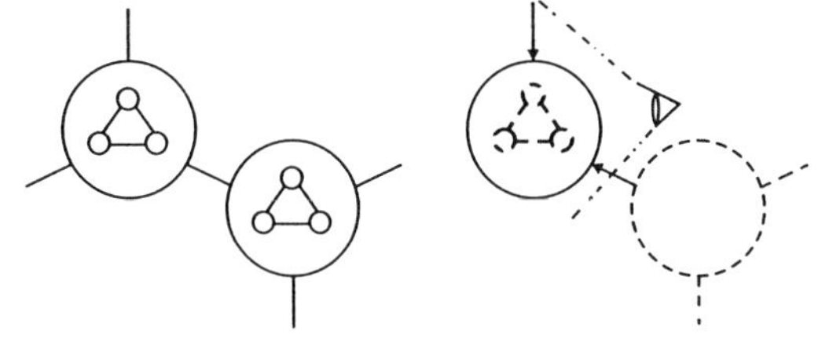

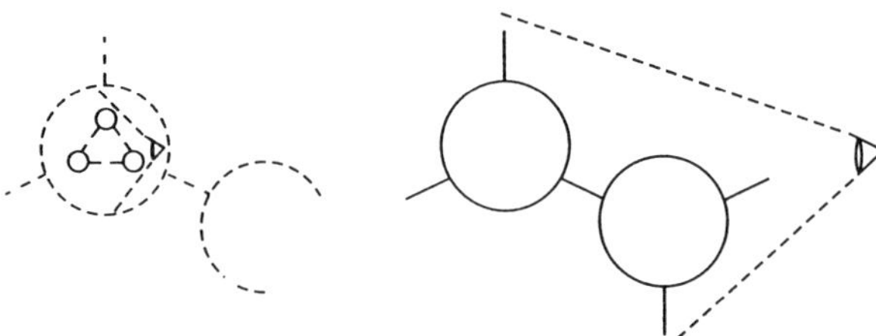

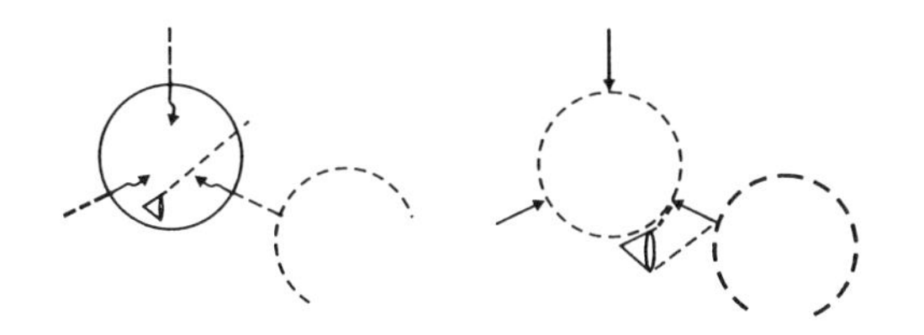

Figure 3.1. The role of the observer and surfaces in observing complex ecological systems.

environmental signals through that boundary. If we put temperature, moisture, and wind sensors both inside and outside a tree, we would find that the inside sensors more moderately and slowly register environmental changes, such as temperature, moisture, and wind outside the tree.

Generally, we are aware of our environment at very selected temporal and spatial scales but not at the larger and smaller scales that also exist. For example, when we developed freon and hydrochloroflurocarbons, we understood they were inert at our scale of the environment, the lower troposphere. We did not consider that there would be a problem using them to cool our refrigerators or vehicles. We have since found that these products were not inert at the larger scale of the stratosphere, becoming quite chemically reactive and altering the protective ozone layer surrounding the earth. The result has been the partial loss of ozone that protects humans from ultraviolet light. Humans have a long history of decisions based on "out of sight is out of mind." We can resolve these problems if the cause and effect are close in space and time, but we need a systems approach to identify cause and effect when time and space differences are large (Senge 1990).

Biophysical Ecology as Ecological Systems

Biophysical ecology can be classified by kinds of ecological systems that include:

- Ecosystems
- Populations
- Landscapes
- Communities (plant and animal)
- Biomes
- Organisms

Kinds of ecological systems are identified, classified, and defined by virtue of their differing processes and states (Allen and Hoekstra 1992). Ecosystems, for example, are characterized by energy states and nutrient cycles with flows and pools of nutrients. Populations are groups of individuals that demonstrate attributes such as birth, growth, and death. As ecological systems, populations are measured, for example, by using the density of individuals and rates of reproduction, mortality, and dispersion of individuals. Other ecological systems, such as communities, have similarly defined structure and process attributes that are described in the extensive literature of this subdiscipline of ecology.

Biophysical ecological systems can be scaled and classified to describe and understand complexes of several different ecological systems superimposed on the land at many different scales (fig. 3.2). Ecological systems can be scaled and classified because each has unique structural components

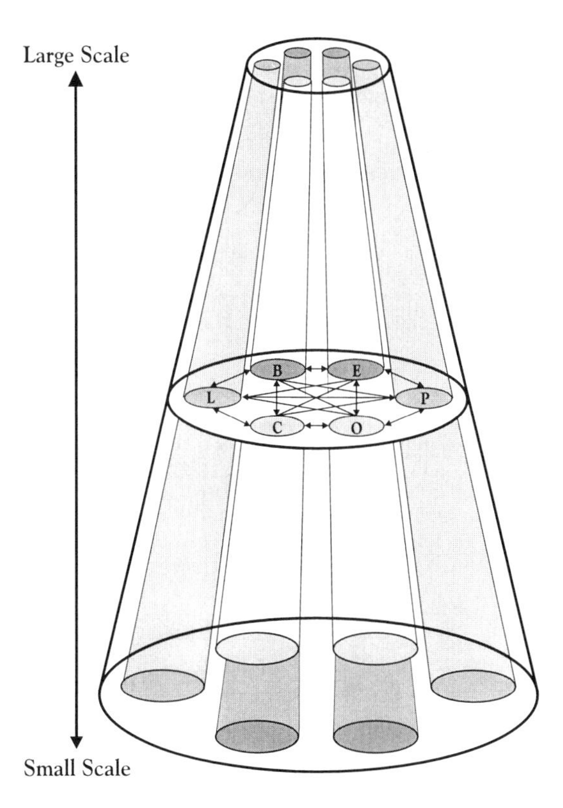

Figure 3.2. The relationship of scale to kinds of ecological systems and their interactions.

and processes and may occur at many different scales. These kinds of ecological systems are not defined by their spatial scale. For example, communities of interacting individuals of various species can be either the context or the content of ecosystems with flows of carbon and nitrogen.

It is also valuable and possible to interpret biophysical ecological systems across scales and kinds of ecological systems because structures and processes of one ecological system are often part of the other ecological systems or processes at other scales (fig. 3.3). Freon is a good example of a structure that can be used for interpreting actions across systems and scales.

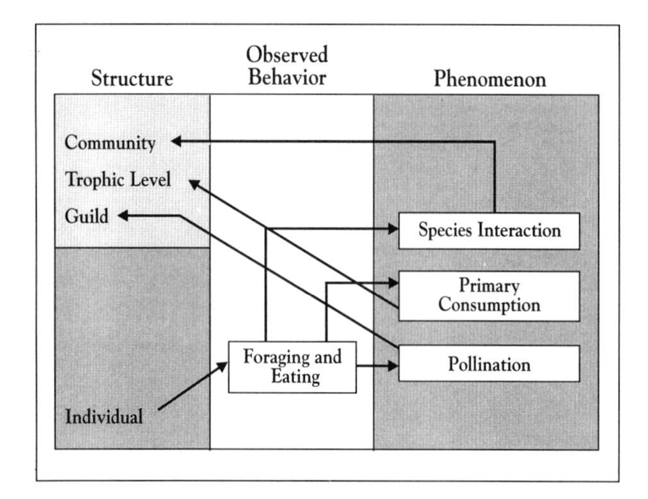

Figure 3.3. The structures at the individual level are observed to change state and so exhibit behavior. Three different phenomena can be chosen on the same observation set, and each links the individual level with a different upper-level structure.

Tradition has it that ecological systems are ordered in a neat nested hierarchy that not only depicts kinds of systems but also the context and content of each system relative to others. In that hierarchy is implicit scaling that severely limits both our full understanding of ecological systems and the utility of our understanding in management. The implicit scaling results when the larger system bounds another system and is the context for the smaller system. The classical hierarchy of organism within population within community within ecosystem, and so on, is a condition that only occurs in rare instances in nature. That restrictive hierarchy is found in many introductory texts on ecology and contributes to the existing misunderstanding of scale and ecological systems in our current management and research. We need to revise our understanding of these ecological systems relative to each other (Allen and Hoekstra 1990). The classic hierarchy that begins with an organism at the lowest level and proceeds through subsequently different systems, such as population, community, ecosystem, and so on, is an inadequate description of what one observes or manages as systems. The fundamental change that we must consider here is to separate the kind of ecological system from its relative scale in time and space. All ecological systems can be described in time and space, and, therefore, only one of their attributes is their time and space scale. However, the scale of ecological systems is not required to describe and define their structural and process attributes. These attributes are necessary to describe the manner in which the system functions. As previously described, a population is generally recognized as being composed of individuals of a single species, occurring at some density, and having some rate of reproduction and mortality. These specific attributes or phenomena of a population are not descriptive of other ecological systems and are descriptive of populations regardless of their temporal and spatial attributes. We can, therefore, define a population as a kind of ecological system and demonstrate that different populations exist at different spatial and temporal scales.† ††

The second key to unlocking the restriction of the classic ecological hierarchy is to relax its requirements that certain ecological systems must be the context for other ecological systems. For example, one can profitably drop the requirement that community ecological systems are always the context for populations. Several examples clarify these issues. First, a population of species A may be the context for a population of species B, such as in the case of an obligate parasite and its host. These populations are descriptive of a tropic hierarchy where different populations occur at different temporal and spatial scales but relate to one another through the food chain. Second, a community is defined by the interactions of individual organisms or small groups, depending on attributes such as the social structure of the species, not population attributes of species in the community. The attributes or phenomena of communities include attributes such as predation, symbiosis, parasitism, mutalism, phytosociology, physiognomy, and so on. These structural and processual relationships among individuals are descriptive of a community, and they are not restricted to a single temporal and spatial scale. These attributes of a community do not describe the relationship between a population of one organism and the populations of another organism, thereby freeing kinds of ecological systems to exist outside of the classical context.† ††

In the case of ecosystems, which are defined by the fluxes of matter and energy, the interactions that are part of communities are not descriptive of what constitutes an ecosystem. Rather, it is the states and processes that are well described, for example, in the carbon cycle, nitrogen cycle, or hydrologic cycle that define the ecosystem. Landscapes are recognized by the size, shape, and distribution of patches of soil, water, and vegetation that result from natural and human-made processes such as rainfall, erosion, frost, farming, and urban sprawl.

If all was ordered according to the classical hierarchy, the task of managers would be somewhat simplified. Unfortunately, as we have shown above, all is not ordered so simplistically. Therein is also a key reason as to why holistic approaches to management are oversimplifications. Forcing the ecological system into the traditional hierarchy keeps us from attaining our objective of managing for sustainable ecological systems.

Sustainability of ecological systems must be based on integrity of the structural and process relationships of each biophysical ecological system and its interaction with other ecological systems. A key point is that the ecological system must be sustainable before consideration is given to concepts such as sustainable development or sustainable management, neither of which can function effectively outside of a sustainable ecological system. The approach that we propose is for managers to consider which types of ecological systems occur at the scale of the proposed management action. The fact that managers must consider several ecological systems is a significant complicating factor for ecological system management. The conceptual structure that should be emerging is that all types of ecological systems occur at many temporal and spatial scales. Therefore, when a manager is considering an action, that action could affect any type of ecological system that functions at the chosen temporal and spatial scale of the management action. Instead of a single hierarchy where all ecological systems are nested one with the next, there are many hierarchies, one for each ecological system.†

Different ecological systems have structures and processes in common with each other. These common structures and processes link kinds of ecological systems in very tangible ways. Consider a community that is linked to populations of plants through the process of dispersion of seeds or the movement of individual organisms such that the community may be invaded by an organism(s) from a population some distance from the current community boundary. In another respect, a community can link with a population through the process of extinction where a population fails to reproduce at a rate that exceeds mortality and the individuals of a population are lost from the community. Dispersion and extinction are only two of the processes that describe the interaction between population and community ecological systems.

For example, an individual that is foraging and eating may be part of several different systems, depending on the phenomenon (activity) it is performing (i.e., it will be part of a guild if it is pollinating, a tropic level if it is a [primary] consumer, a community level if it is interacting with individuals of other species, a population level if it is within the same species).

Human Relationships to Ecological Systems

Humans are an integral component of biophysical ecological systems. We are not only observers and manipulators but are a structural and functional part of the biophysical system. We influence and are influenced by biophysical complexes of ecological systems at multiple scales. In addition, we are sentient beings that attempt to influence how we will be included or excluded from the ecological system functions.

The problem of our relationship to biophysical ecological systems is a "trilemma." First, we have reduced human mortality rates. We will need to expand our stewardship/management role if we continue to expand our population. Second, we may have already overobligated the capacity of some biophysical ecological systems to meet our immediate needs as humans for food and fiber. Third, we may have altered many of the biophysical ecological system components and processes beyond their ecological context, and, therefore, their capacity to be restored to their previous condition is lost.

Thus, human dimensions are an additional and closely related attribute of ecological systems. The point that has become clear is that humans have, in recent times, developed the capacity to alter the earth's ecological systems such that our very existence is threatened. The objective should be clearly focused on the sustainability of the earth's ecological systems with humans as an integral part. Sustainable ecological systems, therefore, become a key concept on which to focus management. Sustainable ecological systems assume that management will be based on the premise that ecological systems have structural and processual integrity or be focused on reestablishing that integrity (Allen and Hoekstra 1995).

Application of the management concept of sustainability needs to be defined on the basis of both an individual ecological system and interaction of ecological systems. Traditional ecological systems include those defined as populations, landscapes, communities, and so on and are identified above as individual ecological systems or elsewhere in this book as simple ecological systems. Ecological system interaction is equally important and represents the notion that there are regular interactions between the same and different kinds of ecological systems; one explanation for this interaction is described in this volume by S. T. A. Pickett and colleagues as metaecological systems (chap. 2). What is needed to sustain a population is not the same as what is needed to sustain an ecosystem or landscape. However, as described previously, failure of one ecological system can have implications for other ecological systems when structures or processes are shared between the systems. The integrity of ecological systems is key to their sustainability. For example, in our definition of sustainable systems, the states of the systems can accumulate or degrade and the processes can be interrupted but only within the capacity of the system to recover. Once the system goes beyond its capacity to recover, a different system takes its place. In addition, the systems are not steady state (i.e., they have and will continue to respond dynamically to many factors including human influences). Sustainable systems that are intensively managed can look and behave much differently than extensively managed systems, but both can be sustainable (Allen and Hoekstra 1994).

Humans fear that they will suffer biologically, economically, and socially from the nonsustainable use of ecological systems. An example of nonsustainable use is the mining of ecological systems for their nutrients or for their animal and plant species. We need to recognize the relationship between the rate at which we alter ecological systems and the rate at which we can respond to restore the system. Ecological systems can change gradually or quickly, but coincident change, whether sudden or gradual, should be expected in the ecological balance between systems. Uriel Safriel's chapter (chap. 8, this volume), for example, discusses the rate of replacement and how this rate cannot increase to replace what is used—such use is nonsustainable.

Nonsustainable use can also include the unplanned contamination of ecological systems with toxins such as pesticides, heavy metals, and radiation. This contamination of air, water, or soil can result in environmental refugees—people forced to move because the ecological system has become uninhabitable.

Our problems of nonsustainable use of ecological systems stem partially from our incomplete knowledge and our single-minded approach. Most past management decisions were based on incomplete social, economic, and ecological information; unfortunately, this is a condition that will not change in the near future; we will never have all the information about the systems that we need to make risk-free decisions. Perhaps, more critically, past efforts have attempted to address human resource needs with independent social, economic, and ecological management considerations, an attribute of decisions we can now change.

Additional problems have come from instances where we have institutionalized scientific thinking in management policies, suppressing the opportunity to let management evolve with science. A salient example here is the use of range condition to assess the health of rangelands in the United States. The concept of range condition was developed using state-of-the-art thinking in community ecology in the mid-1940s. We institutionalized this concept within our land management agency policies. Institutionalizing means we mandated the use of this concept by the agencies in their land management decisions. Range condition was measured and used to adjust the stocking level and season of use of livestock. Once the agencies began to use this measure of range condition, their authority and expertise became tied to the ecological principle associated with the technique (Joyce 1993).

At the same time, scientists began to criticize, test, and refine these ecological principles. Scientists were not bound by agency mandate to use range condition. For some land management agencies, to criticize range condition was to criticize a management structure and authority that was used to make decisions. Research and management had drifted down two very different paths because we had suppressed the opportunity to let management policy evolve with science.

Guiding Principles for Ecological System Management

Seven guiding principles for ecological system management are given in table 3.1. These principles embody concepts from the ecological, economic, and social science disciplines.

Table 3.1. Guiding Principles for Ecological System Management

1. Humans are part of ecological systems and depend upon them for survival and welfare; ecological systems must be sustained for the well-being of humans and other forms of life.
2. All biotic and abiotic elements and processes of an ecological system must be present with sufficient redundancy at appropriate spatial and temporal scales.
3. The integrity of ecological system processes must be present and functioning at appropriate spatial and temporal scales.
4. Human intervention should not impact ecological systems' sustainability by destroying or altering components or processes beyond their restorative capacity.
5. The cumulative effects of human influences, including the production of commodities and services, will require investments of human labor and finances to maintain the integrity of structures and processes of the ecological systems.
6. Management should examine the role that the disturbance patterns played in sustaining the integrity of the ecological system, consider the role that any current subsidy of external inputs has on the ecological system, and determine if restoring or replicating the disturbance patterns would result in sustaining the ecological system.
7. The complexity of biophysical and human dimensions must be reduced to tractable, measurable components and processes.

Principle 1: We manage ecological systems for humans because we are dependent upon the harvestable natural resources produced by ecological systems for food and fiber needs. These systems must be sustained for both our survival and the survival of other forms of life.

Principle 2: One attribute of management for sustainable ecological systems is the need to provide for redundant ecological systems. Redundancy is necessary to reduce the risk of catastrophic loss of a specific kind and scale of a system.

Principle 3: The integrity of ecological system processes must be present and functioning at appropriate spatial and temporal scales. For example, a report released by the National Academy of Science of the United States defined rangeland health as the degree to which the integrity of the soil and the ecological processes of rangeland ecosystems are sustained (NRC 1994). The capacity of rangeland ecological systems to sustain the well-being of humans and other life forms depends on the internal, self-sustaining ecological processes such as soil development, nutrient cycling, energy flow of ecosystems, and the structure and dynamics of plant and animal communities.

Principle 4: Ecological systems are often sustained by using high levels of external inputs, such as irrigation water and fertilizer; the physical environment is modified by tillage, and pests are controlled by applying chemical pesticides. The withdrawal of these external inputs rapidly undermines the sustainability of these human-modified systems. These systems now depend on fertilizer for nutrient inputs and insecticide for control of pests. A question arises about how many external inputs should be used to modify arid and semi-arid systems. In less intensively modified systems, the capacity of rangeland ecosystem systems to produce commodities and other attributes of human value depends on the integrity of nutrient cycles, energy flows, plant community dynamics, intact soil profiles, and stores of nutrients and water. Should the input be a high pulse initially and then no more, or a longer-term lower pulse? What is the implication of altering the dependency of the ecosystem on the inherent nutrient cycles, energy flows, plant community dynamics, and soil processes? How long can we continue to supplement these ecological systems using fixed fossil fuel resources? These are critical questions in light of ecological process integrity.

Principle 5: Human intervention should be very carefully considered when it impacts ecological system sustainability by destroying or altering components or processes beyond their restorative capacity. Ecological systems are dynamic and evolving, but their integrity requires process and state balance. The cumulative effects of human influences, including the production of commodities and services, will require higher levels of investment to maintain the integrity of structures and processes of the ecological systems. Because of the altered states and processes of many ecological systems, humans can no longer take a hands-off approach. We have to use our people and financial resources to sustain the land systems on which we depend.

Principle 6: We have interfered with some processes, like climate, fire, insects, and disease, and the resultant states and processes of ecological systems are substantially altered. Management should examine the role that such processes played in sustaining the integrity of the ecological system, consider the role that any current subsidy of external inputs has on the ecological system, and determine if restoring or replicating the processes could result in sustaining the ecological system.

Principle 7: The total ecology of all but the most simple ecological systems cannot be comprehended let alone managed in their entirety. We need to build our understanding using processes and structures of kinds of ecological systems to the maximum extent possible. Then we need to use science and management interactively to continue to improve our understanding and the results of our efforts. We also can use quantitative methods to help explore and evaluate the complex alternatives at various temporal and spatial scales. Chapters by John Hof and by Geoffrey Pickup and Mark Stafford Smith (chaps. 12 and 11, respectively, this volume) describe some of the alternative quantitative approaches.

These seven guiding principles were drawn from our current ecological understanding of arid and semi-arid systems. In some cases, they reflect new ways of looking at managing these lands, and, in other cases, they reflect new ways of looking at what has been successful about our land management of these areas. The question remains: How do we bring these guiding principles into our management structures? How, for example, would the public management agency bring these concepts into its management process? This is where integrating biophysical ecological system and human needs takes place.†

Ecological System Management: Initial Consideration

The Decision-Making Process

Land management within each of our countries will incorporate a decision-making process that reflects the history of each country's attempt to work with savannas and deserts and arid and semi-arid lands. In this decision-making process, ecological system management specifications need to be incorporated early in that process. For example, first, the ecological systems and their states and processes appropriate to the scale of the management issue need to be identified and state-of-the-knowledge documented as an early step in the decision-making process. Second, all management options should be evaluated based on our current understanding of the system from the ecological, economic, and social perspectives.

We suspect that the decision-making processes in each of our countries would reflect a dominance by one of the three perspectives but inadequate representation of all of them. In the U.S. Forest Service, it is our perception that our former planning process reflects at most consideration of the economic perspective and less so for the ecological perspective, for example (Hoekstra et al. 1990). The Middle East peace talks are an example of an arena where the ecological perspective is now being used to examine policy alternatives that presumably might only have had a social or economic filter applied to them in the past.

Once a policy is set, there is a need to structure a monitoring process to evaluate the results (i.e., the responses of the system to that policy). Land management implementation of a policy is an evolutionary process, and we must find a way to learn from our actions and incorporate the results of that learning into our management. In an article about the Jewish National Fund's land activities, we were encouraged by the description of an area of wetlands (Hula Valley

Project) that had been drained and how later thinking suggested that action might not have been appropriate for the ecology of the situation. It was recognized that there was a need to bring some aspects of the wetlands back into the area. The point is not that we drained the wetlands and are now restoring part of it but that we learned from our land management actions and have adapted our management based on the new knowledge. Incorporating ecological system management concepts into our countries' policy-making processes will undoubtedly require facing some difficult and challenging administrative policy decisions.

Ecological and Administrative Boundaries

Biophysical ecological system processes and structures (and their attendant boundaries) may not be coincident with administrative processes and structures (and their attendant boundaries). Boundaries are the result of ecological processes and states, not the other way around. Air, clean or polluted, does not need a passport to leave any political jurisdiction. It is important to consider the possibility that administrative processes and structures may be easier to change when needed than to try to change the ecological processes and their derived boundaries.

In this regard, the U.S. Forest Service is developing partnerships with its adjacent landowners in a number of areas in the United States, in order to more effectively manage ecological systems. Ecological system management decisions should be made on the basis of ecological system structure and process, regardless of ownership. Ecological systems are dynamic and adaptive, and we need a management structure that also has those characteristics.

Ecological Complexity

Considering the processes and structures of the multiple ecological system hierarchies described above, it is possible to divide the complexity of ecology as a whole into tractable approaches to the management of each different kind of ecological system and begin to describe appropriate management actions. An early, if not the first, step in the management decision-making sequence is to determine the general question or issue and the temporal and spatial scale at which that issue needs to be answered. In addition, the kinds of management actions and desired outcomes or objectives need to be described. Following that step, secondary questions or issues, management actions, and desired outcomes need to be identified, including the scale of each question.

Ecology is complex, and while the subdivision of the subject into tractable subdisciplines provides less ambiguity, it does not eliminate the fundamental complex nature of ecology. In fact, most individuals who have heard earlier presentations of our ideas express frustration about the amount of work ecological system management will likely entail. We do not intend to suggest that ecological system management is easy, but it *is* possible if we begin with tractable definitions and our best understanding of how systems work. It is considerably more work than the management activities of the past that were directed at specific resource-oriented objectives; however, the proposed approach will help reduce past errors in management decisions. An additional consideration is that we do not have the knowledge of most ecological systems that are being managed and should, therefore, restrict our activities to a limited suite of actions where we can estimate the outcome.

An important attribute of ecological system management is the significant change in the rationale or strategy for management. The old strategy was sustained resource-output oriented (i.e., grazing, timber, water, recreation, etc.)—the greatest good for the most people for the longest time. The new strategy is to sustain the ecological system's integrity with resource output as a secondary consideration.

Adaptive Management

We would offer the following assumptions for consideration in using adaptive management to administer ecological system management:

1. Management is an evolutionary process. Our land management needs to be adaptive to changes in the scientific basis for management. Information is doubling every year, and there is a need to incorporate new knowledge into management. Risks and uncertainties are an explicit part of the decision. We will never know everything about systems' states and processes.

2. Management objectives can be quantified. Failure to quantify management objectives reduces our ability to examine our actions rigorously. Monitoring information to compare to our objectives will be useful only if the objectives were well stated at the outset. Otherwise, the comparisons are soft, reducing our ability to learn from experiences.

3. Ecological system structures and processes are an explicit part of the decision. The management decision must primarily focus on ecological system structures and processes and secondarily on resource outputs (e.g., number of livestock, board feet of timber). If we are concerned about the integrity of ecological systems, system processes and structures must form the basis of our considerations.

4. Human dimensions are an explicit ingredient of the decision. We are an integral part of the biophysical system and a key user and impact on ecological systems.

5. Monitoring and evaluation are mechanisms for improving existing management decisions. It is imperative to compare objectives to outcomes to make systematic progress in our understanding of management actions. Monitoring needs must be focused on ecological systems and the attendant human dimensions.

6. The first attempt to structure the management decision is the best possible. The sooner we attempt to structure the management objective and decision, the sooner research and management will share in the process of learning about our arid and semi-arid systems and how to improve our understanding of arid management of these lands.† ††

References

Allen, T. F. H., and T. W. Hoekstra. 1990. The confusion between scale-defined levels and conventional levels of organization in ecology. Journal of Vegetation Science 1:5–12.

———. 1992. Toward a unified ecology. New York: Columbia University Press.

———. 1994. Toward a definition of sustainability. *In* W. W. Covington and L. F. DeBano, tech. coords. Sustainable ecological systems: Implementing an ecological approach to land management, July 12–15, 1993, Flagstaff, Arizona. Gen. Tech. Rep. RM-247. Fort Collins, Colo.: USDA Forest Service, Rocky Mountain Forest and Range Experiment Station.

———. 1995. Sustainability: A matter of human values in a material setting. *In* C. A. Bravo, L. Eskew, C. G. Vicente, and A. B. Villa Salas, eds. and comps. Partnerships for sustainable ecosystem management: Fifth Mexico/U.S. biennial symposium, October 17–20, 1994, Guadalajara, Jalisco, Mexico. Gen. Tech. Rep. RM-266. Fort Collins, Colo.: USDA Forest Service, Rocky Mountain Forest and Range Experiment Station.

Allen, T. F. H., R. V. O'Neill, and T. W. Hoekstra. 1987. Interlevel relations in ecological research and management: Some working principles from hierarchy theory. Journal of Applied Systems Analysis 14:63–79.

Hoekstra, T. W., G. S. Alward, A. A. Dyer, J. G. Hof, D. B. Jones, L. A. Joyce, B. M. Kent, R. Lee, R. C. Sheffield, and R. Williams. 1990. Analytical tools and information. Vol. 4. *In* Critique of land management planning. FS-455. Washington, D.C.: USDA Forest Service, Policy Analysis Staff.

Joyce, L. A. 1993. The life cycle of the range condition concept. Journal of Range Management 46:132–38.

National Research Council, Board on Agriculture, Committee on Rangeland Classification (NRC). 1994. Rangeland health: New methods to classify, inventory, and monitor rangelands. Washington, D.C.: National Academy of Science.

Senge, P. 1990. The fifth discipline. New York: Currency Doubleday.

PART 2 *Ecological Systems and Their Management*

Introduction

The next series of chapters document the existing knowledge of three ecological systems: population, community, and ecosystem. The chapters developed as a result of the workshop by Saltz et al., Archer et al., and Joyce et al. and included individuals from disciplines such as soils, hydrology, plant ecology, animal ecology, and systems ecology. The chapters developed by Sharon E. Nicholson and Uriel N. Safriel were largely individual efforts to synthesize one aspect of the nature of ecological systems and their management. It is interesting that the authors of these chapters have described one ecological system with respect to other ecological systems that are a context and, in some cases, also ecological systems for which that ecological system is the context. This is not unusual as it is very helpful to understand the structure and function of one ecological system in reference to others. These chapters add substantially more detail on the structure and function of ecological systems presented in the previous section on the ecological framework for sustainability.

Nicholson describes the biophysical world from the perspective of an ecosystem and from a large-scale view of the system. Explanation of the ecosystem in her chapter is presented in the context of the climate of arid and semi-arid regions in the Middle East, North America, and Africa. Detailed attention is given to the hydrologic cycle with the energy and water balance processes and the soil-plant water relations process. Implications for sustainability of ecological systems are principally in the form of the climate vegetation feedbacks and the consequences of climate change on sustainability.

A community ecology perspective is provided by S. Archer, W. MacKay, J. Mott, S. E. Nicholson, M. Pando Moreno, M. L. Rosenzweig, N. G. Seligman, N. E. West, and J. Williams through the description of appropriate biophysical processes and management implications. The community ecological system is discussed relative to ecosystem processes (i.e., water flow, nutrients, soil and plant productivity, and landscape processes) influenced by humans (i.e., disturbance, fragmentation, and climate). Community processes that are covered in this chapter include herbivory (grazing), colonization and extinction of species, and succession. The implications of this chapter to the understanding of sustainability concern community structure and function (i.e., competition between shrubs and grasses) and species diversity.

The third chapter of this section by David Saltz, Moshe Shachak, Martyn Caldwell, Steward T. A. Pickett, Jeffrey Dawson, Haim Tsoar, Yoram Yom-Tov, Mark Weltz, and Roger Farrow on ecological systems and their management covers the population ecological system. The context for the discussion includes the ecosystem in the sense of thermal and soil properties and nutrient and soil flows. The landscape context is discussed relative to patchiness, and the community context in terms of interspecies interactions (i.e., herbivory and predation). Several population ecological system processes are discussed including organism and population growth, dispersal and migration, evolution, and population dynamics. The implications for sustainability of the population ecological system are focused on population size (i.e., rare and extinct populations and fragmented populations).

The next two chapters begin to emphasize the management consideration of ecosystem, community, and landscape ecological systems. Chapter 7, by L. A. Joyce. J. J. Landsberg, M. Stafford Smith, J. Ben-Asher, J. R. Cavazos Doria, K. Lajtha, G. E. Likens, A. Perevolotsky, and U. N. Safriel, covers three main topics: management practices, problems arising from management, and scientific issues and research priorities. In that sense, the context for this chapter is the human-induced ecological processes from management actions. The processes themselves include grazing, timber harvesting, human flows from recreation and tourism, conservation management processes, and farming. Sustainability implications for the three ecological systems covered include changes in the landscape pattern (i.e., wetland losses), changes in soil properties (i.e., salinization), changes in biogeochemical cycles, and changes in community ecological system structure.

Safriel's chapter, the last one in this section, focuses on the sustainability of biophysical systems with consideration of the human-economic dimension. The ecological systems of consideration include ecosystems and landscapes. Considerable attention is given to the ecosystem ecological system processes including energy flow and nutrient cycling, landscape processes including patch dynamics, and meta-ecological system processes such as desertification and global change. The economic system processes include the flow of goods and services. The implications for sustainability of the ecological systems of arid and semi-arid lands includes desertification and natural and human-induced local and global environmental changes.

4 The Physical-Biotic Interface in Arid and Semi-Arid Systems: A Climatologist's Viewpoint

Sharon E. Nicholson

The atmosphere is linked to ecosystems in several ways, and the interaction with the land surface has become a current focus in meteorological research. These linkages are apparent in the causes of aridity, the limiting factors in vegetation growth, the exchange of mass, momentum, thermal energy, and moisture between the surface and the atmosphere, the system feedbacks thereby produced, the adaptation of individual species to local climate, the impact of climatic change on ecosystems, and the impact of changing ecosystems on climate.

The key to the sustainability of arid ecosystems is the provision of adequate moisture supply. Prudent management based on sound physical and biological principles can help to maintain the supply of this scarce resource. The major questions for sustainability are how to optimize current resources, how to adapt to short-term fluctuations, and how to prepare for the eventuality of global, human-induced climatic change. Answers to such questions require a detailed understanding of the above interrelationships between the land surface and the atmosphere.†

This chapter describes some of these interrelationships and examines relevant issues concerning the atmosphere and the physical environment, particularly the plant-water relationships. It includes a brief overview of the soil-atmosphere-plant system and a description of the inherent climatic characteristics of arid and semi-arid environments. It then summarizes knowledge and theories concerning the relationship of vegetation characteristics to moisture availability.

The Climatic Environment of Arid and Semi-Arid Regions

The Causes of Dry Climates

The existence and global distribution of arid and semi-arid ecosystems (fig. 4.1) is clearly a function of climate, particularly the amount of precipitation. The causes of the dry climates include insufficient atmospheric moisture, stable air, subsidence (i.e., descending air), divergent patterns of air flow (i.e., air streams spreading apart), and distance from the main tracks of major weather systems.

Except in polar and high altitude deserts, a lack of sufficient atmospheric moisture is generally not the overriding cause of the arid climate. The case of the Sahara desert clearly demonstrates this. Despite the extreme dryness near the surface, the atmosphere above the Sahara in summer is enriched with about as much moisture as over the wetter regions of the southeastern United States.

The other factors, stable and subsiding air and divergent flow, are common to two meteorological situations linked to the occurrence of dry climates: the semipermanent high pressure systems in the subtropics and the rainshadows in the lee of mountains. Consequently, most of the world's deserts are associated with one of these two situations. Semi-arid climates represent the transition between these situations and those which promote precipitation. In these climates lie the tropical savannas and midlatitude grasslands.

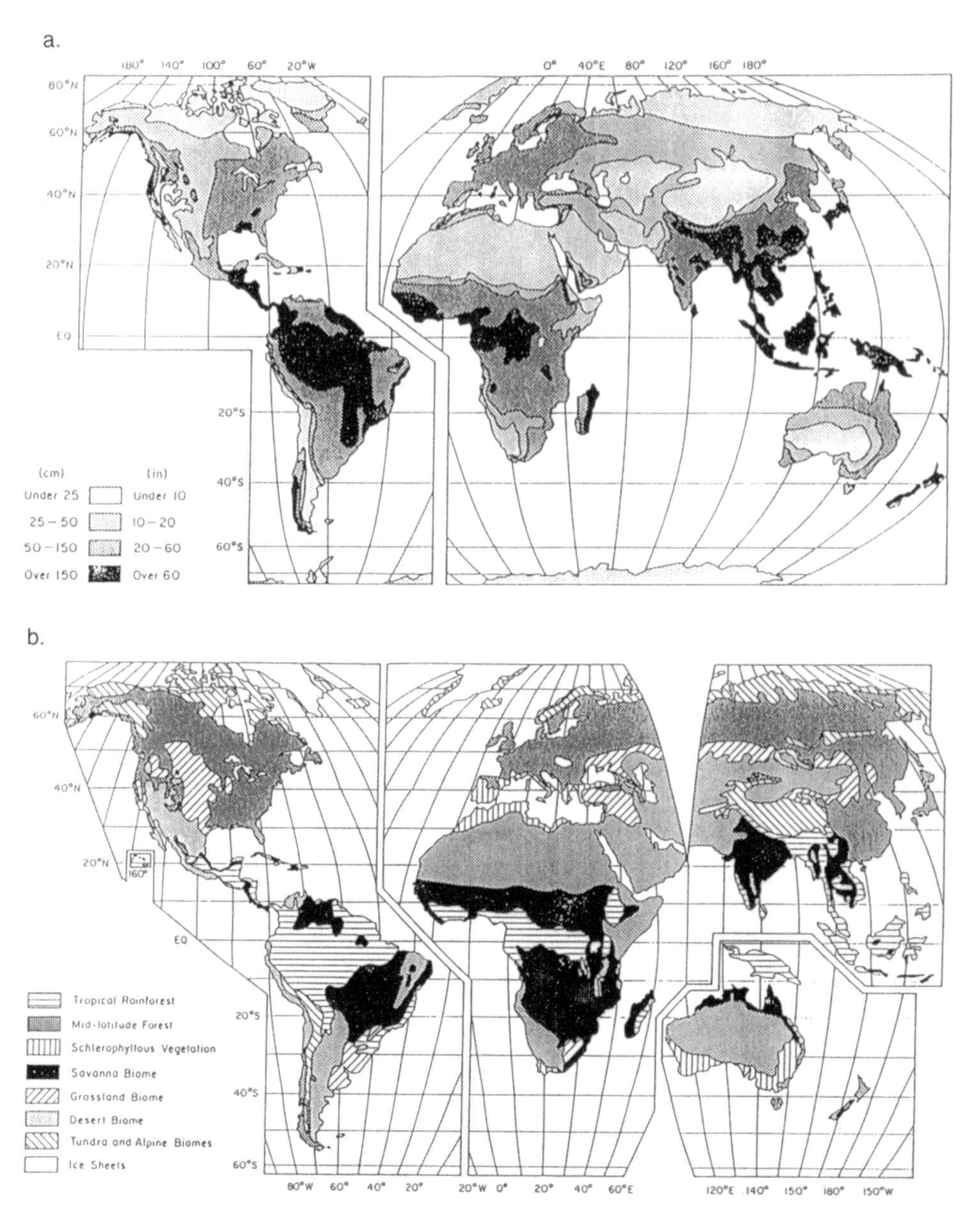

Figure 4.1. (a) Average annual precipitation (based on Strahler and Strahler 1992). (b) Natural vegetation regimes of the world (based on Strahler and Strahler 1992).

General Characteristics of Dryland Climates

The climatic conditions in the world's arid lands are quite diverse. This diversity is recognized in the broad climatic classification scheme of cold deserts, warm deserts, and foggy or coastal deserts (fig. 4.2), but even within these classes generalizations are hard to make. Areas of the Sahara, Namib, and Peru-Atacama deserts are so dry that rainless stretches of ten to fourteen years have been recorded, but few deserts are as dry as those. In the Sonoran, Mojave, Iran, and Arabian deserts and in the Takla-Makan, mean annual rainfall is on the order of tens of millimeters in their driest cores. It exceeds 80–100 mm everywhere in the Thar, Gobi, Patagonian, and Australian deserts and in the Kalahari. In contrast, it scarcely falls below 200 mm in the Chihuahuan Desert and Great Basin of North America. The seasonality of rainfall likewise varies, with some deserts receiving mainly winter or summer rainfall and a few experiencing rainfall during the transition seasons (fig. 4.3).

Rainfall conditions in semi-arid regions are equally diverse. This is particularly true for the savanna environments, because these can result from factors other than climatic aridity. In the savannas, rainfall generally ranges from about 1,000 to 1,500 mm per year, but savannas ex-

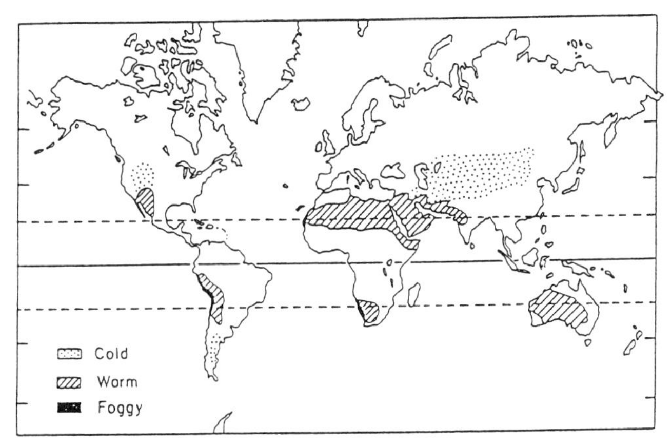

Figure 4.2. Classification of desert regions into cold, hot, and foggy deserts (from Shmida 1985).

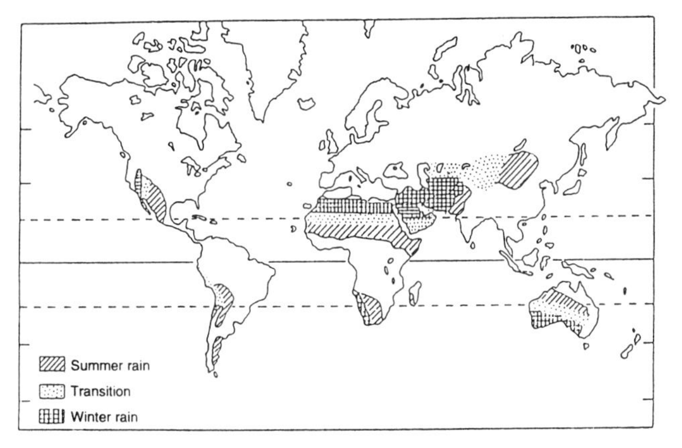

Figure 4.3. Areas of hot deserts with summer rain, winter rain, and rainfall during the transition seasons (from Evenari 1985b, after Shmida).

ist in southern Africa where rainfall is as low as 600–800 mm per year. The midlatitude grasslands tend to occupy areas where mean annual rainfall ranges from 300 mm to 1,000 mm.

Despite this diversity, the dryland regions share many common climatological characteristics. Most are consequences of aridity or of the factors that produce it. Some are consequences of the low-latitude location of most dryland regions. These common characteristics, which are evident to some degree in both the prevailing thermal and moisture regimes, are summarized in this chapter.

Except for coastal deserts, the dryland climates are characterized by:

- low rainfall that is highly variable in time and space;
- severe moisture deficits during some or all of the year;
- localized rainfall events of short duration but high intensity;
- thermal extremes: high temperatures and large diurnal or annual ranges;
- high insolation and low humidity, and therefore, high potential evaporation; and
- relatively strong, turbulent and probably dusty winds.

Semi-arid regions generally share these characteristics but in a less extreme degree. Other commonalities include:

- a pronounced rainfall seasonality;
- abrupt transition between extreme dryness and an often brief rainy season;
- high sensitivity to climatic fluctuations and climatic change; and
- high risk of drought and flood.

Coastal deserts have quite different climatic characteristics because of their proximity to water and the frequent presence of cold offshore currents near these deserts. The maritime influence produces relatively high humidity and a more moderate temperature regime, as the water, with its high thermal capacity, dampens both diurnal and annual fluctuations. In the Namib, for example, the daily and annual ranges are about 6 °C at the coast, compared to 15–20 °C just 100 km inland. A consequence of the cold currents is frequent fog and stratus clouds.

This section will emphasize the climate of arid lands. The characteristics described, particularly the precipitation regime, are nonetheless generally valid for both arid and semi-arid regions, with characteristics being less extreme in the latter. More specific information on the semi-arid savannas and grasslands is included in the "Vegetation and Climate" section.

Thermal Regime

Most deserts are characterized by high surface and air temperatures. In Australia and Asia, absolute maximum air temperatures are on the order of 48–50 °C, but temperatures of 57 and 58 °C have been recorded in the Sahara and at Death Valley, California. Air temperatures as low as -58 °C have been recorded in Asian deserts. Ground temperature is more extreme, with temperatures having exceeded 70 °C at several locations. At Port Sudan on the Red Sea, a sand temperature of 83.5 °C was recorded (Petrov 1975).

The high temperatures result in part from intense insolation, a consequence of the subtropical location of many of the deserts and the scant cloud cover of most. Temperatures are enhanced by the dryness of the soil and the lack of a dense vegetation cover to absorb and redistribute the solar radiation near the ground. Because the ground is dry, very little heat is transported to and stored in deeper layers of soil below the surface. Instead, the heat is concentrated at the surface itself, and temperature decreases rapidly with depth into the ground and with height above the surface. An example from the Kara Kum desert of Asia is shown in figure

4.4. Daytime temperature typically drops over 15 °C within the first few cm and 20–30 °C within the first 10 cm below the surface. Air temperature can drop by more than 10 °C within the first 10 cm above the surface.

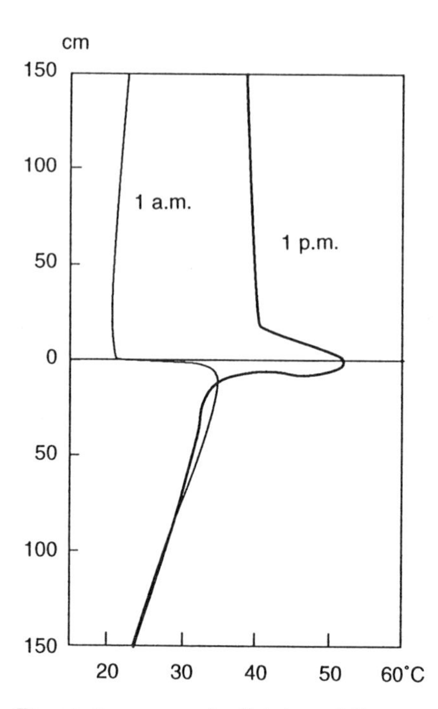

Figure 4.4. Temperature versus depth into and distance above bare ground in the Kara Kum desert (from Petrov 1975).

Since neither the ground nor the sparse vegetation provide a reservoir of heat during the night, the surface cools extremely rapidly and efficiently at night. Then temperature is lowest at the surface and increases sharply with distance from the surface, in both the atmosphere and ground. The usual lack of cloud cover and the dryness of the air near the ground mean that most of the heat accumulated by day escapes to the upper atmosphere by night (90%, compared to 50% in humid regions). This accentuates the nighttime cooling.

The result is a large daily range of both ground and air temperature. The mean daily air temperature range is on the order of 10–20 °C, depending on season and latitude; this is about twice that at humid stations of comparable latitude. The extremes during individual days are more remarkable, with differences of 40–50 °C between maximum and minimum temperatures having been recorded at numerous stations.

The thermal conditions of the semi-arid regions are less extreme than those of arid lands. In the savannas, the annual range is limited by their tropical locations, by the insulating effect of the surface vegetation cover, and by the comparatively moist soils much of the year. These latter two factors also reduce the diurnal temperature range and the near-surface temperature gradients. Mean maximum temperatures in the warmest months are in the range of 30–35 °C near the desert margin, and 25–30 °C near the forest boundary. Temperatures in the cooler months are on the order of 13–18 and 8–13 °C along these same boundaries (Nix 1983). The midlatitude grasslands experience colder winters and larger annual temperature ranges than the savannas, with the short-grass steppes occupying colder environments than the tall-grass prairies.

Precipitation Regime

The hydrologic regime of arid and semi-arid regions is characterized by low rainfall, relatively low runoff (except locally), and a high ratio of evapotranspiration to precipitation. The rainfall is usually concentrated in a brief season, often during the warm season when potential evapotranspiration is high. A dry season prevails much or all of the year, but long dry spells occur within the rainy season. In many true deserts, there is no regular rainy season, and potential evapotranspiration may exceed rainfall throughout the year.

The hydrological regime is most seasonal in the tropical savannas. Total rainfall generally exceeds 600 mm in the wettest 6 months, but falls below 50 mm during the driest 3 months (Nix 1983). The dry season varies in length and intensity, ranging from about 3 or 4 months in the wettest savannas to about 8 or 9 months in the driest. Rainfall is somewhat less seasonal in the midlatitude grasslands, with a dry season ranging from 0 to 8 months and more moderate totals during the wet season. In the drier steppes, potential evapotranspiration exceeds rainfall most of the year, but in the prairies and savannas, long periods of moisture surplus occur within the year.

Other characteristics of the hydrologic regime are strongly dependent on latitude. Tropical rainfall prevails at low latitudes, such as in the southern Sahara or the northern Australian desert. A midlatitude rainfall regime prevails in the higher latitudes, such as the poleward margins of these deserts or in the deserts and grasslands of North America and much of Asia.

Tropical rainfall is generally produced by small-scale convection. Convective rainfall is characterized by high intensity and short duration. The high intensity reduces its effectiveness since the infiltration capacity of the soil is readily exceeded. Thus, much of the rainfall is lost to runoff and little penetrates into the soil. In arid regions, this runoff is generally quite localized and eventually evaporates instead of working its way to streams.

In the midlatitude drylands, the typical character of rainfall depends on its seasonality. In many midlatitude deserts and grasslands, summer rainfall is convective and has the characteristics of tropical rain, particularly its intense, showery nature. Winter rainfall, in contrast, is linked to large-scale warm or cold fronts and tends to be of low to moderate intensity but persistsfor long periods. For these reasons and because potential evapotranspiration is lower in winter than summer, winter precipitation is more effective for vegetation growth.

Other contrasts are apparent between the tropical and midlatitude rainfall regimes. For example, convective rainfall is much more localized than frontal rainfall; it tends to be confined to rain cells on the order of 10–50 km. As a result, the spatial variability of convective rainfall is high, and monthly rainfall at two locations a few kilometers apart might be quite different. In any one year, there are usually a number of areas where rainfall is abnormally low, even if good rainfall prevails overall. The spatial variability tends to be higher during wet years, while drought years tend to be universally dry. Such high spatial variability is not typical in the midlatitude grasslands with winter rainfall or with summer rainfall associated with frontal activity.

Most of the characteristics of convective rainfall become more extreme in the true deserts where rainfall is extremely erratic in both time and space. Rainfall occurs infrequently, generally only a few days per year, and most of the rain that falls occurs within short periods, sometimes as briefly as a few hours. In fact, the mean annual rainfall may occur within a day or even hours. In Helwan, Egypt, where the mean annual rainfall is about 20 mm, seven storms produced a quarter of the rain that fell during an entire twenty-year period. At Lima, Peru, where the mean annual rainfall is 46 mm, 1,524 mm fell during one storm in 1925.

The temporal variability of rainfall is also high in dryland regions. The amount varies tremendously from year to year, especially in the deserts. The rainfall distribution is skewed such that there is an overabundance of subnormal years, the long-term average being inflated by a few years with exceedingly high rainfall. At Biskra, Algeria, where mean annual rainfall is 140 mm, annual totals range from 32 mm to 638 mm (fig. 4.5). The greater the aridity, the greater the interannual variability. The coefficient of variation is on the order of 30–60% in arid regions and 15–30% in semi-arid regions.

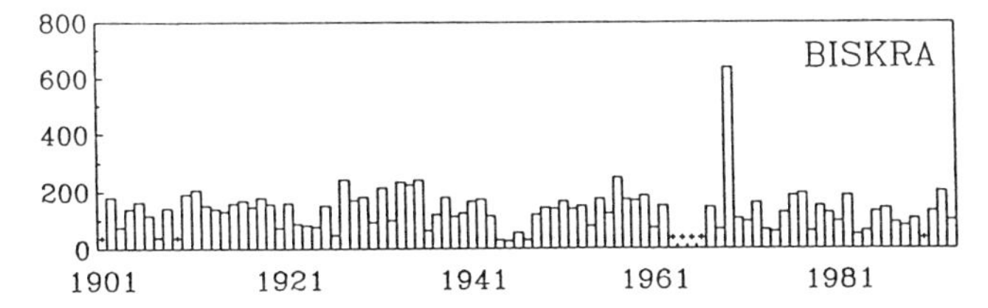

Figure 4.5. Time series of annual rainfall (mm) for Biskra, Algeria, illustrating the high interannual variability of rainfall. The number in upper left is mean annual rainfall.

Microclimates

Many microenvironments of the desert have climatic conditions quite unlike those of the desert at large. For example, streambeds tend to retain moisture near the surface even during long sequences of dry years and can thus support lush stands of vegetation. The same is true for desert oases. Most of the microhabitats tend to ameliorate the extremes of aridity and temperature and, thereby, favor vegetation growth. The importance of the microhabitats is illustrated here with two examples: the sand dune and the Welwitschia plant. In savannas, conditions are more uniform, but the trees themselves produce a microclimate that promotes the grasses of the lower story. This effect helps to sustain the savanna systems. The sand dune is a peculiar microenvironment because its topography plays a large role in distributing radiation, heat, and moisture. The dune contains a number of microhabitats, with climatic conditions differing so markedly that various parts of the dune and the interdune areas have their own unique biota (Robinson and Seely 1980, Seely and Louw 1980, Seely et al. 1990). The differences are apparent in the cycle and intensity of diurnal heating, in the temperature gradient with depth, and in wind conditions. The dune also captures and retains moisture and accentuates year-to-year changes in moisture supply.

A plant modifies the microclimate by intercepting and retaining precipitation and shielding the surface and the interior of the foliage from radiation. It, therefore, ameliorates both the harsh thermal conditions and the moisture stress of the desert, providing a refuge for insects and small animals. The Welwitschia plant of the Namib desert presents a good example (Marsh 1990). Temperature within the leaves and in the litter underneath can be as much as 40 °C cooler than on the surrounding exposed ground. Soil moisture beneath the plant can be twice as high as in the exposed ground.

Surface-Atmosphere Exchange Processes

Between the surface and the atmosphere, there is a constant exchange of gases, water, solutes, particulates, and energy (fig. 4.6). The atmosphere largely prescribes the surface climate and, thereby, influences parameters such as soil mois-

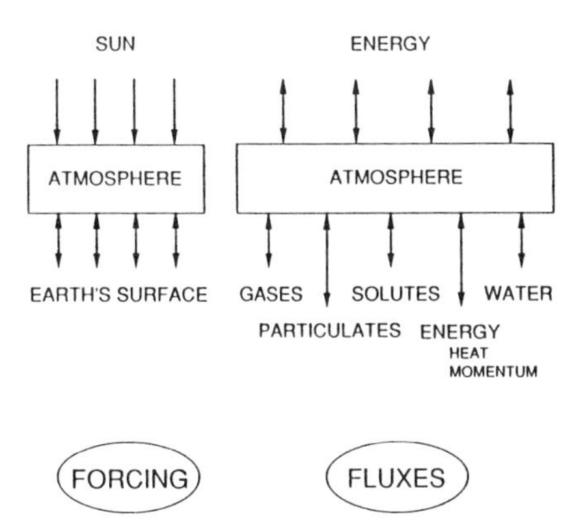

Figure 4.6. Schematic of interactions and fluxes between the land surface and atmosphere. Atmospheric processes are forced via the fluxes of gases, particulates, solutes, energy, and water from the earth and energy from the sun. Double arrows indicate fluxes both to and from the atmosphere.

ture and surface vegetation cover. The surface, in turn, influences the atmosphere through these exchange processes. A number of relevant processes and cycles can be described, such as nutrient flow, carbon cycle, or momentum exchange. For climate, the most relevant are the radiation and energy balance and the water balance.

Radiation and Energy Balance

The net amount of radiant energy available at the surface to drive the climate system, that is, *net radiation,* is defined as:

$$R_{net} = R_{sw}(1 - a_s) + R_{lw}\downarrow - R_{lw}\uparrow \qquad [1]$$

where the subscript *sw* refers to shortwave or solar radiation, a_s is surface albedo, and the subscript *lw* refers to the longwave radiation emitted by the earth, atmosphere, and clouds. $R_{lw}\uparrow$ is that emitted spaceward from the earth's surface; $R_{lw}\downarrow$ is that emitted earthward by clouds and the atmosphere.

The albedo, or reflectivity, is primarily a function of surface moisture and vegetation (table 4.1). It ranges from about 7–12% for tropical rainforests to 15–25% for grasslands. Surface albedo can be as high as 45–50% in extreme deserts with light-colored, sandy, or rock surfaces.

Table 4.1. Typical Albedo for Natural Surfaces

Desert surfaces	35–50%
Sands	25–40
Grassland	15–25
Dry soil	15–25
Wet soil	10
Midlatitude forest	10–20
Tropical forest	7–15

For thermal equilibrium to exist at the surface, the net radiation must be balanced by other forms of heat transfer. This balance is expresssed as:

$$R_{net} = LE + S \qquad [2]$$

where S is sensible heat transfer and LE is latent heat exchange, the product of the latent heat of condensation L and evapotranspiration E. The surface conducts heat to and from the atmosphere and ground so that S includes both heat conducted to or from the subsurface and conduction and convection between the surface and atmosphere.

The partitioning of the surface contribution to atmospheric heating into sensible and latent heat depends on surface moisture and vegetation cover. The wetter the surface, the greater the utilization of heat for evaporation, the lower the temperature and the lower the Bowen ratio (ratio of sensible to latent heating). Water, with its high specific heat and thermal conductivity, also moderates temperature by absorbing radiation with little increase in temperature and by transfering heat to the subsurface. Vegetation similarly moderates temperature by spreading absorbed radiation over a large and distributive surface area and by retarding loss of surface heat.

Water Balance

The latent heat term links the heat balance to another important component of the climate system: the hydrological cycle. The water balance equation states that if there is an equilibrium in the system, the precipitation received (P) must be balanced by the processes of evaporation (E) and runoff (N). Imbalances are manifested as changes in moisture storage. For the surface, this is expressed as:

$$P = E + N + dm/dt \qquad [3]$$

where m is soil moisture stored in the root zone and dm/dt represents changes in storage. E includes both evapotranspiration from plants and direct surficial evaporation from plant, soil, and water surfaces.

Precipitation represents moisture supply, while moisture demand is determined by a parameter termed potential evapotranspiration (PET). It represents the maximum evaporation that can be sustained and is a function of wind, temperature, atmospheric humidity, and insolation. PET is on the order of 2,000–2,500 mm annually in most arid and semi-arid regions but may exceed 6,000 mm annually in some parts of the Sahara. Actual evaporation is determined by a combination of moisture supply and PET, and it is significantly lower than potential most of the year. Thus, aridity at a given location is a function of the relative magnitudes of rainfall and evapotranspiration throughout the course of the year.

Vegetation cover plays a critical but complex role in determining the proportions of precipitation that go into evapotranspiration and runoff. The plants intercept and retain water on their surface, retard evaporation from the soil beneath them, and remove soil moisture via transpiration. On the other hand, vegetation moderates temperatures and reduces surface wind speeds, both effects that reduce evaporation. Both above- and belowground plant materials retard lateral and vertical water movement, thereby reducing runoff, and also alter the soil texture and organic matter content to promote infiltration and soil moisture retention.

Various climate types are characterized by a number of water balance parameters related to the partitioning between evapotranspiration and runoff. These are indicated in table 4.2. In general, the drier the climate the greater the proportion of precipitation that goes into evapotranspiration, the lower the ratio between actual and potential evapotranspiration, and the greater the difference between precipitation and potential evapotranspiration.

Table 4.2. Annual Values of Water Balance Parameters for Various Biomes

	Tundra	Forest	Steppe	Semi-Desert	Desert
R	200–2,000	200–4,000	250–1,500	200–400	< 200
E/R	< 0.30	0.30–0.70	0.70–0.90	0.90–0.97	> 0.97
$R_{net}/(LR)$	< 0.33	0.33–1.00	1.00–2.00	2.00–3.00	> 3.00
N/R	> 0.70	0.30–0.70	0.10–0.30	0.03–0.10	< 0.03
E/E_p	> 0.90	0.70–0.90	0.45–0.70	0.32–0.45	< 0.32
$R\text{-}E_p$	0–2,000	0–3,000	–750–0	–800– –200	< –200

Key: R = annual precipitation; E = actual evapotranspiration; E_p = potential evapotranspiration; R_{net} = net radiation; L = latent heat of vaporization; N = runoff. All parameters expressed in units of mm, except R_{net}, which is in Jm^{-2}, and L, which is in $Jm^{-2}mm^{-1}$.

Source: Based on values in Ripley 1992 and others.

Soil-Plant-Water System

Overview of Water in the Ecosystem

Water is stored at five locations in an ecosystem (fig. 4.7). In the vegetation layer, it is held within plant material and on the plant surfaces that intercept it. Water is also detained on the ground surface in a thin film or as snow or puddles. It is also retained as soil moisture in the porous soil layer with unsaturated flow and as groundwater in the lower soil layer, which is completely saturated and devoid of pores. The water available to the ecosystem is primarily that which infiltrates the soil moisture layer through small pores in the soil.

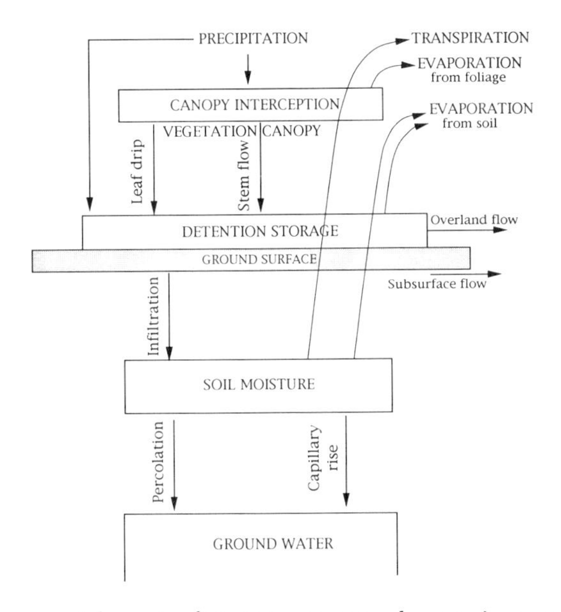

Figure 4.7. Schematic of water movement and storage in an ecosystem. Boxes represent storage reservoirs, and arrows represent fluxes of water in liquid or solid state (straight lines) or vapor state (curved lines) (adapted from Miller 1977). Evapotranspiration is sustained by fluxes from four of these five storage reservoirs.

The water intercepted by plant surfaces (interception water) is derived from precipitation, fog droplets, and surface condensation. The efficiency of interception depends on the nature and amount of precipitation and on vegetation characteristics such as stand architecture and density or surface area of foliage. Many plants in dryland regions have characteristics that tend to maximize the interception of water. Interception is initially high for a dry vegetation layer, but after a saturation threshold of storage is reached, efficiency declines. The excess water falls to the ground as leaf drip or by running down stems.

There are three major paths of removal of water from the ecosystem (fig. 4.7): evapotranspiration to the atmosphere, surface runoff, and downward drainage through the soil layer to the groundwater. The total evapotranspiration is the summation of evaporation of detention water, interception water, and soil moisture plus transpiration through plants. Runoff, or overland flow, is the lateral movement of water in the porous upper few cm of the ground or on the surface itself. Water also flows through large pores directly to the groundwater zone by gravitation, a process called percolation.

The distinction between soil moisture and groundwater is an important one because the utilization of each is quite different. The soil moisture layer essentially constitutes the root zone of plants; the water contained therein is available to plants. Groundwater is long-term storage; some of it flows upward to the soil moisture layer through capillary action, but this is a negligible source of soil moisture in arid and semi-arid regions.

Movement of soil moisture is accomplished in response to four forces: pressure (capillary action), gravity (infiltration and percolation), and the hydraulic and thermal gradients in the soil (Lowry and Lowry 1989). Because these gradients can be extreme in dryland regions, they are important modes of water movement in dryland regions. Capillary action forces water upward, gravity forces it downward. Water (vapor or liquid) flows down the moisture gradient but toward regions of lower temperature. The vapor flow corresponding to a temperature gradient of 1 °C per cm can be as much as 0.4–2 mm per day. This is relatively large compared to evaporation, which is generally on the order of 4–12 mm per day.

The rate of soil water movement is influenced by soil type, ground cover, and topography. Soil type plays a major role because the capillary capacity and the tension forces retarding movement are functions of porosity, texture, and water content of the soil. The latter is measured by a parameter called soil moisture tension, which represents the amount of energy required to extract water from the soil matrix (Dunne and Leopold 1978). The adhesive forces are weakest for large particles, soils of low organic matter content, and wet soils with an open structure, as characteristic of sandy soils. They are greatest for dry, compact soils with small particles and high organic matter, such as clay soils. For a given volume of soil moisture, soil moisture tension is lowest in sand, highest in clays.

For these reasons, soil texture influences the water balance in arid and semi-arid regions and, thus, affects vegetation distribution and growth. Studies by Eagleson (1982a, 1982b) suggest that soil porosity, a parameter closely linked to soil texture, is the most important determinant of optimum vegetation parameters for a given climate. Soil texture influences evapotranspiration and soil moisture infiltration and retention. High soil moisture tension impedes the movement of soil water so that it reduces both infiltration and evapotranspiration and promotes water retention. Thus, both evapotranspiration and infiltration are higher over sand than over clay for any given moisture content (figs. 4.8 and 4.9). Sandy soils retain water the least, clays the most.

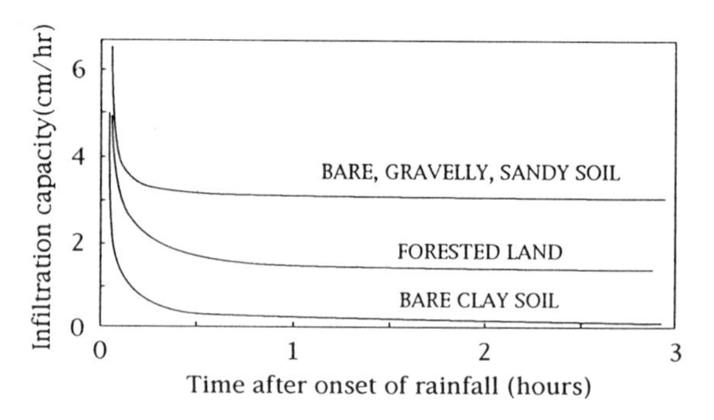

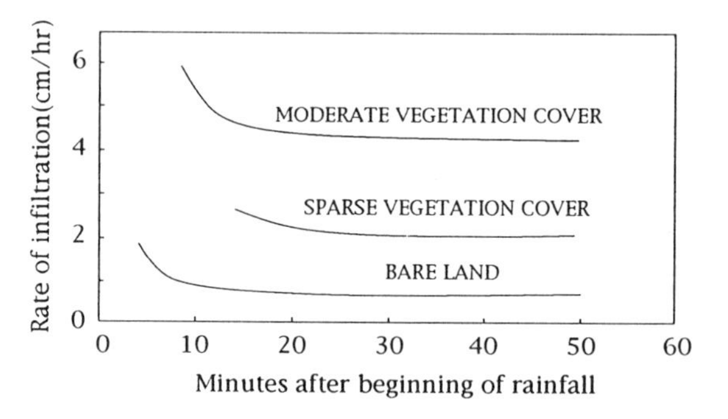

Figure 4.8. Influence of soil texture and vegetation on infiltration capacity of soils (based on Dunne and Leopold 1978, Strahler and Strahler 1992). The vegetation in the bottom diagram is grazed desert shrub. The infiltration capacity is initially high but rapidly decreases as the soil becomes saturated. The actual rate of infiltration is determined by rainfall intensity, as well as vegetation cover and soil texture.

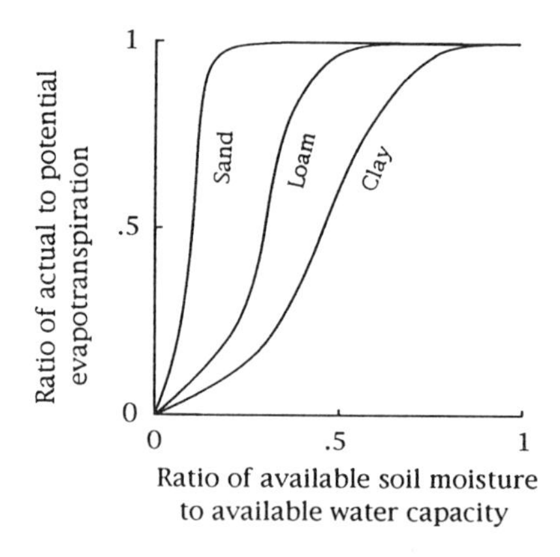

Figure 4.9. Influence of soil texture on evapotranspiration (based on Dunne and Leopold 1978). This represents the percentage saturation of the soil.

Overview of the Soil-Vegetation-Atmosphere Interface

Three extremes of surface types are water, bare dry soil, and complete vegetation canopy. These differ markedly in their effect on the fluxes of radiation, heat, momentum, particulates, moisture, and other gases. In reality, most surfaces present a mixture of the three surface types: a partially vegetated surface overlying moist soil. For the sake of simplicity and brevity, only the limiting cases of bare soil and complete canopy will be considered here, and discussion will focus on fluxes of moisture and latent heat.

Bare Soil

The surface layer of soil interacts directly with the lower atmosphere: it is the direct recipient of precipitation, and its moisture content is the primary determinant of surface temperature and the source of direct surface evaporation. Shallow rooted plants derive their moisture from this layer. The lower layer is a deeper store of moisture and energy, with a delayed and dampened response to surface layer fluxes of moisture and energy. It provides a longer-term source of moisture for deep-rooted plants and soils.

Over bare soil, there are two primary resistances to the exchange of moisture and latent heat (fig. 4.10). A ground resistance (r_g), representing the net contribution of forces affecting subsurface water flow, and an aerodynamic resis-

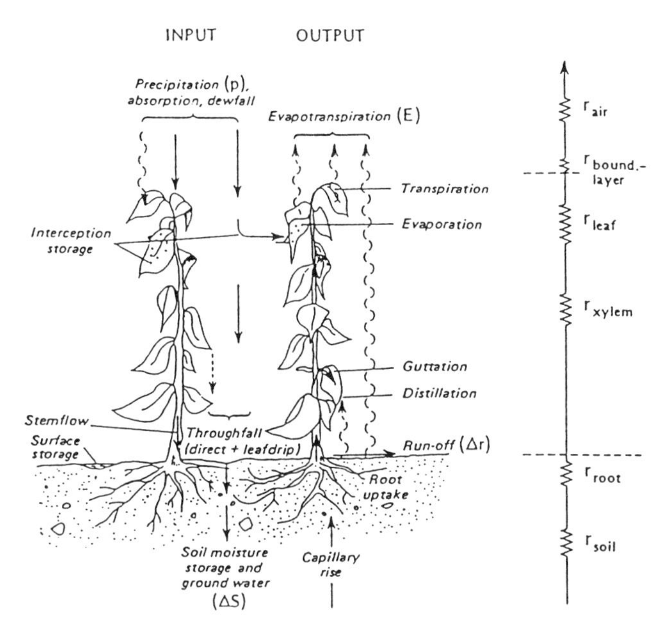

Figure 4.10. Schematic of the various resistances offered by the soil-plant-atmosphere system to water movement and evapotranspiration (from Oke 1978). At each interface is a resistance to the transfer to water. These resistances are assessed in order to calculate evapotranspiration from the system.

tance (*ra*) to passage between the surface and the free atmosphere. Over bare soil, *ra* is a function of the wind regime (surface roughness or turbulence), and r_g is dependent on soil moisture content and soil type.

A Vegetated Surface

In the case of a vegetated surface, the bulk of evapotranspiration is regulated by the soil and air resistances defined above plus the internal resistance the plant also offers to the passage of water from the ground to the air (fig. 4.10). The plant takes in water through the roots. It loses water via evaporation of the interception water on its surface and via transpiration from the internal portions of the leaf. Water transpiring from the leaf takes two pathways: directly through the stomatal openings, which are regulated by two guard cells, and through the epidermis by way of the cuticle. The rate of water flow through the soil depends on the vapor pressure difference between the air and the leaf and the water potential difference between the leaf and the soil.

Both the epidermis and the stomatal pores offer resistance that acts in parallel. The total leaf resistance (*rl*) is calculated as:

$$1/r_l = 1/r_e + 1/r_s \qquad [4]$$

where r_e is cuticular resistance and r_s is stomatal resistance. Minimal stomatal resistance (maximum opening) occurs when conditions for transpiration are optimal; it is usually an order of magnitude smaller than r_e and, under such conditions, is the main contributor to r_l. Controls on stomatal resistance include solar flux, moisture availability (soil water potential and leaf water potential), leaf temperature, leaf resistances, carbon dioxide concentration, and the vapor pressure deficit between the leaf and air.

For a vegetation cover, aerodynamic resistance is a function of wind and canopy structure or leaf area index (LAI). Typical values are on the order of 0.5 s/cm for herbaceous communities 10–100 cm in height but an order of magnitude smaller for forests 10–20 m in height. Where aerodynamic resistance is low, canopy resistance (r_c, the integral of stomatal resistances r_s of the individual leaves) is quite important. Evapotranspiration exhibits less sensitivity to r_c in areas where aerodynamic resistance is high.

Evapotranspiration

The moisture flux *E* from the leaf to the atmospheric boundary layer is calculated as:

$$E = \frac{\rho C_p}{\gamma L} \cdot \frac{\delta e}{r_l + r_a} \qquad [5]$$

where δe is the difference between the saturation vapor pressure at leaf temperature and vapor pressure of the outside air, γ is termed the psychrometric constant (although it is temperature and pressure dependent), ρ is the density of moist air, C_p is the specific heat of air at constant pressure, and r_a and r_l are aerodynamic and leaf resistances, respectively. For an entire canopy, the same equation is assumed to hold with canopy resistance substituted for leaf resistance (Lynn and Carlson 1990, Rutter 1975, Thom 1975).

This equation is often combined with the radiation balance equation:

$$R_{net} - G = LE + H \qquad [6]$$

where R_{net} is the net radiation, *G* is the ground heat flux, *H* is the sensible heat flux, *L* is the latent heat of vaporization, and *E* is evapotranspiration. The result is the Penman-Monteith equation for evaporative flux in the presence of a vegetation canopy:

$$E = \frac{1}{L} \frac{\frac{\Delta}{\gamma}(R_{net} - G) + \frac{\rho L \varepsilon}{P r_a} \delta e}{\frac{\Delta}{\gamma} + 1 + \frac{r_c}{r_a}} \qquad [7]$$

where *P* is atmospheric pressure and *e* is a constant equal to 0.622.

During the course of evapotranspiration, three dynamic processes contribute simultaneously: the absorption of radiation, advection and turbulent exchange, and molecular exchange. These processes are controlled by available energy, water vapor, and wind, respectively; the relevant meteorological variables are net radiation and temperature, the atmospheric humidity and soil moisture content, and wind speed or turbulence. Wind maintains high rates of diffusion between the surface and the ground. When surface water is sufficiently available, evapotranspiration is controlled by the atmospheric variables (the radiation-limited or atmosphere-controlled case). When the surface is relatively dry, evapotranspiration is largely a function of soil moisture (the water-limited/soil-controlled case).

Evaporation over Bare Soil There are three phases in the process of evaporation over bare soil; these vary with the degreee of wetness of the soil. The first phase is that of a wet soil surface. Evaporation is atmosphere limited, and actual evaporation is equivalent to potential. The soil is considered "wet" as long as its moisture content is above field capacity, a threshold defined as the maximum amount of water that the soil can hold against the force of gravity. It ranges from about 25% of saturation in sand to about 90% in clay.

In the second phase, soil moisture is below field capacity. Evaporation is therefore water limited, that is, controlled by the soil moisture content. The rate of evaporation depends primarily on the vertical distribution of soil moisture and the rate at which water evaporated from the surface is replenished from below. The rate can be roughly approximated from the ratio of actual soil moisture (m) to field capacity (m_s) using the relationship

$$m/m_s = E/E_p \qquad [8]$$

and calculating potential evapotranspiration from meteorological variables.

The third phase occurs when the surface is very dry and a second critical level of moisture content is reached. At this point, the soil is so dry that liquid water movement nearly ceases and evaporation consists mostly of vapor from soil pores. Evaporation is small because the available water is below the surface, far from the radiant energy that promotes evaporation. In this phase, the temperature gradient is of primary importance in controlling the rate of evaporation.

Evapotranspiration over a Vegetated Surface The water in four reservoirs (fig. 4.7) sustains the evaporative flux over a vegetated surface. The total evapotranspiration is the summation of evaporation of surface (detention) water, out of soil pores and off the surfaces of plant materials, plus plant transpiration through the stomata. One major distinction between the cases of vegetation and bare soil is that transpiration short-circuits the normal channels of vertical soil moisture transfer by deriving water from the root zone, sometimes well below the surface. Thus, evaporative loss can be greater than over bare soil, especially when the soil is relatively dry.

A "wet" vegetated surface goes through several phases of drying as moisture is depleted. During the first, soil moisture is still plentiful enough to sustain evapotranspiration at the potential rate. In this case, the actual rate is controlled by atmospheric factors and vegetation, as described by equations 5 and 7. This situation is referred to as radiation or energy limited. As the soil dries out, a threshold (roughly field capacity) is reached below which soil moisture becomes the limiting factor. Actual evapotranspiration is generally less than potential and roughly proportional to the available soil moisture, in other words, that above wilting point. In this phase, initially surface evaporation and transpiration from the vegetation both occur. Below a second threshold, the surface dries out and only transpiration continues, with water in the root zone providing moisture. In some cases, this second threshold can occur when moisture in the total soil column is well below field capacity, because the roots extend downward to the moister lower layers of the surface, and evapotranspiration may continue at the potential rate. This could happen when the dominant vegetation has deep roots, as in many dryland regions. Below a third threshold, the wilting point, evapotranspiration also ceases.

The two critical levels of field capacity and wilting point differ greatly among various soil types and are strongly dependent on soil texture. In general, the porosity and hence the wilting point and field capacity increase with clay content (fig. 4.11).

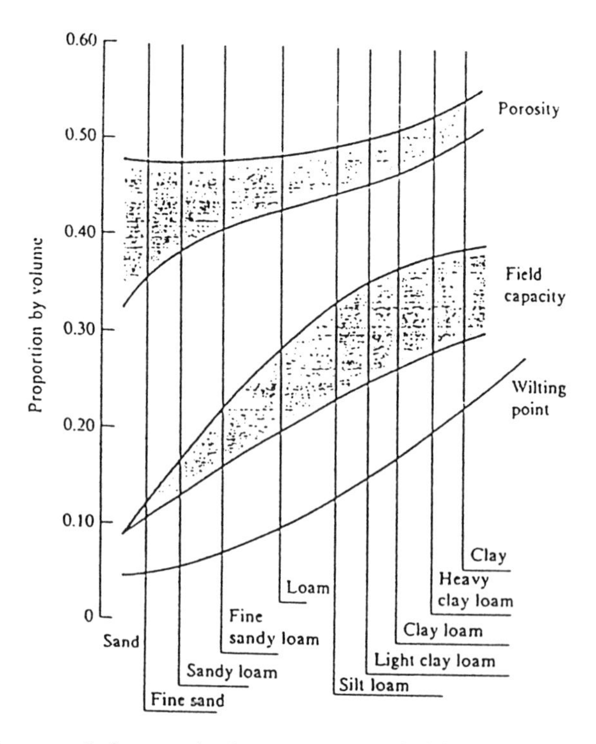

Figure 4.11. Influence of soil texture on water-holding properties of soils (from Dunne and Leopold 1978).

Photosynthesis and Water Use

Green vegetation provides a link between the water, energy, and carbon cycles through its basic growth processes: photosynthesis and respiration. In photosynthesis, water (H_2O) and carbon dioxide (CO_2) are combined to produce carbohydrates ($C_6H_{12}O_6$) and molecular oxygen (O_2):

$$6CO_2 + 12H_2O \xrightarrow[\text{chlorophyll}]{\text{light}} C_6H_{12}O_6 + 6H_2O + 6O_2 \quad [9]$$

Sunlight is required to catalyze the reaction. The opposite occurs in respiration; carbohydrate is broken down and combined with oxygen to yield carbon dioxide and water:

$$C_6H_{12}O_6 + 6H_2O + 6O_2 \rightarrow 6CO_2 + 12H_2O + \text{chemical energy} \quad [10]$$

Net photosynthesis, the net flux of carbon in the system, is the gross photosynthesis minus respiration.

Photosynthesis is affected by a large number of environmental characteristics, the most important being climate. The relevant climatic variables are insolation, wind, temperature, and precipitation; ambient CO_2 levels play a minor role. The portion of the solar insolation that is useful for the process is termed photosynthetically active radiation (PAR) and falls approximately in the wavelength band 0.38–0.71 mm. The amount of PAR intercepted by plants is proportional to the leaf area exposed to sunlight. The area exposed to sunlight ceases to increase once the canopy is so dense that leaves lower in the canopy are completely shielded from solar radiation (Asrar et al. 1984). For this reason, the total leaf area of plants (LAI) is log-linearly related to intercepted PAR (IPAR) (fig. 4.12). Net photosynthesis increases with the LAI until a peak growth rate is achieved and maintained even as the LAI continues to increase (fig. 4.12). The growth rate reaches a maximum because photosynthesis is proportional to the amount of PAR intercepted by plants.

Evapotranspiration and photosynthesis are intimately linked by their common regulatory mechanism: opening and closing of leaf stomata (Rutter 1975, Tucker and Sellers 1986). Therefore, the ratio of stomatal resistances for carbon dioxide and water vapor is the inverse ratio of their diffusion coefficients. This suggests a constant relationship between the rates of photosynthesis dP and evapotranspiration dT,

$$dP/dT = W \quad [11]$$

where W is a constant termed water-use efficiency (Tucker and Sellers 1986) and is species specific. This relationship is often assumed but is not strictly the case, because the behaviors of water vapor and carbon dioxide are not analogous in the passage through the mesophyll or cuticle (Rutter 1975, Tucker and Sellers 1986). Also, some desert species have the ability to regulate transpiration and photosynthesis independently (Evenari 1985a). To some extent, water use efficiency depends on the particular photosynthetic pathway of the plant species. In general, W is higher for C_4 plants (those that begin carbon fixation with 4-carbon compounds) than for C_3 plants (those that begin carbon fixation with 3-carbon compounds). It is highest for CAM (crassulacean acid metabolism) plants, since these can photosynthesize at night and close stomata during the day, thus reducing water loss.

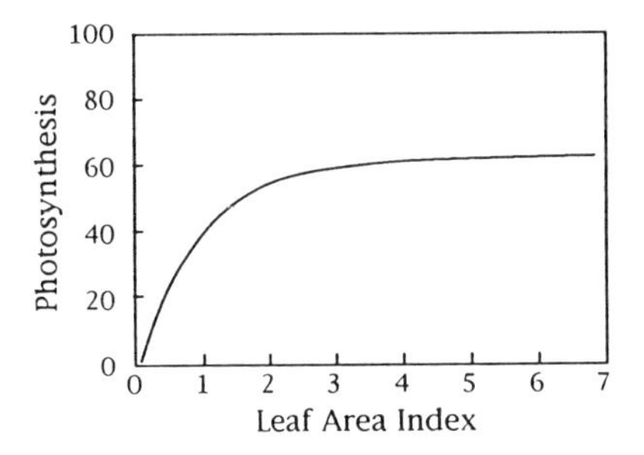

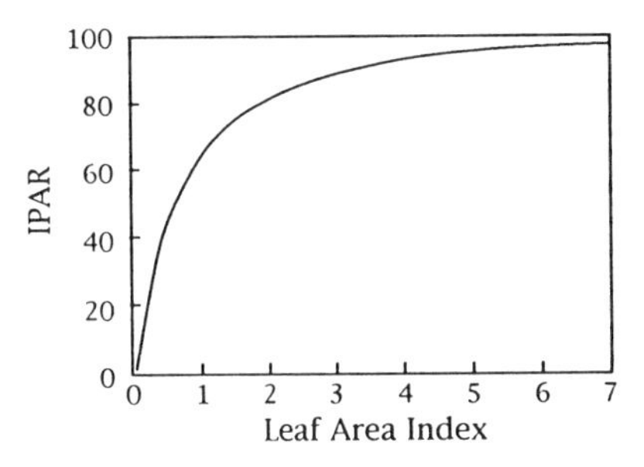

Figure 4.12. Simplified schematic of the relationships among photosynthesis, intercepted photosynthetically active radiation (IPAR), and leaf area index (based on Asrar et al. 1984, Sellers 1987). IPAR is given as a percentage of the incident PAR; units for photosynthesis are mg CO_2 dm^{-1} $hour^{-1}$.

Plants are believed to control the width of stomatal apertures in such a way as to maximize the influx of carbon dioxide while minimizing water loss. When moisture is not a limiting factor, photosynthesis increases almost linearly with the input of photosynthetically active radiation (PAR). At low PAR fluxes, energy is the limiting factor. At high fluxes, the photosynthetic rate changes less rapidly and approaches an asymptotic value; at that point, the plant's internal photosynthetic machinery (e.g., enzymes) is the limiting factor (Tucker and Sellers 1986). This asymptotic value is the photosynthetic capacity and specifies an upper limit of photosynthetic rate for a given PAR flux, assuming all other factors in the process are optimal. It corresponds to minimum values of canopy resistance. The actual photosynthetic rate depends on a number of other factors, including moisture availability.

Vegetation and Climate

General Climatic Controls on Ecosystems

Budyko (1986) suggests that the primary atmospheric control on vegetation can be specified using a simple ratio between energy and precipitation. Termed the dryness ratio, the denominator is precipitation multiplied by the latent heat of condensation and the numerator is net radiation. The ratio compares the energy available to the energy needed to evaporate the "normal" rainfall. Values of 1, 2, and 3 delineate desert, semi-desert, and steppe vegetation, respectively, with savanna and grasslands being distinguished on the basis of the magnitude of net radiation (fig. 4.13).

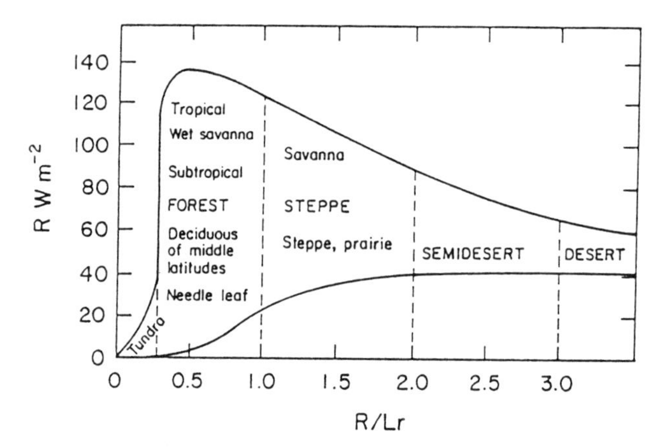

Figure 4.13. Budyko's (1986) concept of geographical zonality: the principal ecosystems as a function of net radiation and the dryness ratio (ratio between annual net radiation and annual average precipitation multiplied by the latent heat of condensation). The upper and lower bounds of the figure represent roughly the limits of net radiation conditions in natural environments.

On a global scale, climate patterns determine to a first approximation the distribution of biomes, and as few as eight climatic variables are needed to specify the major vegetation patterns (Box 1981). On a small scale, biomes can also be delineated when either temperature or net radiation are used in conjunction with rainfall and its seasonality (e.g., Ellery et al. 1991).

In most, but not all, dryland environments, moisture availability limits the type and amount of vegetation (Noy-Meir 1985, Seely and Louw 1980). Water is probably the primary determinant of the rate of photosynthesis in drylands because it is a raw material for the process and is of limited availability in these regions. Plant size, surface cover percentage, and species diversity all roughly increase with rainfall in dryland regions (fig. 4.14). Productivity is also influenced by temperature and wind. Other factors being equal, temperature determines the rate of physiological processes such as photosynthesis. Photosynthesis initially increases with temperature until some species-

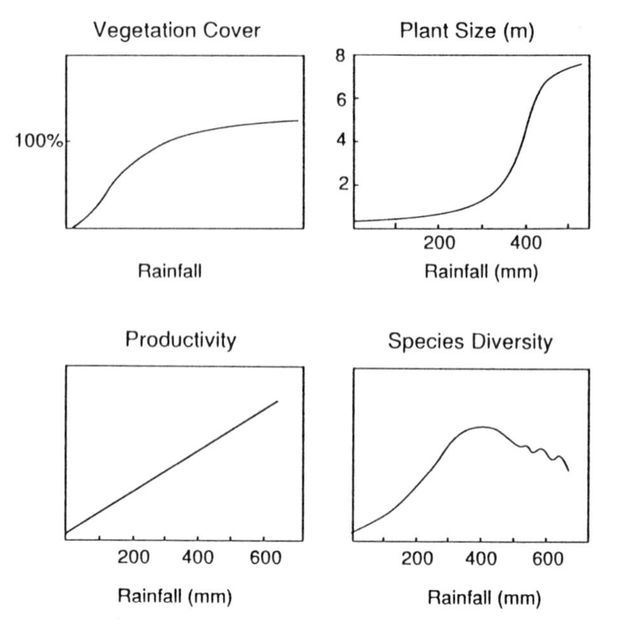

Figure 4.14. Relationship between rainfall and various vegetation characteristics (from Shmida 1985).

dependent optimum is reached, then declines at higher temperatures. Wind promotes both the drying and cooling of vegetation.

Moisture availability represents the interplay of rainfall, evaporation, and the underlying ground surface. Thus, the environmental characteristics determining vegetation cover include the amount and rhythm of rainfall, atmospheric moisture, solar radiation, surface temperature, ground chemistry, and soil moisture conditions. The last factor is strongly dependent on soil type, slope, and topography, the types and characteristics of surface materials, the nature of the drainage system, and the proximity to groundwater, seasonal streams, and exotic streams (Goudie and Wilkinson 1977).

The interplay between rainfall and edaphic factors can be illustrated by the growth cycle using grasslands as an example. When a rain event occurs that is sufficient for growth, the immediate growth is not water limited but proceeds at a rate that is largely dependent on nutrient status. This rate continues (fig. 4.15) until soil water drops below a critical threshold (a function of soil type). The growth rate then continually decreases in proportion to moisture availability. In this scenario (Scholes 1990), water supply (together with soil type) controls the duration of grass production, but nutrients determine the growth rate during productive periods. Similarly, once the soil becomes wet enough during the season that growth is not water limited, grass production will proceed at a nutrient-controlled level until toward or after the end of the rainy season when the soil dries out and reaches the threshold for water-limited growth (Justice and Hiernaux 1986).

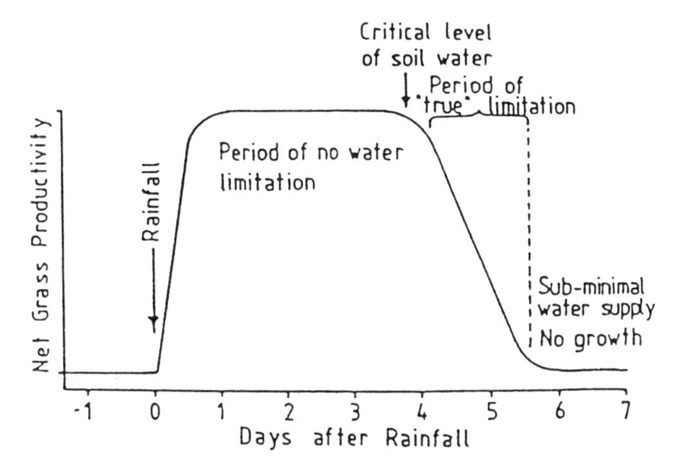

Figure 4.15. Idealized illustration of rainfall and nutrients on production; the peak rate of production is nutrient limited (from Scholes 1990).

Relationships between Vegetation Cover and Moisture Availability

Dryland environments differ greatly in terms of biomass, productivity, characteristic species, variety and richness of these species, surface cover percentage, and size and variety of life forms (Cooke and Warren 1973). This reflects, in part, the diversity of dryland climates. Nevertheless, distinctive characteristics of dryland vegetation emerge because plants must adapt to the conditions of low and variable moisture, extreme temperatures, and high salinity common to arid and semi-arid regions.

The most general characteristics of dryland vegetation are relative scarcity and its variability in time and space (Bourlière and Hadley 1983, Goudie and Wilkinson 1977). Plant cover is concentrated in wetter niches and generally varies from about 0–50%. Both coverage and biomass fluctuate in response to varying moisture during the year or over longer periods. The impact of a wet year can be more dramatic. When a wet year ended a long sequence of dry years in the Namib, desert biomass increased nearly tenfold (Seely and Louw 1980).

Several other characteristics of plants exhibit general relationships to rainfall. Plant size, ground cover percentage, and biomass all tend to increase with rainfall (Shmida 1985), both for ecosystems as a whole and for individual formations or species (fig. 4.14). Some species, particularly cacti, exhibit a remarkable change in size as water becomes more plentiful.

The net production of plant material by photosynthesis is measured in terms of biomass, that is, the dry weight of the organic matter above- and belowground. The net annual increase of biomass, or net primary production, is a measure of the productivity of an ecosystem. In the drier regions, the relationship between net primary production P_r and rainfall P can be generally expressed as

$$P_r = k(P - P_t) \qquad [12]$$

where k is some constant and P_t is a threshold value of rainfall representing the minimum amount of rainfall that is effective for plant growth (Noy-Meir 1985). Relationships of this type are used operationally to estimate net primary production (e.g., Justice and Hiernaux 1986). As with other plant-water relationships, the productivity/rainfall curve varies with the type of soil (fig. 4.16) and vegetation (Davenport and Nicholson 1993, Nicholson and Farrar 1994).

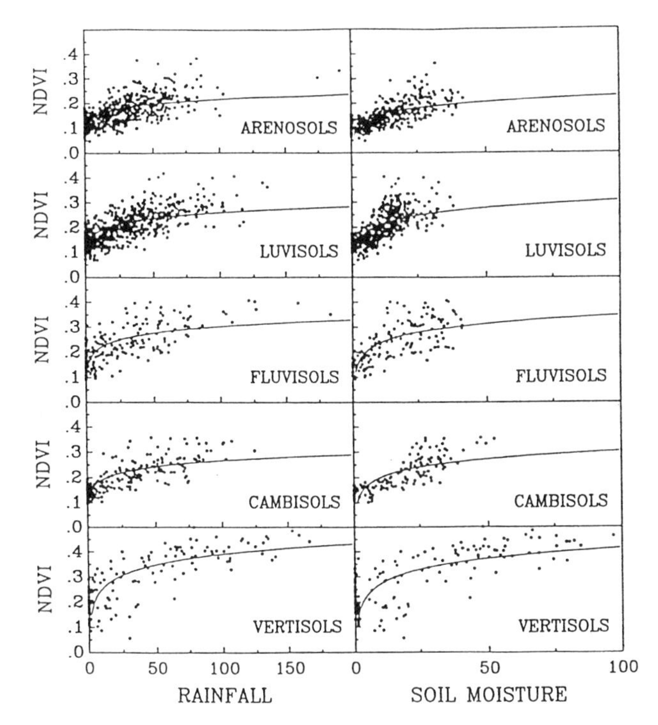

Figure 4.16. Scatter diagrams of monthly normalized difference vegetation index (NDVI) versus rainfall and monthly NDVI versus soil moisture for five soil types (from Nicholson and Farrar 1994). Rainfall is a monthly value averaged for the concurrent plus two previous months; soil moisture is for the month concurrent with NDVI; the line of regression is indicated.

The above relationship (eq. 12) is valid in drier regions, but the rate of increase of P_r decreases and then levels off as rainfall increases and water ceases to be the limiting factor in growth. This happens roughly near the transition to subhumid or humid climates, on the order of 700–900 mm mean annual rainfall (Scholes 1990). The relationship also breaks down at low levels of rainfall. For example, the ratio P_r/P tends to decrease below 100 mm/an. Also, growth may occur below the threshold in microhabitats or when the rainfall is concentrated by runoff patterns (Noy-Meir 1985, Seely 1978).

Noy-Meir (1985) suggests that the ratio of primary production to rainfall (P_r/P) (*rain-use efficiency*) is a better parameter than biomass or productivity for characterizing and comparing different arid regions. It is less variable in time and space and constant over a broad range of moisture conditions. This range can be interpreted as that in which rainfall is the main limiting factor for plant growth. The ratio P_r/P depends on two main variables: the proportion of rainfall returned to the atmosphere by transpiration and the ratio of productivity to transpiration (i.e., the *water-use efficiency*).

Climate-Vegetation Interaction in the Savannas

The savanna environment reflects the intimate link between climate and vegetation. The seasonal rainfall regime, with moisture deficiency and plant stress during the cooler season, is the only commonality of the physical environments of savannas. Savanna environments are extremely diverse, but all share structural and functional characteristics that allow them to tolerant the seasonal drought, and they exhibit a distinctive seasonal cycle of growth in response to the seasonal availability of moisture (Bourlière and Hadley 1983, Cole 1986).

Morphological adaptations include mesophyllous or microphyllous leaves, schleromorphism, and hard, thick, often corky bark (Menaut 1983, Sarmiento and Monasterio 1983). The diminutive leaf size reduces transpirational water loss. The schleromorphism insulates against both thermal and moisture stresses. The bark provides similar protection but also may afford protection against bush fires. Spines and thorns, and sometimes poisonous seed pods, deter predators. Savanna plant species often have extensive root systems to exploit soil moisture and ensure water supply (Menaut 1983).

Many savanna species adapt through rapid growth during the period when moisture is available and through shedding leaves at the end of the dry season. Levels and rates of turnovers of primary production are quite high compared to standing biomass (Bourlière and Hadley 1983, Sarmiento and Monasterio 1983). Production rates are strongly dependent on rainfall.

Climatic, edaphic, and geomorphic factors are important in determining the distribution of savannas (Bourlière and Hadley 1983). Cole's (1986) classification scheme for savanna environments illustrates the relationship with these environmental variables (fig. 4.17). It includes five structural classes differing with respect to the dominance of trees and grasses and the size and spacing of elements: savanna woodlands, savanna parkland, savanna grassland, low tree and shrub savanna. The most important factors controlling the distribution of the categories are water availability (a function largely of rainfall and soil texture) and nutrient availability, but other soil characteristics, such as pH and organic matter content, also play a role.

The broadest distinction is between the savanna woodlands on relatively infertile, highly leached soils at the moist end of the spectrum and the low tree and shrub savanna on

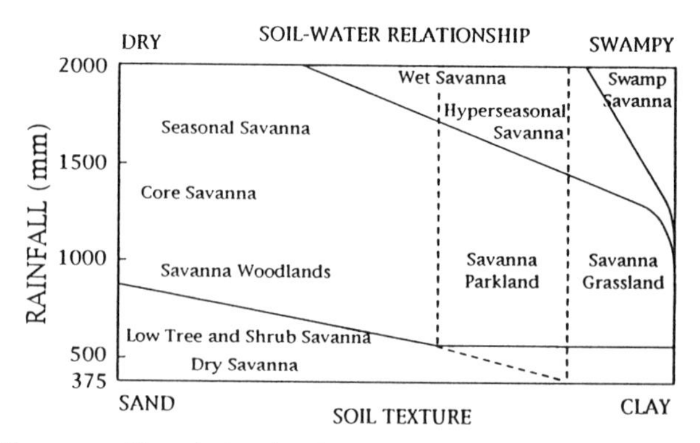

Figure 4.17. The relationship between dominant type of savanna system, rainfall, and soil texture (based on Cole 1986). The higher the rainfall and the finer the soil texture, the greater is the availability of soil moisture and the "wetter" the environment.

unleached, nutrient-rich soils at the more arid margins (Cole 1986). These two extremes are termed moist and dry savannas. Climate, notably the lengthening dry season and increasing extremes of temperature, produces the broad gradation between woodlands and low tree and shrub savanna. For this reason, Scholes (1990) suggests that the distinction between wet and dry savannas be made on the basis of whether or not they are water limited.

Not only the nature of the savanna but also its vegetation associations and species are determined by complex interactive factors in the environment, and overall, the system is in a state of dynamic equilibrium. The core regions of each savanna biome are relatively stable, but the peripheries are *tension zones* (Cole 1986) where the environment shifts markedly in response to drought or exceptional rains or other stresses.

Models have been developed to describe the interactions and the equilibrium states of the savannas and their variability in time and space. As a minimum, a savanna model must include the alternating wet and dry phases, a structure dependent primarily on the competition between woody and herbaceous plants for the available soil moisture, and fire, herbivores, and soil nutrients as principal modifying factors. The model of Walker and Noy-Meir (1982) assumes that grasses have shallow roots drawing water from the upper soil layers and exhausting it quickly through high rates of evapotranspiration, outcompeting trees for the available moisture. Tree species, on the other hand, have longer roots that give them access to the steadier moisture supply of lower layers. The greater the grass cover, the lower the infiltration capacity of the soil, and the less water available to boreal elements. The model suggests dual equilibrium states, one a tree-grass mixture and the other a thicket devoid of grass cover.

Eagleson and Segarra (1985) commenced with the Walker and Noy-Meir (1982) model but simplified it and introduced competition for energy between the two main elements. They also assumed that the grass seeks to minimize moisture stress. Their model shows three equilibrium states: closed forest, grassland, and a tree-grass mixture. Only the last was found to be stable with respect to perturbations in the vegetation components, but metastable with respect to climatic change. Their model was verified with observations of tree-grass savannas in the southern Sudan and in South Africa.

These models suggest a symbiotic relationship between the various system components (Walter 1971). The trees can enrich the soil, prevent erosion, and affect microclimates for the herbaceous layer. Observations after the Sahel drought of the early 1970s demonstrated this (Bourlière and Hadley 1983, Poupon 1979). The drought, which affected flowering and seeds, caused the woody stratum to recede for several years after the drought, with adverse effects on the grass cover as well. Also, the herbaceous vegetation recovered much more quickly in areas with trees and shrubs. These models also describe a delicate equilibrium, sensitive to environmental fluctuations and dependent on maintaining the balance between components and the interactions between them.

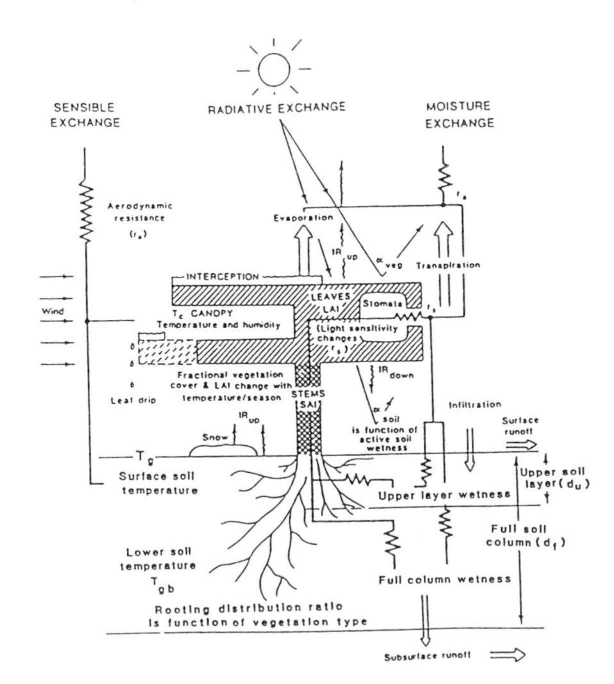

Figure 4.18. Schematic illustrating the features of the land-surface parameterization scheme used in the BATS model (Dickinson et al. 1986).

Vegetation-Climate Interactions

Climatologists are interested in vegetation-atmosphere interactions from the perspective of vegetation as a modifier of climate and climatic variability. Climate is an environmental variable; its changes even on time scales as short as decades or centuries can be sufficiently large to affect the distribution and characteristics of ecosystems. These changes in vegetation in turn exert feedback on the system, modulating the variability of climate.

The feedback mechanism involves the fluxes of energy, moisture, and particulates at the earth's surface. A change of vegetation cover also modifies soil moisture, evapotranspiration, albedo, surface roughness, and soil erodibility. These changes influence momentum and moisture exchange, sensible and latent heating, and the generation of dust. This feedback appears to play a major role in the development of long and severe droughts in continental interiors (Brubaker et al. 1993). It probably played a role in the intensity and unusual duration of drought in the semi-arid Sahel of West Africa (Nicholson 1989).

A current priority in atmospheric science research is modeling of the interactions between the land and atmosphere. Two of the goals of such research are to understand how surface characteristics such as vegetation cover affect climate and to assess the impact of surface modification on climate. Dynamic models of the atmosphere have been developed that explicitly treat land surface processes, using simplified representations of land surface characteristics.††

One such representation is illustrated in fig. 4.18, showing the Biosphere-Atmosphere Transfer Scheme (BATS) developed by Dickinson and colleagues (1986, 1993) for use in atmospheric general circulation models (GCMs). The model consists of two soil layers—an upper layer interacting directly with the surface air and embedded in a total soil column acting as a deeper reservoir of moisture and energy. It has one vegetation layer for which the necessary model paremeters are listed in table 4.3. These relate mainly to fractional coverage, height/roughness length, rooting ratio, albedo (reflectivity), stem area, and leaf area. BATS incorporates eighteen vegetation types based on the work of Kuchler (1983) and Matthews (1983).

Such models have been used to assess the influence of vegetation cover on climate. Bonan and colleagues (1993), for example, investigated the influence of the heterogeneity of cover, concluding that it significantly affects radiation but has a much larger influence on sensible heat transfer and

Table 4.3. Vegetation Parameters in BATS Model

Maximum fractional vegetation cover
Difference between maximum cover and cover at temperature of 269 °K
Roughness length
Depth of the total soil layer
Depth of the upper soil layer
Rooting ratio (upper to total soil layers)
Vegetation albedo for wavelengths < 0.7 mm
Vegetation albedo for wavelengths > 0.7 mm
Minimum stomatal resistance
Maximum LAI
Minimum LAI
Stem (and dead matter) area index
Inverse square root of leaf dimension
Light sensitivity factor

evapotranspiration. The implication of their work and that of others (e.g., Anthes 1984) is that modification of the land surface, even on a relatively small scale, can influence climate. Other models have demonstrated that such heterogeneities can actually induce mesoscale circulations and influence rainfall patterns (Pielke and Avissar 1990, Pielke et al. 1990, 1991).

Several model simulations have evaluated the effects of desertification, a process of great concern in arid and semiarid regions. Early U.N. estimates of desertification's magnitude and extent prompted considerable interest in assessing its potential impact on climate. Charney (1975) calculated that the induced changes in evapotranspiration and albedo in a "desertified" Sahel could reduce rainfall in the region by as much as 40%. More recently, the model of Xue and Shukla (1993) predicted a reduction of 1.5 mm per day, or about 100 mm during a three-month rainy season. Although the early estimates were greatly exaggerated (see, e.g., Hellden 1991, Mainguet 1991), the consensus from modeling efforts suggests that, if desertification becomes sufficiently severe, it could influence climate locally.

It is more difficult to find observational evidence of the influence of land surface changes on climate. However, two case studies are worth noting. Both evaluated grazed and ungrazed sides of international borders: in the Sonaran desert of the United States and Mexico and in the Sinai and Negev. Satellite photos (Otterman 1977, 1981) showed that the soil in the overgrazed Sinai had an average albedo of 0.46, compared to 0.25 in the protected area of the Negev. In the Sonoran case, surface temperature contrasts of 2–4 °C were measured between the more heavily grazed, and warmer, Mexican side of the border versus the U.S. side (Balling 1988, Bryant et al. 1990). This influenced soil moisture and cloudiness, but not precipitation (Balling 1988, Bryant et al. 1990). On the other hand, evidence strongly suggests that irrigated crops can modify the local and mesoscale weather patterns, even to the extent of enhancing rainfall by 15–90% in some months and increasing thunderstorms, hail, and severe weather (Barnston and Schickedanz 1984).

Note

I would like to thank James Shuttleworth and an anonymous reviewer for excellent comments on and critiques of this chapter. Their suggestions have added significantly to the manuscript. I would also like to acknowledge the work of Jeeyoung Kim, Florida State University, and Lee Fortier, National Center for Atmospheric Research, in the production of several of the figures. This work was partially supported by grant ATM-9024340 from the National Science Foundation.

References

Anthes, R. A. 1984. Enhancement of convective precipitation by mesoscale variations in vegetative covering in semiarid regions. Journal of Climate and Applied Meteorology 23:541–54.

Asrar, G., M. Fuchs, E. T. Kanemasu, and J. L. Hatfield. 1984. Estimating absorbed photosynthetic radiation and leaf area index from spectral reflectance in wheat. Agronomy Journal 76:300–306.

Balling, R. C., Jr. 1988. The climatic impact of Sonoran vegetation discontinuity. Climatic Change 13:99–109.

Barnston, A. G., and P. T. Schickedanz. 1984. The effect of irrigation on warm season precipitation in the southern Great Plains. Journal of Climate and Applied Meteorology 23:865–88.

Bonan, G. B., D. Pollard, and S. L. Thompson. 1993. Influence of subgrid-scale heterogeneity in leaf area index, stomatal resistance, and soil moisture on grid-scale land-atmosphere interactions. Journal of Climate 6:1882–97.

Bourlière, F., ed. 1983. Ecosystems of the world. Vol. 13: Tropical savannas. Amsterdam: Elsevier.

Bourlière, F., and M. Hadley. 1983. Present-day savannas: An overview. *In* Bourlière, ed. Ecosystems of the World. 1–17.

Box, E. O. 1981. Predicting physiognomic vegetation types with climate. Vegetatio 45:127–39.

Brubaker, K., D. Entekhabi, and P. Eagleson. 1993. Estimation of continental precipitation recycling. Journal of Climate 6:1077–89.

Bryant, N. A., L. F. Johnson, A. J. Brazel, R. C. Balling, C. F. Hutchinson, and L. R. Beck. 1990. Measuring the effect of overgrazing in the Sonoran desert. Climatic Change 17:243–64.

Budyko, M. I. 1986. The evolution of the biosphere. Dordrecht: Reidel.

Charney, J. G. 1975. Dynamics of deserts and drought in the Sahel. Quarterly Journal of the Royal Meteorological Society 54:642–46.

Charney, J. G., W. J. Quirk, S. H. Chow, and J. Kornfield. 1977. A comparative study of the effects of albedo change on drought in semiarid regions. Journal of the Atmospheric Sciences 34:1366–85.

Cole, M. M. 1986. The savannas. London: Academic Press.

Cooke, R. U., and A. Warren. 1973. Geomorphology in deserts. London: Batsford.

Davenport, M. L., and S. E. Nicholson. 1993. On the relationship between rainfall and the normalized difference vegetation index for diverse vegetation types in East Africa. International Journal of Remote Sensing 14:2369–89.

Dickinson, R. E., A. Henderson-Sellers, and P. J. Kennedy. 1993. Biosphere-Atmosphere Transfer Scheme (BATS) version 1e as coupled to the NCAR Community Climate Model. Tech. Note TN-387+STR. Boulder, Colo.: National Center for Atmospheric Research.

Dickinson, R. E., A. Henderson-Sellers, P. J. Kennedy, and M. W. Wilson. 1986. Biosphere-Atmosphere Transfer Scheme (BATS) for the NCAR Community Climate Model. Tech. Note TN-275+STR. Boulder, Colo.: National Center for Atmospheric Research.

Dunne, T., and L. B. Leopold. 1978. Water in environmental planning. New York: Freeman.

Eagleson, P. S. 1982a. Ecological optimality in water-limited natural soil-vegetation systems. Pt. 1: Theory and hypothesis. Water Resources Research 18:325–40.

———. 1982b. Ecological optimality in water-limited natural soil-vegetation systems. Pt. 2: Tests and applications. Water Resources Research 18:341–55.

Eagleson, P. S., and R. I. Segarra. 1985. Water-limited equilibrium of savanna vegetation systems. Water Resources Research 21:1483–93.

Ellery, W. N., R. J. Scholes, and M. T. Mennis. 1991. An initial approach to predicting the sensitivity of the South African grassland biome to climate change. South African Journal of Science 87:499–503.

Evenari, M. 1985a. Adaptations of plants and animals to the desert environment. *In* Evenari et al., eds. Ecosystems of the world. 79–92.

———. 1985b. The desert environment. *In* Evenari et al., eds. Ecosystems of the world. 1–22.

Evenari, M., I. Noy-Meir, and D. W. Goodall, eds. 1985. Ecosystems of the world. Vol. 12A: Hot deserts and arid shrublands. Amsterdam: Elsevier.

Goudie, A., and J. Wilkinson. 1977. The warm desert environment. Cambridge: Cambridge University Press.

Hellden, U. 1991. Desertification: Time for an assessment? Ambio 20:372–83.

Justice, C. O., and P. H. Y. Hiernaux. 1986. Monitoring the grasslands of the Sahel using NOAA AVHRR data: Niger 1983. International Journal of Remote Sensing 7:1475–98.

Kuchler, A. W. 1983. World map of natural vegetation. Goode's World Atlas. 16th ed. Chicago: Rand McNally. 16–17.

Lowry, W., and P. Lowry. 1989. Fundamentals of biometeorology. McMinnville, Oreg.: Peavine Publications.

Lynn, B. H., and T. N. Carlson. 1990. A stomatal resistance model illustrating plant vs. external control of transpiration. Agricultural and Forest Meteorology 52:5–44.

Mainguet, M. 1991. Desertification: Natural background and human mismanagement. Berlin: Springer.

Marsh, B. A. 1990. The microenvironment associated with *Welwitschia mirabilis* in the Namib desert. *In* M. K. Seely, ed. Namib ecology. Transvaal Museum Monograph No. 7. Pretoria: Transvaal Museum. 149–54.

Matthews, E. 1983. Global vegetation and land use: New high-resolution data bases for climate studies. Journal of Climate and Applied Meteorology 22:474–87.

Menaut, J.-C. 1983. The vegetation of African savannas. *In* Bourlière, ed. Ecosystems of the World. 109–50.

Miller, D. H. 1977. Water at the surface of the earth. International Geophysics Series 21. New York: Academic Press.

Nicholson, S. E. 1989. African drought: Characteristics, causal theories, and global teleconnections. *In* A. Berger, R. E. Dickinson, and J. W. Kidson, eds. Understanding Climate Change. Washington, D.C.: American Geophysical Union. 79–100.

Nicholson, S. E., and T. J. Farrar. 1994. The influence of soil type on the relationships between NDVI, rainfall, and soil moisture in semi-arid Botswana. Pt. 1: Relationship to rainfall. Remote Sensing of Environment 50:107–20.

Nix, H. A. 1983. Climate of the tropical savannas. *In* Bourlière, ed. Ecosystems of the World. 37–62.

Noy-Meir, I. 1985. Desert ecosystem structure and function. *In* Evenari et al., eds. Ecosystems of the world. 93–103.

Oke, T. R. 1978. Boundary layer climates. London: Methuen.

Otterman, J. 1977. Anthropogenic impact on the albedo of the Earth. Climatic Change 1:137–57.

———. 1981. Satellite and field studies of man's impact on the surface in arid regions. Tellus 33:68–77.

Otterman, J., and C. J. Tucker. 1985. Satellite measurements of surface albedo and temperatures in semi-desert. Journal of Climate and Applied Meteorology 24:228–34.

Petrov, M. P. 1975. Deserts of the world. New York: John Wiley and Sons.

Pielke, R. A., and R. Avissar. 1990. Influence of landscape structure on local and regional climate. Landscape Ecology 4:133–55.

Pielke, R. A., G. Dalu, J. S. Snook, T. J. Lee, and T. G. F. Kittel. 1991. Nonlinear influence of mesoscale land use on weather and climate. Journal of Climate 4:1053–69.

Pielke, R. A., G. Dalu, J. Weaver, J. Lee, and J. Purdom. 1990. Influence of land use on mesoscale atmospheric circulations. Fourth conference on mesoscale processes, extended abstracts. Boston: American Meteorological Society. 226–27.

Poupon, M. 1979. Structure et dynamique de la strate ligneuse d'une steppe sahèlienne au nord du Sénégal. Master's thesis. Paris: University of Paris.

Ripley, E. A. 1992. Water flow. *In* R. T. Coupland, ed. Ecosystems of the world. Vol. 8A: Natural grasslands. Amsterdam: Elsevier. 55–73.

Robinson, M. D., and M. K. Seely. 1980. Physical and biotic environments of the southern Namib dune ecosystem. Journal of Arid Environments 3:183–203.

Rutter, A. J. 1975. The hydrological cycle in vegetation. *In* J. L. Monteith, ed. Vegetation and the atmosphere. Vol. 1. New York: Academic Press. 111–54.

Sarmiento, G., and M. Monasterio. 1983. Life forms and phenology. *In* Bourlière, ed. Ecosystems of the World. 79–108.

Scholes, R. J. 1990. The influence of soil fertility on the ecology of southern African dry savannas. Journal of Biogeography 17:415–19.

Seely, M. K. 1978. Grassland productivity: The desert end of the curve. South African Journal of Science 74:295–97.

Seely, M. K., and G. N. Louw. 1980. First approximation of the effects of rainfall on the ecology and energetics of a Namib desert dune ecosystem. Journal of Arid Environments 1:117–28.

Seely, M. K., D. Mitchell, and K. Goelst. 1990. Boundary layer microclimate and *Angolosaurus skoogi* (Sauria: Cordylidae) activity on a northern Namib dune. *In* M. K. Seely, ed. Namib ecology. Transvaal Museum Monograph No. 7. Pretoria: Transvaal Museum. 155–62.

Sellers, P. J. 1987. Canopy reflectance, photosynthesis and transpiration. Pt. 2: The role of biophysics in the linearity of their dependence. Remote Sensing of Environment 21:143–83.

Shmida, A. 1985. Biogeography of the desert flora. *In* Evenari et al., eds. Ecosystems of the world. 23–77.

Shuttleworth, W. J., and J. S. Wallace. 1985. Evaporation from sparse crops: An energy combination theory. Quarterly Journal of the Royal Meteorological Society 111:839–55.

Strahler, A. H., and A. N. Strahler. 1992. Modern physical geography. 4th ed. New York: John Wiley and Sons.

Thom, A. S. 1975. Momentum, mass, and heat exchange in plant communities. *In* J. L. Monteith, ed. Vegetation and the atmosphere. Vol. 1. New York: Academic Press. 57–110.

Tucker, C. J., and P. J. Sellers. 1986. Satellite remote sensing of primary production. International Journal of Remote Sensing 7:1395–1416.

Walker, B. H., and I. Noy-Meir. 1982. Aspects of the stability and resilience of savanna ecosystems. *In* B. J. Huntley and B. H. Walker, eds. Ecology of tropical savannas. New York: Springer. 556–90.

Walter, H. 1971. Ecology of tropical and subtropical vegetation. London: Longman.

Xue, Y., and J. Shukla. 1993. The influence of land surface properties on Sahel climate. Pt. 1: Desertification. Journal of Climate 6:2232–45.

5 Arid and Semi-Arid Land Community Dynamics in a Management Context

S. Archer, W. MacKay, J. Mott, S. E. Nicholson, M. Pando Moreno, M. L. Rosenzweig, N. G. Seligman, N. E. West, and J. Williams

Community and Resource Management

Resource Integrity and Management Goals The rich habitat diversity of arid and semi-arid lands creates opportunities for manipulation, exploitation, and conservation. In addition to traditional economic production activities associated with grazing, cultivation, irrigation, and water harvesting, other potential activities include recreation, tourism, landscape enhancement, and conservation of physical and biological resources. In less-developed arid regions, the agropastoral sector is still the main economic activity, but, in the more affluent arid regions of the world, settlement and development are based on strong physical and socioeconomic links with the rest of society and with the economy outside the arid zone. These links promote urbanization of the region and create socioeconomic values that tend to overshadow the traditional primacy of the agropastoral sector. Environmental and "nonproduction" aspects of land use become important issues, especially around areas of human abode and activity.

Communities are set in distinct units of landscape and provide a framework for spatial evaluation of management impacts on it. Land management cascades through a hierarchy of administrative units from the state, region, local unit (farm, park, mine, nature reserve), and on to subunits (field, paddock, block, site). A management unit normally has cadastral boundaries that encompass a number of plant and animal communities set in distinct landscape units with different characteristics and different relevance to the management goals. Attempts to achieve a balance between exploitation and conservation of the communities that constitute the resource is complicated by the conflicting nature of these goals. Maintaining the integrity and richness of the resource is an ecological challenge that must be met within a prevailing socioeconomic context.

The workshop that provided the basis for this book was concerned with updating the information that land managers in dry regions require to make decisions that promote socioeconomic and ecological goals. In this chapter, we will address some current issues of dryland community ecology that relate mainly to the use of the vegetation as rangeland and to the conservation of its biological resources.

Management Options Along the continuum between the semi-arid and arid communities, potential vegetation density decreases from closed canopy to sparse open stands while the ability to maintain cultivation and sedentary populations of large animals decreases until migration, transhumance, and nomadism become essential strategies for survival of the larger mammals and people dependent on them. Management can impact arid land vegetation communities by imposition of very different measures: utilization by animals, removal or addition of selected species, manipulation of moisture and nutrients, or even nonintervention. The nature and intensity of these practices depend not only on the management goals and resource constraints but also on the availability of appropriate techniques. Command over the necessary techniques is a function of the knowledge and experience of the manager. Application then depends not only on the availability of the necessary labor, capital, and information about market and environmental trends but also on social and administrative constraints. These factors together determine *management maneuverability* among specific management options, ranging from intensive land and livestock management to subsistence pastoralism. The more extensive options are not always open because they often involve substitution of labor for capital. Where labor costs become too high or where access to more favored areas is restricted, even extensive solutions like transhumance or nomadism become unfeasible.†

With the development of a more urbanized, industrial, service, and leisure-based society, the arid regions become increasingly dependent on material, economic, and demographic flows from other regions. These developments are facilitated by improved communications and greater accessibility. Recreation and environmental values become increasingly important. The need for shade and softening of the harshness of the arid environment motivates landscape enhancement projects, often on a fairly large scale; intrusion of heavy engineering earthworks for highway construction, mining, and settlement create vandalized landscapes that need restoration, often dictated by law. The conservation of the genetic resources of arid regions can be threatened by these developments even as management for conservation commands increasing recognition in many societies. Land management is becoming more complex and capable of inflicting greater disturbance than ever before. Understanding the implications of management interventions for the constituent plant and animal communities is an urgent scientific, administrative, and political issue.

Some Basic Concepts

Communities In its widest sense, communities are any combination of two or more species populations in a given land area. Consequently, definitions of *community* are scale dependent. Some communities occur in recognizable patterns over large areas and are ecologically significant in that they reflect interrelations between the species and their dependence on topography, soils, microclimate, geomorphology, and past disturbance or history. As such, "the community and its environment are inseparable" (Daubenmire 1959), especially when we recognize the fauna, domesticated and wild, micro and macro, as part of the environment. The interdependence among the constituents of the community is not as tight nor as obligative as in an organism, and consequently, plant communities can be heterogeneous and have indeterminate boundaries. In addition, they are dynamic organizations that change with variations in climate, management, and disturbance.

The populations that constitute communities include a few species that are common in the community and many that are much rarer and more spottily distributed (Hanski et al. 1993). Which group of species should be the focus of management depends on management's goals. If those are agriculture or pastoralism, management has to focus on the common species that constitute the bulk of the biomass because usually these serve most or conflict most with the objectives of management. However, if biodiversity or ecotourism is the goal, management must focus on the rarer or more spectacular species. Moreover, these are sometimes endangered species in need of special care. In any case, a focus on the goals of management and the recognition of the properties and responses of key species to environmental change and management manipulation are necessary to make both the management and research agenda more tractable and more relevant. Problems arise when these goals conflict and challenge managers, scientists, and policy makers.

Soil surface properties are often no less important than vegetation in assessing or monitoring arid and semi-arid communities. Information provided from both these components can indicate whether perceived change can be ascribed to interactions between land-use practices and vegetation alone or whether the soil as habitat for plants has itself been altered. Protocol for assessing vegetation properties is relatively advanced, and potentially useful procedures for assessing changes in soil surface conditions have recently been developed (Tongway 1994).

Functional Groups Species in a community constitute *functional groups* or *guilds* when they have ecologically similar responses to and effects on the environment or when they have similar physiological adaptations. Plants within functional groups have similar rates of resource acquisition, similar effects on resource supply, energy balance, and water balance, and similar responses to certain types of disturbance and climatic fluctuation (Chapin 1980, Grime 1977, Keddy 1992b). C_3 or C_4 perennial grasses that respond in a similar manner to grazing can be valid functional groups. Pollinator guilds share the same plant species; rodent guilds consume the same food resources; soil invertebrates can be grouped according to their role in detrital food webs and bacteria according to their function in a system context (e.g., nitrifiers, methanogens). No two species have identical physiological traits, so when all ecosystem processes are considered, the number of functional groups may approach the number of species. However, only a few processes are normally considered for any particular purpose in management or research so that, when there are clear correlations among response variables, meaningful functional groups can be defined (Chapin 1993, Keddy 1992b). Definition of criteria for functional classification of plants dates back to Alexander von Humboldt in the last century (Smith et al. 1993). Refinements that include features such as plant stature and longevity, response to extreme temperatures, photosynthetic pathway, leaf longevity, size, shape, and consistency have facilitated the study of regional and global plant-climate relationships (Box 1981, Emanuel et al. 1985a, 1985b, Neilson et al. 1992, Rizzo and Wiken 1992, Running et al. 1994, Shugart 1990).

Indicator Species Some species are indicators of specific conditions related to edaphic properties, climatic conditions, depth to groundwater, presence of pathogens, and disturbance and management history. Inconspicuous lichens, mosses, and algal crusts can significantly influence soil nitrogen, soil surface stability, and water infiltration in dryland ecosystems and, thus, may warrant closer monitoring than has traditionally been the case. At times, minor species can have disproportionately large effects on habitat structure, species composition, and biogeochemical processes. The kangaroo rat (*Dipodomys* spp.), a native, fossorial, nocturnal, granivorous rodent of southwestern North American deserts, is one example. Changes in its distribution or abundance appear to have greater control over desert-grassland transitions than the more conspicuous livestock that graze these areas (Brown and Heske 1990). Identification and monitoring of such species are critical for understanding and effective management of arid lands communities for both exploitation and conservation.

Major Factors

Climate

Climate and Desertification Climate is a major factor in the development of the biophysical environment, particularly of the dominant vegetation. Climate is not, however, an environmental constant; it fluctuates on a spectrum of time scales from years to millennia. The environment responds to these fluctuations with various intensities and time lags. Superimposed on the climate-environment interaction is human intervention, often in response to climate fluctuations and weather phenomena, such as droughts and floods.

These complex interactions are particularly apparent in semi-arid regions where grazing pressure exacerbates the change in vegetation accompanying drought. In the early 1970s, a number of scientists suggested that these forces were leading to irreversible desertification, to "the expansion of desert-like conditions and landscapes to areas where they should not occur climatically" (Graetz 1991a). The Sahel drought, which commenced in the late 1960s, was seen as an outstanding example of this process. Biswas and Biswas (1980) estimated that 9 million km² of drylands were desertified and that virtually all arid and semi-arid regions (35% of the earth's surface) are at risk. Researchers now recognize that these statistics are greatly exaggerated and that no serious scientific study was behind their production (e.g., Hellden 1991, Mainguet 1991, Nicholson 1990, Rhodes 1991, Verstraete 1986). The figures in West Africa, for example, were based on vegetation changes in one or two locations during a time period in which mean rainfall had decreased by 50% (Nicholson 1990). More recently, Tucker and colleagues (1991) have demonstrated that the Sahelo-Saharan boundary, rather than advancing continuously, responds nearly linearly to the interannual fluctuations of rainfall in the region.

This does not mean that desertification is not a real problem. Certainly, land degradation seriously affects many arid and semi-arid regions. However, the paradigm of the advancing desert is misleading. The degree and extent of degradation is not well known and its physical manifestations are not well established. Answers to these questions are a prerequisite to understanding critical issues, such as the feedbacks among climate, vegetation and grazing, the degree of amplification of environmental disturbances, and the resilience of the environment (Graetz 1991b, Schlesinger et al. 1990, Verstraete 1986).

These issues have been recognized and incorporated into a recent national scale evaluation of rangeland condition in Australia by Tothill and Gillies (1992) in an effort to determine whether range rehabilitation would occur on its own at the end of drought conditions and with an improvement in weather conditions or whether major noneconomic land management inputs would be needed.

Climatic Variability and Forecasting Studies of productivity and water availability, based on the remotely sensed normalized difference vegetation index (NDVI), indicate that the impact of climate variability depends on the degree of aridity. At relatively low rainfall amounts, annual productivity increases roughly linearly with rainfall in excess of a minimum unavoidable loss by evaporation until nutrients or radiation become limiting factors. In southern Africa, annual productivity increases progressively less with rainfall up to a threshold of about 600–800 mm per year. In eastern and western Africa, this threshold is higher at 1,000–1,200 mm per year (Nicholson et al. 1990). This means that, in drier environments, the interannual variability of productivity is closely correlated with the interannual variability of rainfall. In wetter semi-arid regions, the correlation is weaker and the impact of drought on productivity will depend on the degree to which rainfall during the drought year falls below the threshold.

Long-term climatic forecasting has become more reliable with the greater understanding of the El Niño-Southern Oscillation (ENSO). ENSO is a periodic disruption of global ocean temperature and wind patterns that has a major impact on rainfall in many semi-arid regions (Nicholson 1994, Ropelewski and Halpert 1987, 1989). It is tracked by way of an atmospheric pressure index between the South Pacific and Indian oceans called the southern oscillation index (SOI). Negative indices of the SOI below -5 (termed El Niño events) are associated with low rainfall in eastern Australia, Indonesia, some parts of India, and eastern and southern Africa. There are good examples of the effect in both Africa and Australia. In northern Kenya, rainfall is below average during the first half of an ENSO event but above average during the "little rains" of main ENSO years; the opposite sequence is evident in Botswana. On the eastern Australian seaboard, the SOI has had a dramatic effect on the climate and agricultural production of Queensland (e.g., McKeon and White 1992). About half the droughts (rainfall less than 3 decile) have been associated with changes in the SOI.

In both Australia and southern Africa, the relationship between changes in the SOI and major weather patterns has led to an appreciation of their potential use in predicting weather patterns (McKeon and White 1992). In Queensland, there is increasing use of SOI with remotely sensed real time weather data and simulation models of agricultural and seasonal climate forecasts to provide a spatial predictive analysis for weather changes within the state (Brook et al. 1992). However, recent studies (Russell et al. 1992) have shown that the use of the SOI alone is limited because it is a global estimate. Much more regional specificity and accuracy can be obtained by analyzing SOI data with sea surface temperature (SST) data. The use of process models, such as general circulation models (GCMs), may also eventually improve the resolution and reliability of forecasting skill.

Water and Primary Production

Water, by definition, is the primary determinant of plant growth and other biological functions in dry regions. Once water is available, nutrients and other factors come into play (van Keulen and Seligman 1992). Availability of water in these regions is, however, a particularly complex phenomenon because rainfall distribution and redistribution by runoff and run-on interact strongly with evaporation from the soil surface to create extreme variability both within and among growing seasons. Estimating the terms of the water balance in dry regions has, therefore, attracted research attention throughout this century and is the basis of many models of plant production (e.g., McCown and Williams 1990, van Keulen 1975).

The Water Balance The interaction of rainfall with soil surface properties determines the partitioning of water among entry, transmission, storage in the profile, generation of overland flow, and redistribution elsewhere in the landscape. The evaporative demand of the atmosphere is determined by the combined influences of solar radiation, wind flow, and water vapor pressure. The resultant balance between the input of rainfall and the output of soil and plant evaporation are the two most significant determinants that must be quantified in order to predict the water status of a site in a semi-arid region. The water balance can be written as:

$$P + Es + T + RO + DD + S = 0 \qquad [1]$$

where P is precipitation as rainfall or snow, Es is soil evaporation, T is transpiration, RO is runoff (or run-on), DD is deep drainage beneath the root zone, and S is the change in the quantity of water stored in the soil profile.

Water Use and Plant Production Early work by Briggs and Shantz (summarized by Shantz and Piemeisel 1927) set a foundation for predicting plant production as related to water stress by demonstrating a strong, consistent relationship between transpiration (T) and plant biomass. This relationship results from the close linkage between photosynthesis and transpiration as shown by the theoretical work of Penman (1948) and later by the experimental work of Boyer and Macpherson (1975). De Wit (1958) examined the relationships between accumulated transpiration and plant production. In particular, he and many others (e.g., Fisher and Turner 1978, Hanks et al. 1969, Tanner and Sinclair 1982) have shown that, in arid and semi-arid environments where productivity is water limited, plant biomass production is some function of T/Eo where T is transpiration accumulated over the observed growing period and Eo is the average daily potential evaporation or evaporative demand of a free water surface during that period.

Physical Processes and Water Supply to the Plant The terms of the water balance can be determined by appropriate models based on soil physics, hydraulics, and plant growth processes that include canopy development. Water entry and runoff are important components of such models because overland flow is a dominant feature of arid lands hydrology and is controlled by surface hydraulic properties, vegetation cover, litter residues, and surface roughness. Physically based models that can simulate the influence of soil surface condition on infiltration and runoff can improve our understanding of the impact of grazing pressure on soil hydrology and on plant population dynamics. They are also necessary to facilitate prediction of erosion, nutrient, and organic carbon loss from arid soils.

Soil Water Infiltration and Movement The physical processes that govern and describe water entry and movement through soil and to the plant root are well understood and the mathematics is well developed. The influence of surface roughness and surface mulches on soil evaporation and pore flow theory are areas, however, that need further attention.

Storage Overflow Models In the storage overflow models, the soil parameter is simply a measure of the size of the soil profile store or bucket. This is usually estimated from depth of rooting and the upper and lower limits of available water for each horizon of the profile. It is presented as a single value and ranges from 50 mm in shallow, light textured soils up to 400–500 mm for deep, heavier textured soils that are fully exploited by deep-rooting grasses, shrubs, and trees (Williams 1983). There have been many variations in this form of model where the soil profile is subdivided to represent layers or horizons (de Wit and van Keulen 1972). One form that can be useful in evaluating semi-arid regions is where the surface detention store and the interception loss in the plant canopy are simulated as two additional stores that receive rainfall and then lose it either to the soil profile or directly to the atmosphere as evaporation.

The capability of these storage overflow models to simulate the runoff in arid and semi-arid environments has been shown to be poor under high intensity rainfall, as is often the case in these climates (Winkworth 1969). Prediction of water supply for the plant is significantly more accurate than observation.

Interception and Surface Redistribution Where rainfall arrives in many showers of small magnitude, interception by vegetation and subsequent evaporation can be very important (Slatyer 1967). This water, while never entering the soil store, can represent a significant part of the semi-arid water balance. Surface detention of water is determined by the roughness of the soil and vegetation. There is much theoretical and experimental evidence that increased surface roughness as well as increased plant density, including microphytes, reduces the volume and velocity of runoff, increases its depth of water penetration, and so protects the soil surface from erosional loss (Dunne et al. 1991, Williams and Bonell 1988). The redistribution of water from runoff areas to run-on areas as a consequence of heterogeneity in microrelief, surface roughness, and soil hydraulic properties has a major influence on plant productivity in semi-arid lands (Williams et al. 1994). Seed dispersal and germination are influenced by these aspects of the water distribution: run-on areas, where there is usually soil deposition and higher infiltration rates, provide favorable niches for seed germination and plant vascular establishment. Management of the Mulga, Mallee, and Chenopod shrublands of arid and semi-arid Australia is based on these patterns of heterogeneity in water distribution in the landscape. Spatial models that can treat these issues have yet to be developed, although some of the tools are becoming available.

Soil Evaporation and Deep Drainage The water evaporated from the soil surface (Es) is a significant component of the water balance in semi-arid regions. If plant biomass is predicted by using the water use efficiency methodology, then the separation of *Es* from transpiration (*T*) is most important (Amir and Sinclair 1991, van Keulen 1975). For semi-arid regions in southern Australia, the soil evaporation in pastures and rangelands can exceed 150 mm for a 500 mm annual rainfall. The soil surface water and temperature status, that are critically important to seed germination and plant establishment, are largely determined by soil evaporation. Litter and surface mulches not only modify soil evaporation and soil temperature but also provide niches for soil animals.

Deep drainage is that part of the water that enters the soil profile and moves beneath the root zone toward the local or regional groundwater aquifer. In one sense, this water is lost to the plant system but in many landscapes it moves laterally and may contribute to seeps and soaks at other locations in the landscape. In semi-arid lands, this subsurface redistribution of water can play an important part in plant and animal production. Regions of recharge and discharge are features of most semi-arid landscapes. This deep drainage component of the water balance is affected by changing vegetation patterns through land clearing and grazing and is responsible for extensive salinization in semi-arid regions of Australia (Peck and Hurle 1973).

Nutrients

Water and Nutrient Limitations to Productivity Primary production and plant community composition in semi-arid regions is usually viewed as water limited but, in fact, nutrients also frequently limit production (Ludwig 1987, van Keulen 1975, West 1991). Plant community composition significantly influences nutrient cycling in semi-arid environments. This influence is particularly noticeable when the proportions of plants with differing life forms/life histories are altered (Archer 1994b, West 1991). The opening of nutrient cycles due to conversion to annually tilled farmlands are most dramatic (Hobbs 1993), but the more subtle chronic disturbances to wildlands also have consequences.

Plant Form and Nutrient Flow Semi-arid wildland plant communities are usually dominated by a mix of herbaceous and woody species that appears well adapted to the variable climate in which some years are favorable for grassland species and others for shrubs. The higher density of fine, fibrous roots of grasses, concentrated in the surface soils, facilitates extraction of water and nutrients following rains that only infiltrate the upper soil profile. Shrubs and small trees extract water and nutrients from much greater soil depths. With this deeper soil moisture extraction, they can respond to much earlier infiltration events and maintain activity after the grasses and forbs are effectively drought and/or cold limited. Furthermore, shrubs and trees of semi-arid zones may lift water at night during late summer and so provide additional soil water for their near surface roots, microorganisms, and nearby grasses (Caldwell 1990, Caldwell et al. 1990).

Considerable differences among species can be noted in their ability to compete for space, water, and nutrients (Tilman and Wedin 1991a, 1991b). Annual grasses are particularly plastic in their responses to moisture and nutrient inputs. The only nutrient translocation to surviving parts that annuals make before death is to the seeds. Perennials, however, typically translocate most critical nutrients from senescing to longer-lived parts before shedding them. Shrubs are particularly adept at translocating nutrients from senescing parts and retaining critical nutrients in continuing phytomass (West 1991). Furthermore, shrub roots possess much higher cation exchange capacities than grass roots (Woodward et al. 1984). Largely because of the nutrient translocation before litter drop, the litter from shrubs and trees is of much lower quality for microinvertebrate and decomposer activity than is grass litter (Whitford 1986). Evergreen shrubs, so common in shrub steppes, and trees in savannas typically have lower growth rates and lower nutrient demands because of less need for foliar replacement and, thus, lower maintenance costs than deciduous plants (Goldberg 1982). Relative to herbs, woody growth forms may be better suited to sites with inherently low potential (e.g., shallow soils) or sites with deeper soils from which nutrients have been depleted or that have been degraded through accelerated soil erosion (Archer 1994b). Removal of nutrients from herbs and grasses by grazing can be critical in determining their ability to persist in relation to the shrubs.

Microphytic soil crusts are another common feature of the vegetation of semi-arid regions. They respond to very light rains and even dew (Metting 1991, West 1990). Some of the crust organisms, particularly the cyanobacteria, can contribute to nitrogen input (Harper and Pendleton 1993).

Vegetation Change and Nutrient Implications The native vegetation of semi-arid regions is subject to continuous change. Some factors of change are direct and intentional (e.g., clearing of trees and shrubs by wood gathering, chaining, cabling, roller-chopping, with herbicides, or root ploughing to enhance herbaceous forages for livestock). Other factors are indirect and inadvertent (e.g., alteration of fire regime by excessive grazing, global climatic change, fertilization effects of pollutants, invasion of exotics, etc.). Disturbances on wildlands that have led to loss of perennials and encouragement of annuals have usually led to changes in fire regimes. Frequent and spatially continuous fires can eventually eliminate all woody plants. In North America, the spread of cheatgrass (*Bromus tectorum* L.), an exotic annual, has caused much more frequent fires earlier in the season, eventually eliminating the native shrubby vegetation. Its litter also inhibits nitrification, but the cheatgrass itself prefers NH_4^+ to NO_3^- (Wilkeem and Pitt 1982). The net result is biotic pauperization and microenvironmental homogenization and simplification of many U.S. intermountain systems where this annual has come to dominate. The implications for nutrient cycling are serious but have scarcely been addressed.

Nutrients and Sustainability Soils, not vegetation, are the major store of nutrients in most semi-arid environments, although in tree-dominated systems, a higher fraction of the total nutrient pools, particularly nitrogen, can be found in the phytomass than in the case of grasslands (Whitford 1992). The long residence times of some vital nutrients in humic materials and fixed to clays means that nutrient depletion could be buffered by soil organic matter over very long periods. Extensive, low-level nutrient transfers by migratory animals (McNaughton 1988, McNaughton et al. 1988) and nomads in subsistence economies may not permanently degrade a region (Ellis and Swift 1988). However, when cash economies depend heavily on exports of plant and animal products, nutrient limitations begin to arise (Ellis and Swift 1988). Loss of soil and organic carbon by erosion are also major outflows of nutrients that cannot be replaced easily.

Fertilizer is commonly applied to cultivated land to redress nutrient loss or deficiency. However, fertilizers can have undesirable side effects on natural plant communities. Enhanced productivity is generally accompanied by reduced biodiversity (e.g., Carpenter et al. 1990, Cornelius and Cunningham 1987, Goldberg and Miller 1990, Tilman 1987, Tilman and Pacala 1993). This is caused by the intense competition imposed by aggressive species that respond differentially to increased availability of nutrients (Ellenberg 1985, Grime 1979, Wisheu and Keddy 1992). The phenomenon was reported over a hundred years ago and is virtually ubiquitous (Huston 1979). It has also been shown that there are fewer animal species in very productive situations (Rosenzweig and Abramsky 1993). Annuals in particular are favored by fertilization as well as by overgrazing and mechanical disturbance. As a rule, fertilizer application for higher pasture production in dry regions also increases the interannual variability of production because plant response is greatest in favorable years and poorest in drought years (Seligman 1995). Fertilizers are also expensive, and widespread application on extensive rangelands is seldom economically feasible. Except for nitrogen, which is biologically renewable, all other mineral nutrient losses can only be replaced by slow geochemical processes. Consequently, productivity of dryland communities cannot be sustained under conditions of accelerated erosion and organic matter depletion. The transient benefits of fertilizer application in these situations underline the importance of management for resource conservation. Fertilizers come into their own when conditions are much more favorable and pasture can be rotated with cultivated crops, as in the wheat belt in Australia.

Rangeland Communities: Aspects of Rangeland Ecology

Climatic Complexities

Climate, Community Processes, and Human Activities These factors are interdependent and interact upon each other (fig. 5.1). Climatic processes and events directly influence human activities and community processes via extreme events such as drought, floods, hurricanes, and tornadoes. In other instances, climate influences human activity indirectly via community processes that control productivity, species abundance, and distribution. At the same time, human activities may influence climatic processes ei-

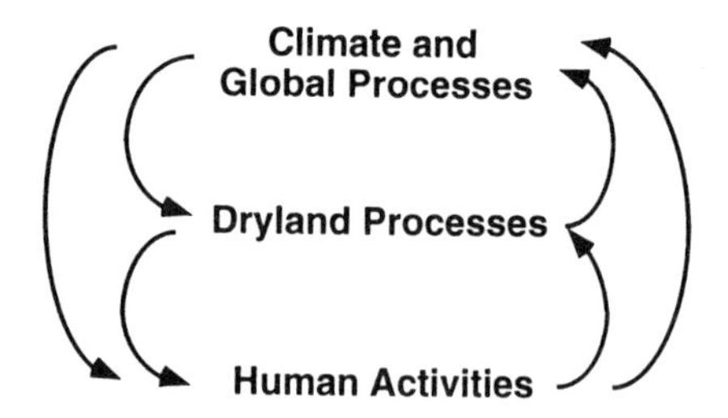

Figure 5.1. Interactions and feedbacks among climatic and atmospheric processes, dryland ecosystem processes, and human activities (modified from Archer 1994a).

ther directly or indirectly. Industrial and land-use practices have increased concentrations of greenhouse gases (CO_2, CH_4, N_2O, CFCs) that have the potential to cause climatic change (Houghton et al. 1992).

Climate-vegetation classifications, such as that developed by Holdridge (1964), have been used to construct maps to display the potential impact of climatic changes predicted under future atmospheric CO_2 concentrations (Emanuel et al. 1985a, 1985b, Rizzo and Wiken 1992, Shugart 1990). However, extrapolation of relationships derived from present-day climate-vegetation correlations to climates with atmospheric CO_2 concentrations without present-day analog is suspect. Moreover, climate-vegetation relationships are typically based on a region's "potential natural vegetation," and this may have little bearing on actual land cover that has been shaped by anthropogenic activities. In addition, static, time-independent, vegetation-climate models predict only the vegetation in equilibrium with regional climate. Dynamic vegetation models explicitly representing key ecological processes, such as competition, mortality, and recruitment, are more likely to work under novel environmental conditions not represented in empirical classification models, but their utility for assessing vegetation dynamics at regional and global scales is currently constrained by numerous factors (Prentice and Solomon 1991, Shugart 1990, Walker 1994). Plant production models for the main native pasture systems in northern Australia have been developed to predict the probable impact of a range of climate scenarios on rangeland productivity and to suggest property management options to accommodate these changes (Howden et al. 1993, McKeon et al. 1990, White et al. 1993).

Elusive ANPP Correlative and mechanistic models highlight precipitation as a key climatic variable regulating community structure and function in arid and semi-arid regions (Hunt et al. 1991, Parton et al. 1988, Sala et al. 1988, 1992). Despite this recognition, several factors limit the relevance of this knowledge. First, the relationships between precipitation and aboveground net primary production (ANPP) derived from regional data sets differ from those produced using temporal data sets (Lauenroth and Sala 1992). Models developed from regional analyses of climate-ANPP relationships (often the only option for areas where long-term data sets are unavailable) may thus contain critical errors. Second, estimates of grassland ANPP, typically based on changes in aboveground biomass, may significantly underestimate total biomass production, especially in tropical and subtropical environments (Long and Hutchin 1991). Apparent changes in ANPP may be real or may reflect a change either in the turnover rate of plant parts or in root/shoot allocation. Finally, the effects of seasonality and intensity of rainfall, which may change under future climates (Pittock 1993), are only poorly understood.

Multiple Scale Disturbances Understanding community structure and function and community dynamics requires a working knowledge of the degree of interaction between plant, animal, and environmental components across an array of spatial and temporal scales (Collins 1987, Collins and Barber 1985, Loucks et al. 1985). On a large scale, precipitation and temperature regulate vegetation dynamics in arid and semi-arid systems (Austin et al. 1981, MacMahon 1980, Sala et al. 1988). Climate also influences vegetation indirectly via its effects on soil development (Jenny 1980) and in creating conditions contributing to natural disturbances such as fire (Baisan and Swetnam 1990, Clark 1988, Swetnam and Betancourt 1990), floods, and windthrow. However, most plant communities and landscapes are extremely patchy (Belsky 1983), and broad-scale climatic factors cannot account for the existence of these small-scale patterns. As spatial and temporal frames of observation diminish and resolution increases, edaphic heterogeneity and disturbances (e.g., grazing, fire, cropping, flooding) assume greater importance in determining community structure and function. Frequent small-scale perturbations, such as ant, termite, or rodent mounds or patchy grazing, occur within the context of larger-scale, less frequent disturbances, such as fires, floods, and wind storms, to produce complex disturbance regimes (Collins 1987) that shape community structure and function.

Episodic Triggers Interpreting the role of climate in shaping ecological systems is particularly difficult in dryland regions where rare or infrequent (episodic) events may do more to shape patterns of plant distribution and abundance than mean conditions (Beatley 1974, Chew 1982, Griffin and Friedel 1985, MacMahon 1980, Turner 1990). These episodic events cause changes in community structure that are stepwise rather than linear (fig. 5.2). Thus, an assemblage of plants may be rather stable and resistant to disturbance or climatic change up to certain thresholds. Beyond these threshold levels, changes can be rapid, dramatic, and potentially irreversible over time frames relevant to management (Griffin and Friedel 1985, Johnstone 1986). Climate and/or disturbance may generate episodes of plant establishment

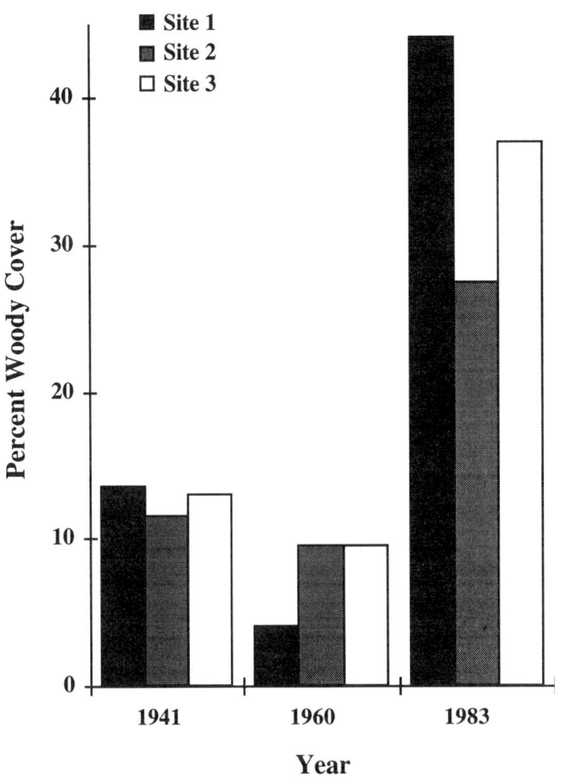

Figure 5.2. Changes in woody plant cover in a grazed subtropical savanna, 1941–93 (from Archer et al. 1988). The 1941–60 period was characterized by a severe drought, whereas the 1960–83 period received normal to above-normal annual rainfall. It appears that drought conditions during the 1950s may have created conditions conducive to rapid expansion of woody vegetation in the subsequent pluvial period.

or mortality whose impacts on community structure are maintained long after the triggering event has passed. Past or future changes in the frequency and/or magnitude of extreme events (either favorable or unfavorable) may thus be more important than gradual shifts in mean values (Katz and Brown 1992).

Directional Change and Fluctuations Climatic variability also makes it difficult to distinguish between directional change and fluctuation in community structure. Fluctuation represents reversible changes in dominance within a stable species composition, whereas succession and retrogression are directional changes in composition and dominance (Rabotnov 1974). Chronic disturbance or changes in environmental conditions can cause retrogression, which typically leads to a loss of diversity, net primary production, and ground cover. During the course of retrogression, site processes become increasingly coupled to and regulated by abiotic factors. This, in turn, may accentuate fluctuation. Progressive directional changes occur in the opposite direction and represent the recovery of community structure following biotic or abiotic disturbance. As plant diversity, production, and ground cover increase through time, the plants themselves influence microclimate, energy flow, nutrient cycling, and species interactions, thus potentially dampening fluctuation associated with oscillation of weather and abiotic factors. Rates and patterns of fluctuation or directional change may be further influenced by the life history traits of species involved (Huston and Smith 1987).

Database Limitations While the concepts of directional change versus fluctuation are relatively straightforward, climate and vegetation databases typically do not cover sufficient time periods to enable making these critical distinctions. Rainfall records in excess of a hundred years at many sites in northern Australia have been used to assess probabilities of drought occurrence and to identify unusual conditions in the past that triggered events that have shaped present-day community structure (McKeon and White 1992). Even when such fairly detailed long-term records are available, our understanding of what causes some sites to fluctuate while others change in a directional fashion may be limited. Collins and colleagues (1987) analyzed 39 years of vegetation data from two sites grazed by cattle. Each site contained an exclosure to prevent cattle from grazing portions of each pasture. The vegetation in the grazed areas of both pastures exhibited shifting patterns of abundance rather than sequential species replacement (i.e., fluctuation), as did one of the exclosures. The other exclosure exhibited a directional change from dominance by annuals to dominance by perennial grasses. Reasons for the contrasting patterns in these pastures were not clear.

Lags and Habitat Change

Biological Inertia The extent to which shifts in vegetation structure lag behind the climatic changes that drive them and the extent to which vegetation can ever be said to be in equilibrium with climate are not easily determined (Davis 1982). This makes it difficult to establish cause-and-effect relationships. For example, patterns of plant recruitment and productivity in one year may be dependent, in part, on climatic conditions of the previous year or years. Vegetation established from seed under one climatic regime may survive under a subsequent regime in a vegetative state. This phenomenon, whereby perennial plants persist for periods of tens to hundreds, even thousands, of years under conditions very different from those under which they initially became established, represents biological inertia (Cole 1985, Lewin 1985). In these instances, correlations between recent changes in vegetation and short-term climatic variables or disturbances may be spurious or low. If the present climatic conditions are such that the dominant plants cannot successfully reestablish from seed with sufficient frequency to maintain the population, community composition is destined to change. Neilson (1986) hypothesized that "the pristine vegetation of the Chihuahuan Desert recorded 100 years ago was a vegetation established under and adapted to 300 years of 'little ice age' and is only marginally supported under the present climate." If this hypothesis is correct, the succession from desert grassland to shrubland reported for this region may have been under way at the time of settlement and was only augmented or accelerated by anthropogenic activities, such as livestock grazing.

Grazing, Climate, and Habitat Degradation In many parts of the world, anthropogenic activities are thought to have caused widespread changes in vegetation and even desertification. However, it can be difficult to assess the extent to which disturbances have influenced communities relative to climatic factors that may be operating simultaneously (Foran 1986, Herbel et al. 1972, McNaughton 1983). As a result, changes in community structure related to climate may be mistakenly attributed to other factors. Western and van Praet (1973) document a case of regional mortality of fever tree (*Acacia zanthophloea*) woodlands in the Amboseli Basin of East Africa. The decline of the woodlands was accompanied by a marked shift toward a more arid habitat, with pastoralist activities and excessive grazing commonly regarded as the principal cause. Subsequent research has demonstrated that livestock grazing was concentrated around the perimeter of the basin where tree mortality was low. Toward the interior of the basin where tree mortality approached 100%, livestock were absent. Additional data indicated fever trees to be highly intolerant of salts and that tree stands in an advanced stage of decline were associated with saline soils. Evidence was presented to show that the fever tree woodlands had developed over a period of decades when the water table was low. Later, during a series of high rainfall years, the water table rose and introduced salts into the tree-rooting zones, causing their demise. In this instance, long-term, climate-induced oscillations in the water table, not pastoralist activities or livestock grazing, appear to have been the proximate driving force for vegetation change.

Retrogression or desertification can thus be natural (Haynes 1982), human-induced (Gornitz and NASA 1985, Owen 1979), or a combination of the two (Hastings and Turner 1965, Verstraete 1986).

Herbivory

Vertebrate Fauna In many ecosystems, herbivory by native or introduced animals can have a profound influence on community structure and function. In locations where the climate and soils might support forest or woodland, utilization of woody plants by browsers and insects can create and/or maintain a grassland or savanna (Belsky 1984, Berdowski 1987, Naiman 1988, Yeaton 1988). Preferential utilization of seeds by granivorous rodents can determine whether short- or tall-grass growth forms predominate in desert grasslands (Brown and Heske 1990). Widespread eradication of a burrowing, colonial prairie rodent (prairie dogs, *Cynomys* spp.) from North American grasslands may have contributed to the encroachment of woody vegetation into many areas (Weltzin 1990). During the same period, direct and indirect effects of grazing by large numbers and high concentrations of livestock appear to have facilitated encroachment of unpalatable woody plants into many communities, thus triggering succession from grassland and savanna to shrublands and woodlands (Archer 1989, Buffington and Herbel 1965, Madany and West 1983, Skarpe 1990, 1991). In Australia, water developments for livestock have contributed to increased abundances of various kangaroo species and, thus, to an increase in overall grazing pressure (Edwards et al. 1995, Wilson 1991). In these situations, management of livestock numbers to promote sustainability of grazing lands may be of little consequence, unless kangaroo densities can be simultaneously controlled. This is difficult to achieve, as actions taken on one property may only serve to increase kangaroo densities on adjoining properties. In addition, regulation of kangaroo numbers is an economic cost to the land manager and fosters controversy with animal rights and conservation groups.

Invertebrate Fauna Although most research has focused on effects of aboveground grazers on vegetation, belowground herbivores such as root-feeding nematodes, larvae, and grubs may actually consume more plant material than aboveground herbivores and may have a proportionally greater impact on total primary production or species composition than would be predicted on their consumption rates (Coleman et al. 1976, Detling et al. 1980). Unfortunately, annual estimates of the numbers or biomass of native animals in ecosystems and the importance of fauna in regulating plant productivity and composition are seldom available. It is, therefore, difficult to gauge their importance relative to more conspicuous domestic herbivores or to climatic events. This also makes it difficult to develop and implement resource management plans.

Invertebrate populations are strongly influenced by the water, temperature, aeration, and strength status of the surface horizons, particularly the litter layers. Sheep grazing in western Australia results in an increase in bare ground and a reduction in litter mass, canopy cover, and soil moisture (Abensperg-Traun 1992). Invertebrate species richness and abundances are usually diminished in grazed areas compared to ungrazed controls. This is thought to be due, in part, to the reduction in architectural diversity (Morris 1973, 1990). Web spider species diversity is highly correlated with vegetation structural diversity and is unaffected by prey availability (Greenstone 1984). In other instances, individuals of some species may actually increase in abundance after grazing due to the changes in microclimate and habitat structure (Thomas 1983). Siepel (1990) and Siepel and van Wieren (1990) have documented the effects of cattle grazing on microarthropod detritivores. The species abundances and compositions were different in grazed and ungrazed plots. Succession in the microarthropods occurred faster in grazed plots. Decomposition rates in grazed plots were also higher, presumably due to the higher temperatures in plots where vegetation was removed.

System Processes

Thresholds and Feedbacks A community may be rather stable and resistant to stress and chronic disturbance up to a certain threshold. Beyond this threshold, changes can be rapid, dramatic, and potentially irreversible over time frames relevant to ecosystem management (Westoby et al. 1989). Thresholds are unstable states in the sense that, when a community is near a threshold, small disturbances can trigger transition from one stable domain to another. The conceptual model in figure 5.3 postulates thresholds in community composition related to the direct and indirect effects of grazing. In grasslands and savannas, thresholds may exist between transitions from one herbaceous assemblage (state) to another, from a herbaceous assemblage to various woody assemblages, and from stable to degraded soils (Friedel 1991). Community structure and function in a given state will depend on the characteristics of the soil, seedbank, and vegetative regenerative potential of the vegetation. Change

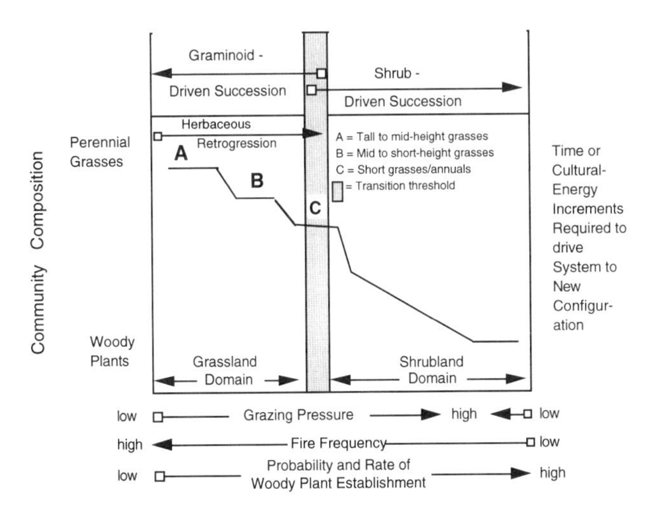

Figure 5.3. Conceptual model of changes in community structure as a function of interactions between grazing and fire (from Archer 1989). Several herbaceous and woody states are possible, and transitions between them are potentially mediated by other disturbances and climate. Recovery to preceding states or transition to an alternate state will depend on species life history traits, the extent to which soils and seedbank have been modified, and climatic factors. Management intervention may be required to augment desirable transitions or prevent undesirable transitions.

can be gradual, cumulative, and reversible until at a certain point a threshold is crossed when some species are reduced to extinction or when habitat is depleted to the extent that it can no longer maintain previous levels of productivity. In semi-arid communities, episodic or catastrophic events, such as drought, torrential rain, or a hot fire, commonly trigger threshold transitions by inducing widespread plant mortality and soil erosion. Systems subjected to chronic stress, such as continuous, heavy livestock grazing, may be more susceptible to these extreme events. On the other hand, establishment of some perennial species, especially long-lived trees and shrubs, can be dependent on very infrequent climatic and environmental combinations that are favorable for seedling establishment.

Whether a community will cross a threshold or remain within a domain of stability depends on the comparative strengths of both positive feedbacks that reinforce deviations from a current state and negative feedbacks that halt or reverse movement away from that state. Schlesinger and colleagues (1990) maintain that positive feedback is operational in desertification. In their proposed scenario, long-term grazing of semi-arid grasslands by livestock produces spatial and temporal heterogeneity of soil resources such as nitrogen and water that promotes invasion by desert shrubs. This leads to a further localization of soil resources under shrub canopies and the creation of barren areas between shrubs. Soil fertility is further reduced by erosion and gaseous emissions. A greater percentage of the soil surface is exposed, and soil surface and air temperatures increase and produce a hotter, drier microclimate that favors drought-tolerant shrubs over grasses.

In other cases, the appearance of a *keystone species* may initiate a positive feedback that dramatically alters community composition and function. Establishment of leguminous shrubs such as *Myrica fava* (Vitousek and Walker 1989), *Prosopis glandulosa* (Archer et al. 1988), and *Mimosa pigra* (Lonsdale 1993) alter patterns of soil nutrients, microclimate, and seed dispersal to eventually control the successional dynamics of whole communities. Introduced grasses such as *Bromus tectorum* (Billings 1990) and *Eragrostis lehmanniana* (Anable et al. 1992) in North America and buffelgrass (*Cenchrus* spp.) in Australia (Walker et al. 1981) have contributed to increased fire frequencies in a self-reinforcing manner, consequently reducing the diversity of native plants and animals (D'Antonio and Vitousek 1992). Once such species establish, reach a critical mass, and put a community into a positive feedback mode, it may be difficult to halt or reverse progression to alternate states. This could be used to advantage in the case of restoration or reclamation efforts by identifying and using desirable species that can initiate positive feedback to a more advantageous state.

State and Transition The existence of such thresholds and asymmetry of transitions between them forms the basis for a *state-and-transition approach* to vegetation monitoring and management (Laycock 1991, Westoby et al. 1989). Knowledge of climatic influences and plant life history attributes, disturbance impacts, and successional processes can be used to identify circumstances whereby desirable transitions can be augmented or facilitated and undesirable transitions mitigated or avoided. In systems where climatic variability is the rule rather than the exception, situations conducive to vegetation improvement or deterioration may arise infrequently and unexpectedly. Failure to recognize and respond to either situation constitutes a missed opportunity. If the potential for transition to undesirable states is ignored, long-lasting potentially irreversible impacts can result. Conversely, progressive and flexible management schemes that can capitalize on infrequent windows of opportunity for vegetation improvement may realize long-term benefits.

Unfortunately, we know little of the ecological attributes of communities that may predispose them to abrupt change. For example, it is generally assumed that heavily grazed areas are more prone to woody plant invasion than are lightly grazed areas, other factors being equal. Yet, little is known about the relationships between probabilities of woody plant seedling establishment and grazing intensity. There are few guidelines for determining when or how often to use prescribed fire to effectively maintain a grass-woody plant balance. If the goal of sustainable management is to anticipate critical thresholds and adjust land-use practices so as not to exceed them, then the challenge for research is to identify reliable plant, soil, and climatic indicators that can be monitored.††

Shrubs and Grasses

Among the plant community phenomena on arid and semi-arid rangeland of direct interest to management, one that has aroused worldwide concern is the status of shrubs vis-à-vis the herbaceous vegetation.† ††

Woody Plant Encroachment In many arid and semi-arid ecosystems throughout the world, the density, biomass, and stature of unpalatable trees and shrubs has increased over the past two hundred years (Archer 1994b). As a rule, these changes in vegetation structure are undesirable because they have reduced the carrying capacity of the land for livestock, contributed to soil erosion and reductions in stream flows, altered wildlife habitat, and threatened pastoralist or ecosystem sustainability. Consequently, shrub encroachment has attracted much research and political attention because of conflicting environmental and production interests. Grazing has been implicated in the spread of bush in Africa, desert and thorn scrub in North and South America, and *Acacia* and *Eucalyptus* woodlands in Australia. Even though our understanding of the mechanisms that control the rates, dynamics, and extent of shrub encroachment has increased in the recent past, control over grass-shrub relations is limited by the constraints of extensive management, especially in dry regions.

Factors regulating the balance between graminoid and woody plant life-forms include climate, soils, disturbance (e.g., grazing, fire), and their interactions. Changes in one or more of these factors may enable woody plants to increase in abundance. A shift from grassland or savanna to shrub or woodland may result if (1) climate or disturbance regimes change to enable native woody species to extend their geographic range or increase in stature and density within their historic ranges or (2) introduced woody species successfully establish and reproduce. Characteristics commonly shared by woody species that increase in grazed environments include (1) high seed production, (2) seeds that persist in soil for many years, (3) ability to disperse over long distances, (4) ability to sprout following top removal, (5) tolerance to low levels of water and nutrients, and (6) low palatability.

Bush Encroachment and Climatic Change The fact that grasslands and savannas have given way to shrublands and woodlands over similar time frames and over such broad geographic areas constitutes one line of evidence that broad-scale factors like climatic change may have been operating. Local and regional studies offer some evidence that seasonal patterns of rainfall and temperature may have changed in recent history to favor woody plants over grasses in portions of North America (Hastings and Turner 1965, Neilson 1986). Grasslands and savannas, which established under previous climatic regimes, may have been only marginally supported by the recent climate and were perhaps prone to woody plant invasion (Neilson 1986, 1987). In addition, oscillations among different climatic regimes in recent history (Mitchell 1980) may have affected shifts in plant recruitment patterns to promote episodes of woody plant seed production and seedling establishment but not necessarily their local extinction (Neilson and Wullstein 1985).

Changes or fluctuations in broad-scale climatic regimes or atmospheric CO_2 levels cannot, however, explain how grasslands and savannas have persisted on some sites within a climatic zone but not others. In cases where shifts from grass-to-shrub domination have been well documented, there is little evidence to support the climatic change hypothesis as a proximate cause for bush encroachment. At the Jornada Experimental Range in southern New Mexico, an analysis of climatic data from 1914 to 1984 failed to demonstrate any consistent statistical rationale for invoking climate as a causal mechanism in the observed vegetation changes (Conley et al. 1992).

Aggressive Exotics In the Mitchell Grasslands of Australia, a leguminous arborescent from Africa, prickly acacia (*Acacia nilotica*), was introduced along water courses on some properties in the 1940s to provide shade for livestock. Since that time, distinctive fenceline contrasts have developed as prickly acacia has spread throughout the pastures where it was introduced and has formed dense stands. Where prickly acacia was not introduced, pastures have been maintained as grassland. The fact that the spread of the plant has been limited by fencelines suggests that livestock rather than native fauna are the principle agents of seed dispersal. However, seed production and seedling establishment of prickly acacia away from watercourses is episodic, occurring only during periods of sufficient rainfall that are infrequent and widely spaced (Burrows et al. 1990).

Fire and Grazing Interactions On a local scale, fire, grazing, and soil properties interact within a variable climate to determine the balance between community components. Temporal patterns of fire and soil moisture are the primary factors influencing woody plant versus grass abundance in the semi-arid regions of southeastern Australia. Episodic rainfall events that trigger woody plant seedling establishment also stimulate grass production (Burrows et al. 1990). Fire (either natural or prescribed) becomes a possibility that, if realized, kills most juvenile and many adult woody plants and promotes grassland vegetation (Harrington and Hodgkinson 1986). Where livestock or high concentrations of grazing wildlife are prevalent, the capacity of grasses to competitively exclude shrub seedlings is reduced, fine fuels that would normally accumulate are consumed, and fire is eliminated as a mortality factor. Climate-fire interactions that maintain grassland or savanna are disrupted, and succession to woodlands dominated by unpalatable trees and shrubs may occur quickly.

Figure 5.2 illustrates how woody plant abundance increased over a 42-year period in savannas of southern Texas. In this system, changes in woody plant abundance have been punctuated and abrupt, not gradual or linear. However, because of the interaction of numerous factors, one can only speculate about causes for the observed change. This site had been fire free and heavily grazed by livestock since the early 1900s. The 1941–60 period was characterized by severe drought, whereas the 1960–83 period received normal to above-normal rainfall. The data suggest that drought may have predisposed the system for rapid rates of woody plant invasion in the postdrought period. Would a change of this magnitude have occurred if livestock had not been grazing the site and (1) accentuating drought stress on grass plants, (2) spreading seeds of the dominant woody species (Brown and Archer 1989), and (3) preventing fine fuels from accumulating and fire from occurring? Conversely, would these activities of livestock have produced this change even if the drought had not occurred?

In a unique field study in Utah, Madany and West (1983) have documented a case in which a savanna protected from cattle grazing was maintained, whereas nearby edaphically similar sites subjected to cattle grazing changed from savanna to dense woodland soon after the introduction of livestock in the late 1800s. Low frequencies of fire were documented on the savanna site protected from grazing, indicating that frequent fire was not required to maintain the grass-tree balance in this system. Both sites experienced the same climate, yet one changed dramatically over the past hundred years whereas the other did not. Such data indicate that livestock grazing has been the proximate cause of vegetation change on this site, not changes in fire or climatic regimes. If climatic fluctuations or atmospheric CO_2 enrichment in recent history were necessary to cause the change on the grazed site, they were not sufficient to cause change on the ungrazed site.

Grazing and Shrub Dominance Where persistence of perennial grasses is sensitive to grazing, conditions can be favorable for shrub encroachment. Woody plant recruitment can proceed in numerous, self-reinforcing ways:

1. Livestock may effectively disperse woody plant seeds, particularly those of some leguminous shrubs and arborescents.
2. Utilization of grasses increases light levels at the soil surface and may increase chances for germination and early establishment of woody seedlings.
3. Concomitant reductions in transpirational leaf area, root biomass, and root activity associated with grazing of grasses can:
   a. increase superficial soil moisture to enhance woody seedling establishment and growth of shallow-rooted woody species;
   b. increase the amount of water percolating to deeper depths and benefit established woody species with deep root systems;
   c. increase nutrient availability to woody plants; and
   d. release suppressed populations of established tree or shrub "seedling reserves."
4. Grazing increases mortality rates and decreases plant basal area, seed production, and seedling establishment of palatable grasses (O'Connor 1991, O'Connor and Pickett 1992).
5. Grazing may also increase susceptibility of grasses to other stresses such as drought (Clarkson and Lee 1988, Herbel et al. 1972, Nelson 1934, Paulsen and Ares 1962). These factors would combine to increase the rate of gap formation and available area for woody plant seedling establishment, especially in postdrought periods.
6. Herbaceous species may be replaced by assemblages that compete less effectively with woody plants.
7. Reduction of the biomass and continuity of fine fuel may reduce fire frequency and intensity (Savage and Swetnam 1990).
8. Invading woody species are often unpalatable relative to grasses and forbs and are thus not browsed with sufficient regularity or severity to limit establishment or growth.

9. Lower soil fertility and alterations in physicochemical properties occur with loss of ground cover and subsequent erosion. This favors N_2-fixing woody plants (e.g., *Prosopis, Acacia*) and evergreen woody plant growth forms that are tolerant of low nutrient conditions (Bush and Van Auken 1989, Cohn et al. 1989, Van Auken and Bush 1989).

Role of Shrubs in Grassland Communities As a rule, woody vegetation is poor forage for domestic livestock and inferior to the better pasture grasses. Nevertheless, it is important not to lose sight of the fact that, in many cases, woody species play an important role in the functioning of arid and semi-arid communities, often with significant benefits for pastoralism and conservation:

1. Woody plants may be best adapted to the prevailing biotic and abiotic conditions on some landscape units. On such sites, it may not be realistic to expect management or manipulation to enhance herbaceous production.

2. Woody plants do not always reduce herbaceous production. Effects of woody plants on herbaceous production varies from positive to negative:

a. Stimulation beneath tree or shrub canopies relative to interstitial zones has been reported for poplar box (Christie 1975), for algarrobo (Barth and Klemmedson 1978), for huisache (Scifres et al. 1982), for *Acacia karroo* (Stuart-Hill et al. 1987), for various hot desert shrubs (Ludwig et al. 1988), and for umbrella thorn and baobab (Belsky and Amundson 1989, Weltzin and Coughenour 1990).

b. No effect was observed by Harrington (1973) for *Acacia hockii.*

c. Decreased standing crop was recorded for pines (Jameson 1987), for poplar box and narrow-leaved iron bark (Walker et al. 1986), for pinyon-juniper woodlands (Clary 1987), and for subtropical thorn shrubs (Scanlan 1988).

d. Conflicting patterns of overstory-understory relationships that were observed in *Eucalyptus crebra* woodlands in Australia (Burrows et al. 1990) add to the difficulties of generalizing about such relationships. The effect of woody plants on herbaceous production may depend upon site (topoedaphic features), climate (especially the temperature and rainfall regime), species of woody plant and its growth form (i.e., single vs. multiple stemmed), canopy architecture, size, and density [Aucamp et al. 1983, Belsky and Amundson 1989, Burrows et al. 1990, Frischknecht 1963, Scifres et al. 1982, Walker et al. 1986]).

Herbaceous species respond differently to the removal of woody plants. Mesquite savannas in the Rolling Plains of Texas were characterized by C_3 grasses beneath mesquite canopies versus C_4 grasses in zones between trees. Following herbicide application that top-killed the mesquite, C_4 grasses displaced C_3 grasses on the mesquite-influenced microsite (Heitschmidt et al. 1986). As a result of reducing the C_3 grass component, spring forage production was delayed and greatly reduced, necessitating extended supplemental feeding.

The presence of woody plants may:

1. reduce grazing pressure on grasses and provide refugia for heavily utilized herbaceous species that might otherwise be grazed to local extinction (Davis and Bonham 1979, Jaksic and Fuentes 1980, Welsh and Beck 1976);
2. reduce supplemental feed requirements during cold or dry periods and aid in meeting animal nutritional requirements (Cook 1972);
3. enhance soil nutrient status and water infiltration;
4. provide shade that helps improve animal energy balance and, hence, increase weight gain and reproductive output; and
5. improve habitat for game and nongame wildlife.

Bush Clearing Woody plants are often the target of vegetation manipulation technologies aimed at improving livestock carrying capacity or facilitating livestock handling. However, the encroachment of unpalatable woody vegetation into grasslands and savannas is often the result of inappropriate grazing management and the elimination of fire as a key management tool. In management plans aimed at manipulating vegetation, symptoms (the undesirable woody plants) are often treated without addressing underlying factors that caused the transition to brush domination. Many woody plants of rangelands are long-lived (decades to centuries) and capable of regenerating via seed or vegetative propagation following natural or anthropogenic disturbance. As a result, it is difficult and expensive to limit their density and stature once established on a site. Therefore, when assessing whether to invest in efforts to reduce woody plant cover or density, the following points should be considered:

1. Will woody manipulation stimulate herbaceous production and increase livestock carrying capacity sufficiently to offset treatment costs? If so, how much time should elapse before administering a follow-up treatment? Will treating one problem perhaps create another that may be worse (i.e., replacement of evenly spaced, single-stemmed, old decadent plants by vigorously growing, younger multiple-stemmed plants that may create dense thickets; replacement of nonsprouting species like big sagebrush with sprouting species like rubber rabbitbrush). Answers to these questions vary according to species and sites (Scifres et al. 1988), the type, timing, and sequencing of vegetation manipulation technologies (Scifres et al. 1983), and site potential with respect to soils and rainfall. For example, in dry (510 mm annual rainfall) *Eucalyptus populnea* woodlands, increases in herbage production were not observed until stands were thinned by 70%; in contrast, herbage production in higher rainfall (750 mm) zones increased linearly with reductions in *E. crebra* density (Walker et al. 1986). The cost of effective treatment, rate of return on the investment, and level of risk would differ markedly in these systems.

2. Desirable woody plants that provide cover and palat-

able forage for livestock and wildlife may be replaced by other woody species after application of nonselective brush manipulation practices (Fulbright and Beasom 1987). When woody plants are sprayed with herbicides, forbs utilized by wildlife (e.g., deer and quail) may also be lost.

3. How will such manipulation affect aesthetic value? This is becoming an increasingly important consideration as many segments of society regard shrubby species as attractive elements of the landscape and have come to view bush clearing unfavorably, particularly where chemicals are used or where wildlife habitat is lost.

Grasslands Induced by Grazing The tendency for grazing to favor shrub encroachment is more characteristic of dry regions than of more humid regions. Many grasslands in the world are induced and maintained by grazing—often heavy grazing. An outstanding example is New Zealand where brush control is accomplished by high stocking rates to maintain a dense sward of perennial grass and prevent establishment of woody plant seedlings. To maintain pasture productivity at a level that can sustain such high stocking rates, plant nutrition must be boosted with adequate fertilizer applications (Clark and Lambert 1986). Another example is shrub encroachment on alpine grasslands that has traditionally been controlled by heavy grazing. Recent reductions of stock numbers have begun to create an environmental problem, not only because grasslands are a national heritage and regarded as an integral part of the Swiss landscape, but also because tall shrubs and trees on alpine grasslands are undesirable when, in winter, the grasslands become ski slopes.

A classic example of control of woody species by grazing is cited by Crawley (1983, p. 303):

> On an acid heath in Surrey where only small isolated clumps of mature Scots pine trees could be seen, a fence was erected to exclude cattle. Almost immediately pine saplings sprang up everywhere in abundance. Intrigued by this, Darwin (1859) looked more closely at the treeless heath and [noted,] "... in one square yard, at a point some hundred yards distant from one of the old clumps, I counted thirty-two little trees; and one of them, judging from the rings of growth, had during twenty six years tried to raise its head above the stems of the heath, and had failed. No wonder that, as soon as the land was enclosed, it became thickly clothed with vigorously growing young firs."

Similar situations are common in Mediterranean woodland communities, as evidenced by the rapid development of the oak woodlands in central and northern Israel after cessation of the intensive grazing that was practiced in the area for thousands of years (Seligman and Perevolotsky 1994). In another example, reported by Quinn (1986), introduced exotic livestock severely reduced woody vegetation cover under high stocking rates that developed on islands near California. Constant grazing pressure caused attrition of shrub cover because seedling establishment was inhibited. When grazing ceased, the chaparral communities regenerated rapidly.

From this wider perspective, one can conclude that, as a rule (to which there are many outstanding exceptions), where productivity is high, grass dominance over shrub encroachment is dependent on high stocking rates. Where productivity is low, heavy grazing is conducive to shrub dominance. However, in any specific case, site conditions and interactions with weather and anthropogenic factors can have decisive effects on the general trend.

Marshaling Information for Range Management

Managing Rangeland Communities Management to achieve desirable effects on the structure and function of rangeland communities involves decisions that have to be made under conditions of uncertainty and incomplete knowledge of the relevant processes and of the possible consequences of intervention (or nonintervention). Where theory and scientific understanding of critical processes can be shown to reduce uncertainty, they will probably be welcomed by managers. It would help to have a method for analysis of vegetation dynamics that has some theory to give it generality, situation-specific parameters to allow application to "reality," and implications that are relevant to the needs of management and can be related to the appropriate scale and dimensions of the "management space." Integration of geographic information systems and heuristic approaches in expert systems can draw from a large data and information base to provide useful guidance for management (Buchanan and Duda 1983, Folse et al. 1990, Loehle 1987, Loh and Rykiel 1992, Plant and Stone 1991, Rauscher and Cooney 1986, Starfield et al. 1993). Such systems for arid and semi-arid range management also require basic data on community structure and function as well as a useful management-oriented approach to extract maximum benefit from the data. Is there a pragmatic, commonsense program that can improve communication between up-to-date community ecology and range management? The following outline is proposed as an indication of a possible approach.†

Much of the recent discussion on the theory of ecological change as related to arid and semi-arid rangeland systems has revolved around the question of how to define and understand the different states that a rangeland community might go through in response to grazing, fire, and cultivation and "natural undisturbed" ecological processes (Friedel 1991, George et al. 1992, Laycock 1991, Westoby et al. 1989). The relationships among the different (actual and potential) states create the conceptual basis for the definition of *range condition,* a problem-oriented concept that is the basis for many management and political decisions. This concept is based on Clementsian succession theory and has guided range management since the end of the 1940s (Westoby et al. 1989). However, the simple elegance of considering grazing by domestic livestock as the central driving force in a linear progression of succession and retrogression in rangeland could not cope with many special situations that required theoretical adaptations. The large number of exceptions has provoked a reexamination of the relevance of Clementsian equilibrium theory to rangeland dynamics (e.g., Walker 1988, Westoby 1979–80); a number of revised concepts have crystallized out as the state-and-transition model (Westoby et al. 1989). By examining these concepts more closely, we may be able to expand this essentially descriptive approach to community change into a template for management-oriented organization of ecological information.

State of a Community By *state* we usually mean a recognizable and distinctive set of species that represents the community that dominates a given habitat at a particular point in time. In this sense, state and community are equivalent terms. However, it is seldom that all the component species are given equal weight, especially when used as cues for management. Even in the most "objective" and comprehensive ordination methods, the individual species can be represented in ways that give very different weights to a species. Species richness can be emphasized or dominant species can be emphasized; records can be presence/absence data or different types of quantitative measures. Thus, depending on the method of recording, some species can be distinctive in one view and redundant in another (Noy-Meir 1973a). Seemingly objective multivariate analyses are dependent on subjective choice of recording format and on whether and how the data are transformed (Seligman 1973). Consequently, any definition of a *state of vegetation* is a statement of a point of view. This standpoint can reflect a rancher interested in forage value of the vegetation, a forester concerned with woody species, an environmentalist interested in "pristine" condition, a conservationist concerned with rare and endangered species, or an ecologist interested in species diversity, energy flow, or nutrient cycling. The possibilities are not infinite, but they are many. Whenever vegetation on a specific site is analyzed, it is for a specific purpose. Therefore, the purpose should determine the method. Appropriate definition of a state then comes down to listing, in an appropriate form, the main species that are relevant to the problem. Assembly and response rules have been proposed to explain the species composition of a community (Keddy 1992a), but on rangelands, particularly in the arid zone, the main interest to management is often only two or three species, such as "creosote bush and tarbush . . . on tobosa grasslands, or velvet mesquite or honey mesquite on black grama grasslands in New Mexico" (Laycock 1991). In some cases, a group of species (e.g., "hemicryptophytic thistles," "annual grasses," or "native perennial grasses") is the subject of concern. Such functional groups have proved useful in vegetation analysis (Keddy 1992b). It is not that all the other species or functional groups on the site are unimportant or uninteresting, only that, as a cue for management, most of them have little information value, at least at the time of decision.

Stability and Transition If management is to be based on a few critical or key species (or functional groups of species), then we need to know their autecology, "vital statistics," and life histories in order to determine whether the state that they define is "stable" or transient. *Stability* in the context of constantly changing biological systems is a relative term because, if a "stable state" means the continued, viable presence of the dominant species on a site, the question then is: continued presence for how long? Stability can mean the continued viable presence of an individual (tree, shrub, grass), its regenerating organs, or its offspring at least until the relevant management horizon. Depending on the management problem, that could be anywhere between one and thirty or more years. Stability depends on the response of longevity and reproductive behavior of the key species to chronic or episodic disturbance that can include herbivory by domestic and wild animals, fire, chemical control, inundation, cultivation, or any other relevant disturbance. Where establishment or extinction of key species is dependent on "episodic windows" (unusual or rare weather conditions and/or interactions among factors like weather, fire, and grazing), it would be helpful to know the frequency or periodicity of these events and to recognize the conditions that preceded them.

In a management context, *transition* from one stable state to another means replacement of one or a number of key species or functional groups by another or others. As a rule, because the issue is problem oriented, this replacement has a value judgment attached to it. We regard it either as good or bad, desirable or undesirable in relation to our chosen criteria. However, the manager, as a rule, has multiple objectives even without the additional objectives imposed by other parties with interests in the rangeland community. The determination of desirable or undesirable status must then involve some compromise between conflicting interests.

Determinants of Stability These are the factors in a problem situation that allow some species to flourish and others to linger and die. Factors for continued survival success (or stability) include:

- physical capture of territory and efficient exclusion of competitors for the available resources;
- ability to withstand fluctuating environmental stress by maintaining essential (vegetative or sexual) regenerative organs that can effectively exploit opportunities of release from stress;

- lack of efficient predators (including herbivores) or efficient means of warding them off;
- resistance to disease, including soilborne pathogens; and
- "competitive advantage," in relation to the other critical species in the set, that may include allelopathy or be the result of relative differences in the previously listed factors.

Causes of failure to persist and flourish are basically the opposite of the causes of survival:

- capture of the territory by other species that effectively exclude establishment;
- predator pressure (including herbivores) that reduces regenerative recovery below survival levels;
- inability to accommodate fluctuating environmental stress with the result that there are insufficient essential residual organs (including seeds) to enable recovery when stress is released;
- susceptibility to disease, including soilborne pathogens; and
- "competitive disadvantage," in relation to the other critical species in the suit, that may be the result of relative differences in the above factors.

Environmental stress includes all perturbations from mild to wild—fluctuations due to weather, fire, lightning, inundation, cultivation, chemical additions, trampling by animals or recreation vehicles, and so on. The information necessary to decide on a course of action (that does not become a curse of action!) is that which includes the specific responses of the critical species to each relevant factor and combination of factors. In some regions, much of the relevant information is available, even if from descriptive or anecdotal sources. But much more must be done to create a reliable and comprehensive database for rangeland analysis.

A Management Information Template The information should be organized so as to facilitate conclusions and guidelines for management action. While there can be many potential states and transitions on a specific site, at a given point in time a current state may be "desirable" or "undesirable" from a management viewpoint. The management objective is to maintain a desirable state or to move from an undesirable to a desirable state. Whether the objectives can be achieved or not depends on the characteristics of the species that distinguish between these two states, the cost of management resources, and other constraints on management. A list that summarizes the available information (and the lack of essential information) should define the conditions necessary to achieve the management objective and to conclude whether a transition is feasible within the management horizon. Structuring the necessary management information can then provide a framework for decision making in a specific management situation and in determining research objectives where information is inadequate.†

An example or two may help to illustrate both the possibilities and limitations of this approach for defining the state of the relevant information (tables 5.1 and 5.2). These examples are not rigorously defined and should not be construed as being in any way authoritative. They are only meant to illustrate one way of marshaling information to guide management of some rangeland situations.

Table 5.1. Stability and Transition Factors in an Arid Mediterranean Hilly Rangeland with Mixed Dwarf Shrub-Herbaceous Vegetation

	Desirable Species[a]	Undesirable Species[b]
Presence	Moderate	Moderate
Longevity	Annual and perennial	Medium ca. 25 yrs
Sensitivity of persistence to:		
Grazing	Low	Insensitive
Browsing	Low	Low
Fire	Low	Moderate
Competition	High	Low
Cultivation	Not relevant	Not relevant
Chemical control	Low	Moderate to high
Drought	Low	Low
Inundation	Not relevant	Not relevant
Nutrient enhancement	High	Insensitive
Nutrient depletion	Moderate	Low
Establishment ability:		
Existing seed stocks	High	Low to moderate
Seed production potential	High	High
Reseeding	Not relevant	Not relevant
Vegetative	No	Low
Reestablishment potential	Very high	Low to moderate
Other factors:		
Soil erodibility	Low	Low

Conclusions

Stability:

1. The undesirable state is very stable but may revert to the desirable state if the nutrient level of the habitat becomes more favorable or if recurrent fire is feasible.
2. Desirable state probably stable over a long period as dwarf shrubs slowly recolonize the site.

Conditions for transition:

1. Possibly enhancement of soil nutrient status.
2. Chemical control.

Costs of transition:

1. Moderate to expensive with nutrient enhancement.

Risks:

1. Possibly poor response of dwarf shrubs to chemical control or to nutrient enhancement.

a. Annual and perennial pasture species (e.g., grasses, legumes, forbs).
b. Unpalatable dwarf shrubs (e.g., *Sarcopoterium* sp.).

Table 5.2. Stability and Transition Factors in a Semi-Arid Mediterranean Cultivable Rangeland with High Cover of Undesirable Shrubs Unpalatable for Domestic Livestock

	Desirable Species[a]	Undesirable Species[b]
Presence	Low	High
Longevity	Annual	Medium < 30 yrs
Sensitivity of persistence to:		
Grazing	Low	Insensitive
Browsing	Low	Low
Fire	Low	Very high
Competition	High	Low
Cultivation	Low	Very high
Chemical control	Low	Moderate
Drought	Low	Low
Inundation	Not relevant	Not relevant
Nutrient enhancement	High	Low
Nutrient depletion	Moderate	Low
Establishment ability:		
Existing seed stocks	Low	High
Seed production potential	High	High
Reseeding	Possible	Not relevant
Vegetative	No	No
Other factors:		
Soil erodibility	Low	Low

Conclusions

Stability:

1. The undesirable state is stable over long periods and, because of longevity and competitive advantage of the Cistus and high seed stocks, it is not likely to revert to the desirable state unless subjected to practices to which it is highly sensitive.
2. Desirable state probably transient, maintenance possible only with periodic, possibly infrequent intervention.

Conditions for transition:

1. Grazing alone ineffective.
2. Recurrent fire or cultivation until depletion of Cistus seed stocks.
3. Enhancement of soil nutrient status.

Costs of transition:

1. Low with fire.
2. Moderate with cultivation, but crop may cover the costs.
3. Nutrient enhancement may raise cost unrealistically.

Risks:

1. Low soil erosion risk with cultivation.

a. Annual pasture species (e.g., grasses, legumes, forbs).
b. Inedible shrubs (e.g., *Cistus* spp.).

Conservation of Biodiversity

Principles and Problems

Diversity in Dry Biomes Surprisingly, diversity is extraordinarily high in arid and semi-arid biomes (Mares 1992). There is evidence that some important measures of mammal diversity actually peak in landscapes characterized by 150–300 mm precipitation (Abramsky and Rosenzweig 1984, Rosenzweig 1992, 1995, Rosenzweig and Abramsky 1993). Western (1991) has concluded that even large mammal diversity peaks in semi-arid areas of Africa. This is probably related to the type of variability in semi-arid zones where both very dry and very wet seasons are not unusual (Seligman et al. 1992). Even where diversity is not particularly high, the special adaptations of plants and animals to dryness and unpredictability make for unusual biotas of great appeal.

The diversity of arid and semi-arid zone flora has already been tapped for practical uses. Its species have been a source of major cultivated cereals in both the New and Old Worlds. Pharmaceuticals, such as ephedrine, have also emerged. In addition, the arid zone has special value as a biological laboratory in which environmental principles have been developed that are applicable in many other ecosystems beyond the bounds of water-poor lands. Because of intense human pressure on these other environments, they may never be able to sustain such research themselves. Arid systems harbor communities that are easier and cheaper to study than any others for several reasons. The arid landscape is more two-dimensional and open than others. Equipment lasts longer and requires less attention in low humidities. Finally, the low productivity and consequent low level of human intervention in many arid regions provide the investigator with larger expanses of land, functioning more naturally than perhaps any other extensive terrestrial environment. It is, therefore, hardly accidental that so many basic ecologists focus their research on arid or semi-arid environments.

The future maintenance of biodiversity in the arid zone cannot be taken for granted because pressures ranging from urban development to excessive grazing, tourism, and recreational vehicles are constantly increasing in both developed and less developed countries. Conservation of diversity requires consideration in planning and implementation of development projects in arid regions no less than in more severely threatened zones.

Inventory: Necessary but Insufficient Taking inventory is rarely a problem in arid and semi-arid lands. Their floras and faunas are much better known than those of some tropical regions. But no less than elsewhere (and perhaps even more), the inventory is dynamic. Many species become scarce or abundant without warning. Some of the most remarkable examples come from Australia's avifauna; from time to time, certain parrots and honeyeaters suddenly appear in vast numbers in an arid community and then utterly vanish, not to be seen again perhaps for decades. Such species may somehow specialize in erratic population size or continent-scale migration (Rhoades 1985). In any case, their preservation requires developing tactics to recognize and deal with extreme dynamism.

The great variability of populations in arid and semi-arid regions highlights the limitations of the emphasis many scientists place on the location of diversity *hot spots*—core areas with unusually high diversity. The prevalent attitude is that, once such areas have been identified, it is up to managers and conservation authorities to set the areas aside and then the species they contain will be preserved. But inventories change dramatically over short periods. Moreover, as we shall see, until we understand what produces a hot spot, we cannot know if it will remain hot as conditions change.

Determinants of Biodiversity Two primary problems in the maintenance of biodiversity can be derived from basic ecology. First, diversity depends on available area at several scales. Second, many, if not all, communities have "tolerant" species that depend for their continued existence on species-poor locations (Wisheu and Keddy 1992).

Area The oldest known pattern in ecology is that larger areas harbor more species than do smaller ones. Moreover, we know also that in general the pattern fits a simple logarithmic equation:

$$\ln S = c + z \ln A$$

where S is the number of species, A is the area, and c and z are constants. Most investigators have found that the value of z that most closely predicts the loss is about 0.25–0.30 (Rosenzweig 1995). The appropriate value of c depends on the taxon.

This equation states that diversity is a mass effect dependent on large areas. It also shows why deserts and semi-arid regions have so many species: they cover large areas. When recreational windows are opened to ecotourists in particularly diverse or beautiful locations, they may seem to use only a tiny fraction of the area available. But that is an illusion. These windows depend for their diversity on the much larger areas around them. If we reduce area substantially, we lose significant numbers of species. The species/area equation allows us to estimate what fraction of diversity will disappear in a few generations after any proportional loss of area. The following sample calculations, based on a z-value of 0.25, indicate the loss in the number of species that are represented as the area of conservation reserves is reduced:

Land in conservation reserves (%)	50	30	20	10	5
Remaining species (%)	84	74	67	56	47

Tolerant Species Ecologists have spent decades and learned much from investigating how species avoid extinction. One important lesson is that most species in natural communities do not have a habitat in which they excel. They stay alive despite their inability to exclude all their competitors from some specialized habitat. They stay alive by tolerating poorer habitats (Rosenzweig 1991, Wisheu and Keddy 1992). Richer habitats have more species, not necessarily as a consequence of higher productivity, but because they can harbor tolerant species as well as intolerant ones, albeit as sink species (Kadmon and Shmida 1990, Pulliam and Danielson 1991). To do their job well, biodiversity reserves must include both rich and poor habitats.

Evidence is less satisfactory regarding temporal variations in environment. Some species require unusual times (the specialized intolerants), and some manage well enough every year (the tolerant ones). We need to ensure that we do not prevent the occurrence of the rare events that support those species.

Fragmentation As the landscape grows wetter, pastoralism and agriculture become more profitable and settlement more practical. Proportionally less land will remain available for biodiversity uses, making loss of area an important concern. In semi-arid regions, the land becomes fragmented and takes on the appearance of a mosaic comprising pastoralism, settlements, perhaps some dry farms, and some biological reserves. Biodiversity needs to find a secure place in the semi-arid mosaic because many species found in semi-arid communities do not occur in more xeric ones. The fragmented landscape brings added questions about dynamics and viability that are hotly debated by ecologists today. Should reserves be combined into one large area or will there be more species if the same area is subdivided into several small ones? The topic is often called the SLOSS controversy (single large or several small). The issues are:

1. *Dynamic persistence:* Does subdividing an area reduce the probability of extinction? Many believe that if population dynamics exhibit chaos, cyclicity, or even a considerable amount of stochastic variation, then subdivision into a number of weakly linked populations improves the ability of each species to persist. If disaster and extinction hits one population, the others are saved.

2. *Gene pool size:* What is the effect on genetic diversity of having a set of small weakly linked populations? Does such a metapopulation adapt as readily as a single large one? More readily? Most population geneticists traditionally say that metapopulations have more variability because they experience many semi-independent genetic drift and founder effects. But an opposing view (Bowers et al. 1973, Brown 1957, Greenbaum et al. 1978) suggests that small populations cannot adapt to new circumstances. The manager of semi-arid communities will be keeping close watch on this controversy.

3. *Edge effects:* Will species from surrounding communities swamp those we are trying to preserve? They might do so by migrating as sink species from the edge of very small preserves. Sink species may not be able to persist without continual immigration into a reserve, yet they do utilize its resources. As the ratio of reserve edge to area grows, such species may greatly reduce the populations of the source species that the reserve is meant to harbor. Those source species may fall below minimum viable population size.

4. *Principles of island biogeography:* Fragmentation turns mainland areas into islands. That is why they tend to fit species-area curves with z-values near 0.25. Evidence strongly suggests that the reason for the long lifespan of the SLOSS controversy is that fragmentation itself makes little or no

difference. The species-area curve that is fitted to the separate islands of an archipelago also predicts the archipelago's total diversity (Rosenzweig 1995). Yet, one sort of species may well be harmed by fragmentation. If a species is widespread and rare, it may need the larger area of the single reserve to have a minimum viable population anywhere. Populations of this sort often occur among predators high in the trophic chain. Thus, fragmentation threatens some of the most charismatic species.

Managers will have to anticipate this problem by regular censuses of threatened species and exchange of their individuals among the reserve systems' fragments. Leaving them to their own devices will promote their loss. A corollary of this management need is that conservation cannot totally abandon its efforts to save individual species. The point is nowhere more valid than in facing the danger to species in high trophic levels.

5. *Practical problems:* To these questions we must add several less biological but still quite practical ones: given a fixed amount of money to invest in reserves, will more area be preserved if we do not insist on contiguity? Will it be placed in reserve sooner? Is a set of small reserves easier to manage? Or is it easier and more economical to manage a single large preserve? Some of these questions do not yet have answers, but all may affect managers' treatment of preserves in semiarid regions.

Support Requirements

Education Public attitudes are an important factor in determining the priority of conservation. On the whole, people are willing and interested to learn about their natural heritage. Most appreciate the stark beauty of arid landscapes but are unaware of their spectacular biological diversity. The University of Arizona's public television station, KUAT, has for many years produced a local introductory series about the Sonoran desert ("The Desert Speaks"). General distribution of this series (and similar ones) to people who live in more mesic regions should improve understanding, advance curiosity, and foster interest in desert conservation and arid lands ecotourism.

In addition, appealing instructional programs must be designed to explain ecological tolerance, sink species, and the ecological importance of large areas. When people understand these concepts, they will be better able to discriminate between a zoo and a recreational window on the natural world. They will appreciate that it takes vast, apparently unexploited areas to support the survival of the species they enjoy seeing in relatively tiny areas.

Israel's popular network of field schools are perhaps misnamed—they are really unique, nature-oriented vacation facilities often booked years in advance. They already offer minicourses and on-site lectures in geology, archaeology, and natural history of their specific locales. With the addition of similar experiences involving more general principals of ecology, these schools could offer the world a successful template to adopt and modify as local conditions require.

Infrastructure Education without parallel construction and development of recreational infrastructure could lead to irresponsible recreation. Appropriate recreational windows must be identified and facilities developed. A good example is the Ularu National Park in Northern Territories, Australia. Facilities are located in a large Aboriginal land reserve. An airfield and a good road system move people in and out comfortably and efficiently. Modern hotels, living quarters for service personnel, and camping facilities are concentrated in a small settlement far from the chief scenic areas. These are accessible in a variety of ordinary ways. Even more could be done to guide people to a better appreciation of the flora and fauna in the park. People tolerate being restricted to a small recreational window because they respect Aboriginal rights and culture. Ularu could serve as a model in many other situations.

Integration Conservation of diversity can coexist with many other uses of an area. Nevertheless, these uses need to be integrated and prioritized. Economic opportunity will no doubt engender conflicts of interest. The relative roles of private and public organizations will probably vary from country to country and even from time to time, but it seems difficult to imagine accomplishing coordination without an objective public body supervising and refereeing disputes over whose ox has to bleed for the sake of the common good.

Research Lengthy, carefully conceived studies require a combination of tactics. But most certainly they will not happen by accident. Thus, research use of arid and semi-arid lands requires planning and coordination. There are complementary advantages to empowering either resident or visiting scientists as planners. Residency brings intimate knowledge of the particular locale but also tends to produce intellectual isolation and obsolescence, while visiting scientists bring objective perspective but also a lack of understanding of the local situation. Perhaps a combination of resident and visiting scientists would avoid those problems.

Central bodies should assiduously resist the temptation to dictate local research, but they should prescribe and then facilitate its broad goals. This probably means setting aside research areas, funding essential research on the special problems of preserving local biodiversity, and providing infrastructure including housing. For example, each recreational window will need permanent monitoring of some of its species. Initially, ecologists will need to find out how the local communities are organized. What habitats do various taxa recognize? Which species are sink species? Which species are tolerant or intolerant? Over the longer term, ecologists will discover the population-dynamic behavior of the species and whether any appear to specialize on rare events. Their deepening acquaintance with local flora and fauna will enrich the recreational experiences of the ecotourists.

Management Aspects

Habitat Planning Small areas lose diversity, in part, because they constitute an incomplete sample of habitats (Rosenzweig 1995, Williams 1943). Knowing this, the designer and the manager of a fragmented reserve system can ameliorate the problem. Purchase of a smaller patch of habitat not yet included in the system may be considerably more useful than adding a large patch of already represented habitat. The same applies to allocating a fixed sum for reserve acquisition and development. Getting the most land area may not do the most good.

Once acquired, a system's habitats need to be maintained. This will probably require active local management. Fires need to come at an acceptable frequency. Soil needs to be protected. Successional seres need to be inventoried and perhaps rotated from place to place. In some habitats, grazing pressure (from native or exotic grazers) needs to be regulated so as to maintain the habitat type. Such management is facilitated by mangers' knowledge of the habitats in the system and their role in protecting diversity. Then, continued censuses of key populations will warn of the need to fine-tune management policies.

Corridors What can be done when a large reserve is needed to preserve a species but only a set of small ones is available? Some ecologists have suggested interconnecting the small ones with corridors of land that can support movement of individuals among fragments. The land would not necessarily have to lie fallow but would be restricted to uses that did not interfere with movement. This tactic may work for single species, but the evidence is as yet unclear. Even if it does succeed, species have various abilities to move and various requirements for traveling. Thus, a corridor for one species may be a barrier to another unless no competing uses are tolerated within it. As conservation moves away from concentrating on preserving single species to preserving ecosystems, corridors may become less and less practical.

Species with Source Populations outside Semi-Arid Reserves One of the more recent principles of community and geographical ecology is that species abundance and range are positively correlated (Hanski et al. 1993). It is likely that tolerant species are those that tend to be dense and widely distributed, whereas intolerant species tend to be rare and local. Managers can use these properties of species to further maximize the work done by fragmented reserves.

Some species, probably tolerants, will be found coexisting as source populations outside reserves. Yet, these same species will most likely also have source populations inside reserves. Those reserve populations will be taking resources from species that cannot coexist with pastoralism, settlement, or agriculture. This presents reserve managers with a great challenge. If the common species can be eliminated (or substantially reduced) from a reserve, all its area can be devoted to preserving the rare local ones. Some of these should cross the threshold of viability and persist. But eliminating common species is not easy. Because they are likely to be the tolerants, any adverse impact on the habitat is more likely to hurt the rare, intolerant species. Biological control directed at common species can be hazardous. Specific parasites and diseases for biological control of tolerants are treacherous. They do not stay within the preserve boundaries, and no species, no matter how common, is safe from decimation and even extinction in the face of a new exploiter (Rosenzweig 1995). Indeed, exploiters are one of the main factors that can make species rare.

Genetic Sampling Managers should conduct occasional sampling of the gene pool in metapopulations of rarer species whose individuals are not being interchanged. In the absence of firm conclusions about the genetic effects of fragmentation, this is about the only way to guard against a genetic disaster. Should a population suffer a serious loss of heterozygosity, exchange of individuals with other populations should begin (at least for a time).

Arid Zone Ethnobotany

The flora of arid zones is a largely untapped source of useful plants. Mexico is an interesting example because of the traditional importance of many native species that are used for food, medicinal, industrial, and ritual purposes. Some of these plants are the only source of monetary income for thousands of families who inhabit the arid regions. However, only a few species are exploited commercially. Among these are *Euphorbia antisyphilitica* Zuco., which is harvested to obtain a white wax; *Parthenium argentatum* Gray, whose stems contain latex used to produce rubber; jojoba (*Simmondsia chinensis*), which produces oils; and lecheguilla (*Agave lecheguilla* Torr.), which is harvested to extract hard fibres called "ixtle."

There are many studies that deal with the ethnobotany of Mexican desert plants (e.g., Cepeda 1949, Chapa 1956, Marroquin et al. 1981, Sandoval 1980, Sheldon 1980). However, most of this research is descriptive and focuses on taxonomy, use, and brief social and economical analyses. Very few provide details on species or population management (examples of some that do not include Briones and Valiente-Banuet 1993, Eufracio 1992, Valiente-Banuet and Ezcurra 1989). Consequently, there is little agreement on appropriate management of even the more widely used species.

One example is lecheguilla, which has been an important item in the economy of Mexico's dryland inhabitants for millennia (Sheldon 1980). Some maintain that populations of lecheguilla have been decreasing due to overexploitation for fiber (Maldonado 1979), while others argue that if the part of the plant used to obtain fiber (the *cogollo*) is periodically removed, apical dominance is released, and the number of new suckers (*hijuelos*) increases so that the plant lives longer (Sheldon 1980). If this is correct, then exploitation is the best conservation strategy for this species. However, the physiology of lecheguilla has barely been studied and only one article (Eufracio 1992) has treated the response of the lecheguilla populations to harvesting. Even this study was incomplete since it covered only three years of evaluation. The physiology, ecology, and sustainable management of these useful plants is a research challenge that still has to be met.

Issues and Challenges

Arid land ecology has received much research attention over the past half century (Evenari et al. 1985, Noy-Meir 1973b, 1974) and a wide theoretical base has developed. However, the relevance of the theory to management of arid lands communities has not always been clear. Recent evidence even suggests that some theories were more dogma than science. So, for instance, desertification and overgrazing became an inseparable pair even where the evidence was very scanty. Recent studies support the view that desertification over vast areas has been largely climate induced (Nicholson et al. 1990). Specific cases of rangeland habitat degradation have been shown to have had no relation to grazing (Western and van Praet 1973). While there are certainly cases of major community change where grazing is the proximate cause of degradation, it is by no means true in all cases (Archer 1994b). Recognizing the true causes of community change and habitat degradation is a challenge that community ecology must face if it is to be relevant to management. However, identifying the real causes of change in many cases has been an intriguing and often frustrating quest with many surprises on the way as well as many dead ends. Often, causes were overlooked because of either theoretical misconceptions or because the proximate cause of change was either episodic or transient and occurred in the unretrievable past or was due to a seemingly minor factor (Weltzin 1990).

A more realistic view of livestock grazing as one ecological factor among many has surfaced in the state-and-transition model of Westoby and colleagues (1989). This approach recognizes the complexity of ecological change and allows for any observable transition, including orderly Clementsian succession as a special case. It provides an opportunity to break loose from the bonds of a one-dimensional view but gives little solace for range managers because the loss of theoretical security is replaced by the need for multidimensional site-specific information, much of which is unavailable. "Opportunistic management" puts the brunt of decision making on the shoulders of managers who must become experts at recognizing both episodic events with ominous portent and "windows of opportunity" for desirable change. Nevertheless, it stimulates a fresh mind-set that recognizes the need for more sophisticated, nonsimplistic approaches to community dynamics in arid and semi-arid rangelands.

Urbanization and major demographic change in the arid lands of developed countries has brought changes in the values that society attaches to biological resources in these lands. Conservation of biodiversity has become more than a buzz word in many countries, and concern for the "health" of the extensive arid lands communities has encouraged greater scientific involvement in the management of these resources. The field is wide open and the need for comprehensive monitoring and effective action to ensure conservation of the more endangered areas and species is widely recognized, even if not widely implemented. Conflicting interests, budgetary constraints, public indifference, and unresolved theoretical questions are all factors that make conservation a continuing issue.

The arid lands management problems in developed countries are a source of controversy and concern, but they are tractable in the sense that there are feasible management options. The situation is much more serious in developing countries where rural populations are high and growing and where pressures of survival leave few options other than maximum exploitation of all available local resources. Under such conditions, community ecology has to take a backseat. How human ecology will respond to impending disaster can be too frightening to contemplate. Although global solidarity may seem a utopian concept in the late 1990s, it is an ecological imperative no less in arid lands than in more densely populated parts of the world.

Exploiting and conserving the biological and ecological riches of arid lands will continue to challenge both scientists and managers. Open discussion of the options for management and their ecological implications is one way to arrive at a clearer view of what must and can be done.

References

Abensperg-Traun, M. 1992. The effects of sheep-grazing on the subterranean termite fauna (*Isoptera*) of the western Australian wheatbelt. Australian Journal of Ecology 17:425–32.

Abramsky, Z., and M. L. Rosenzweig. 1984. Tilman's predicted productivity-diversity relationship shown by desert rodents. Nature 309:150–51.

Amir, J., and T. R. Sinclair. 1991. A model of water limitation on spring wheat growth and yield. Field Crops Research 28:59–69.

Anable, M. E., M. P. McClaran, and G. B. Ruyle. 1992. Spread of introduced Lehman lovegrass *Eragrostis lehmanniana* Nees. in southern Arizona, USA. Biological Conservation 61:181–88.

Archer, S. 1989. Have southern Texas savannas been converted to woodlands in recent history? American Naturalist 134:545–61.

———. 1994a. Regulation of ecosystem structure and function: Climatic versus non-climatic factors. *In* J. Griffiths, ed. Handbook of agricultural meteorology. Oxford: Oxford University Press. 245–55.

———. 1994b. Woody plant encroachment into southwestern grasslands and savannas: Rates, patterns, and proximate causes. *In* M. Vavra, W. Laycock, and R. Pieper, eds. Ecological implications of livestock herbivory in the West. Denver, Colo.: Society of Range Management. 13–68.

Archer, S., C. J. Scifres, C. R. Bassham, and R. Maggio. 1988. Autogenic succession in a subtropical savanna: Rates, dynamics and processes in the conversion of a grassland to a thorn woodland. Ecological Monographs 58:111–27.

Aucamp, A. J., J. E. Danckwerts, W. R. Teague, and J. J. Venter. 1983. The role of *Acacia karroo* in the False Thornveld of the Eastern Cape. Proceedings of the Grassland Society of South Africa 18:151–54.

Austin, M. P., O. B. Williams, and L. Belbin. 1981. Grassland dynamics under sheep grazing in an Australian Mediterranean-type climate. Vegetatio 47:201–12.

Baisan, C. H., and T. W. Swetnam. 1990. Fire history on a desert mountain range: Rincon Mountain Wilderness, Arizona, USA. Canadian Journal of Forestry Research 20:1559–69.

Barth, R. C., and J. O. Klemmedson. 1978. Shrub induced spatial patterns of dry matter, nitrogen, and organic carbon. Journal of the Soil Science Society of America 42:804–9.

Beatley, J. C. 1974. Phenological events and their environmental triggers in Mojave desert ecosystems. Ecology 55:856–63.

Belsky, A. J. 1983. Small-scale pattern in grassland communities in the Serengeti National Park, Tanzania. Vegetatio 55:141–55.

———. 1984. Role of small browsing mammals in preventing woodland regeneration in the Serengeti National Park, Tanzania. African Journal of Ecology 22:271–79.

Belsky, A. J., and R. G. Amundson. 1989. The effects of trees on their physical, chemical, and biological environments in a semi-arid savanna in Kenya. Journal of Applied Ecology 26:1004–24.

Berdowski, J. J. M. 1987. Transition from heathland to grassland initiated by the heather beetle. Vegetatio 72:167–73.

Billings, W. D. 1990. *Bromus tectorum,* a biotic cause of ecosytem impoverishment in the Great Basin. *In* G. M. Woodel, ed. The earth in transition: Patterns and processes of biotic impoverishment. Cambridge: Cambridge University Press. 301–22.

Biswas, M. K., and A. K. Biswas, eds. 1980. Desertification. Oxford: Pergamon Press.

Bowers, J. H., R. J. Baker, and M. H. Smith. 1973. Chromosomal, electrophoretic and breeding studies of selected populations of deer mice (*Peromyscus maniculatus*) and black-eared mice (*P. melanotis*). Evolution 27:378–86.

Box, E. O. 1981. Macroclimate and plant forms: An introduction to predictive modeling in phytogeography. The Hague: Dr. W. Junk Publishers.

Boyer, J. S., and H. G. Macpherson. 1975. Physiology of water deficits in cereal crops. Advances in Agronomy 27:1–23.

Briones, O., and A. Valiente-Banuet. 1993. Importancia e impacto de los estudios de comunidades en los ecosistemas áridos de México. Memorias del XII Congreso Mexicano de Botánica. Mérida, Yuc., México.

Brook, K. D., J. O. Carter, T. J. Danaher, G. M. McKeon, N. R. Flood, and A. Peacock. 1992. The use of spatial modeling and remote sensing for the monitoring and forecasting of drought-related land degradation events in Queensland. Sixth Australasian Remote Sensing Conference 1:140–49.

Brown, J. H., and E. J. Heske. 1990. Control of a desert-grassland transition by a keystone rodent guild. Science 250:1705–7.

Brown, J. R., and S. Archer. 1989. Woody plant invasion of grasslands: establishment of honey mesquite (*Prosopis glandulosa* var. *glandulosa*) on sites differing in herbaceous biomass and grazing history. Oecologia [Berlin] 80:19–26.

Brown, W. L., Jr. 1957. Centrifugal speciation. Quarterly Review of Biology 32:247–77.

Buchanan, B. G., and R. O. Duda. 1983. Principles of rule-based expert systems: Advances in computers. Orlando, Fla.: Academic Press.

Buffington, L. D., and C. H. Herbel. 1965. Vegetational changes on a semidesert grassland range from 1958 to 1963. Ecological Monographs 35:139–64.

Burrows, W. H., J. O. Carter, J. C. Scanlan, and E. R. Anderson. 1990. Management of savannas for livestock production in north-east Australia: Contrasts across the tree-grass continuum. Journal of Biogeography 17:503–12.

Bush, J. K., and O. W. Van Auken. 1989. Soil resource levels and competition between a woody and herbaceous species. Bulletin of the Torrey Botanical Club 116:22–30.

Caldwell, M. M. 1990. Water parasitism from hydraulic lift: A quantitative test in the field. Israel Journal of Botany 39:395–402.

Caldwell, M. M., J. H. Richards, and W. Beyschlag. 1990. Hydraulic lift: Ecological implications of water afflux from roots. *In* D. Atkinson, ed. Plant root systems: The effect on ecosystem composition and structure. Oxford: Basil Blackwell. 423–36.

Carpenter, A. T., J. C. Moore, E. F. Redente, and J. C. Stark. 1990. Plant community dynamics in a semi-arid ecosystem in relation to nutrient addition following a major disturbance. Plant and Soil 126:91–99.

Cepeda, R. 1949. La fibra de lechuguilla. Ph.D. diss. Buenavista, Saltillo, Coahuila, Mexico: Universidad Autonoma Agraria Antonio Narro.

Chapa, R. M. T. 1956. Estudio morfologico de la candelilla *Euphorbia* spp. Ph.D. diss. Saltillo, Coahuila, Mexico: Escuela Superior de Agricultura Antonio Narro.

Chapin, F. S., III. 1980. The mineral nutrition of wild plants. Annual Review of Ecology and Systematics 11:233–60.

———. 1993. Functional role of growth forms in ecosystem and global processes. *In* J. R. Ehleringer and C. B. Field, eds. Scaling physiological processes: Leaf to globe. San Diego: Academic Press. 287–311.

Chew, R. M. 1982. Changes in herbaceous and suffrutescent perennials in grazed and ungrazed desertified grassland in southeastern Arizona, 1958–1978. American Midland Naturalist 108:159–69.

Christie, E. K. 1975. A study of phosphorus nutrition and water supply on the early growth and survival of buffelgrass grown on a sandy red earth from south-west Queensland. Australian Journal of Experimental Agriculture and Animal Husbandry 15:239–49.

Clark, D. A., and M. G. Lambert. 1986. Pasture improvement and animal production using low inputs on New Zealand Hills. *In* F. M. Borba and J. M. Abreu, eds. Grasslands facing the energy crisis. Proceedings of the eleventh general meeting of the European Grasslands Federation. Portugal: European Grasslands Federation. 471–78.

Clark, J. S. 1988. Effect of climate change of fire regimes in northwestern Minnesota. Nature 334:233–35.

Clarkson, N. M., and G. R. Lee. 1988. Effects of grazing and severe drought on a native pasture in the Traprock region of southern Queensland. Tropical Grasslands 22:176–83.

Clary, W. P. 1987. Herbage production and livestock grazing on pinyon-juniper woodlands. Proceedings, pinyon-juniper conference. Gen. Tech. Rep. Ogden, Utah: USDA Forest Service. 440–47.

Cohn, E. J., O. W. Van Auken, and J. K. Bush. 1989. Competitive interactions between *Cynodon dactylon* and *Acacia smallii* seedlings at different nutrient levels. American Midland Naturalist 121:265–72.

Cole, K. 1985. Past rates of change, species richness and a model of vegetation inertia in the Grand Canyon, Arizona. American Naturalist 125:289–303.

Coleman, D. C., R. Andrews, J. E. Ellis, and J. W. Singh. 1976. Energy flow and partitioning in selected man-managed and natural ecosystems. Agroecosystems 3:45–154.

Collins, S. L. 1987. Interaction of disturbances in tallgrass prairie: A field experiment. Ecology 68:1243–50.

Collins, S. L., and S. C. Barber. 1985. Effects of disturbance on diversity in mixed-grass prairie. Vegetatio 64:87–94.

Collins, S. L., J. A. Bradford, and P. L. Sims. 1987. Succession and fluctuation in *Artemisia* dominated grassland. Vegetatio 73:89–99.

Conley, W., M. R. Conley, and T. R. Karl. 1992. A computational study of episodic events and historical context in long-term ecological processes: Climate and grazing in the northern Chihuahuan desert. Coenoses 7:55–60.

Cook, C. W. 1972. Comparative nutritive values of forbs, grasses, and shrubs. *In* C. M. McKell, J. P. Blaisdell, and J. R. Goodin, eds. Wildland shrubs: Their biology and utilization. Gen. Tech. Rep. Ogden, Utah: USDA Forest Service. 303–10.

Cornelius, J. M., and G. L. Cunningham. 1987. Nitrogen enrichment effects on vegetation of a northern Chihuahuan desert landscape. *In* E. F. Aldon, C. E. G. Vincent, and W. H. Moir, eds. Strategies for classification and management of native vegetation for food production in arid zones. Fort Collins, Colo.: USDA Forest Service, Rocky Mountain Forest and Range Experiment Station. 112–16.

Crawley, M. J. 1983. Herbivory: The dynamics of animal-plant interactions. Berkeley: University of California Press.

D'Antonio, C. M., and P. M. Vitousek. 1992. Biological invasions by exotic grasses, the grass/fire cycle, and global change. Annual Review of Ecology and Systematics 23:63–87.

Darwin, C. 1859. The origin of the species. London: John Murray.

Daubenmire, R. F. 1959. Plants and environment. 2d ed. New York: John Wiley and Sons.

Davis, J. H., and C. D. Bonham. 1979. Interference of sand sagebrush canopy with needle and thread. Journal Range Management 32:384–86.

Davis, M. B. 1982. Quaternary history and the stability of forest communities. *In* D. C. West, H. H. Shugart, and D. D. Botkin, eds. Forest succession: Concepts and application. New York: Springer-Verlag. 132–53.

Detling, J. K., D. J. Winn, C. Proctor-Gregg, and E. L. Painter. 1980. Effects of simulated grazing by belowground herbivores on growth, CO_2 exchange, and carbon allocation patterns of *Bouteloua gracilis*. Journal of Applied Ecology 17:771–78.

de Wit, C. T. 1958. Transpiration and crop yields. Agricultural Research Reports 64.8. Wageningen, The Netherlands: Pudoc.

de Wit, C. T., and H. van Keulen. 1972. Simulation of transport processes in soils. Wageningen, The Netherlands: Pudoc.

Dunne, T., W. Zhang, and B. F. Aubry. 1991. Effects of rainfall and vegetation on infiltration and runoff. Water Resources Research 27:2271–85.

Edwards, G. P., T. J. Dawson, and D. B. Croft. 1995. The dietary overlap between red kangaroos (*Macropus rufus*) and sheep (*Ovis aries*) in the arid rangelands of Australia. Australian Journal of Ecology 20:324–34.

Ellenberg, H. H. 1985. Veränderungen der Flora Mitteleuropas unter dem einfluß von Düngen und Immisionen. Schweizer Zeitschrift für Forstwissenschaft 136:19–39.

Ellis, J. E., and M. D. Swift. 1988. Stability of African pastoral systems: Alternate paradigms and their implications for development. Journal of Range Management 41:450–59.

Emanuel, W. R., H. H. Shugart, and M. Stevenson. 1985a. Climatic change and the broad-scale distribution of terrestrial ecosystem complexes. Climatic Change 7:29–43.

———. 1985b. Response to comment: Climatic change and the broad-scale distribution of terrestrial ecosystem complexes. Climatic Change 7:457–60.

Eufracio, O. 1992. Respuesta de una poblacion natural de lechuguilla (*Agave lecheguilla* Torr.): A diferentes sistemas de aprovechamiento. Ph.D. diss. Facultad de Ciencias Forestales, Universidad Autonoma de Nuevo Leon, Mexico.

Evenari, M., I. Noy-Meir, and D. W. Goodall, eds. 1985. Ecosystems of the world. Vol. 12A: Hot deserts and arid shrublands. Amsterdam: Elsevier.

Fisher, R. A., and N. C. Turner. 1978. Plant productivity in the arid and semi-arid zones. Annual Review of Plant Physiology 29:277–317.

Folse, L. J., H. E. Mueller, and A. D. Whittaker. 1990. Object-oriented simulation and geographic information systems. AI Applications in Natural Resource Management 4(2):41–47.

Foran, B. D. 1986. The impact of rabbits and cattle on an arid calcareous shrubby grassland in central Australia. Vegetatio 66:49–59.

Friedel, M. H. 1991. Range condition assessment and the concept of thresholds: A viewpoint. Journal of Range Management 44:422–26.

Frischknecht, N. C. 1963. Contrasting effects of big sagebrush and rubber rabbitbrush on production of crested wheatgrass. Journal of Range Management 16:70–74.

Fulbright, J. E., and S. L. Beasom. 1987. Long-term effects of mechanical treatment on white-tailed deer browse. Wildlife Society Bulletin 15:560–64.

George, M. L., J. R. Brown, and W. J. Clawson. 1992. Application of non-equilibrium ecology to management of Mediterranean grasslands. Journal of Range Management 45:436–40.

Goldberg, D. E. 1982. The distribution of evergreen and deciduous trees relative to soil type: An example from the Sierra Madre, Mexico, and a general model. Ecology 63:942–51.

Goldberg, D. E., and T. E. Miller. 1990. Effects of different resource additions on species diversity in an annual plant community. Ecology 71:213–25.

Gornitz, V., and NASA. 1985. A survey of anthropogenic vegetation changes in West Africa during the last century: Climatic implications. Climatic Change 7:885–325.

Graetz, R. D. 1991a. Desertification: A tale of two feedbacks. *In* H. A. Mooney et al., eds. Ecosystem experiments. New York: John Wiley and Sons. 59–87.

———. 1991b. The nature and significance of the feedback of changes in terrestrial vegetation on global atmospheric and climate change. Climatic Change 18:147–73.

Greenbaum, I. F., R. J. Baker, and P. R. Ramsey. 1978. Chromosomal evolution and the mode of speciation in three species of *Peromyscus*. Evolution 32:646–54.

Greenstone, M. H. 1984. Determinants of web spider species diversity: Vegetation structural diversity vs. prey availability. Oecologia 62:299–304.

Griffin, G. F., and M. H. Friedel. 1985. Discontinuous change in central Australia: Some implications of major ecological events for land management. Journal of Arid Environments 9:63–80.

Grime, J. P. 1977. Evidence for the existence of three primary strategies in plants and its relevance to ecological and evolutionary theory. American Naturalist 111:1169–94.

———. 1979. Plant strategies and vegetation processes. Chichester: John Wiley and Sons.

Hanks, R. J., H. R. Gardner, and R. L. Florian. 1969. Plant growth-evapotranspiration relations for several crops in the central great plains. Agronomy Journal 61:30–34.

Hanski, I., J. Kouki, and A. Halkka. 1993. Three explanations of the positive relationship between distribution and abundance of species. *In* R. Ricklefs and D. Schluter, eds. Species diversity in ecological communities: Historical and geographical perspectives. Chicago: University of Chicago Press. 108–16.

Harper, K. T., and R. L. Pendleton. 1993. Cyanobacteria and cyanolichens: Can they enhance availability of essential minerals for higher plants? Great Basin Naturalist 53:59–72.

Harrington, G. N. 1973. Brush control: A note of caution. East African Agricultural Forest Journal 39:95–96.

Harrington, G. N., and K. C. Hodgkinson. 1986. Shrub-grass dynamics in Mulga communities of eastern Australia. *In* P. J. Joss, P. W. Lynch, and O. B. Williams, eds. Rangelands: A resource under siege. Proceedings of second International Rangeland Congress. Canberra: Australian Academy of Sciences. 26–28.

Hastings, J. R., and R. L. Turner. 1965. The changing mile: An ecological study of vegetation change with time in the lower mile of an arid and semi-arid region. Tucson: University of Arizona Press.

Haynes, C. V., Jr. 1982. Great Sand Sea and Selima Sand Sheet, eastern Sahara: Geochronology of desertification. Science 217:629–33.

Heitschmidt, R. K., R. D. Schultz, and C. J. Scifres. 1986. Herbaceous biomass dynamics and net primary production following chemical control of honey mesquite. Journal of Range Management 39:67–71.

Hellden, U. 1991. Desertification: Time for an assessment? Ambio 20:372–83.

Herbel, C. H., F. N. Ares, and R. A. Wright. 1972. Drought effects on a semi-desert grassland. Ecology 53:1084–93.

Hobbs, R. J. 1993. Effects of landscape fragmentation on ecosystem processes in the western Australian wheat belt. Biological Conservation 64:193–201.

Holdridge, L. R. 1964. Life zone ecology. San Jose, Costa Rica: Tropical Science Center.

Houghton, J. T., B. A. Callander, and S. K. Varney. 1992. Climate change: Supplementary report to the IPCC Scientific Assessment. Cambridge: Cambridge University Press.

Howden, S. M., G. M. McKeon, J. C. Scanlan, J. O. Carter, and D. H. White. 1993. Changing stocking rates and burning management to reduce greenhouse gas emissions from southern Queensland grasslands. Proceedings of seventeenth International Grassland Congress. Palmerston North, New Zealand. 1203–6.

Hunt, H. W., M. J. Trlica, E. F. Redente, J. C. Moore, J. K. Detling, T. G. F. Kittel, D. E. Walter, M. C. Fowler, D. A. Klein, and E. T. Elliott. 1991. Simulation model for the effects of climate change on temperate grassland ecosystems. Ecological Modeling 53:205–46.

Huston, M. 1979. A general hypothesis of species diversity. American Naturalist 113:81–101.

Huston, M., and T. Smith. 1987. Plant succession: Life history and competition. American Naturalist 130:168–98.

Jaksic, F. M., and E. R. Fuentes. 1980. Why are native herbs in the Chilean matorral more abundant beneath bushes: Microclimate or grazing? Journal of Ecology 68:665–69.

Jameson, D. A. 1987. Climax or alternative steady states in woodland ecology. Proceedings, pinyon-juniper conference. Gen. Tech. Rep. Ogden, Utah: USDA Forest Service.

Jenny, H. 1980. The soil resource, origin and behavior. New York: Springer-Verlag.

Johnstone, I. M. 1986. Plant invasion windows: A time-based classification of invasion potential. Biological Review 61:369–94.

Kadmon, R., and A. Shmida. 1990. Spatiotemporal demographic processes in plant populations: An approach and a case study. American Naturalist 135:382–97.

Katz, R. W., and B. G. Brown. 1992. Extreme events in a changing climate: Variability is more important than averages. Climatic Change 21:289–302.

Keddy, P. A. 1992a. Assembly and response rules: Two goals for predictive community ecology. Journal of Vegetation Science 3:157–64.

———. 1992b. A pragmatic approach to functional ecology. Functional Ecology 6:621–26.

Lauenroth, W. K., and O. E. Sala. 1992. Long-term forage production of North American shortgrass steppe. Ecological Applications 2:397–403.

Laycock, W. A. 1991. Stable states and thresholds of range condition on North American rangelands. Journal of Range Management 44:427–33.

Lewin, R. 1985. Plant communities resist climatic change. Science 228:165–66.

Loehle, C. 1987. Applying artificial intelligence techniques to ecological modeling. Ecological Modeling 38:191–212.

Loh, D. K., and E. J. Rykiel. 1992. Integrated resource management systems: Coupling expert systems with geographic information systems and database management systems. Journal of Environmental Management 16(2):167–77.

Long, P., and R. P. Hutchin. 1991. Primary production in grasslands and coniferous forests with climate change: An overview. Ecological Applications 1:139–56.

Lonsdale, W. M. 1993. Rates of spread of an invading species: *Mimosa pigra* in northern Australia. Journal of Ecology 81:513–21.

Loucks, O. L., M. L. Plumb-Mentjes, and D. Rogers. 1985. Gap processes and large-scale disturbances in sand prairies. *In* S. T. A. Pickett and P. S. White, eds. The ecology of natural disturbance and patch dynamics. Orlando, Fla.: Academic Press. 71–83.

Ludwig, J. A. 1987. Primary productivity in arid lands: Myths and realities. Journal of Arid Environments 12:1–7.

Ludwig, J. A., G. L. Cunningham, and P. D. Whitson. 1988. Distribution of annual plants in North American deserts. Journal of Arid Environments 15:221–27.

MacMahon, J. A. 1980. Ecosystems over time: Succession and other types of change. *In* R. Waring, ed. Forests: Fresh perspectives from ecosystem analyses. Corvallis: Oregon State University Press. 27–58.

Madany, M. H., and N. E. West. 1983. Livestock grazing–fire regime interactions within montane forests of Zion National Park, Utah. Ecology 64:661–67.

Mainguet, M. 1991. Desertification: Natural background and human mismanagement. Berlin: Springer-Verlag.

Maldonado, J. L. 1979. Uso multiple de los recursos naturales de las zonas aridas. Bulletin Ciencia Forestal [Instituto Nacional de Investigaciones Forestales (INIF), Mexico] 4(17):12–20.

Mares, M. A. 1992. Neotropical mammals and the myth of Amazonian biodiversity. Science 255:976–79.

Marroquin, J., G. Borja, R. Velazquez, and J. A. de la Cruz. 1981. Estudio ecologico dasonomico de las zonas aridas del Norte de Mexico. Mexico: Instituto Nacional de Investigaciones Forestales (INIF).

McCown, R. L., and J. Williams. 1990. The water environment and implications for productivity. Journal of Biogeography 17:513–20.

McKeon, G. M., K. A. Day, S. M. Howden, J. J. Mott, D. M. Orr, W. J.

Scattini, and E. J. Weston. 1990. Northern Australian savannas: Management for pastoral production. Journal of Biogeography 17:355–72.
McKeon, G. M., and D. H. White. 1992. El Niño and better land management. Search 23:197–200.
McNaughton, S. J. 1983. Serengeti grassland ecology: The role of composite environmental factors and contingency in community organization. Ecological Monographs 53:291–320.
———. 1988. Mineral nutrition and spatial concentrations of African ungulates. Nature 334:343–45.
McNaughton, S. J., R. W. Recess, and S. W. Weagle. 1988. Large mammals and process dynamics in African ecosystems. BioScience 38:794–800.
Metting, B. 1991. Biological surface features of semiarid lands and deserts. *In* J. Skujins, ed. Semiarid lands and deserts: Soil resource and reclamation. New York: Marcel Dekker. 257–93.
Mitchell, J. M. 1980. History and mechanisms of climate. *In* H. Oeschger, B. Messerli, and M. Svilar, eds. Das klima-analysen und modelle, geschichte und zukunft. Berlin: Springer-Verlag. 31–42.
Morris, M. G. 1973. Chalk grassland management and the invertebrate fauna. *In* A. C. Jermy and P. A. Stott, eds. Chalk grassland: Studies on its conservation and management in south-east England. Maidstone: Kent Trust for Nature Conservation Special Publications. 27–34.
———. 1990. The effects of management on the invertebrate community of calcareous grassland. *In* S. H. Hillier, D. W. H. Walton, and D. A. Wells, eds. Calcareous grasslands: Ecology and management. Huntingdon: Bluntisham Books. 128–33.
Naiman, R. J. 1988. Animal influences on ecosystem dynamics. BioScience 38:750–52.
Neilson, R. P. 1986. High resolution climatic analysis and Southwest biogeography. Science 232:27–34.
———. 1987. Biotic regionalization and climatic controls in western North America. Vegetatio 70:135–47.
Neilson, R. P., G. A. King, and G. Koerper. 1992. Toward a rule-based biome model. Landscape Ecology 7:27–43.
Neilson, R. P., and L. H. Wullstein. 1985. Comparative drought physiology and biogeography of *Quercus gambelii* and *Quercus turbinella*. American Midland Naturalist 114:259–71.
Nelson, E. W. 1934. The influence of precipitation and grazing upon black grama grass range. Tech. Bull. 32. N.p.: USDA Forest Service.
Nicholson, S. E. 1990. The need for a reappraisal of the question of large-scale desertification: Some arguments based on consideration of rainfall fluctuations. Report of the SAREC-Lund International Meeting on Desertification, December 1990. Lund, Sweden.
———. 1994. Recent rainfall fluctuations in Africa and their relationship to ENSO. Proceedings of eighteenth annual climate diagnostics workshop. Oklahoma City: National Oceanic and Atmospheric Administration. 58–61.
Nicholson, S. E., M. L. Davenport, and A. R. Malo. 1990. A comparison of the vegetation response to rainfall in the Sahel and East Africa, using normalized difference vegetation index from NOAA AVHRR. Climatic Change 17:209–41.
Noy-Meir, I. 1973a. Data transformations in ecological ordination. Pt. 1: Some advantages of non-centering. Journal of Ecology 61:329–41.
———. 1973b. Desert ecosystems: Environment and producers. Annual Review of Ecology and Systematics 4:25–41.
———. 1974. Desert ecosystems: Higher tropic levels. Annual Review of Ecology and Systematics 5:195–214.
O'Connor, T. G. 1991. Local extinction in perennial grasslands: A life history approach. American Naturalist 137:753–73.
O'Connor, T. G., and G. A. Pickett. 1992. The influence of grazing on seed production and seedbanks of some African savanna grasses. Journal of Applied Ecology 29:247–60.
Owen, D. F. 1979. Drought and desertification in Africa: Lessons from the Nairobi Conference. Oikos 33:139–51.
Parton, W. J., J. W. B. Stewart, and V. C. Cole. 1988. Dynamics of C, N, P, and S in grassland soils: A model. Biogeochemistry 5:109–31.
Paulsen, H. A., Jr., and F. N. Ares. 1962. Grazing values and management of black grama and tobosa grasslands and associated shrub ranges of the Southwest. Tech. Bull. 56. N.p.: USDA Forest Service.
Peck, A. J., and D. H. Hurle. 1973. Chloride balance of some farmed and forested catchments in southwestern Australia. Water Resources Research 9:643–537.
Penman, H. L. 1948. Natural evaporation from open water, bare soil, and grass. Proceedings, Royal Society, Series A 193:120–46.
Pittock, A. B. 1993. A climate change perspective on grasslands. Proceedings of seventeenth International Grassland Congress. Palmerston North, New Zealand. 1053–50.
Plant, R. E., and N. D. Stone. 1991. Knowledge-based systems in agriculture. New York: McGraw-Hill.
Prentice, C., and A. M. Solomon. 1991. Vegetation models and global change. *In* R. S. Bradley, ed. Global changes of the past. Boulder, Colo.: UCAR/Office for Interdisciplinary Earth Studies. 365–83.
Pulliam, H. R., and B. J. Danielson. 1991. Sources, sinks, and habitat selection: A landscape perspective on population dynamics. American Naturalist 137:S50–S66.
Quinn, R. D. 1986. Mammalian herbivory and resilience in Mediterranean-climate ecosystems. *In* B. Dell, A. J. M. Hopkins, and B. B. Lamont, eds. Resilience in Mediterranean-type ecosystems. Dordrecht: Junk. 113–28.
Rabotnov, T. A. 1974. Differences between fluctuations and successions. *In* R. Knapp, ed. Vegetation dynamics. The Hague: Dr. W. Junk Publishers. 19–24.
Rauscher, H. M., and T. M. Cooney. 1986. Using expert system technology in a forestry application: The CHAMPS experience. Journal of Forestry 84:14–17.
Rhoades, D. F. 1985. Offensive-defensive interactions between herbivores and plants: Their relevance in herbivore population dynamics and ecological theory. American Naturalist 125:205–38.
Rhodes, S. L. 1991. Rethinking desertification: What do we know and what have we learned? World Development 19.
Rizzo, B., and E. Wiken. 1992. Assessing the sensitivity of Canada's ecosystems to climatic change. Climatic Change 21:37–55.
Ropelewski, C. F., and M. S. Halpert. 1987. Global and regional precipitation associated with the El Niño/Southern Oscillation. Monthly Weather Review 115:985–96.
———. 1989. Precipitation patterns associated with the high index phase of the Southern Oscillation. Journal of Climate 2:268–84.
Rosenzweig, M. L. 1991. Habitat selection and population interactions: The search for mechanism. American Naturalist 137:S5–S28.
———. 1992. Species diversity gradients: We know more and less than we thought. Journal of Mammalogy 73:715–30.
———. 1995. Patterns of diversity in space and time. Cambridge: Cambridge University Press.
Rosenzweig, M. L., and Z. Abramsky. 1993. How are diversity and productivity related? *In* R. Ricklefs and D. Schluter, eds. Species diversity in ecological communities: Historical and geographical perspectives. Chicago: University of Chicago Press. 52–65.
Running, S. W., T. R. Loveland, and L. L. Pierce. 1994. A vegetation classification logic based on remote sensing for use in global biogeochemical models. Ambio 23:77–81.
Russell, J. S., N. McLeod, and M. B. Dale. 1992. Combined Southern

Oscillation index and sea surface temperature as predictors of seasonal rainfall. Proceedings of International COAD Workshop. Boulder: University of Colorado Press. 145–56.
Sala, O. E., W. K. Lauenroth, and W. J. Parton. 1992. Long-term soil water dynamics in the shortgrass steppe. Ecology 73:1175–81.
Sala, O. E., W. J. Parton, L. A. Joyce, and W. K. Lauenroth. 1988. Primary production of the central grassland region of the United States. Ecology 69:40–45.
Sandoval, G. 1980. Algunas consideraciones sobre *Yucca schidigera* y su aprovechamiento. Memorias de la Primera Reunión Nacional sobre Ecología, Manejo y Domesticació de las Plantas Utiles del Desierto. 139–54.
Savage, M., and T. W. Swetnam. 1990. Early nineteenth-century fire decline following sheep pasturing in a Navajo ponderosa pine forest. Ecology 71:2374–78.
Scanlan, J. C. 1988. Spatial and temporal vegetation patterns in a subtropical *Prosopis* savanna woodland, Texas. College Station: Department of Rangeland Ecology and Management, Texas A&M University.
Schlesinger, W. H., J. F. Reynolds, G. L. Cunningham, L. F. Huenneke, W. M. Jarrell, R. A. Virginia, and W. G. Whitford. 1990. Biological feedbacks in global desertification. Science 247:1043–48.
Scifres, C. J., W. T. Hamilton, J. M. Inglis, and J. R. Conner. 1983. Development of integrated brush management systems (IBMS): Decision-making processes. *In* K. W. McDaniel, ed. Proceedings of brush management symposium. Lubbock: Texas Tech Press. 97–104.
Scifres, C. J., W. T. Hamilton, B. H. Koerth, R. C. Flinn, and R. A. Crane. 1988. Bionomics of patterned herbicide application for wildlife habitat enhancement. Journal of Range Management 41:317–21.
Scifres, C. J., J. L. Mutz, R. E. Whitson, and D. L. Drawe. 1982. Interrelationships of huisache canopy cover with range forage on the coastal prairie. Journal of Range Management 35:558–62.
Seligman, N. G. 1973. A quantitative phytosociological analysis of the vegetation of the golan. Ph.D. diss. Hebrew University of Jerusalem.
———. 1995. Management of Mediterranean grasslands. *In* J. Hodgson and A. W. Illius, eds. Ecology and management of grasslands. United Kingdom: CABI.
Seligman, N. G., and A. Perevolotsky. 1994. Has intensive grazing by domestic livestock degraded Mediterranean basin rangeland? *In* M. Arianoutsou and R. H. Groves, eds. Plant-animal interactions in Mediterranean-type ecosystems. Dordrecht, The Netherlands: Kluwer Academic Publishers. 93–103.
Seligman, N. G., H. van Keulen, and C. J. F. Spitters. 1992. Weather, soil conditions, and the interannual variability of herbage production and nutrient uptake on annual Mediterranean grasslands. Agricultural and Forest Meteorology 57:265–79.
Shantz, H. L., and L. N. Piemiesel. 1927. The water requirements of plants at Akron, Colo. Journal of Agricultural Research 34:1093–90.
Sheldon, S. 1980. Ethnobotany of *Agave lecheguilla* and *Yucca carnerosana* in Mexico's Zona Ixtlera. New York: University Bonaventure. 378–87.
Shugart, H. H. 1990. Using ecosystem models to assess potential consequences of global climatic change. Trends in Ecology and Evolution 5:303–7.
Siepel, H. 1990. Decomposition of leaves of *Avenella flexuosa* and microarthropod succession in grazed and ungrazed grasslands. Pt. 1: Succession of microarthropods. Pedobiologia 34:19–30.
Siepel, H., and S. E. van Wieren. 1990. Decomposition of leaves of *Avenella flexuosa* and microarthropod succession in grazed and ungrazed grasslands. Pt. 2: Chemical data and comparison of decomposition rates. Pedobiologia 34:31–36.
Skarpe, C. 1990. Shrub layer dynamics under different herbivore densities in an arid savanna, Botswana. Journal of Applied Ecology 27:873–85.
———. 1991. Impact of grazing in savanna ecosystems. Ambio 20:351–56.
Slatyer, R. O. 1967. Plant water relations. New York: Academic Press.
Smith, T. M., H. H. Shugart, F. I. Woodward, and B. J. Burton. 1993. Plant functional types. *In* A. M. Solomon and H. H. Shugart, eds. Vegetation dynamics and global change. New York: Chapman and Hall. 272–92.
Starfield, A. M., D. H. M. Cummings, R. D. Taylor, and M. S. Quading. 1993. A frame-based paradigm for dynamic ecosystem models. AI Applications 7:1–13.
Stuart-Hill, G. C., N. N. Tainton, and H. J. Barnard. 1987. The influence of an *Acacia karroo* tree on grass production in its vicinity. Journal of the Grassland Society of South Africa 4:83–88.
Swetnam, T. W., and J. L. Betancourt. 1990. Fire–Southern Oscillation relations in the southwestern United States. Science 249:1010–20.
Tanner, C. B., and T. R. Sinclair. 1982. Efficient water use in crop production: Research or re-search. *In* H. M. Taylor, W. R. Jordan, and T. R. Sinclair, eds. Limitations of efficient water use in crop production. Madison, Wis.: American Society of Agronomy Monographs. 1–27.
Thomas, J. A. 1983. The ecology and conservation of *Lysandra bellargus* (Lepidoptera: Lycaenidae) in Britain. Journal of Applied Ecology 20:59–83.
Tilman, D. 1987. Secondary succession and the pattern of plant dominance along experimental nitrogen gradients. Ecological Monographs 57:189–214.
Tilman, D., and S. Pacala. 1993. The maintenance of species richness in plant communities. *In* R. Ricklefs and D. Schluter, eds. Species diversity in ecological communities: Historical and geographical perspectives. Chicago: University of Chicago Press. 13–25.
Tilman, D., and D. Wedin. 1991a. Dynamics of nitrogen competition between successional grasses. Ecology 72:1038–49.
———. 1991b. Plant traits and resource reduction for five grasses growing on a nitrogen gradient. Ecology 72:685–700.
Tongway, D. 1994. Rangeland soil condition assessment manual. Canberra, Australia: CSIRO Publications. 1–69.
Tothill, J. C., and C. Gillies. 1992. The pasture lands of northern Australia. Occasional Publications No. 5. St. Lucia, Queensland: Tropical Grassland Society of Australia.
Tucker, C. J., H. E. Dregne, and W. W. Newcomb. 1991. Expansion and contraction of the Sahara desert from 1980 to 1990. Science 253:299–301.
Turner, R. M. 1990. Long-term vegetation change at a fully protected Sonoran desert site. Ecology 71:464–77.
Valiente-Banuet, A., and E. Ezcurra. 1989. Aspectos causales de la asociación *Neobuxbaumia tetetzo:* Plantas nodrizas en el Valle de Zapotitlán de las, Salinas, México. Centro de Ecología, UNAM. Memorias de la Reunión sobre Líneas de Investigación Ecológica en Zonas Aridas. SEDUE, UAM, U.S. Fish and Wildlife Service and International Association of Fish and Wildlife Agencies. Washington, D.C.
Van Auken, O. W., and J. K. Bush. 1989. *Prosopis glandulosa* growth: Influence of nutrients and simulated grazing of *Bouteloua curtipendula.* Ecology 70:512–16.
van Keulen, H. 1975. Simulation of water use and herbage growth in arid regions. Wageningen, The Netherlands: Pudoc.
van Keulen, H., and N. G. Seligman. 1992. Moisture, nutrient availability, and plant production in the semiarid region. *In* Th.

Alberda, H. van Keulen, N. G. Seligman, and C. T. de Wit, eds. Food from dry lands: An integrated approach to agricultural development. Dordrecht, The Netherlands: Kluwer Academic Publishers. 25–81.

Verstraete, M. M. 1986. Defining desertification: A review. Climatic Change 9:5–18.

Vitousek, P. M., and L. R. Walker. 1989. Biological invasion by *Myrica faya* in Hawaii: Plant demography, nitrogen fixation, and ecosystem effects. Ecological Monographs 59:247–65.

Walker, B. H. 1988. Autecology, synecology, climate, and livestock as agents of rangeland dynamics. Australian Rangeland Journal 10:69–75.

———. 1994. Landscape to regional-scale responses of terrestrial ecosystems to global change. Ambio 23:67–73.

Walker, J., R. W. Condon, K. C. Hodgkinson, and G. N. Harrington. 1981. Fire in pastoral areas of poplar box (*Eucapyptus populnea*) lands. Australian Rangeland Journal 3:12–23.

Walker, J., J. A. Robertson, L. K. Penridge, and P. J. H. Sharpe. 1986. Herbage response to tree thinning in a *Eucalyptus crebra* woodland. Australian Journal of Ecology 11:135–40.

Welsh, R. G., and R. F. Beck. 1976. Some ecological relationships between creosotebush and bush muhly. Journal of Range Management 29:472–75.

Weltzin, J. F. 1990. The role of prairie dogs (*Cynomys ludovicianus*) in regulating the population dynamics of the woody legume *Prosopis glandulosa*. Master's thesis. College Station: Texas A&M University.

Weltzin, J. F., and M. B. Coughenour. 1990. Savanna tree influence on understory vegetation and soil nutrients in northwestern Kenya. Journal of Vegetation Science 1:325–34.

West, N. E. 1990. Structure and function of microphytic soil crusts in wildland ecosystems of arid to semi-arid regions. Advances in Ecological Research 20:179–223.

———. 1991. Nutrient cycling of soils of semiarid and arid regions. *In* J. Skujins, ed. Semiarid lands and deserts: Soil resource and reclamation. New York: Marcel Dekker. 295–332.

Western, D. 1991. Climatic change and biodiversity. *In* S. H. Ominde and C. Juma, eds. A change in the weather: African perspectives on climatic change. Nairobi: African Centre for Technology Studies Press. 87–96.

Western, D., and C. van Praet. 1973. Cyclical changes in the habitat and climate of an East African ecosystem. Nature 241:104–6.

Westoby, M. 1979–80. Elements of a theory of vegetation dynamics in rangelands. Israel Journal of Botany 28:169–94.

Westoby, M., B. H. Walker, and I. Noy-Meir. 1989. Opportunistic management for rangelands not at equilibrium. Journal of Range Management 42:266–74.

White, D. H., D. Collins, and S. M. Howden. 1993. Drought in Australia: Prediction, monitoring, management, and policy. *In* D. A. Wilhite, ed. Drought assessment, management, and planning: Theory and case studies. Dordrecht, The Netherlands: Kluwer Academic Publishers. 213–36.

Whitford, W. G. 1986. Decomposition and nutrient cycling in deserts. *In* W. G. Whitford, ed. Pattern and process in desert ecosystems. Albuquerque: University of New Mexico Press. 93–117.

———. 1992. Effects of climate change on soil biotic communities and soil processes. *In* R. L. Peters and T. E. Lovejoy, eds. Global warming and biological diversity. New Haven, Conn.: Yale University Press. 124–36.

Wilkeem, S. J., and M. D. Pitt. 1982. Soil nitrogen gradients as influenced by sagebrush canopy in southern British Columbia. Northwest Science 56:276–86.

Williams, C. B. 1943. Area and the number of species. Nature 152:264–67.

Williams, J. 1983. Soil hydrology. *In* Division of Soils, Commonwealth Scientific and Industrial Research Organization, ed. Soils: An Australian viewpoint. London: CSIRO; Melbourne: Academic Press. 507–30.

Williams, J., and M. Bonell. 1988. The influence of scale of measurement on the spatial and temporal variability of the Philip infiltration parameters: An experimental study in an Australian savanna woodland. Journal of Hydrology 104:33–51.

Williams, J., K. R. Helyar, R. S. B. Greene, and R. A. Hook. 1994. Soil characteristics and processes critical to the sustainable use of grasslands in arid, semi-arid, and seasonally dry environments. Proceedings of seventeenth International Grassland Congress. 1335–50. Wellington, New Zealand: SIR Publishing.

Wilson, A. D. 1991. The influence of kangaroos and forage supply on sheep productivity in the semiarid woodlands. Australian Rangeland Journal 13:69–80.

Winkworth, R. E. 1969. The soil water regime of an arid grassland *Eragrostis eripoda* in central Australia. Agricultural Meteorology 7:387–94.

Wisheu, I. C., and P. A. Keddy. 1992. Competition and centrifugal organization of plant communities: Theory and tests. Journal of Vegetation Science 3:147–56.

Woodward, R. A., K. T. Harper, and A. R. Tiedemann. 1984. An ecological consideration of the significance of cation exchange capacity of roots of some Utah range plants. Plant and Soil 79:169–80.

Yeaton, R. I. 1988. Porcupines, fires, and the dynamics of the tree layer of the *Burkea africana* savanna. Journal of Ecology 76:1017–29.

6 The Study and Management of Dryland Population Systems

David Saltz, Moshe Shachak, Martyn Caldwell, Steward T. A. Pickett, Jeffrey Dawson, Haim Tsoar, Yoram Yom-Tov, Mark Weltz, and Roger Farrow

A General Approach to Studying and Managing Populations

The study and management of population dynamics is concerned with species. Each species demonstrates a unique distribution and abundance pattern. It is impossible to study and manage the population dynamics of all species in all of the habitats within their range of distribution. For successful management, we have to select sample populations that represent the range of population dynamics in the region. Decisions should be made on the following considerations: (1) Which species should be studied and managed? (2) Where along their distribution should they be studied and managed? (3) Which factors controlling the population dynamics should be studied and managed? The landscape within which a population exists and the population's distribution and abundance along environmental gradients within the landscape are the basic elements by which answers to these questions will be determined.†

Drylands comprise a significant proportion of the earth's landmass, yet existing definitions for drylands are vague and boundaries are often nebulous (McGinnes 1979). Nevertheless, it is widely accepted that drylands are characterized by low levels of precipitation varying considerably in time and space (Safriel et al. 1989, Shreve 1934). Evenari (1981) termed this the *rainfall instability syndrome* (RIS) and considered it the most important element affecting desert life. Thus, in terms of the organisms living within them, deserts are patchy and unpredictable. If we define a population as a collection of individuals of a single species or subspecies bound together by interactions (dynamics, competition, mutualism, behavior), then desert populations are those that are adapted (in terms of these interactions) to the dry and unpredictable desert environments. This does not necessarily mean that these populations could not exist in other environments. However, if they did, their intrinsic rate of natural increase (r) would be lower. The lower r may result because their specific adaptations to deserts make them less adapted to other environments or to competition with other populations that are better adapted to nondesert environments.

To understand and manage populations, it is necessary to work within the context of the factors that affect their distribution and abundance (environmental or otherwise). We term these factors *controllers,* and a population together with its controllers is called a *population system.* The changes in distribution and abundance over time are defined as *flow.* Dryland population systems are typified by limited water and patchiness, which are important controllers on the flow of organisms.†

The first step in the study and management of dryland populations is developing a distribution and abundance curve. If the target population is indeed a dryland population, then the second step is addressing population processes, specifically those factors such as water, patchiness, and others that affect population flow at the local level.

Since we are dealing with population systems, we can use system language consisting of state and flow variables and controllers on the flow variables (fig. 6.1). The population system deals with questions concerning the factors that brought about the changes in population abundance (flow) between times t and $t + 1$. The problem in this case is how to identify and select the factors controlling the system. Since population systems are associated with community, ecosystems, and landscape systems, we can select the controllers according to the kind of organization:

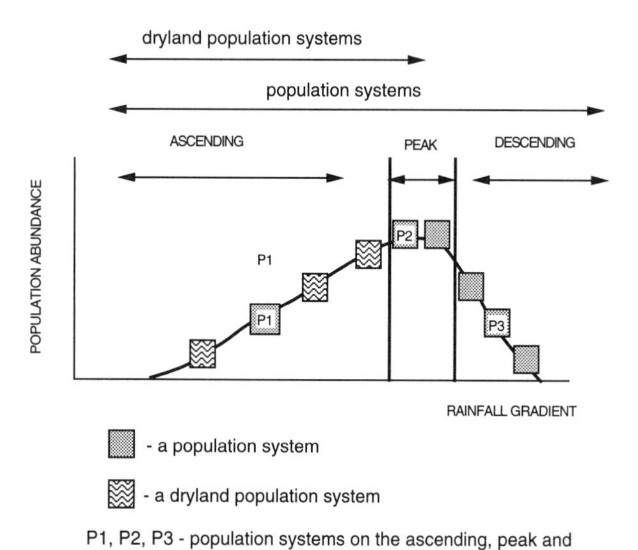

Figure 6.1. Dryland population systems of a species along the rainfall gradient.

1. How do the other systems (e.g., communities, ecosystems, and landscapes) affect population dynamics? We may ask how patchiness, which is a landscape phenomenon, affects population dynamics. How does nutrient flux, an ecosystem phenomenon, affect population dynamics? Or how do the interactions between populations, a community question, affect population dynamics?

2. How do individual animal traits and population-level processes affect population dynamics? The influence of physiological traits on population dynamics may be considered as the effect of individuals on the population dynamics. The effect of density on population dynamics is a feedback of the population within its own level of organization.

In addition to kind of system, human activities affect population dynamics. Human activities as controllers of population dynamics may be divided into two categories: (1) activities not directed at the population but nevertheless affecting it and (2) activities directed at the population itself. The first category is often associated with economic development, including factors such as urbanization, agriculture, mining, and so on, and impacts populations by affecting their controllers (e.g., loss of habitat, fragmentation, etc.). The second category includes activities such as consumption (harvesting), pest control, and conservation and entails management steps such as regulation and law enforcement, habitat improvement, reintroduction, and culling.

The third step is to identify similarities and differences among species. Specifically, what commonalities and differences do populations in the same location have in population dynamics? Since it is not possible to study all species in a given locale, we should attempt to classify the species into aggregates. We can aggregate species according to different criteria. The criteria may be: (1) physiological (homeotherms, poikilotherms, etc.), (2) behavioral (diurnal, nocturnal, etc.), (3) trophic levels (predators, prey, herbivores, parasites, or detritivores), or (4) abundance (i.e., rare and common species). The criteria define functional groups. After aggregating the species into functional groups, we are able to approach a more specific question: What are the similarities and differences in population processes within and between functional groups?

The purpose of this chapter is to describe the attributes characterizing the major controllers of populations in drylands. Based on these attributes, we discuss the important aspects that, in our opinion, should be considered in management of dryland populations. This chapter is not a review of population management techniques in arid lands; it is concerned with the traits of the ecosystem, communities, and populations that should be taken into account when dealing with population management strategies.†

The chapter has four sections. The first two sections deal with the ecosystem traits and the nature of primary producers that should be considered when developing and implementing management strategies. The third section deals with the known and required knowledge for animal population management and with the general approach that we believe should be taken toward managing arid lands animal populations. We do not address the issue of managing plant populations because plant management is most often studied and implemented at the community level. In those cases where plants are managed as individual populations, many of the principals and concerns outlined for managing animal populations in this chapter are applicable. The last section is a summary to assist management personnel who are dealing with a particular species in a given area in identifying the specific problem in the wider context. For this reason, we concentrated all the questions, functional groups, and flows described in the first three sections. We hope that readers will be able to choose from all these questions, groups, and flows those relevant to their own interest. We feel that such a process will allow the formation of a specific management strategy based on general principles.

The Environmental Context for Population Management

The existence of organisms in a saline environment marked by high temperatures and limited water availability depends on their ability to maintain thermal equilibrium within their environment and to attain resources necessary for reproduction. Maintaining thermal equilibrium and attaining resources depend, in part, on four components of the physical environment: substrate, water, heat flow, and nutrients.

Deserts are typified by several major substrates: rocky with soil patches, loessial, and sand dunes. Different substrates have different properties that determine their ability to retain water and heat. The rules that determine this ability are the same for all substrates and, in many situations, the different substrates may have somewhat similar problems. Sand dunes, which are common in deserts, present a substrate convenient for understanding the relationships among the thermal regime, water, and salinity that create the surroundings in which the populations are active. We will use sand dunes as a case study for substrate and devote the first part of this section to exploring the dunes as an example for analyzing the characteristics of the surroundings and their influence on the vegetation and animals. It is important to note that, while we devote the first part of this section to sand dunes, the rest of this chapter addresses deserts in general and not necessarily sandy environments.

In the second part of this section, we present the relationships between recycling of elements and populations. An environment poor in water and high in salinity limits primary and secondary productivity. One of the coordinating factors for water, salinity, and biomass is the recycling of elements. We feel that it is difficult to manage desert populations without understanding that the plants and animals live in an environment limited in nutrients.

Surface Properties and Population: Dune Sand as a Case Study

About 20% of the world's arid and extremely arid zones are covered by aeolian sand. In total, they cover an area of more than 5,000,000 km^2 (Pye and Tsoar 1990). Dune sand is usually considered devoid of any positive ecological characteristics because it has low fertility and low water-retaining capacity. Indeed, aeolian sand encroachment has become the image of desiccation and desertification in drylands (Hellden 1988, Lamprey 1975). Regardless of this popular concept, sand dunes have an extensive cover of vegetation and support a wide variety of fauna. In addition to human impact, three major physical attributes affect dune flora and fauna: thermal properties, sand texture, and sand moisture regime.

Thermal Properties of Dune Sand When sand is devoid of or has only meager vegetative cover, its texture plays an important role in determining the thermal properties of its environment. Heat is absorbed from solar short-wave radiation and concentrated in the sand's surface to about 1 mm thickness (Van Wijk and de Vries 1963). The transfer of this heat energy from the hot sandy surface into the cooler, underlying sand strata is known as heat conduction. Thermal conductivity measures the rate at which heat passes from a warmer area of sand into a cooler one. Specific heat measures the ratio of the total amount of heat needed to increase the temperature of a unit mass of sand by one degree to the total amount of heat required to raise the same mass of water at 15 °C by one degree.

The rate at which temperature modification takes place in below-surface sand is in direct proportion to its ability to transmit heat (thermal conductivity) but inversely proportionate to the amount of heat required to achieve the modification of temperature (specific heat). Quartz, the main component of sand, is a very poor conductor of heat (Janza 1975). On the other hand, sand has a relatively high specific heat. Thus, sand is typified by a very slow penetration of surface temperature changes.

These thermal properties of sand result in three significant diurnal patterns in the temperature at deeper levels of dune sand: the maximal temperature is lessened with depth, the diurnal amplitude is dampened with depth, and there is a time lag between the daily minimum and maximum surface temperature and those at various depths.

Water requires a very large input or output of heat to increase or decrease its temperature relative to air. The result is that dry sand is much more sensitive to thermal changes than wet sand (Satoh 1967). Since sandy soil with low field capacity contains little soil water, wet sandy soils are warmer during daytime than wet heavy soils, which have higher field capacity. Hence, in wintertime, the minimum surface temperature of wet subtropical desert sand reaches temperatures above freezing regardless of the low atmosphere relative humidity that increases the infrared outgoing radiation.

The high range of diurnal sand temperature variations is reduced by vegetation cover. Birds nesting on small shrubs on the sand can experience a maximal temperature of 43 °C, which is lower than the surface temperature (58 °C) but higher than the shade temperature (35 °C) (Williams 1954).

Ecosystem Properties: Texture, Water, and Salt The texture of sand is characterized by: (1) low field capacity, (2) low value for soil moisture availability for plants, and (3) high rate of permeability and leaching. The latter results in the washing away of nutrient elements necessary for plant growth. The extreme mobility of loose sand particles allows them to be easily swept away by the wind in a leaping movement described as a ballistic path very close to the bed (White 1982). This process, known as saltation, is detrimental to plants, since the grains bombard the plants causing damage and making the plants vulnerable to disease.

Dust does not easily accumulate together with sand unless the latter is covered by vegetation that serves as a trap for the suspended particle load in the lower atmosphere. Biogenic crust is formed on sand dunes when vegetation spreads and reduces the sand movement. Filamentous cyanobacteria (blue-green algae) are known to form a crust in the northwestern Negev sand dunes. This crust serves as a trap for atmospheric dust. There is a feedback mechanism between the amount of dust that is deposited on the surface and the mass of the cyanobacteria. A significant correlation has been found in the northwestern Negev between the percentage of fines (silt + clay) and the percentage of organic matter in the crust. For about 1.5% of organic matter, there are more than 30% of fines.

Sand dunes are known to have high rates of infiltration because of their relatively large pore spaces and low field capacity. The rate at which water flows through a permeable medium is known as hydraulic conductivity, a measure of a medium's ability to transmit water and a dimension of percolation velocity. On well-graded dune sand consisting of particle size of 0.15 mm, 1 mm of rain penetrates to an average depth of 7 mm; whereas, on coarser sand of 0.3 mm, 1 mm of rain achieves an average of 20 mm penetration (Dincer et al. 1974). Finer soils with a better field capacity than sand reduce percolation to much smaller depths. One mm of rainfall wets loessial soil to a depth of 5 mm (Orev 1984) and finely granulated clay to 2 mm (Walter 1973).

The salt in the desert soil originates from rainwater and aeolian dust. The rate at which salt is leached into the ground depends on the hydraulic conductivity of the soil. Loamy soil, with 20% clay, has the highest salt content (up to 800 ppm of Na^+), mostly at depths of 35–40 cm. The sandy loam and sandy soils are much less salty (about 100–150 ppm of Na^+) throughout the soil depth. In summer and autumn, salt concentration in loamy soil at depths of 0–15 cm increases (from 80 ppm in winter to 400 ppm in autumn) because of the capillary movement of soil moisture to the surface. Such changes do not occur in sandy loam and sandy soils because of low capillarity.

The biogenic crust formed on top of the dune has a lower hydraulic conductivity than dune sand because of the high proportion of fines. A crust of 1–2 mm thickness that is composed of up to 79% fines induces runoff over the dune during rain intensity of more than 10 mm hr^{-1} for a duration of three minutes (Yair 1990).

Vegetation in Sand Dune Landscape: Plant Response to Texture, Shape, and Moisture Because water easily percolates in sand, soil moisture availability to plants tends to be low. The volumetric available water content for sand usable by plants (which is between the wilting point and the field capacity) is about 3–9% (Brady 1974, Noy-Meir 1973). Dune sand has 1.28% moisture content at wilting point (Chepil 1956). Exact measurements of field capacity are practically impossible. The advantage of dune sand is in deserts where rain is meager. A small amount of water input brings sand moisture to values above the wilting point, while the same amount of water in loamy soils is still below the wilting point and is not available for plants. The soil moisture measure for optimum plant growth is near the field capacity. This range in moisture content for sand is very narrow (Brady 1974); therefore, plants on sand need smaller but more frequent inputs of water to the soil than those on loamy soils, in order to suck water rapidly. This is why desert sand is considered a mesic habitat and may have as much as five times more primary production than loamy soils when rain is well distributed throughout the rainy season (H. Tsoar, personal communication, 1994).

The advantage of coarse sand texture has its most important manifestation in extreme arid regions. The minimum isolated rain shower required to produce some signs of vegetation on sandy soils is on the order of 10–15 mm. Such meager precipitation has no effect on loam soils (Le Houerou 1986). The breakpoint between the advantage of coarse textured sand (low evaporation) as compared with fine textured soil, versus its inherent disadvantage (high hydraulic conductivity), lies somewhere between 300 mm and 500 mm mean annual precipitation (Noy-Meir 1973).

Sandy textured soil has more biomass in arid areas than loamy textured soil also because there is less concentration of salt in sand and low evaporation of deep soil moisture. The high rate of hydraulic conductivity on dune sand offers the feature of deep percolation where moisture is protected from evaporation during the long dry periods. Moisture retention is assured in desert sand below the upper 30–60 cm. Hence, desert sand may allow for relatively more vigorous and more perennial mesophytic shrubs than on loamy textured desert soils (Alizai and Hulbert 1970, Chadwick and Dalke 1965).

In arid areas of the Negev, Egypt, and Saudi Arabia, sand retains a certain moisture content just above the wilting point down to a depth of several meters. At the end of the long dry summer, the upper dry layer reaches down to a depth of 30–60 cm. This favors perennial plants rather than ephemeral ones (Dincer et al. 1974, Migahid 1961, Tsoar and Zohar 1985). For that reason, most seedlings do not fare well on desert sand dunes, as they are unable to reach the moisture below the dry surface layer (Bowers 1986).

Areas undergoing a high rate of erosion show little or no vegetation because plant roots are denuded and soil moisture is lost. In dune sands, erosion, not lack of moisture, is the primary limiting factor for vegetation (Tsoar 1990). The rate of sand erosion depends on wind magnitude, frequency, and dune shape. Dune fields in areas where wind velocities above the threshold are very frequent and multidirectional are under a continued processes of erosion that hinders any development of vegetation growth. Sand dunes in arid areas with low and infrequent wind energy are stabilized completely by microphytes and macrophytes, as in the northwestern Negev, where the average amount of rainfall is 100–150 mm (Tsoar et al. 1994).

Animals in Sand Dune Landscape The two most important factors affecting animals living in sand habitats are the thermal qualities that cause surface temperatures to increase frequently up to 70 °C and the low cohesiveness of the sand particles. The most common animal adaptation to high temperatures is avoidance: most animals living in sand dune habitats in deserts are nocturnal or are active during the early morning or late afternoon when soil temperature decreases to tolerable temperatures. During the rest of the day, most of these animals hide below the sand surface either in their own burrows or in holes dug by other animals. This behavior is common among small- to medium-sized animals, such as insects and other invertebrates, but also among reptiles and small mammals such as rodents.

In areas without sand deposition, moisture is always found in the sand below the upper 30–60 cm. Temperature decline with depth is logarithmic. Consequently, the most favorable conditions for the biota in the summer are at depths of 20–30 cm during the daytime. Burrows, therefore, provide a microclimate of relatively low temperature and high relative humidity, preventing overheating and excessive water loss.

Some animals, mostly medium to large, do not hide in burrows; they avoid hot surface temperatures by hiding in the shade of bushes. Some animals are less sensitive to high temperatures than others and are diurnal. Such animals can tolerate air temperatures of about 40 °C but not long exposure to higher surface temperatures. To avoid overheating, they minimize the time they spend on the hot surface by frequently hiding in the shade or in burrows or decrease their contact with the surface by raising one or more of their legs while standing on the hot sand. The temperature gradient above the surface is very steep (when surface temperature is 65 °C, air temperature 2 cm above it is only 45 °C). Many of the diurnal animals have relatively long legs (e.g., lizards and beetles) and can elevate themselves high above the hot surface, thus keeping their bodies at relatively low temperatures while only the soles of some of their feet are in contact with the hot sand.

Low cohesiveness is another character of the sand that affects animals because it makes movement (particularly walking and running) difficult. The most common adaptation by animals evolved to the low cohesiveness of sand dunes is increased surface area of the soles and fingers. This is achieved by enlarged or additional scales on the fingers (as have many lizards and skinks) or having stiff hairs on the sole (as have many rodents and carnivores). The other technique to overcome low cohesiveness of the soil is sand swimming: some animals, particularly skinks and some insects, avoid walking on the surface. These animals have relatively long and narrow bodies and short legs that help them move while winding their bodies between the sand particles.

The Impact of Humans Desert sand dunes have long attracted the attention of humans. Dunes carry a richer biomass than other desert terrain types and are, therefore, attractive for grazing. Soil on stabilized dunes is more easily cultivated and less prone to salinization than loamy soils.

Grazing is undeniably one of the principal agents of biological stress and landscape changes on sand dunes. Other human activities on sand dunes include collection of firewood or building material, use of off-road vehicles, and agricultural cultivation, all of which result in elimination of the biogenic crust, destruction of vegetation, the concatenated increase of sand movement, deflation of suspendable fines, and the decrease of soil moisture. Many sand dune areas of the world have borne the consequences of these desertification processes (Danin 1987, Tsoar and Moller 1986). In some cases, the extirpation of wild herbivores and absence of domestic grazing pressure has led to stabilization of dunes that were characteristically mobile.

Nutrient Flux and Population Systems

Dryland populations of plants and animals tend to be limited in occurrence and growth by both the availability and mobility of nutrients. Nutrient flow is linked closely to water availability and movement. Water is necessary as a medium for soil microbial activity central to nutrient cycling. Water also serves as a solvent in which mineral nutrients move in soil ecosystems and in which they are taken up and translocated by plants. Plant water deficit is the most important factor limiting carbon fixation, plant growth, and net primary productivity on a global basis (Boyer 1982, Schultze 1986). Aridity limits the rate and amount of nutrient movement among organic and mineral pools in soil even where soils are fertile.

Adaptation of Organisms to Low Nutrients Environment Populations of microorganisms, plants, and animals of drylands possess adaptations that allow them to obtain nutrients limited in availability and mobility by aridity. Populations of dryland organisms are composed of individuals that can tolerate or avoid prolonged drought and can conserve scarce nutrients and opportunistically exploit periods and spaces of water and nutrient availability. Individuals in plant populations may possess chemical and physical defenses against herbivory or predation or have the ability to overcome the defenses of their food source. The populations themselves will exhibit dynamics characteristic of drylands and reflective of individuals within the species population. Populations may increase rapidly during periods of nutrient and water availability and, subsequently, maintain a sufficient number of individuals during periods of prolonged drought, thus persisting in the ecosystem through time.

Nutrient Flux and Population Dynamics Oases, streams, lakes, and subsurficial water available to plants represent zones of potential high-nutrient flux in deserts. Their role in supporting populations of plants, microbes, and animals beyond their fringes is undoubtedly important but not well defined scientifically. A more reliable supply of water from year to year in these restricted moist zones can sustain small areas with consistent annual plant and animal production within larger regions characterized by infrequent precipitation and extreme temporal variability in plant productivity and associated nutrient flow. Nutrient recycling via turnover of plant organic material and return of animal tissue and wastes to soil tends to be concentrated in areas of relatively high primary productivity, indicating available forage, browse, water, and shelter.

The plant macronutrients nitrogen, phosphorus, and potassium can be limiting to plant growth in drylands. Nitrogen and phosphorus contents of soil are often correlated with organic matter. Wind erosion and drifting sand can mix and dilute organic matter in lighter arid soils. On ancient soils, phosphorus can be limiting because of loss by leaching over millennia. Sandy soils with low cation exchange capacity tend to be particularly low in potassium. Heavier clay soils with higher cation exchange capacities tend to be richer in nutrients in arid regions. On such relatively fertile soils in arid regions, plant water deficit rather than plant nutrients limits primary productivity.

In northern Australia's arid regions, Mitchell Grasslands are dominated by *Astrebla* spp. on the fertile heavy clay soils, coarse spinifex grasslands dominated by *Trioidia* spp. occur on the low fertility sandy soils, and large expanses of mulga with shrubby, nitrogen-fixing *Acacia* spp. support an understory of mixed perennial and annual grasses (Mott and Reid 1985). Forage quantity from year to year can vary by a factor of 5 (Christie and Hughes 1983). The best forage is derived from the nutrient-rich and digestible ephemerals (Lorimer 1978). Protein content can be above 9% in most vegetation types (Wilson and Harrington 1984) with ephemerals preferred over perennial grasses that are, in turn, preferred over browse by sheep. The browse species may have protein contents of 9% or greater but digestibility is less than 50%, limiting its value for supporting animal populations (Wilson 1977). Low density of plants in these arid lands, extreme variability in annual phytomass production, and low digestibility of perennial grasses and browse probably influence animal reproductive success and carrying capacity more than nutrient levels in individual plants.

Symbiosis, Nutrient Flux, and Population Dynamics Competitive advantages in drylands may accrue to plants that have a symbiotic relationship with mycorrhizae and nitrogen-fixing bacteria, allelochemicals, abilities to withstand fire, advantageous rooting habit, efficient pollination systems, prolonged seed variability, high water use efficiency, the ability to fix more carbon per unit of mineral nutrient, or other functional attributes. One example of a plant possessing some of these characteristics is desert she oak, *Allocasuarina decaisneana*. This tree grows to 15 m in height in the arid climatic zone of central Australia (Doran and Hall 1983). She oak is one of the largest trees in Australia's western field, often attaining tree stature while the largest associated plants are shrubs. It typically occurs in sandy soils on the lower slopes of dunes, in interdunal depressions, and along drainage lines in plains of low relief. The 50 percentile annual rainfall is 200–250 mm with potential annual evaporation of 2,500–3,300 mm. Heavy rainfall occurs sporadically with occasional tracking of monsoons from the north across central Australia. Annual patterns of rainfall vary greatly, with a recorded minimum value for annual rainfall of 40–65 mm.

Among the arid zone's adaptive characteristics of she oak are the ability to tolerate fires, fix nitrogen symbiotically with its nodule-forming, actinomycetous symbiont, *Frankia,* develop sinker roots that presumably descend to great soil depths, produce needle-like cladodes that minimize the surface-to-volume ratio of photosynthetic organs, and discourage browsing by producing as fodder young cladodes that are sharp, hard, and unpalatable and low in overall nutritive value (Doran and Hall 1983). Near the surface, lateral roots are sparse and it is difficult to find nodulated roots. Considering the desert she oak's large stature, its occurrence in depressions between dunes and along drainage lines, its association with sandy soils, and its development of prominent sinker roots indicates that these trees are able to exploit deep stores of soil moisture.

Nodulation may also occur more abundantly on roots in deep, moist soil strata. *Prosopis glandulosa* of the Sonoran desert of North America has been found to produce nodules and fix atmospheric nitrogen on deep roots near the water table where a distinct rhizobial population exists (Jenkins et al. 1987). In the Australian Capital Territory, the capacity of *Casuarina* spp. to nodulate was greatest in soils from deeper strata nearest to rivers and least in surficial soils of adjacent uplands (Dawson et al. 1989). Nodulation and subsequent fixation of atmospheric nitrogen by *Casuarina* spp. and woody legumes can also be inhibited where available soil phosphorus is low (Redell et al. 1988, Sylvester-Bradley et al. 1980), illustrating how macronutrient availability can limit production and movement of another nutrient. The distribution of plant species able to obtain nitrogen on infertile sites, but unable to compensate for low phosphorous, would be limited by soil phosphorous availability in soils across the landscape.

Fire, Water, Nutrients, and Populations Fires in deserts are rare but do occur. Although considered fire resistant, the *Allocasuarina decaisneana* groves in central Australia burn. Monsoons sweep through these deserts in most years, resulting in a flush of ephemeral plants that may aid in carrying the fire. Fires have been recorded in the Sonoran desert near Tucson, Arizona, where invading exotic grasses carried the fire that damaged old saguaro cactus stands (Dawson personal observation). Like central Australia, this desert is typified by seasonal monsoons.

Fires in arid ecosystems can, in many instances, release

mineral nutrients from organic matter more rapidly than microbial activity, making them available for plant uptake when soil moisture allows. High temperature fires will, however, volatilize nitrogen. Nitrogen-fixing plants will often be important in desert ecosystems where fires occur frequently, perhaps because they have the advantage of nodular, nitrogen-fixing symbiosis, affording them independence from soil nitrogen. Both cycads possessing a blue-green algal nitrogen-fixing nodular symbiont and shrubby rhizobially nodulated acacias are components of Australian ecosystems subject to periodic fires (Dawson 1983). The role of these plants in the nitrogen economy and nutrient dynamics of important ecosystems is worthy of increased scientific attention.

Within the family Casuarinaceae, species of *Allocasuarina* generally occur on drier harsher sites than species of *Casuarina,* which tend to be riparian or coastal in distribution. For example, on Australia's southern coast, *Allocasuarina littoralis* is found on steep slopes with shallow soil exposed to salt spray from the ocean. On these sites *A. littoralis* trees are nodulated by both *Frankia* and mycorrhizae. *Allocasuarina* spp. are less often found to be nodulated in their natural habitats than *Casuarina* spp. (Lawrie 1982), suggesting that genetic or environmental differences influence nodulation and nitrogen fixation of these two genera. These characteristics certainly play a role in the distribution of populations of Casuarinaceae in arid regions.

Salt tolerance varies among species in the family Casuarinaceae with some species able to tolerate sodic soils characteristic of coastal and arid region soils (El-Lakany and Luard 1982). Salt tolerance, drought tolerance, and physiological adaptations to osmotic stress vary also among strains of *Frankia* (Burleigh and Dawson 1991, 1994, Dawson and Gibson 1987). The role these differences play in geographic distribution of plant and microbe populations and their role in nutrient flows have not been determined. Salt tolerance together with a degree of drought tolerance have made *C. equisetifolia* one of the most commonly planted trees for windbreaks in tropical and subtropical coastal areas.

Cyanobacteria occur in desert soils and may be important to the nitrogen economy of some arid regions. For example, *Nostoc flagelliforme* from China can survive two-year droughts in hot arid deserts, then take up water rapidly on rewetting and resume metabolic activity, including nitrogen fixation, during brief irregular periods of rainfall (Scherer et al. 1984). Lichens with cyanobacterial symbionts can similarly withstand drought, activate metabolically upon rewetting, and fix nitrogen in crusts formed at the surface of desert soils (Rogers et al. 1966).

The fact that many of the processes, especially those important to nutrient cycling and flux in desert ecosystems, occur belowground and often probably much deeper underground than in moister climates has been a major factor limiting subterranean biological investigations and progress in understanding nutrient dynamics in these ecosystems.

The Nature of Plant Population Systems in Arid Lands

In this section, we address the features of plant populations and how they respond to dryland conditions. Demographic and adaptive responses of plant populations in dryland habitats and their implications for management will be covered. The population biology of plants is based, in part, on the ecophysiological characteristics of individual plants in addition to the population processes such as recruitment, mortality, natality, dispersal, and so on. Population biology, in turn, relates to community, ecosystem, and landscape ecology and has many implications for management.

Dispersal and Establishment

The dispersal of individual seeds and how populations of seeds are distributed in the landscape are of particular importance in the responses of plant populations to the spatial and temporal heterogeneity of resources in drylands and how such populations can recover from disturbance events. The fate of individual seeds during dispersal has been pursued for some species (e.g., the exotic invasive annual grass, *Bromus tectorum*) (Kelrick 1988). For populations, the fate of seeds, consumption by granivores, and dispersion have been examined for several dryland species (e.g., Crist and MacMahon 1992, Kelrick et al. 1986, Marlette and Anderson 1986, Pierson and Mack 1990a, Pyke 1990). However, much less information exists on seedbanks of dryland species (e.g., Osman et al. 1987), in part perhaps because of the tedious work involved in collecting such information.

How and where seeds enter seedbanks and whether seeds are transported to different components of seedbanks are open questions in many habitats (Leck et al. 1989). Seedbanks are unevenly distributed in the landscape. Not only the location of parent plants but also how seed movements are affected by wind, water movement, and animal activity will finally determine the spatial heterogeneity in seedbank distribution. Although some of this is obvious, quantification of these factors and their effect on seedbanks as well as the management implications of these processes are needed. Patchiness of seedbank distribution occurs at several scales: seed deposition in the immediate proximity of parent plants will determine one scale of seedbank aggregation, and longer distance distribution of seed by wind and water movement will determine spatial patchiness at larger scales.

Seedling survival of dryland species has received much more attention (e.g., Aguilera and Lauenroth 1993, Bassiri et al. 1988, Booth 1990, Brown and Archer 1989, Friedman and Orshan 1975, Khan and Wilson 1984, Owens and Norton 1989, Pierson and Mack 1990b). Detailed demography in which the fates of individual plants have been followed in the field over considerable periods of time (Chambers and

Norton 1993, Mack and Pyke 1983, 1984, West and Rea 1979) is particularly valuable for constructing models that assess the invasion of competitive exotics (e.g., Moody and Mack 1988). In one sense, such information is descriptive in nature as it does not address the mechanisms of why seedlings of some invasive species can compete so effectively; however, it is necessary for a quantitative understanding of a species' success or failure under varying circumstances. The differential demographic responses along gradients or in contrasting patch types is key information. Emergence may often be restricted to patches in which the microphytic crust is disturbed (Boeken and Shachak 1994). Such disturbance can permit concentration of water in the soil and penetration of roots into deeper soil layers.

Much of the competition for seedling establishment is belowground, as has been shown in root exclusion studies with semi-arid species (Cook and Ratcliff 1984, 1985, Reichenberger and Pyke 1990). The importance of rapid root elongation in establishing seedlings of semi-arid species in order to reach retreating soil moisture during drying cycles has also been shown (Harris 1967, Wilson 1984a). However, it should not be assumed that competition between establishing seedlings with roots of surrounding vegetation is only for water. Much more needs to be learned about how seedlings establish a foothold in the community, especially for invasive plants. Some invaders do not require disturbance in order for seedlings to be successful (e.g., *Isatis tinctoria* in western North American rangelands) (Farah et al. 1988). The detection of characteristics that may generate contrasting functional groups with different potentials for invasiveness may be useful for management.

Basic Growth Processes

The understanding of processes such as light interception by canopies, photosynthesis, respiration, and growth of dryland plant species is reasonably well developed, at least for representatives of important growth forms. Models of light interception by vegetation, though very well developed for the continuous vegetation canopies that occur in many ecosystems, are only now being developed for the discontinuous canopies and isolated plants that are characteristic of arid and semi-arid vegetation (Ryel et al. 1994). Such models permit a more accurate characterization of light interception by individual plants and can be used for any species as long as basic plant structure information is available. They also allow assessments of the potential for light competition among plants that occurs in groups close to one another. Often shrubs and other plant lifeforms occur in clumps or islands, in which case competition for light can develop. Such clumping is an important source of patchiness in drylands. Some populations may experience severe competition in the physically favorable microclimate generated by shrub canopies or plant clumps. Directly analogous to assessing canopy distributions on radiation harvesting by plants and competition for light, models mentioned above can be used to analyze the beneficial effects of canopies for shade and long-wave radiation loss. The so-called nurse plant effect is, in part, a matter of larger plants protecting smaller plants and seedlings from severe radiation conditions. For example, tree canopies can protect seedlings both from photoinhibition due to excessive short-wave radiation and from frost damage due to long-wave radiation loss. Thus, models that deal with isolated plants and discontinuous canopies can be employed effectively in quantifying the nurse plant effects on the radiation microenvironment.

Solar radiation models can be coupled with basic models of photosynthesis, respiration, and growth. Representatives of most of the major lifeforms in dryland vegetation have been studied in the field with regard to photosynthesis and respiration and how these processes are influenced by environmental factors such as temperature and water stress (e.g., Caldwell et al. 1981, DePuit and Caldwell 1975, Lange et al. 1974, Monson et al. 1986, Nowak et al. 1988, Smith and Nowak 1990, Wan et al. 1993a). While the majority of such gas exchange studies have been conducted with individual leaves or branches, a few have involved whole plant gas exchange measurements (e.g., Caldwell et al. 1981). Many of these studies have followed plants through much of the growing season. As in most ecosystems, extending models of gas exchange to plant growth is less well developed, especially in any mechanistic sense. Plant growth must necessarily involve allocation and senescence to be realistic, and such processes are understood primarily in an empirical sense. It is unlikely that a basic understanding of allocation will develop soon to the extent of practical utility; thus, an empirical approach will probably be needed in the coming few years at least.

Differential growth and allocation within populations is one of the main sources of population structure. Interaction with grazers or browsers can also affect growth and allocation. Of course, one of the greatest areas of uncertainty involves belowground growth. Estimates of allocation to belowground biomass are available but usually not much beyond this.

Belowground Activity

Root to total biomass ratios of vegetation in dryland communities are high, on the order of 0.75:0.95 (Drobrowolski et al. 1990). The reasons for this are not clear. While the investment in root systems can be massive, the functional significance of root systems in dryland plants has only been investigated to a limited extent (as in most ecosystems). Obviously, roots extract soil moisture and nutrients, but they can also have unexpected functions. For example, in shrubs and perennial grasses in semi-arid and arid lands, root systems can transfer soil moisture from one soil horizon to another in a process known as hydraulic lift (Caldwell and Richards 1989, Richards and Caldwell 1987, Wan et al. 1993b). Recent descriptions of hydraulic lift in herbaceous and woody plants suggest a means by which plants can overcome the limitations of low soil water content to surficial root growth and nutrient uptake (Dawson 1993, Richards and Caldwell 1987). In the process of hydraulic lift, negative water potentials in shoots and surficial roots develop during the daytime with evapotranspiration. This can create a steep water potential gradient in the plants where deeper roots occur in soil strata that are moist relative to surficial soils. Not only does this process provide water to surficial roots in dry soil, allowing root survival and growth even when soil water is unavailable in surface strata, it may also be important *ex planta*. Water from roots in deeper, moister soil strata has been shown to hydrate the rhizospheres of surficial roots during nighttime adjustment of water potential within the plant. This rhizosphere hydration allows nutrients close to roots to become soluble and taken up by plants in arid regions. Because nutrient uptake occurs against diffusion gradients at the expense of metabolic energy, well-aerated surficial roots able to access oxygen for aerobic respiration are generally more important in nutrient uptake than deeper roots. The occurrence and importance of this process to nutrient dynamics in plants of arid lands merits priority for future survey and study.

Much of the competition among plants in dryland systems occurs belowground. The process and mechanisms of this root competition are slowly being investigated in dryland systems. For example, dual-radioisotope studies have indicated that species differ substantially in their ability to extract nutrients when in competition, and this is not necessarily proportional to plant investment in root density or mycorrhizae (Caldwell et al. 1985, 1987, 1991b).

Soil nutrients and moisture are often very heterogeneously distributed in soil, and effective exploitation of these resources can involve foraging activities of roots. The proliferation of roots in resource-rich soil microsites is well known. In semi-arid species, this can vary considerably among species and can be modulated in accord with the degree of enrichment of the microsite (Jackson and Caldwell 1989). Root proliferation can also be extremely rapid in some circumstances (e.g., within twenty-four hours) (Jackson and Caldwell 1989). When roots of more than one species encounter an enriched microsite, the degree of root proliferation can be influenced differently according to the presence of roots of other species (Caldwell et al. 1991a). Much is yet to be learned about how roots of different species interact in the same soil spaces. Such root behavior may have a considerable bearing on plant competition. In addition to root proliferation, root physiological uptake capacity has been shown to increase in roots encountering enriched patches (Jackson et al. 1990), and this can occur in older roots already present in the microsite if nutrients are added subsequently (Jackson and Caldwell 1991). The possibility exists that functional groups can be differentiated on the basis of root foraging capacities or patterns.

On a larger scale, the natural abundance of stable isotopes have been used to determine where different dryland species obtain soil moisture. In particular, the ability of different species to utilize summer precipitation varies considerably among co-occurring species (Flanagan et al. 1992). Application of stable isotopes with natural abundance should have many powerful applications in future research in dryland systems including, for example, aspects of nitrogen cycling (Evans and Ehleringer 1993). Distributions of root systems belowground and shoot mass aboveground are not necessarily correlated spatially. Also, different functional groups of plants may utilize resources at different depths and at different times of the year.

As in any ecosystem, the greatest gaps in our knowledge tend to be with the belowground system. While some interesting progress has been made in recent years in the investigation of dryland systems, much more is needed. The nature and role of belowground patchiness remains an unknown influence in dryland plant populations.

Population Responses to Disturbance

Disturbance can take many forms and occur at different spatial and temporal scales. Soil disturbance can be a matter of animal activity on rather small scales (e.g., porcupines in the Negev desert of Israel or gophers in North America) or on a somewhat larger scale, human activity, such as military operations, mining, and recreation. Disturbances involving soil disruption may require longer periods for recovery and may open areas to invasion by less desirable plant species. This depends on the particular context of the disturbance. Burrowing activity of animals can promote plant species diversity and contribute to seedbank dynamics in several ways. Large-scale human disturbances can leave lasting imprints on the landscape and often require intense mitigation and restoration, especially in arid areas.

Unlike in arid ecosystems, fire can play an important role in some semi-arid ecosystems. Some species are specifically deterred and some promoted by fire. For example, in semi-arid systems of North America, the shrubs are, for the most part, usually killed and unable to sprout following fire, whereas invasive annuals such as the exotic *Bromus tectorum* are less affected. *Bromus* is thought to accelerate the fire cycle since it is highly flammable in the dried state. The seeds easily survive fire. Although more carefully controlled research is needed, it appears that fire in fact enhances the persistence of this species.

At the population level, if plants are largely killed by a fire event, reestablishment depends on the presence of a viable seedbank or dispersal of new seeds into the disturbed area and propagule germination and establishment of new plants. As mentioned above, information on seedbanks in dryland systems needs more attention. Much of the information on seed longevity to date is anecdotal in nature. Dispersal of propagules is also in need of more study. In the case of disturbances such as fire, the scale of disturbance becomes an issue. Large patches of decimated vegetation obviously demand effective long-distance dispersal. Dryland populations may commonly have a metapopulation structure based either on pronounced physical heterogeneity or disturbance-generated patchiness. Genetic differentiation is one possible result. Little is known in this area.

Other episodic events apart from fire can also cause significant mortality, such as significant insect damage (e.g., Eroga moth devastation of *Artemisia tridentata* or factors such as anoxia of soils during periods of unusual wetness) (Ganskopp 1986). Often, large population die-off events are difficult to attribute to specific causes (Nelson et al. 1989). In order adequately to anticipate responses to general climate change in dryland systems, it is important to understand how disturbance events cause population perturbations and affect the recovery process.

Whether or not prolonged drought and severe overgrazing are disturbances is, in large part, a matter of scale. Both are obviously components of dryland ecosystems at large spatial scales, and, over long periods of time, plant populations have evolved under these pressures. Unusually heavy grazing by livestock or unusually severe drought might be considered as disturbances. Much is known of dryland plant population response to water limitation, at least for normal annual cycles (Dobrowolski et al. 1990, Evenari et al. 1983, Smith and Nowak 1990). However, information on population responses to unusually severe drought is very limited. Episodic mortality in long-lived plants resulting from drought can affect both population and community structure (Walker 1993).

The timing of precipitation can also play a large role in the balance between cryptogams and higher plants. For example, rainfall distributed in a series of very small events (< 1 mm) may promote cryptogams at the expense of higher plants as opposed to situations with fewer precipitation events but with storms of greater magnitude. Soil cryptogams also interact indirectly with higher plants by altering soil infiltration characteristics.

Responses to actual or simulated herbivory has received considerable attention in many important semi-arid species (Briske and Richards 1994). In the last decade, some general concepts of plant response to grazing have been questioned and some alternative concepts introduced. For example, for decades, considerable attention has been paid to cycles of soluble carbohydrate concentrations in plant tissues as indicators of plant susceptibility to grazing, the idea being that these carbohydrates represented a critical reserve to support tissue regrowth following defoliation (Caldwell 1984). This has subsequently been shown not to be particularly useful, at least in some species, since carbohydrate concentrations are not indicators of regrowth capacity. Much of the regrowth is actually supported by concurrent photosynthesis of the regrowing tissues (Richards and Caldwell 1985).

The effect of growth morphology and the locations and activity of meristems on herbivory tolerance has received revived attention in the past decade (Briske and Richards 1994, Richards et al. 1988). Equally important are belowground responses to grazing above ground (Hodgkinson and Becking 1977, Richards 1984). Species that do not sensitively adjust their allocation belowground when grazed may be at a severe disadvantage (Richards 1984). Generally, plants in a resource-poor environment are expected to invest more in antiherbivore defense or have higher root/total biomass ratios (Coley et al. 1985, Meijden et al. 1988, Rosenthal and Kotanen 1994).

Depending on the geographic area, grazing reaches different degrees of severity. In areas of the Middle East and Africa, some regions have been under such grazing pressure that the only remaining species are annuals, geophytes, or perennials that are heavily defended chemically, by thorns, or by other mechanisms. Cryptogams and cryptogamic crusts are also affected by grazing activity both directly, such as snail grazing of cryptogams in the Negev desert (Shachak et al. 1976), or indirectly, such as that resulting from trampling activity of large herbivores (West 1990). Such areas require very different consideration than areas with palatable perennials where grazing optimization schemes are attempted. The population level implications are obviously quite different.

At the population level, responses of species to interacting factors such as fire and water stress (Hodgkinson 1992), grazing, and drought (Chambers and Norton 1993) are not well understood. Although the importance of competitive pressure on a plant's ability to recover from defoliation has been known for some years (e.g., Mueggler 1972), it is too often ignored in studies of recovery from defoliation (Caldwell 1984).

Patchiness and Population Management

The patterns of response to patchiness can be a major feature of dryland plant populations. Fixed patchiness on ecological time scales can reflect the substrate heterogeneity, sites for dew deposition, sinks for water redistribution, and so on (Nicholson chap. 4, this volume). Patches to which plant populations respond can also result from the physical or chemical effects of plants. Shade, litter, allelochemicals, nutrients, and symbionts may reflect the presence or performance of plants. Whether the existence of "islands of fertility" (Garcia-Moya and McKell 1970) is a general phenomenon in drylands needs to be documented.

Patchiness can vary in time. Patches generated by the growth, death, or spread of plants or by the activity of animals are likely to exhibit changing spatial patterns. Such patch dynamics are a changing template that can affect the distribution and abundance of plant populations in drylands. Temporal variance of environmental quality within patches is a subtle component of patch dynamics. Annual variations in rainfall, for example, can affect the degree of contrast between patch types.

Management and restoration can effectively exploit the existence and dynamics of patches. Fire, seedbed preparation, provision of nurse sites, dispersal foci (e.g., perches), extirpation of competitive exotics, and direct manipulation of mutualist or antagonistic fungi, bacteria, or animal populations are all mechanisms of altering the existence or quality of patches on which dryland plants depend.

Climate Change and Population Responses

Much attention has been focused on plant response to elevated CO_2 and temperature levels in studies of plants from several ecosystems. Little has been done with elevated CO_2 and increased temperature on dryland species, especially for populations. Of even greater significance are likely alterations in the seasonality and variance in precipitation. In dryland ecosystems, the variance in precipitation is already great, and a well-developed understanding of how plant populations respond to the variance does not exist. Whether the variance may be increased or decreased in future climates is difficult to predict. Changes in seasonality of precipitation are a more consistent prediction of climate models. For example, in the intermountain region of North America, increased summer precipitation is considered likely in future climates. One might speculate about changes in populations that would accompany the increased warm season precipitation. For example, plants with C_4 photosynthesis may benefit at the expense of plants with C_3 photosynthesis since the former are more active in growth during the summer. However, controlled research to test population responses to altered seasonality of precipitation is largely lacking for dryland plant species.

The Study and Management of Animal Population Systems in Arid Lands

The Study of Animal Populations

There have been several stages in developing an understanding of animal populations in deserts. The stages include (1) forming an inventory, (2) accumulating knowledge of the natural history, including intraspecific interactions of individual species, and (3) studying physiological and behavioral mechanisms of survival. Only the accumulation of sufficient knowledge in these areas enables moving to the next stage: the understanding of population systems. Such knowlege is fundamental to the sound management of populations.

The publication of *Deserts of the World* (McGinnes et al. 1968) was a major advance toward achieving the first stage (i.e., forming an inventory of animals living in arid environments). All deserts have been visited by specialists of various kinds and, at least for vertebrates, there are almost complete lists of the animals inhabiting those areas, although some groups (e.g., bats) are less well known than others. However, this cannot be said with the same certainty about other taxonomic groups of animals where much remains to be discovered. Hence, one of the future tasks of desert biologists is to complete the gaps in our knowledge of desert animals by studying and describing unknown desert animals, particularly invertebrates.

From previous sections of this chapter, it is evident that desert organisms exist in an uncertain environment where conditions vary greatly between extremes over time and space. These organisms must cope with high surface temperatures that occasionally exceed 60 °C, limited water and nutrient availability that may change abruptly over space and time, and low primary productivity. Many plant species are unpalatable or have most of their biomass underground. The problem of organisms' adaptation to such an environment has prompted, in the last three decades, a large number of studies on the natural history of desert animals and their behavioral and physiological adaptations (e.g., Louw and Seely 1982).

Most natural history studies were carried out on vertebrates, chiefly on four-legged reptiles (Mayhew 1968), birds, and rodents (Abramsky et al. 1990, Brown 1987). Much less was done on other vertebrates, and little is known about snakes, insectivores (although some work has been done in southern Africa), bats (a group that composes a third of terrestrial mammals), ungulates, and carnivores. Research has been done on several groups of invertebrates such as ants (Morton and Davidson 1988), snails (Yom-Tov 1970), isopods, scorpions, spiders, and beetles (Edney 1974). Most studies have been short term and have spanned less than 2–3 generations of the studied species. They provide us with some knowledge of the life cycle of desert species. Nevertheless, basic biological knowledge, even for those groups that were the subject of more intense studies, remains exiguous. For example, although the mean clutch size of most desert birds is known, there is as yet no information on the number of clutches laid per annum or on survival rates of these birds. Further studies on the natural history of desert animals are clearly necessary.

Many behavioral (Kotler and Brown 1988, Rubenstein 1989, Wilson 1984b) and physiological (Hadley 1975, Schmidt-Nielsen 1979) studies were carried out on desert animals, particularly on reptiles, rodents, and ungulates. Animals inhabiting arid zones may compensate for limited water and nutrients and high temperatures in various ways, behavioral and physiological. The seasonal migration of large herbivores for water and food and the nocturnal habits of smaller vertebrates are some examples of behavioral adaptations to temporal and spatial variability in the availability of water and plant nutrients. The ability to manipulate body temperature and concentrate urine are examples of physiological adaptation. From the perspective of ecologists, such studies can answer the question on the importance of water in relation to other factors, such as nutrient cycling and soil properties, in determining life history in the desert. Although water seems to be the most important direct factor in deserts, its indirect effect on animal life through energy availability is apparently of similar importance (Shkolnik 1992).

The gaps in our knowledge of desert animal populations require answers from long-term studies. Such studies are essential for understanding and managing population systems. Only a few studies of desert animals have been long-term (Abramsky et al. 1990, Brown 1987, Shachak and Steinberger 1980). For example, one of the characteristics of deserts are large annual fluctuations in water availability; the effects of such fluctuations on population size, reproduction, and mortality can be determined only during long-term studies that include several cycles of water availability. Similarly, there is limited knowledge on how desert populations are controlled. The most elementary data needed for such assessments (e.g., survival curves, reproductive success, and how these vary) are unknown for most desert species. Because desert populations are naturally fragmented, immigration and emigration are vital to the genetic makeup and survival of the metapopulation. All of these data are essential if we are to predict how populations respond to perturbation, how they recover from such perturbation, and what factors affect the recovery.

Most studies of desert animals are observational. They may indicate key factors affecting population parameters but do not determine their relative importance. Hence, the determination of which factors are responsible for population structure, their parameters, and mode of operation should be examined by experimental studies carried out in the field with laboratory control (Abramsky et al. 1994, Yom-Tov 1970).

The Management of Animal Populations

Wild populations are managed for one of three reasons (Caughley 1976): conservation, consumption, or pest control (with management strategies differing accordingly). When managing for conservation, the main goal is ensuring the species' existence by halting population decline or increasing population density or range. Management, in this case, is geared toward increasing production and reducing mortality. Populations managed for consumption are viewed as a renewable natural resource, where the goal is achieving a maximum sustainable yield (MSY). Similar to conservation, treatment is implemented to maximize productivity. In contrast to conservation, reducing mortality is directed only at factors that compete with harvesting. Pest control is often implemented by reducing population density either by amplifying mortality or reducing recruitment. The three categories are not necessarily mutually exclusive, and management, therefore, is not always straightforward. For example, conflicts caused when endangered species are pests or are an integral part of the local human population's livelihood result in complex, delicate situations. These often require interdisciplinary efforts with considerable, and sometimes prohibitive, costs in terms of funding for research and implementation.

Scale, Rarity, and Management of Animal Populations Low levels of precipitation with high variance in time and space are unique to arid environments (Safriel et al. 1989). This results in low-average, high-variance primary productivity, making desert habitats patchy in time and space for animals (Garcia-Moya and McKell 1970). Consequently, desert wildlife populations are typified by low average densities and reproduction and by slow recovery from perturbation. Temporal and spatial scales of patchiness play an important role in the dynamics and survival of a species, with the persistence of patches over time being more important than changes in their population size (Fahrig 1992).

In comparison to other species, those inhabiting drylands have received little attention. This may be due to the relative rareness of desert species. Most ecological research has concentrated on common species, with rare species receiving less attention (Kunin and Gaston 1993). Kunin and Gaston (1993) suggest this is because rare species are either considered "less important" or are logistically harder to study. Where management-oriented questions are concerned, a third, fundamental factor—economics—has further limited the amount of attention paid to desert species. In two out of the three management categories (consumption and pest control), management is driven solely by economic factors. Rarity yields marginal economic value. Lower densities mean fewer hunters and minor impacts as pests. As a result, there has been little incentive to study and invest in management techniques of desert populations.

Their rareness and slow rate of increase make desert species susceptible to overharvesting and rapid decline due to loss of habitat. Thus, many desert animals have become endangered with the advent of human expansion and modern firearms (Kaul and Thalen 1979, Le Houerou 1979). As interest in conservation has increased in recent years, more attention has been paid to the ecology of small and endangered populations (Olney et al. 1994, Scott et al. 1994). A considerable body of theoretical work exists on the probability of extinction and its causes, with specific emphasis on carrying capacity (Goodman 1987) and the impact of environmental, demographic, and genetic stochastic processes on small populations (Nunney and Campbell 1993, Shaffer

1981). Conservation measures, such as protection laws, captive breeding, and reintroduction programs (Gipps 1991, Olney et al. 1994, Saltz and Rubenstein 1995, Soule 1986, Stanley Price 1989), are first steps in reestablishing viable wildlife populations. However, such procedures do not address the major long-term problem of loss of habitat due to human activity. Ensuring adequate habitat is probably the most important factor in minimizing the impact of stochastic factors and risk of extinction.

Animal-habitat relationships are an integral component in managing game and other vertebrate populations (Verner et al. 1984). However, specific literature on managing wildlife in desert habitats is scarce compared to other environments. The recent techniques manual published by the Wildlife Society (Bookhout 1994) includes chapters devoted to wetlands, farmlands, rangelands, and forestlands but none to drylands. Since deserts are characterized by low primary production, slow recovery after perturbation, patchy habitats, and low abundance of wildlife, the management of desert wildlife populations may differ considerably from that of nondesert species. This is especially true in view of eminent loss of desert habitat due to human activity.

Standing biomass in deserts fluctuates considerably over short-term (seasonal) and long-term (annual) spans with an overall low average. Desert herbivores exist in low densities and are often regulated by density-dependent reproductive success. Mass mortalities are most often caused by drought (a density-independent factor), while die-offs from density-dependent factors, such as epizootics, are uncommon (Young 1994). While epizootics can be controlled by harvesting or immunization (Winkler and Bogel 1992), droughts are unpredictable and their impacts cannot be buffered. The fragmentation characteristic of desert populations, which could prevent the spread of a disease, provides little protection against drought. Thus, such natural catastrophes may greatly reduce the viability of desert populations.

Deserts are typified by an overall low primary productivity concentrated in islands. Consequently, although overall density of herbivores and the carnivores dependent on them is low, they may be abundant locally. The short-term fluctuations in vegetation cause fluctuations in annual reproductive success and postnatal survival and may result in certain age groups being scarce (Owen-Smith 1990). In deserts, this has been shown to generate marked long-term trends in herbivore density, even when weather has no time trend (Caughley and Gunn 1993).

Management of Populations in Heterogeneous Environments Rabinowitz and colleagues (1986) defined three basic kinds of rarity: (1) geographic range—the size of the area in which the species can be found, (2) habitat specificity—whether the species can exist in variable habitats or is limited only to a certain type, and (3) local population size—how dense the species is in the areas in which it occurs. Most desert mammals and reptiles exhibit rarity patterns of types 2 and 3—restricted to islands where water and/or food are available—and occur at overall low densities throughout their range. The patchiness of desert habitats dictates the patchiness of its wildlife populations (Stafford Smith and Pickup 1990, Yom-Tov 1993).

Human intervention in the past century (mostly in the form of hunting and competition with domestic stock) has caused the range of many desert species to shrink considerably. Loss of habitat due to human development ensures that, even though conservation steps are taken (such as hunting regulations, law enforcement, and range management), the former range can never be reestablished. It is within this context that desert species should be studied and managed.

Deserts are a series of islands among which the larger animals move and through which the smaller animals are connected by narrow veins of genetic flow. The islands may be completely isolated or connected by narrow corridors (dry riverbeds). While many plants and birds are good colonizers, mammals and reptiles are not. Mammals, especially, face a range of problems stemming from their being homeotherms. The impact of habitat isolation and fragmentation disturbs them more than any other class of animals (Cody 1986, Morton 1990). Consequently, the management of desert wildlife should be based on our knowledge of island biogeography (MacArthur and Wilson 1967), drawing on theoretical and empirical studies of metapopulations (Craig 1994), corridors (Beier 1993), SLOSS theory (single large or several small?) (Diamond 1975), rarity (Rabinowitz et al. 1986), and sink-source theory (Pulliam 1988). Fragmentation must be considered for each species in its own appropriate scale (Kotliar and Wiens 1990). The key questions when managing for an individual species are therefore: (1) How much fragmentation can the species endure (distance between islands, fineness of the matrix, etc.)? (2) How many islands are needed to ensure species survival? (3) What is the minimum size of an island that will support a population? (4) How are stochastic processes affected?

Theoretical models concerning patchy environments are numerous (Karieva 1990). Several researchers have shown that a species is more likely to survive in a single large reserve than in a series of smaller reserves with an overall similar area (Burkey 1989) and that clumping enhances the persistence of the population (Adler and Nuernberger 1994). There is some empirical support for these models (Fahrig and Merriam 1985, Hanski 1994). However, the "single large" category from SLOSS is not a relevant option for desert species already existing in a naturally fragmented environment. Little work has been done on how changes in patch structure in a fragmented environment affect species survival or on the effects of fragmentation in dryland environments (Andren 1994). Neither has the large body of work concerning minimum viable populations (MVP) (Simberloff 1988) and demographic, environmental, and genetic stochasticity (Nunney and Campbell 1993, Shaffer 1981) addressed these issues in terms of desert species existing in a naturally fragmented environment.

Changes in overall patch area may take several forms parallel to the kinds of rarity discussed by Rabinowitz and colleagues (1986): (1) patches at the peripheries of the species range may disappear, maintaining average patch size and distance between patches but reducing the overall range of the species, (2) patches and corridors may disappear randomly throughout the range, maintaining range and average size but increasing average distance between patches, or (3) average patch size may decline.

Models suggest that a reduction in the number of islands may be less harmful than a reduction in their average area and that maintaining corridors for migration is vital (Bleich et al. 1990, Burkey 1989). However, the theoretical work of Fahrig (1992) has demonstrated that persistence of patches over time is more important to species survival than patch size. The scale at which patch clustering occurs is probably critical, but there is no good empirical data on the matter (Doak et al. 1992). More theoretical work and empirical studies over larger areas and longer time spans are needed if we are to make the correct decisions for the development of arid environments while maintaining wildlife populations within them (Fahrig 1992, Saunders et al. 1991).

Less preferred areas may appear of marginal importance but may not in fact be so. Many desert species are migratory, moving during the dry season to areas where water is available (Ghobrial 1974). Migration routes may traverse ranges of little value in terms of primary productivity, but disrupting these habitats may block migration routes, thus leading to the demise of complete populations (Moody and Alldredge 1986). For example, isolated water points can be critical, serving as "stepping stones" between patches. Thus, the elimination of a small and apparently insignificant water source may, in effect, obliterate a corridor and isolate populations (Bleich et al. 1990). Similarly, areas between the islands may be regarded as sinks (Pulliam 1988) and of little value to the population, but their destruction may limit movement between islands, thus indicating their importance for the survival of the metapopulation (Howe et al. 1991).

Degradation of desert landscapes due to overgrazing is often equated with soil erosion and coupled with sharp declines in herbivorous populations (Friedel et al. 1990) and other species dependent on them. Recovery in arid lands from overgrazing damage is a very slow process and sometimes not feasible without active, costly, and time-consuming management (Friedel et al. 1990, Williams and Roe 1975). The damage is most expected around water sources (Berger 1977, Stafford Smith and Pickup 1990), disrupting the social fabric of the populations (Rubenstein 1989) and, at times, causing the virtual collapse of the population. Consequently, resources in the vicinity of water must be continuously and closely monitored so management measures (such as controlled hunting) can be deployed prior to damage.

Ecosystem Processes and Population Management Desert animals evolved special behavioral and physiological adaptations for existing in the desert (Rubenstein 1989, Schmidt-Nielsen 1979, Ward and Saltz 1994, Wilson 1984b). However, all large mammals, even those considered especially well adapted to desert conditions, require free water at some stage (Ghobrial 1970, Jhala et al. 1992, Stanley Price 1989, Williamson and Delima 1991). Even species that were shown to be independent of free water in laboratory conditions may require free water to reproduce or even survive in the wild (Robbins 1983). Thus, water sources are the most critical factor for the existence of desert wildlife, and their loss due to development for tourism, pumping, or isolation from other resources could lead to sharp declines and even the disappearance of entire populations of wild animals (Zhirnov and Ilyinsky 1986). When managing for a specific species, water sources must be the underlying foundation on which the management program should be built. In many of the desert mammals, lactation continues well into the dry season. Free water is, therefore, a critical component in reproductive success and distribution of wildlife in deserts. Habitat requirements should be estimated based on the limiting season by assessing resource availability within the range of the animals' daily movements around the water source.

Climate, Stochastic Processes, and Population Management Desert animals are adapted to withstand long periods of seasonal drought (Grenot 1992). Long-term drought spells may dry up water holes and springs, increasing densities above carrying capacity around the few remaining water sources. The high densities could eventually lead to severe overgrazing, erosion, and eventual collapse of the population and system surrounding the water source. Populations would then be established over time through immigration. With the disappearance of corridors due to human encroachment, the probability of a successful reestablishment is low. Consequently, management steps should be taken to prevent such collapses. Such steps can be in the form of artificial water sources away from the natural ones (Kie et al. 1994) and spreading the grazing pressure over larger areas. Harvesting could perhaps be implemented. Artificial water sources may also increase accessibility to habitats formerly inaccessible to certain species and, in this manner, increase overall carrying capacity.

Small populations are susceptible to extinction due to

genetic, demographic, and environmental stochasticity. Loss of genetic variability is thought to impact populations that are smaller than five hundred individuals (Simberloff 1988), but this may be an underestimate (Lande 1988). In deserts, populations are subdivided into small units with subpopulations expected to have naturally low genetic variance and few deleterious genes (Craig 1994). Thus, genetic impacts of fragmentation in desert species should be less than in other species. However, because desert subpopulations are small, demographic and environmental stochasticity will cause frequent extinction and recolonization, making corridor blockage in desert environments more harmful than fragmentation.

Human Activity and Population Management The use of arid lands for grazing domestic stock is common throughout the world. However, current and past management practices are far from achieving sustainable, let alone optimal, utilization (Fuls 1992). Although it is well known that herbivores use different parts of their range disproportionately (Stoddart et al. 1975), this is rarely taken into account in patchy environments such as deserts. The complexity and importance of correct assessment of stocking rates in patchy desert environments and the catastrophic outcomes of overstocking have been pointed out by several investigators (Fuls 1992, Kellner and Bosch 1992, Morton 1990). In Australia, for example, the low-mean/high-variance precipitation results in long stretches of drought that are detrimental to an overgrazed habitat. Overgrazing in arid woodlands has given a competitive advantage to woody plants following periods of rainfall, causing a considerable reduction in carrying capacity (Friedel et al. 1990). In shrubland, overstocking caused die-offs of saltbush and soil erosion following drought years (Friedel et al. 1990). Thus, not only has long-term sustainable utilization by domestic livestock not been realized, but many wild species have been driven to extinction in the process. Of the thirty-eight Australian terrestrial mammals considered threatened or extinct, twenty-three (60%) inhabit (or inhabited) arid zones (Morton 1990).

The transportation of water over large distances has enabled the development of large human populations and intensive agriculture in arid areas. The waste generated by humans (in the form of garbage dumps) and the green agricultural islands affect the distribution and behavior patterns of local fauna (Saltz and Alkon 1989). These species may eventually become a pest problem. Open disposal sites enable the existence of relatively dense populations of canids and other predators, causing a proliferation of rabies outbreaks. Oral vaccination has recently been used successfully to immunize wild canids and control the spread of rabies (Winkler and Bogel 1992). The agricultural fields attract various herbivores that do considerable damage. The pest species are likely to be labeled endangered or protected, requiring the use of nonlethal methods for control. Fencing is often the only effective nonlethal, although costly, method for controlling damage to agricultural crops. Care should be exercised so that fences extending over some distance do not block important wildlife corridors (Moody and Alldredge 1986).

Some Generalizations for Study and Management of Population Systems

The diverse surroundings and interactions in deserts enable coexistence of a wide variety of populations associated with many taxonomic groups. The populations are not isolated entities. They function and interact within the community, ecosystem, and landscape contexts. The dynamics of each population are a complex process and are determined by the species characteristics and its specific interactions with the biotic and abiotic surroundings.

In the study and management of a specific population system, there are inherent problems related to the general and specific knowledge needed. The general knowledge entails information attained by the community of investigators and managers. The specific knowledge refers to the gaps in general knowledge that should be filled in order to solve a particular problem. The best way to obtain general knowledge is by referring to a general theory. Unfortunately, the scientific community has not yet generated a general theory on population systems. General knowledge is even less organized for dryland population systems. Within this framework, we suggest an approach for advancing toward the development of some generalizations for study and management of specific dryland population systems.

In our presentation, we identified three components that are necessary in order to develop general applications. The first component is general questions related to dryland population systems. If investigators or managers have at their disposal a pool of agreed-upon general questions, they can then select and modify questions relevant to the specific study. To illustrate the concept of general questions, we developed a set of questions classified into four categories (table 6.1). The categories are based on the types of interactions that may affect population systems. The first category is within the population, such as the importance of life history, behavior, physiology, and intraspecific interactions. The next three categories focus on interactions with other ecological systems: community, ecosystem, and landscape. The fifth category relates to population management. These questions may be adopted and modified by investigators or managers when dealing with population research or management problems in arid areas. Most of the questions are still open for basic research and management in drylands.

Table 6.1. General Questions Related to the Study and Management of Dryland Population Systems, Classified by Type of Interaction

Within Populations

How does the natural history of the target species affect its population dynamics?

- What are the main physiological mechanisms that affect the survival and reproduction of the species?
- What are the adaptations of plants and animals that enable them to obtain nutrients limited in availability and mobility?
- How do desert populations conserve scarce nutrients?

What are the physical and chemical defenses of populations to conserve nutrients against herbivory and predation?

- What are the long-term trends in population dynamics?
- What are the relative importances of life history, behavior, physiology, and intraspecific interactions for population structure and function?

Population-Community

What are the interspecific interactions among organisms that affect their demography?

Population-Ecosystem

What is the effect of thermal properties of the environment on populations?

- What is the effect of soil texture and moisture regime on populations?
- What are the effects of soil salinity on populations?
- How does soil texture affect water and nutrient availability for organisms?
- How do both availability and mobility of nutrients affect population dynamics?
- How does the limited rate and amount of nutrient movement among organic and mineral pools in soil affect plant population dynamics?
- What is the effect of wind erosion on nutrients and population dynamics?
- How does variation in nutrient availability (annuals vs. perennials) affect grazing and herbivory in general?
- What are the relative effects of low density of plants, extreme variability in phytomass production, low digestibility of perennial grasses, and nutrient levels on the carrying capacity of animal populations?
- What is the relationship between animal migration and nutrient availability?
- What is the effect of N fixing shrubs on associated organisms?
- How do symbiotic relationships with mycorrhizae affect population dynamics?
- What are the relationships among nitrogen fixation, nitrogen availability, phosphorus availability, fire, and population dynamics?

How do the short periodic increases in nutrient and water availability affect population dynamics?

- What is the relationship between short-term fluctuations of the producers and long-term fluctuations of the consumers?
- What is the relationship between soil erosion and desert populations?

Population-Landscape

- How do small patches rich in water and nutrients (oases, streams, lakes, and surficial water) affect population dynamics in the overall landscape?
- How do temporal scales affect the populations?
- What are the effects of geographic range, habitat specificity, and local population size on population dynamics?

Table 6.1. (cont.)

- What is the effect of patchiness on the distribution and abundance of organisms inhabiting deserts?
- What is the effect of isolation and fragmentation of habitats on the distribution and abundance of organisms inhabiting deserts?
- What is the relationship between free water distribution in the landscape and population dynamics in the desert?
- How do long stretches of drought affect desert populations?

Population Management Dynamics

- What are the effects of development on population dynamics?
- How does waste disposal influence populations and nutrient availability?
- How do pollutants influence nutrient flow and populations of plants and animals?
- How do urbanization and recreation impact nutrient dynamics and populations of desert plants and animals?
- Should the management policies differ for rare and for common species?
- What are the effects of overgrazing and desertification on the population dynamics?
- What are the effects of hunting on the population dynamics?
- Is burning a feasible management option? What impact will it have on desert populations?

Note: These questions were generated in the present essay.

The second component refers to functional groups (table 6.2). Populations from different taxonomic groups may possess similar physiological traits, behavioral patterns, life history strategies, and functions in the community, ecosystem, and landscape. We define a collection of populations that show similarity in one of these aspects as functional groups. Our basic assumption is that populations of the same functional group may have similarities in their dynamics, and we feel that linking a local and general problem may be accomplished by understanding population dynamics in the context of functional groups. This methodology has been useful in studying such phenomena as succession (Bazzaz 1987) and environmental gradients (Keddy 1991, Wisheu and Keddy 1992).

The third component deals with flows and controllers in population systems (fig. 6.1). We identified several flows in relation to dryland populations. Since we are dealing with population systems, our main concern (focal flow) is the flow of population size from time t to $t+1$. Often, water will be the main controller in dryland population systems. It is necessary to distinguish between physiological and ecosystem water flows. The physiological implies input, usage, and output of water by the individual. The ecosystem flow relates to the water regime in the organism's environment, which performs many functions in addition to the physiological ones.

In addition to water, there are other flows that may control the flow of individuals. We identified several, including species, energy, heat, nutrients, soil, wind, and patches. Species flow is the change in species diversity that affects the flow of individuals of the target species. Energy flow

Table 6.2. Functional Groups Related to the Study and Management of Dryland Population Systems

Classification Criteria	Functional Groups
Physiological	Homeotherms, poikilotherms, diurnal, nocturnal, aestivators, hibernators
	Free water drinkers, dew and humidity utilizers, metabolic water utilizers
	Salt-tolerant plants, salt-sensitive plants
	N fixers, nonfixers
	Generalists, specialists
	C_4,C_3—CAM plants
Behavior	Migrators, permanent residents, life cycle
	Perennial, ephemeral
	Herbaceous, woody plants
	Deciduous, evergreen plants
Population	Rare, common species
	Gender, age groups, cohorts
	Adult, seedling
Function in the community	Predators, prey, herbivores, parasites, symbionts
	Mycorrhizal symbionts, absence of mycorrhizal symbionts and ecosystem
Landscape	Microphytes, macrophytes
	Understory plants, open space plants
	Subterranean organisms, aboveground organisms

Note: These functional groups were generated in the present essay.

through the population is the energy acquired by the individuals as part of the food web. Heat flow must be differentiated into physiological and ecosystem flows. Physiological heat flow is individual heat balance, while ecosystem flow is the transfer of heat energy in the environment. Nutrient flow, as explained in this chapter, may be crucial in the understanding of population systems. Soil flow alters the structure and resource base of the surroundings. Wind flow affects the system directly by dispersing seeds and organisms and indirectly by disturbances and transportation of materials. The last flow we described was that of patches. The changes in patchiness of a landscape through time alters the resource base in space and time for populations.

To sustain viable populations in drylands, human-induced controllers must be integrated with the natural ones. To do so, we must first understand the natural flows and controllers of the populations. Then, we must establish the flows and controllers that were, are, and will be introduced into the system by human activity and how these impact the natural flows and controllers. Finally, we must buffer those controllers that have a negative impact on the population and, at the same time, establish new controllers that will prevent extinction of populations while minimizing the conflict with human activities, allowing their sustained use.

Although considerable advances have been made in our understanding of and skills in managing wild populations, our ability to predict population response to perturbation remains limited. Even more limited is our ability to predict the impact of such perturbation on the system within which the population exists. Much work is still needed before population management can be viewed as standard procedure. Currently, all management must be viewed as experiments with a predicted, albeit indefinite, outcome. Thus, postimplementation monitoring must be standard protocol in population management in order to verify the consequences of our activities and adjust the management tactics.

References

Abramsky, Z., O. Offer, and M. L. Rozenzweig. 1994. The shape of a *Gerbillus pyramidum* (Rodentia, Gerbillinae) isocline. Oikos 69:318–26.

Abramsky, Z., M. L. Rozenzweig, B. Pinshow, J. S. Brown, B. P. Kotler, and W. A. Mitchell. 1990. Habitat selection: An experimental field test with two gerbil species. Ecology 71:2358–79.

Adler, F. R., and B. Nuernberger. 1994. Persistence in patchy irregular landscapes. Theoretical Population Biology 45:41–75.

Aguilera, M., and W. K. Lauenroth. 1993. Seedling establishment in adult neighborhoods: Intraspecific constraints in the regeneration of the bunchgrass *Bouteloua gracilis.* Journal of Ecology 81:253–61.

Alizai, H. U., and L. C. Hulbert. 1970. Effects of soil texture on evaporative loss and available water in semi-arid climates. Soil Science 110:328–32.

Andren, H. 1994. Effects of habitat fragmentation on birds and mammals in landscapes with different proportions of suitable habitat: A review. Oikos 71:355–66.

Bassiri, M., A. M. Wilson, and B. Grami. 1988. Dehydration effects on seedling development of four range species. Journal of Range Management 41:383–86.

Bazzaz, F. A. 1987. Experimental studies on the evolution of niche in successional plant populations. *In* A. J. Gray, M. J. Crowley, and P. J. Edwards, eds. Colonization, succession, and stability. Boston: Basil Blackwell. 245–72.

Beier, P. 1993. Determining minimum habitat areas and habitat corridors for cougars. Conservation Biology 7:94–108.

Berger, J. 1977. Organizational systems and dominance in feral horses in the Grand Canyon. Behavioral Ecology and Sociobiology 2:131–46.

Bleich, V. C., J. D. Wehausen, and S. A. Holl. 1990. Desert-dwelling mountain sheep: Conservation implications of a naturally fragmented distribution. Conservation Biology 4:383–90.

Boeken, B., and M. Shachak. 1994. Desert plant communities in human-made patches: Implications for management. Ecological Applications 4:702–16.

Bookhout, T. A., ed. 1994. Research and management techniques for wildlife and habitats. Bethesda, Md.: Wildlife Society.

Booth, D. T. 1990. Seedbed ecology of winterfat: Effects of mother-plant transpiration, wind stress, and nutrition on seedling vigor. Journal of Range Management 43:20–24.

Bowers, J. E. 1986. Seasons of the wind. Flagstaff, Ariz.: Northland Press.

Boyer, J. S. 1982. Plant productivity and environment. Science 218:443–48.

Brady, N. C. 1974. The nature and properties of soils. New York: Macmillan.
Briske, D. D., and J. H. Richards. 1994. Physiological responses of individual plants to grazing: Current status and ecological significance. *In* M. Vavra, W. A. Laycock, and R. D. Pieper, eds. Ecological implications of livestock herbivory in the West. Denver, Colo.: Society for Range Management. 147–76.
Brown, J. H. 1987. Variation in desert rodent guilds' patterns, processes and scales. *In* J. H. R. Gee and P. S. Gilles, eds. Organization of communities: Past and present. Oxford: Basil Blackwell.
Brown, J. R., and S. Archer. 1989. Woody plant invasion of grasslands: Establishment of honey mesquite (*Prosopis glandulosa* var. *glandulosa*) on sites differing in herbaceous biomass and grazing history. Oecologia 80:19–26.
Burkey, T. V. 1989. Extinction in nature reserves: The effects of fragmentation and the importance of migration between reserve fragments. Oikos 55:75–81.
Burleigh, S. H., and O. J. Dawson. 1991. Effects of sodium chloride and melibiose on the in-vitro growth and sporulation of *Frankia* strain HFPCcI3 isolated from *Casuarina cunninghamiana*. Australian Journal of Ecology 16:531–35.
———. 1994. Desiccation tolerance and trehalose production in *Frankia*. Soil Biology and Biochemistry 26:593–98.
Caldwell, M. M. 1984. Plant requirements for prudent grazing. *In* National Academy of Sciences, ed. Developing strategies for rangeland management. Boulder, Colo.: Westview Press. 117–52.
Caldwell, M. M., D. M. Eissenstat, J. H. Richards, and M. F. Allen. 1985. Competition for phosphorus: Differential uptake from dual-isotope-labeled interspaces between shrub and grass. Science 229:384–86.
Caldwell, M. M., J. H. Manwaring, and S. L. Durham. 1991a. The microscale distribution of neighbouring plant roots in fertile soil microsites. Functional Ecology 5:765–72.
Caldwell, M. M., J. H. Manwaring, and R. B. Jackson. 1991b. Exploitation of phosphate from fertile soil microsites by three Great Basin perennials when in competition. Functional Ecology 5:757–64.
Caldwell, M. M., and J. H. Richards. 1989. Hydraulic lift: Water efflux from upper roots improves effectiveness of water uptake by deep roots. Oecologia 79:1–5.
Caldwell, M. M., J. H. Richards, D. A. Johnson, R. S. Nowak, and R. S. Dzurec. 1981. Coping with herbivory: Photosynthetic capacity and resource allocation in two semi-arid *Agropyron* bunchgrasses. Oecologia 50:14–24.
Caldwell, M. M., J. H. Richards, J. H. Manwaring, and D. M. Eissenstat. 1987. Rapid shifts in phosphate acquisition show direct competition between neighbouring plants. Nature 327:615–16.
Caughley, G. 1976. Wildlife management and the dynamics of ungulate populations. *In* T. H. Coaker, ed. Applied biology. Vol. 1. New York: Academic Press. 183–246.
Caughley, G., and A. Gunn. 1993. Dynamics of large herbivores in deserts: Kangaroos and caribou. Oikos 67:47–55.
Chadwick, H. W., and P. D. Dalke 1965. Plant succession on dune sands in Fremont County, Idaho. Ecology 46:765–80.
Chambers, J. C., and B. E. Norton. 1993. Effects of grazing and drought on population dynamics of salt desert shrub species on the Desert Experimental Range, Utah. Journal of Arid Environments 24:261–75.
Chepil, W. S. 1956. Influence of moisture on erodibility of soil by wind. Proceedings of the Soil Science Society of America 20:288–92.
Christie, E. K., and P. G. Hughes. 1983. Interrelationships between net primary production, ground storey condition and grazing capacity of the *Acacia aneura* rangelands in semi-arid Australia. Agricultural Systems 12:191–211.
Cody, M. L. 1986. Diversity, rarity, and conservation in Mediterranean-climate regions. *In* M. E. Soule, ed. Conservation biology: The science of scarcity and diversity. Sunderland, Mass.: Sinauer Associates. 122–52.
Coley, P. D., J. P. Bryant, and F. S. Chapin III. 1985. Resource availability and plant antiherbivore defense. Science 230:895–99.
Cook, S. J., and D. Ratcliff. 1984. A study of the effects of root and shoot competition on the growth of green panic (*Panicum maximum* var. *trichoglume*) seedlings in an existing grassland using root exclusion tubes. Journal of Applied Ecology 21:971–82.
———. 1985. Effects of fertilizer, root, and shoot competition on the growth of siratro (*Macroptilium atropurpureum*) and green panic (*Panicum maximum* var. *trichoglume*) seedlings in a native speargrass (*Heteropogon contortus*) sward. Australian Journal of Agricultural Research 36:233–45.
Craig, J. L. 1994. Metapopulations: Is management as flexible as nature? *In* P. J. S. Olney, G. M. Mace, and A. T. C. Feistner, eds. Creative conservation: Interactive management of captive and wild animals. New York: Chapman and Hall. 50–66.
Crist, T. O., and J. A. MacMahon. 1992. Harvester ant foraging and shrub-steppe seeds: Interactions of seed resources and seed use. Ecology 73:1768–79.
Danin, A. 1987. Impact of man on biological components of desert in Israel. Proceedings of the annual meeting of the Israeli Botanical Society, Beer Sheva. 6–7.
Dawson, J. O. 1983. Dinitrogen fixation in forest ecosystems. Canadian Journal of Microbiology 29:979–92.
Dawson, J. O., and A. H. Gibson. 1987. Sensitivity of selected *Frankia* isolates from *Casuarina, Allocasuarina* and North American host plants to sodium chloride. Physiology Plantarum 70:272–78.
Dawson, J. O., D. G. Kowalski, and P. J. Dart. 1989. Variation with soil depth, topographic position, and host species in the capacity of soils from an Australian locale to nodulate *Casuarina* and *Allocasuarina* seedlings. Plant and Soil 118:1–11.
Dawson, T. E. 1993. Hydraulic lift and water use by plants: Implications for water balance, performance and plant interactions. Oecologia 95:565–74.
DePuit, E. J., and M. M. Caldwell. 1975. Stem and leaf gas exchange of two arid land shrubs. American Journal of Botany 62:954–61.
Diamond, J. M. 1975. The island dilemma: Lessons of modern biogeographic studies for the design of natural reserves. Biological Conservation 7:129–45.
Dincer, T., A. Al-Mugrim, and U. Zimmerman. 1974. Study of the infiltration and recharge through the sand dunes in arid zones with special reference to the stable isotopes and thermonuclear tritium. Journal of Hydrology 23:79–109.
Doak, D. F., P. C. Marino, and P. M. Karieva. 1992. Spatial scales mediate the influence of habitat fragmentation on dispersal success: Implications for conservation. Theoretical Population Biology 41:315–36.
Dobrowolski, J. P., M. M. Caldwell, and J. H. Richards. 1990. Basin hydrology and plant root systems. *In* C. B. Osmond, G. Hidy, and L. Pitelka, eds. Plant biology of the basin and range. Berlin: Springer-Verlag. 243–92.
Doran, J., and N. Hall. 1983. Notes on fifteen Australian *Casuarina* species. *In* S. J. Midgely, J. W. Turnbull, and R. D. Johnson, eds. *Casuarina* ecology, management, and utilization. Melbourne: Commonwealth Scientific and Industrial Research Organization. 19–52.
Edney, E. B. 1974. Desert arthropods. *In* G. W. Brown, ed. Desert biology. Vol. 2. New York: Academic Press. 312–84.
El-Lakany, M. H., and E. J. Luard. 1982. Comparative salt tolerance of selected *Casuarina* species. Australian Forest Research 13:11–20.
Evans, R. D., and J. R. Ehleringer. 1993. A break in the nitrogen cycle in arid lands? Evidence from EB15N of soils. Oecologia 94:314–17.

Evenari, M. 1981. Ecology of the Negev desert: A critical review of our knowledge. *In* H. Shuval, ed. Developments in arid zone ecology and environmental quality. Philadelphia: Balaban International Science Services. 1–33.

Evenari, M., L. Shanan, and N. Tadmor. 1983. The Negev: Challenge of a desert. Cambridge, Mass.: Harvard University Press.

Fahrig, L. 1992. Relative importance of spatial and temporal scales in a patchy environment. Theoretical Population Biology 41:300–314.

Fahrig, L., and G. Merriam. 1985. Habitat patch connectivity and population survival. Ecology 66:1762–68.

Farah, K. O., A. F. Tanaka, and N. E. West. 1988. Autecology and population biology of Dyers Woad (*Isatis tinctoria*). Weed Science 36:186–93.

Flanagan, L. B., J. R. Ehleringer, and J. D. Marshall. 1992. Differential uptake of summer precipitation among co-occurring trees and shrubs in a pinyon-juniper woodland. Plant, Cell and Environment 15:831–36.

Friedel, M. H., B. D. Foran, and D. M. Stafford Smith. 1990. Where the creeks run dry or ten feet high: Pastoral management in arid Australia. Proceedings of the Ecological Society of Australia 16:185–94.

Friedman, J., and G. Orshan. 1975. The distribution, emergence, and survival of seedlings of *Artemisia herba-alba* Asso in the Negev desert of Israel in relation to distance from the adult plants. Journal of Ecology 63:627–32.

Fuls, E. R. 1992. Semi-arid and arid rangelands: A resource under siege due to patch-selective grazing. Journal of Arid Environment 22:191–93.

Ganskopp, D. C. 1986. Tolerances of sagebrush, rabbitbrush, and greasewood to elevated water tables. Journal of Range Management 39:334–37.

Garcia-Moya, E., and C. M. McKell. 1970. Contribution of shrubs to nitrogen economy of a desert-wash plant community. Ecology 51:81–88.

Ghobrial, L. I. 1970. The water relations of the desert antelope *Gazella dorcas*. Physiological Zoology 43:249–56.

———. 1974. Water relation and requirements of dorcas gazelle in Sudan. Mammalia 38:88–101.

Gipps, J. H. W. 1991. Beyond captive breeding: Reintroducing endangered mammals in the wild. Symposia, Zoological Society of London, No. 62. Oxford: Clarendon Press.

Goodman, D. 1987. Consideration of stochastic demography in the design and management of biological reserves. Natural Resource Modeling 1:205–34.

Grenot, C. J. 1992. Ecophysiological characteristics of large herbivorous mammals in arid Africa and the Middle East. Journal of Arid Environments 23:125–55.

Hadley, N. F., ed. 1975. Environmental physiology of desert organisms. Stroudsberg, Pa.: Dowden, Hutchinson, and Ross.

Hanski, I. 1994. Patch-occupancy dynamics in fragmented landscapes. Trends in Ecological Evolution 9:131–35.

Harris, G. A. 1967. Some competitive relationships between *Agropyron spicatum* and *Bromus tectorum*. Ecological Monographs 37:89–111.

Hellden, U. 1988. Desertification monitoring: Is the desert encroaching? Desertification Control Bulletin 17:8–12.

Hodgkinson, K. C. 1992. Water relations and growth of shrubs before and after fire in a semi-arid woodland. Oecologia 90:467–73.

Hodgkinson, K. C., and H. G. B. Becking. 1977. Effect of defoliation on root growth of some arid zone perennial plants. Australian Journal of Agricultural Research 29:31–42.

Howe, R. W., G. J. David, and V. Mosca. 1991. The demographic significance of "sink" populations. Biological Conservation 57:239–55.

Jackson, R. B., and M. M. Caldwell. 1989. The timing and degree of root proliferation in fertile-soil microsites for three cold-desert perennials. Oecologia 81:149–54.

———. 1991. Kinetic responses of *Pseudoroegneria* roots to localized soil enrichment. Plant and Soil 138:231–38.

Jackson, R. B., J. H. Manwaring, and M. M. Caldwell. 1990. Rapid physiological adjustment of roots to localized soil enrichment. Nature 344:58–60.

Janza, F. J. 1975. Interaction mechanisms. *In* R. G. Reeves, A. Anson, and D. Landen, eds. Manual of remote sensing. Falls Church, Va.: American Society of Photogrammetry. 75–179.

Jenkins, M. B., R. A. Virginia, and W. M. Jarrell. 1987. Rhizobial ecology of the woody legume mesquite (*Prosopis glandulosa*) in the Sonoran desert. Applied Environmental Microbiology 53:36–40.

Jhala, Y. V., R. H. Giles Jr., and A. M. Bhagwat. 1992. Water in the ecophysiology of blackbuck. Journal of Arid Environments 22:261–69.

Karieva, P. 1990. Population dynamics in spatially complex environments: Theory and data. Proceedings, Royal Society of London, Series B 330:175–90.

Kaul, R. N., and D. C. P. Thalen. 1979. South-west Asia. *In* D. W. Goodall and R. A. Perry, eds. Arid-land ecosystems: Structure, functioning, and management. Cambridge: Cambridge University Press. 213–72.

Keddy, P. A. 1991. Working with heterogeneity: An operator's guide to environmental gradients. *In* J. Kolasa and S. T. A. Pickett, eds. Ecological heterogeneity. New York: Springer-Verlag. 181–201.

Kellner, K., and O. J. H. Bosch. 1992. Influence of patch formation in determining the stocking rate for southern African grassland. Journal of Arid Environments 22:99–105.

Kelrick, M. 1988. The dispersion pattern of annual plants: Factors affecting the seeds of shrub-steppe annuals. Ph.D. diss. Logan: Utah State University.

Kelrick, M. I., J. A. MacMahon, R. R. Parmenter, and D. V. Sisson. 1986. Native seed preferences of shrub-steppe rodents, birds and ants: The relationships of seed attributes and seed use. Oecologia 68:327–37.

Khan, S. M., and A. M. Wilson. 1984. Nonstructural carbohydrates and dehydration tolerance of blue grama seedlings. Agronomy Journal 76:637–42.

Kie, J. G., V. C. Bleich, A. L. Medina, J. D. Yoakum, and J. W. Thomas. 1994. Managing rangelands for wildlife. *In* Bookhout, ed. Research and management techniques for wildlife and habitats. 663–88.

Kotler, B. P., and J. S. Brown. 1988. Environmental heterogeneity and the coexistence of desert rodents. Annual Review of Ecological Systematics 19:281–307.

Kotliar, N. B., and J. A. Wiens. 1990. Multiple scales of patchiness and patch structure: A hierarchical framework for study of heterogeneity. Oikos 59:253–60.

Kunin, W. E., and K. J. Gaston. 1993. The biology of rarity: Patterns and consequence. Trends of Ecological Evolution 8:298–301.

Lamprey, H. F. 1975. Report on the desert encroachment reconnaissance in northern Sudan, 21st October to 10 November 1975. UNESCO/UNEP. Mimeo.

Lande, R. 1988. Genetics and demography in biological conservation. Science 241:1456–60.

Lange, O. L., E. D. Schulze, M. Evenari, L. Kappen, and U. Buschbom. 1974. The temperature-related photosynthetic capacity of plants under desert conditions. Pt. 1: Seasonal changes of the photosynthetic response to temperature. Oecologia 17:97–110.

Lawrie, A. C. 1982. Field nodulation of nine species of *Casuarina* in Victoria. Australian Journal of Botany 30:447–60.

Leck, M. A., V. T. Parker, and R. L. Simpson, eds. 1989. Ecology of soil seedbanks. San Diego, Calif.: Academic Press.

Le Houerou, H. N. 1979. North Africa. *In* D. W. Goodall and R. A. Perry, eds. Arid-land ecosystems: Structure, functioning, and management. Cambridge: Cambridge University Press. 357–84.

———. 1986. The desert and arid zones of Northern Africa. *In* B. M. Evenari, I. Noy-Meir, and D. W. Goodall, eds. Ecosystems of the world. Vol. 12A: Hot deserts and arid shrublands. Amsterdam: Elsevier. 101–47.

Lorimer, M. S. 1978. Forage selection studies. Pt. 1: The botanical composition of forage selected by sheep grazing, *Astrebla* spp. pastures in N.W. Queensland. Tropical Grasslands 12:97–108.

Louw, G., and M. K. Seely. 1982. Ecology of desert organisms. London: Longman.

MacArthur, R. H., and E. O. Wilson. 1967. The theory of island biogeography. Princeton, N.J.: Princeton University Press.

Mack, R. N., and D. A. Pyke. 1983. The demography of *Bromus tectorum:* Variation in time and space. Journal of Ecology 71:69–94.

———. 1984. The demography of *Bromus tectorum:* The role of microclimate, grazing, and disease. Journal of Ecology 72:731–48.

Marlette, G. M., and J. E. Anderson. 1986. Seedbanks and propagule dispersal in crested wheatgrass stands. Journal of Applied Ecology 23:161–76.

Mayhew, W. W. 1968. Biology of desert amphibians and reptiles. *In* G. W. Brown, ed. Desert biology. Vol. 1. New York: Academic Press. 196–356.

McGinnes, W. G. 1979. Arid-land ecosystems: Common features throughout the world. *In* D. W. Goodall and R. A. Perry, eds. Arid-land ecosystems: Structure, functioning, and management. Cambridge: Cambridge University Press. 299–316.

McGinnes, W. G., B. J. Goldman, and P. Paylore, eds. 1968. Deserts of the world. Tucson: University of Arizona Press.

Meijden, E. van der, M. Wijn, and H. J. Verkaar. 1988. Defense and regrowth: Alternative plant strategies in the struggle against herbivores. Oikos 51:355–63.

Migahid, A. A. 1961. The drought resistance of Egyptian desert plants. Arid Zone Research 16:12–233.

Monson, R. K., M. R. Sackschewsky, and G. J. Williams III. 1986. Field measurements of photosynthesis, water-use efficiency, and growth in *Agropyron smithii* (C_3) and *Bouteloua gracilis* (C_4) in the Colorado shortgrass steppe. Oecologia 68:400–409.

Moody, D. S., and A. W. Alldredge. 1986. Red Rim: Mining, fencing, and some decisions. Proceedings of the Pronghorn Antelope Workshop 12:57.

Moody, M. E., and R. N. Mack. 1988. Controlling the spread of plant invasions: The importance of nascent foci. Journal of Applied Ecology 25:1009–21.

Morton, S. R. 1990. The impact of European settlement on the vertebrate animals in arid Australia: A conceptual model. Proceedings of the Ecological Society of Australia 16:201–13.

Morton, S. R., and D. W. Davidson. 1988. Comparative structure of harvester ant communities in arid Australia and North America. Ecological Monographs 58:19–38.

Mott, J. J., and R. Reid. 1985. Forage and browse: The northern Australian experience. *In* G. E. Wickens, J. R. Goodin, and D. V. Field, eds. Plants for arid lands. London: Allen and Unwin. 145–61.

Mueggler, W. F. 1972. Influence of competition on the response of bluebunch wheatgrass to clipping. Journal of Range Management 25:88–92.

Nelson, D. L., K. T. Harper, K. C. Boyer, D. J. Weber, B. A. Haws, and J. R. Marble. 1989. Wildland shrub die-offs in Utah: An approach to understanding the cause. Proceedings, symposium on shrub ecophysiology and biotechnology. Gen. Tech. Rep. INT-256. Ogden, Utah: USDA Forest Service. 119–35.

Nowak, R. S., J. E. Anderson, and N. L. Toft. 1988. Gas exchange of *Agropyron desertorum:* Diurnal patterns and responses to water vapor gradient and temperature. Oecologia 77:289–95.

Noy-Meir, I. 1973. Desert ecosystems: Environment and producers. Annual Review of Ecological Systematics 4:25–51.

Nunney, L., and K. A. Campbell. 1993. Assessing minimum viable population size: Demography meets population genetics. Trends of Ecological Evolution 8:234–39.

Olney, P. J. S., G. M. Mace, and A. T. C. Feistner. 1994. Creative conservation: Interactive management of captive and wild animals. New York: Chapman and Hall.

Orev, Y. 1984. Sand is greener. Teva-va-Aretz 26:15–16. [in Hebrew]

Osman, A., R. D. Pieper, and K. C. McDaniel. 1987. Soil seedbanks associated with individual broom snakeweed plants. Journal of Range Management 40:441–43.

Owens, M. K., and B. E. Norton. 1989. The impact of available area on *Artemisia tridentata* seedling dynamics. Vegetatio 82:155–62.

Owen-Smith, N. 1990. Demography of a large herbivore, the greater kudu *Tragelaphus strepsiceros,* in relation to rainfall. Journal of Animal Ecology 59:893–913.

Pierson, E. A., and R. N. Mack. 1990a. The population biology of *Bromus tectorum* in forests: Distinguishing the opportunity for dispersal from environmental restriction. Oecologia 84:519–25.

———. 1990b. The population biology of *Bromus tectorum* in forests: Effect of disturbance, grazing, and litter on seedling establishment and reproduction. Oecologia 84:526–33.

Pulliam, H. R. 1988. Sources, sinks, and population regulation. American Naturalist 132:652–61.

Pye, K., and H. Tsoar. 1990. Aeolian sand and sand dunes. London: Unwin Hyman.

Pyke, D. A. 1990. Comparative demography of co-occurring introduced and native tussock grasses: Persistence and potential expansion. Oecologia 82:537–43.

Rabinowitz, D., S. Cairns, and T. Dillon. 1986. Seven forms of rarity and their frequency in flora of the British Isles. *In* M. E. Soule, ed. Conservation biology: The science of scarcity and diversity. Sunderland, Mass.: Sinauer Associates. 182–204.

Redell, P., P. A. Rosbrook, G. D. Bowen, and D. Gwaze. 1988. Growth responses in *Casuarina cunninghamiana* plantings to inoculation with Franki. Plant and Soil 108:79–86.

Reichenberger, G., and D. A. Pyke. 1990. Impact of early root competition on fitness components of four semi-arid species. Oecologia 85:159–66.

Richards, J. H. 1984. Root growth response to defoliation in two *Agropyron* bunchgrasses: Field observations with an improved root periscope. Oecologia 64:21–25.

Richards, J. H., and M. M. Caldwell. 1985. Soluble carbohydrates, concurrent photosynthesis and efficiency in regrowth following defoliation: A field study with *Agropyron* species. Journal of Applied Ecology 22:907–20.

———. 1987. Hydraulic lift: Substantial nocturnal water transport between soil layers by *Artemesia tridentata* roots. Oecologia 73:486–89.

Richards, J. H., R. J. Mueller, and J. J. Mott. 1988. Tillering in tussock grasses in relation to defoliation and apical bud removal. 62:173–79.

Robbins, C. T. 1983. Wildlife feeding and nutrition. New York: Academic Press.

Rogers, R. W., R. T. Lange, and D. J. D. Nicholas. 1966. Nitrogen fixation by lichens of arid soil crusts. Nature 209:96–97.

Rosenthal, J. P., and P. M. Kotanen. 1994. Terrestrial plant tolerance to herbivory. Trends of Ecological Evolution 9:145–48.

Rubenstein, D. I. 1989. Life history and social organization in arid adapted ungulates. Journal of Arid Environments 17:145–56.

Ryel, R. J., W. Beyschlag, and M. M. Caldwell. 1994. Light field heterogeneity among tussock grasses: Theoretical considerations of light harvesting and seedling establishment in tussocks and uniform tiller distributions. Oecologia 98:241.

Safriel, U., Y. Ayal, B. P. Kotler, Y. Lubin, L. Olsvig-Whittaker, and B. Pinshow. 1989. What's special about desert ecology: Introduction. Journal of Arid Environments 17:125–30.

Saltz, D., and P. U. Alkon. 1989. Characterizing the spatial behavior of Indian crested porcupines from radiotracking. Journal of Zoology [London] 217:255–66.

Saltz, D., and D. I. Rubenstein. 1995. Population dynamics of a reintroduced Asiatic wild-ass (*Equus hemionus*) herd. Ecological Applications 5:327.

Satoh, I. 1967. Studies on some peculiar environmental factors related to cultivation in sand dune fields. Journal of Faculty Agriculture [Tottori University, Japan] 5:1–41.

Saunders, D. A., R. J. Hobbs, and C. R. Margules. 1991. Biological consequences of ecosystem fragmentation: A review. Conservation Biology 5:18–32.

Scherer, S., A. Ernst, T. W. Chen, and P. Boger. 1984. Rewetting of drought-resistant blue-green algae: Time course of water uptake and reappearance of respiration, photosynthesis, and nitrogen fixation. Oecologia 62:418–23.

Schmidt-Nielsen, K. 1979. Desert animals: Physiological problems of heat and water. New York: Dover.

Schultze, E. D. 1986. Carbon dioxide and water vapor exchange in response to drought in the atmosphere and in the soil. Annual Review of Plant Physiology 37:247–74.

Scott, J. M., S. A. Temple, D. L. Harlow, and M. L. Shaffer. 1994. Restoration and management of endangered species. *In* Bookhout, ed. Research and management techniques for wildlife and habitats. 531–39.

Shachak, M., E. A. Chapman, and Y. Steinberger. 1976. Feeding, energy flow, and soil turnover in the desert isopod, *Hemilepistus reaumuri*. Oecologia 24:57–69.

Shachak, M., and Y. Steinberger. 1980. An algae desert food chain: Energy flow and soil turnover. Oecologia 46:402–11.

Shaffer, M. L. 1981. Minimum population sizes for species conservation. BioScience 31:131–34.

Shkolnik, A. 1992. The black Bedouin goat. Bielefelder Okologische Beitrage 6:53–60.

Shreve, F. 1934. The problems of the desert. Scientific Monthly 40:199–209.

Simberloff, D. 1988. The contribution of population and community biology to conservation science. Annual Review of Ecological Systematics 19:473–511.

Sinclair, J. G. 1922. Temperature of the soil and air in a desert. Monthly Weather Review 50:142–44.

Smith, S. D., and R. S. Nowak. 1990. Ecophysiology of plants in the intermountain lowlands. *In* C. B. Osmond, L. F. Pitelka, and G. M. Hidy, eds. Plant biology of the basin and range. New York: Springer-Verlag. 179–241.

Soule, M. E., ed. 1986. Conservation biology: The science of scarcity and diversity. Sunderland, Mass.: Sinauer Associates.

Stafford Smith, D. M., and G. Pickup. 1990. Pattern and production in arid lands. Proceedings of the Ecological Society of Australia 16:195–200.

Stanley Price, M. R. 1989. Animal reintroduction: The Arabian oryx in Oman. New York: Cambridge University Press.

Stoddart, L. A., A. D. Smith, and T. W. Box. 1975. Range management. New York: McGraw-Hill.

Sylvester-Bradley, R., L. A. DeOliveira, J. A. DePodesta Filho, and T. V. St. John. 1980. Nodulation of legumes, nitrogenase activity of roots, and occurrence of nitrogen-fixing *Azospirillum* spp. in representative soils of central Amazonia. Agro-Ecosystems 6:249–66.

Tsoar, H. 1990. The ecological background, deterioration, and reclamation of desert dune sand. Agriculture, Ecosystems, and Environment 33:147–70.

Tsoar, H., V. Goldsmith, S. Schoenhaus, K. Clarke, and A. Karnieli. 1994. Reversed desertification on sand dunes along the Sinai/Negev border: From satellite imagery. *In* V. P. Tchakerian, ed. Desert aeolian processes. New York: Chapman and Hall.

Tsoar, H., and J. T. Moller. 1986. The role of vegetation in the formation of linear sand dunes. *In* W. G. Nickling, ed. Aeolian geomorphology. Boston: Allen and Unwin. 75–95.

Tsoar, H., and Y. Zohar. 1985. Desert dune sand and its potential for modern agricultural development. *In* Y. Gradus, ed. Desert development. Dordrecht, The Netherlands: Reidel. 184–200.

Van Wijk, W. K., and D. A. de Vries. 1963. Periodic temperature variations in a homogeneous soil. *In* W. K. Van Wijk, ed. Physics of plant environment. Amsterdam: North-Holland. 102–43.

Verner, J., M. L. Morrison, and C. J. Ralph. 1984. Wildlife 2000: Modeling habitat relationships of terrestrial vertebrates. Madison: University of Wisconsin Press.

Walker, L. R. 1993. Nitrogen fixers and species replacements in primary succession. *In* J. Miles and D. W. H. Walton, eds. Primary succession on land. Boston: Blackwell Scientific Publications. 249–72.

Walter, G. 1973. Vegetation of the earth. London: English University Press.

Wan, C. G., R. E. Sosebee, and B. L. McMichael. 1993a. Broom snakeweed responses to drought. Pt. 1: Photosynthesis, conductance, and water-use efficiency. Journal of Range Management 46:355–59.

———. 1993b. Does hydraulic lift exist in shallow-rooted species? A quantitative examination with half-shrub *Gutierrezia sarothrae*. Plant and Soil 153:11–17.

Ward, D., and D. Saltz. 1994. Foraging at different spatial scales: Dorcas gazelles foraging for lilies in the Negev desert. Ecology 75:48–58.

West, N. E. 1990. Structure and function of microphytic soil crusts in wildland ecosystems of arid and semi-arid regions. Advances in Ecological Research 20:180–223.

West, N. E., and K. H. Rea. 1979. Plant demographic studies in sagebrush-grass communities of southeastern Idaho. Ecology 60:376–88.

White, B. R. 1982. Two-phase measurements of saltating turbulent boundary-layer flow. International Journal of Multiphase Flow 8:459–73.

Williams, C. B. 1954. Some bioclimatic observations in the Egyptian desert. *In* J. L. Cloudsley-Thompson, ed. Biology of deserts. London: Institute of Biology. 18–27.

Williams, O. B., and R. Roe. 1975. Management of arid grasslands for sheep: Plant demography of six grasses in relation to climate and grazing. Proceedings of the Ecological Society of Australia 9:143–56.

Williamson, D. T., and E. Delima. 1991. Water intake of Arabian gazelles. Journal of Arid Environments 21:371–78.

Wilson, A. D. 1977. The digestibility and voluntary intake of leaves of trees and shrubs by sheep and goats. Australian Journal of Agricultural Research 28:501–8.

Wilson, A. D., and G. N. Harrington. 1984. Grazing, ecology, and animal production. *In* G. N. Harrington, A. D. Wilson, and M. Young. Management of Australian rangelands. Melbourne: CSIRO. 63–78.

Wilson, A. M. 1984a. Leaf area, nonstructural carbohydrates, and root growth characteristics of blue grama seedlings. Journal of Range Management 37:514.

Wilson, M. F. 1984b. Vertebrate natural history. Philadelphia: Saunders College Publishing.

Winkler, W. G., and K. Bogel. 1992. Control of rabies in wildlife. Scientific American 266:56–62.

Wisheu, I., and P. A. Keddy. 1992. Competition and centrifugal organization of plant communities: Theory and tests. Journal of Vegetation Science 3:147–56.

Yair, A. 1990. Runoff generation in a sandy area: The Nizzana sands, western Negev, Israel. Earth Surface Proceedings of Landforms 15:597–609.

Yom-Tov, Y. 1970. The effect of predation on population density of some desert snails. Ecology 51:907–11.

———. 1993. The importance of stopover sites in deserts for Palearctic migratory birds. Israel Journal of Zoology 39:271–73.

Young, T. P. 1994. Natural die-offs of large mammals: Implications for conservation. Conservation Biology 8:410–18.

Zhirnov, L. V., and V. O. Ilyinsky. 1986. Wild-ass. *In* V. Y. Sokolov, ed. The Great Gobi National Park: A refuge for rare animals of the central Asian deserts. N.p. 68–79.

7 Ecosystem-Level Consequences of Management Options

L. A. Joyce, J. J. Landsberg, M. Stafford Smith, J. Ben-Asher, J. R. Cavazos Doria, K. Lajtha, G. E. Likens, A. Perevolotsky, and U. N. Safriel

Land use on arid and semi-arid lands has intensified over the thousands of years that humans have drawn from their wealth. Some of these land uses have never been assessed critically from an ecosystem perspective. The productivity of arid and semi-arid lands is often more dependent on the inherent ecosystem processes than the intensively managed lands in mesic environments. Consequently, the disruption of these processes imperils ecosystem sustainability. Links between ecosystem science and "on-the-ground" management are tenuous at best because of lack of awareness, poor training, insufficient time, different goals, and poor communication between researchers and managers. In this chapter, we assess the consequences of various management practices in terms of their impact on ecosystem processes such as the flows of energy and materials (nutrients, soil, water), the production of plant biomass, and the changes in ecosystem states.

The evolution of land use on arid and semi-arid lands has been driven by human population densities, affluence, and available technology (Graetz 1994). Earliest uses were by hunter-gatherers whose impact may have been changes in fire frequency. Subsistence pastoralists herding domesticated herbivores displaced hunter-gatherers from the most productive grasslands. In arid and semi-arid lands, traditional common-land pastoralists developed seasonal migration practices to search for the best available feed across large land areas. With access to a cash or market system, the availability of technology increased and the focus of animal husbandry shifted from survival to the production of a regular or increasing income (Graetz 1994). Similarly, the harvest of fuelwood or other products may have increased with access to a market or larger populations (Mercer and Soussan 1992). Increasing availability of technology also allowed intensification of grazing uses and consideration of uses such as dryland cropping or irrigated agriculture. With increasing affluence, demands for both recreation and conservation have increased. With increased population density, human demands have broadened to include waste disposal and some activities, such as military training, that require large unsettled areas. Increasing populations have also pressured the conversion of semi-arid lands to cropland, pushing grazing uses to the more arid lands. This evolution of land use over time is seen across the globe today with intensity of land use increasing as a function of available capital, knowledge, and technology.

Management usually implies planning (selecting combinations of practices for a desired goal), operating (implementing the practices), and monitoring (evaluating the results) (Stuth et al. 1991). The production of these desired goals or resource outputs requires the selection of a suite of management practices to bring the desired goal to fruition. These desired goals could be obtained without management planning with potential unanticipated consequences at the ecosystem level. The basis for selecting management practices varies from the use of informal assessments by individuals to the use of relatively sophisticated mathematical models by resource management staffs. Diminishing resources (e.g., fuelwood) are often an incentive to develop sustainable management practices.

In the following sections, we outline how resource production from different land uses interacts with energy flow and nutrient cycling, biomass production and removal, and changes in ecosystem states and how, without proper management, these ecosystem processes can be disrupted with long-term consequences. Finally, we assess the scientific knowledge needed to understand the ecosystem-level consequences with the objective of developing management practices for sustainable ecosystems. The land uses we consider span the traditional uses of grazing and the selective harvesting of unmanaged species, including fuelwood, to more recent uses such as recreation and conservation management, to more intensive uses such as dryland (arable) agriculture, waste disposal, and military activities. Land uses in three focus areas of the workshop (Australia, southwestern United States/northern Mexico, and the Middle East) are discussed in detail in Fisher and colleagues' article (chap. 9, this volume).† ††

Management Options and Ecosystem Processes

Grazing

Grazing of domestic herbivores is the historical, traditional use of arid and semi-arid lands and, because of this long history, much experience and knowledge of the defoliation effects on plants is available. Ecosystem-level consequences of grazing have only been examined within the last few decades. Grazing management includes the selection of type and number of animals, the timing of grazing, the spatial distribution of animals, and under more intensive operations, the modification of plant species present, the modification of soil surface, and the input of subsidies to enhance the grazing value of the land (Heady 1975, Heitschmidt and Stuth 1991, Senft et al. 1987). The suite of management practices associated with grazing management can be generalized into either extensive or intensive management.

Extensive management manipulates the temporal and spatial distribution and the numbers of herbivores (Briske and Heitschmidt 1991). It is based on small inputs of external resources (capital, supplemental feeds) and high inputs of internal resources (labor, natural resources such as grasslands). Output is moderate and risk is low. Extensive management occurs in areas where the plant productivity is low or variable and where ecosystem types are diverse. Management practices may be controlled in the sense of fixed grazing patterns within fenced or well-bounded areas or nomadic, as in the case of the Bedouins whose sheep and goats graze across a large landscape. The production of livestock under extensive management is intimately related to the ecosystem processes that sustain plant production. The effects of extreme events, such as drought, on ecosystem processes can have disastrous consequences on animal production. Improper management can also exacerbate these ecosystem processes with similar dire effects on livestock production.

Intensive management is based on high investment of external resources in order to obtain high production for export. Output is high and risk is also high. Intensive management occurs in the semi-arid lands where climate, soils, and topography are more favorable than in the extensively managed areas. Fertilization, land clearing of woody plants, and the introduction of forage species are examples of energy-intensive inputs that have been used on some semi-arid lands (Briske and Heitschmidt 1991). Because intensive management supplements some ecosystem processes, the dependency on these processes is less than in extensive management. For example, fertilization lessens the system's dependency on soil mineralization rates to release nutrients that limit plant production. Understanding the interactions between these subsidies and ecosystem processes is important to avoid unplanned consequences such as excessive nitrates in the groundwater.

Because herbivores harvest plant biomass, the grazing activities of domestic livestock and wildlife can influence the production of and composition of vegetation and, hence, have ecosystem-level effects on plant production, nutrient cycling, and ecosystem states. Plant production is a function of the physiological processes within plants, competition for resources (water, nutrients, light) among plants, and the availability of nutrients and moisture within the soil. The removal of plant biomass, typically the photosynthetically active plant parts, reduces the plant's conversion of energy from the sun, the first step in the energy flow of ecosystems. Photosynthesis is the process by which the plant produces simple food materials for use in root replacement, regeneration of leaves and stems after dormancy, respiration during dormancy, bud formation, and regrowth after top removal (Holechek et al. 1995). The coevolution of plants and grazing animals has resulted in physiological and morphological traits that increase the plant's resistance to grazing; for example, resistant traits in grasses can include higher proportion of culmless shoots, delay in elevation of the apical buds above the soil surface, ready sprout from basal buds after defoliation, and higher ratio of vegetative to reproductive stems. Plants have also adapted to resist grazing through mechanical and biochemical mechanisms such as spines and toxic secondary compounds. This coevolution influences the amount of energy that passes from plants to herbivores—unless humans intervene to enhance this energy flow to domestic herbivores.

Vallentine (1990) described three opportunities to enhance the energy efficiency in the soil-forage-ruminant system:

1. increase the conversion of solar energy by photosynthesis (e.g., enhance the quantity and quality of forage produced);
2. increase the consumption of forage over space and time through optimal management of grazing animals; and
3. increase the conversion of ingested energy by the grazing animal into products directly useful to humans through improved animal genetics, nutrition, and health.

The production of plant biomass influences the diversity of plant and microbe species and, through that diversity, the lifeform dominants and invasion of nonnative plants. Extensive management typically does not add plant or microbial species to arid and semi-arid lands; rather, the adaptations of the species present are manipulated to produce the desired outputs. Because herbivores have evolved to prefer one or more of the lifeform groups of grasses, forbs, or shrubs (Stuth 1991), the mix of species grazing can influence vegetation composition. Light and moderate grazing can influence competition among plants and, thus, the diversity of arid and semi-arid vegetation. However, heavy grazing generally lowers species diversity. Species diversity influences the availability of nutrients such as nitrogen through nitrogen-fixing plants, the accumulation of minerals within plants, and, at the landscape scale, the spatial patterns of soil nutrients and energy flow. Species diversity may play a critical role in the ecosystem's ability to withstand stresses such as drought, insects, or disease.

For ecosystems, recommendations have been made on the percentage of biomass that can be safely harvested without impairing the vegetation. Grazing systems seek to maximize the harvest of plant biomass within or between years, with nongrazing periods sufficient for plants to recover from defoliation. Though natural vegetation patchiness and the tendency for some livestock to concentrate on stream- and riverbanks influence livestock distribution on the landscape, management can direct the distribution through herding or watering points and salt licks, thereby harvesting more of the vegetation across the landscape. Management practices seek to optimize the amount of energy that the herbivores can harvest from the plants without impairing the ecosystem's ability to respond to biomass removal and its associated impacts.

This energy captured in photosynthesis and converted into plant biomass fuels the trophic structure of the ecosystem. In some areas, this trophic structure has been altered. In the Old World, most of the wildlife are extinct due to heavy continuous hunting and competition from domestic livestock (e.g., Perevolotsky and Baharav 1991). In some arid regions, the only native herbivores of importance may be granivores (seed eaters) (Noy-Meir 1985).

The movement of nutrients through arid and semi-arid ecosystems—the biogeochemical cycle—is influenced by abiotic processes, biological processes (plants, soil microbes), and the physical characteristics of the soil. Their degree of importance varies by nutrient. Nitrogen cycling tends to be more dependent on the biological (microbial, plant) in contrast to the cycling of sulfur, which is more influenced by the abiotic processes such as weathering. Plants take up nutrients primarily from the soil, store these nutrients in tissue, and release them in dying tissue. Cryptogamic crusts on the soil surface can be the dominant source of nitrogen inputs to some arid lands (Evans and Ehleringer 1993). Nutrients such as carbon and nitrogen can be tied up for long periods in the woody parts of plants or in the soil humus. Nutrients can be released from plant tissue or soil through fire or removed from the ecosystem through harvest. Biomass removal, as in grazing or selective harvest, removes nutrients but, under proper management, not in sufficient quantities to disrupt the cycle. Typically, extensive management does not add large quantities of water or fertilizer to these lands; thus, the integrity of soil processes is fundamental to the ecosystem's ability to withstand wind and water erosion events.

Intensive management may involve modification of the plant species present to channel production into species that are considered suitable forage. Such modification may involve the removal of species by physical means such as fire or chaining (e.g., pinyon-juniper or mesquite in the southwestern United States), chemical removal by herbicides, introduction of additional species to supplement the pasture (e.g., legumes in northern Australia, grasses in northern and central Australia, chenopod shrubs in shrub steppe formation in southern Australia) (Harrison 1986), or removal of all vegetation and planting of introduced species (e.g., crested wheatgrass in western United States).

Both extensive and intensive management must consider the variability of the climates in the arid and semi-arid lands. A review of desertification caused by grazing suggested that there were few documented cases where grazing alone had caused desertification (Dodd 1994). More often, these cited cases of desertification represented the variability of plant production in areas of highly erratic precipitation.†

Fuelwood Use and the Selective Harvesting of Unmanaged Species

Woody species, harvested for fuelwood use and charcoal production, are the energy source for cooking and heating for over 30% of the world's population (Williams 1994); in many parts of the world, alternatives to fuelwood for cooking and heating are not easily available, placing much pressure on the available woody species. Lack of fuelwood is greatest in Africa where, in some savannas, the rate of removal greatly exceeds the rate of growth (Williams 1994). In some areas, such as the Middle East, decisions to harvest for fuelwood or other uses probably were made thousands of years ago and have created the current treeless situation (Perevolotsky chap. 15, this volume). Unlike grazing, the harvesting of fuelwood is little regulated. Only recently have management practices been developed because of concerns about overharvesting preferred species. For example, the collection of firewood or fuelwood is now regulated by permit within the Australian Capital Territory, on forest lands managed by the federal government in the United States, and in Mexico by the federal government.

The selective harvesting of other plant species has occurred for traditional uses such as medicinal, commercial (e.g., income enhancement in northern Mexico, native plant landscaping, wood products such as *Acacia aneura* for artifacts, and sandalwood for export in Australia), or the illegal collecting of rare species (cacti in southwestern United States). The identification of species to be harvested has evolved with the uses of these arid and semi-arid lands. The Bedouin territoriality rules (in the Sinai) forbid members of other tribes to dig wells, establish gardens, or build permanent stone houses within the territory of another tribe. However, a tribesman may freely erect his tent, graze his herd, gather firewood, and prepare charcoal on a neighbor's land (Perevolotsky 1987). More than 20,000 years ago, the survival of hunting and gathering communities in Mexico was related to identifying useful species for shelter, food, medicine, and ornament (Maldonado 1985). With increased human population pressure, material demands increased and the pressures on species and ecosystems broadened to include, for example, construction materials such as timber for houses and fences (Manzanilla 1994).

Management decisions involve the selection of species to be harvested, the manner of harvesting, and the timing and rate of harvesting. However, management "decisions" are usually the result of individual choices by the person harvesting the plant (e.g., Bormann et al. 1991). The method of harvest varies by species and typically does not involve removing large amounts of biomass from the ecosystem. Fuelwood harvests near population centers may include both living and dead trees, whereas in rural areas only the dead tree is typically harvested (Williams 1994). Large individuals and standing dead trees contribute litter and snag inputs to the forest floor. Small mammals and soil organisms, in addition to the deep rooted woody species, influence the vertical and horizontal distribution of soil nutrients. Vertical and horizontal concentrations under arid shrub canopies result from peripheral roots searching at increasing depths and lateral spread into shrub interspaces (Charley and West 1975). Direct nutrient losses from fuelwood harvesting are likely to be small if the wood harvested is mainly stems because most plant nutrients are in the leaves and small shoots.

Horizontal patterns of soil nutrients reflect the topographic variability and deposits by wind and water. Woody species may have large canopies that mitigate the impact of rainfall on the soil surface and limit the soil surface exposure to wind. Woody species within savannas or woodlands typically have deep roots accessing sufficient water for transpiration throughout the year. Removal of the woody species potentially alters the relationship between the vegetation and the water table.

The production of plant biomass and the cycling of nutrients may not be greatly affected by species additions or removal (Johnson and Mayeux 1992, Vitousek 1990). Harvesting dead material or cutting larger branches or boles of living trees has relatively little impact on the rooting system or deep soil. The harvesting of medicinal or commercial species may involve the destructive uprooting of the plant, as is the case with candelilla (*Euphorbia antisyphilitica*) in the Chihuahuan desert, but this effect is limited to the scale of the plant. The outputs from these ecosystems and their value to humans are clearly altered in species shifts, such as the reduction of a commercially important species.†

Recreation and Tourism

Recreation and tourism is a growing land use throughout the world and, in many areas, represents a major hope for income generation in dry regions. The low human population density of arid and semi-arid lands provides a sharp contrast to the human congestion of the more densely settled mesic areas. Tourism may be usefully divided into commercial tourism that tends to focus on a small number of intensively used locales and is, therefore, readily manageable in principle and private tourism that is often dispersed over large areas and, in arid areas, often based on access by four-wheel drive vehicles, therefore requiring very broadscale management or community education.

Commercial tourism involves management of intensively used locales where an infrastructure surrounding the site is used to manage access. Landscape enhancement may involve creating water bodies and local woodlots, as in Israel where green belts are planted around towns in the arid zone. These are localized and often involve high levels of investment of time and money. Biomass may be removed and replaced with other vegetation, such as conversion from grass and shrubs to trees. Typically, this conversion is subsidized with fertilizer and water, temporarily removing the dependency on the mineralization rates of the soil to supply nutrients and the annual precipitation to supply moisture. The fact that many desert plants and animals occur at low densities, due to the low inherent productivity of the landscapes, and are inactive for extensive dry periods has led to calls for "ecosystem engineering" to artificially trigger growth responses to assure visitors of seeing at least small areas of the deserts in bloom.

Private tourism involves individuals touring the landscape using some form of transportation such as hiking, off-road vehicles, or a road system. Typically, little biomass is removed under private tourism management. Management may emphasize the faunal attractions where the openness of the arid and semi-arid lands makes wildlife more visible compared to other environments. Many countries encourage the use of wildlife for recreation, hunting, or passive viewing, with a strong focus on large mammals. Managing these animals and their predators to ensure adequate visitor contact times may result in large concentrations of animals in small areas. In arid lands, wildlife are easily attracted to water bodies or "feeding restaurants." The phenomenon of bird migration is easily observed in desert oases and likewise attracts ecotourists (e.g., Elat in Israel).

The underlying features of arid areas—low productivity and variability in space and time—draw tourists seeking the "wilderness" experience in areas less densely populated. However, the landscape-scale spatial heterogeneity tends to be expressed in terms of large expanses of arid vegetation and limited areas that are attractive for visitation, such as scenic vistas, water bodies and riparian zones, or known rare species populations. For example, within the large expanses of arid land, recreationists, native fauna, and domestic livestock may concentrate in riparian zones. Wetlands are important not only aesthetically, as the refuge for a wide range of interesting flora and fauna, but also because they provide the control systems for stream and river flows, build streambanks, serve as filters for sediments, and capture and break down nutrients and water pollutants. Water can be stored in wetlands and released slowly, thereby extending the supply of water through the summer season.

Both commercial and private tourism development requires road construction to develop a transportation network. Such a network may remove small amounts of biomass, often replaced with roadside plantings, as in Israel. Runoff diversion caused by roads may be the largest interaction with the ecosystem processes. Runoff often provides additional moisture to plants and can be enhanced by careful shaping to focus the runoff on roadside plantings. Highway plantings may be nonnative species (e.g., in Colorado, heightening the contrast between the native short-grass prairie and the introduced mesic mid-grasses along the roadside). From the landscape perspective, the road structures can alter overland flow of water and materials, altering the vegetation type by either restricting the former flows and, hence, decreasing water and nutrient inputs, or trapping the flows and, thereby, increasing moisture and nutrients.

Conservation Management On and Off Reserve

Conservation management is aimed at conserving or preserving the ecological integrity of ecosystems. While various international agreements, such as the Biodiversity Convention, mandate some aspects of conservation, in practice, conservation management is driven mainly by perceptions of use and existence values, often the concerns of affluent urban populations in developed countries (Fisher et al. chap. 9, this volume). Use values are dominated by tourism and recreation but also include genetic resources for pharmaceuticals and other uses as well as the protection of water catchments.

Conservation in reserves is traditionally based on theoretical concepts, such as island biogeography, that help determine appropriate sizes of reserves (Margules 1989). Choice of reserve area is driven mainly by a balance among biological, political, and recreational concerns and is rarely optimal for conservation (Morton et al. 1995). Management will vary depending upon the objectives of the reserve (e.g., a single species focus, watershed protection for drinking water, or ecosystem protection). Typically, biomass removal is minimal although management may seek to restore the dependency on ecosystem processes in resurrecting the historical fire frequency. Management practices may contrast with the surrounding areas when protected species are regarded as pests on adjoining lands (e.g., kangaroos and dingos in some areas of Australia, prairie dogs in the United States) or species cultivated for grazing or agriculture in adjoining lands may be weeds from the conservation viewpoint (e.g., buffel grass in central Australia) (Griffin 1993). Competition for water may also result.

Reserve management has raised considerations about the role of landscape processes. Reserve sizes determined by compromise appear to be smaller than the migration needs of large mammals, insufficiently large to capture the nature of fire frequency or water and nutrient redistribution within the watershed, or too small to support sufficient numbers of populations to survive the highly variable climate. Especially in arid areas, the patchy nature of the landscape and the unreliable persistence of wet areas over time render conservation untenable without allowing for extinction and reinvasion processes (cf. Morton 1990). Management to conserve the landscape matrix depends on a good understanding of source and sink areas for populations, rates of extinction and immigration, and the effects of different land uses on these processes and the biogeochemical cycles.

Dryland Arable Agriculture

Cropping Cereals are the most widely grown crops in arid and semi-arid regions; some of the world's great wheat growing regions are semi-arid. In North America, winter wheat is grown largely on water stored in the soil from winter precipitation. This system, therefore, relies both on the occurrence of winter precipitation and on adequate soil water holding capacity. Winter wheat in Australia also relies on winter precipitation, although stored water is less important: winters are not generally cold enough to halt growth totally; the wheat develops slowly through late winter and spring. In Israel, nomadic pastoralists may grow rain-fed barley following traditional ways (Marx 1967). The land may be tilled by draft animals so that compaction and the effects on soil structure are different from the winter wheat areas; timing of operations may also be different, being governed to some extent by the condition of the animals, which in turn depends on the availability of forage. The areas involved are also much smaller.

Management decisions focus on whether to plant or leave an area fallow and on the crop variety to use. Fallow cropping allows storage of enough water for one crop. In the winter wheat areas, the cropping practices are now based on well-established principles and reliable information. Nevertheless, cropping in semi-arid regions is risky from the point of view both of productivity (the risk of crop failure because of drought) and of the land (sustainability). The American dust bowl of the 1930s demonstrated the disastrous consequences of misunderstanding semi-arid ecosystems.

The economic viability of semi-arid cropping systems often depends on extensive subsidy, as in the case of crop supports in the United States. In Israel, as part of a national policy to encourage agricultural settlements in the semi-arid region, a system of drought compensation to cereal farmers has been implemented based on a study of climate-production variability (Zaban 1981).

Agroforestry Agroforestry generally consists of tree-forage systems that may be designed to improve grazing, enhance water utilization by forage or by trees, or perhaps enhance tourism through shade, campsites, and aesthetic considerations. The options available to managers are, therefore, determined by the benefits that may be obtained, in any particular area, for some combination of trees and forage and by the ecological consequences of the practice. One advantage of agroforestry is that the annual crop removed is very small in terms of total biomass, thus reducing the risks of soil erosion.

Benefits of agroforestry are determined by the growth patterns of both components of the system and their effects on each other. Interactions may be positive, such as the effects of nitrogen fixation (e.g., modest proportions of the introduced *Acacia nilotica* in Mitchell Grass pastures in Australia promotes overall productivity through N-fixation) (Carter et al. 1989), or negative, such as competition (in the same example, excessive invasion by *Acacia nilotica* can result in loss of grass productivity through competition for light and water) (see also Walker 1985, Walker et al. 1989, and Werner 1991 on models of interactions; Scanlan 1992, Scanlan and Burrows 1990, Walker et al. 1986 for interactions).

The hydrology of a tree-forage system also depends on the architecture of the tree crowns and the extent to which they intercept rainfall, as well as the characteristics of the rainfall itself and the way the tree and forage root systems interact and compete for water (Walker et al. 1989). In areas where precipitation occurs as frequent light showers, interception losses are proportionally higher than in areas where rainfall tends to occur in heavy storms.

Runoff Farming The essence of run-off farming is the collection of water from areas where plant growth is unlikely to occur and its concentration in areas where additional water can be used by plants. The collecting area is normally much larger than the receiving area.

Decisions concerning the collection of runoff are normally made based on experience. Structures are built and the runoff areas modified—for example, by stripping vegetation to increase the amount of water collected in the receiving areas. Given rainfall pattern and the surface and hydraulic characteristics of the runoff areas (*microcatchments*), it is possible to calculate the runoff from specified areas and, hence, the amount of water that could be collected from any given microcatchment (Yair and Shachak 1987).

As a farming technique, runoff farming was used only in Israel in ancient times (100 B.C.E. to 600 C.E.). One site has been reconstructed and has served for many years as a research and extension station (Evenari et al. 1971). Currently, runoff farming is mainly used in Israel for the establishment of trees in the Negev desert. The Savannization Project is an example of such efforts in Israel (Sachs and Moshe chap. 17, this volume).

On a larger scale, managers in Israel have created structures called limans, which are earthen dams constructed across wadis. Water accumulates behind the dams, infiltrates, and is stored in the soil. These limans are usually planted to trees that provide shade for stock and tourists and improve the landscape from an aesthetic viewpoint. Whether water collected in the limans is effective for plant growth depends on the soil depth and water storage characteristics. Comparable measures are used for landscape rehabilitation in Australia (Purvis 1986; for effects on nutrient capture, see Tongway 1990) and for local production of supplementary "ponded pastures" in northern Australia where seasonal rainfall is more reliable than further south (Stafford Smith 1994).

Runoff harvesting is used widely in Australia to provide water for stock. The collecting dams are often dug down to relatively impermeable clay layers so that the water does not infiltrate the soil; the ratio of surface area to depth is also reduced to minimize evaporation losses. Because most of the Australian rangeland areas are flat, the water catchment areas for these dams are often poor, and they rely on the occasional heavy rains to fill them. However, with minor replenishment, they can store water for several years.

Runoff harvesting is also widely used in Mexico and the western United States. In the arid and semi-arid northern lands where annual mean precipitation ranges from 120 mm to 550 mm, water catchments provide water not only for stock but also for growing subsistence crops and for plantations of trees and shrubs.

Waste Disposal

Activities such as waste disposal are increasingly impinging on low-population areas such as most arid and semi-arid lands. Modern cities in all climate zones generate vast quantities of waste that are rapidly becoming unmanageable particularly in the more densely settled areas. When the waste is nutrient-rich sewage sludge with low levels of industrial pollutants or toxins, an increasingly popular solution to the disposal problem is land application—the controlled spreading of sewage sludge onto and into the soil surface (Harris-Pierce 1994). Municipal sewage sludge has also been used to amend sites for reclamation of coal-strip mine spoils, gravel spoils, coal refuse, and sites devastated by toxic fumes (Sopper 1993). Storage for nuclear or medical waste involves long-term isolation of the waste from soil processes and plant growth.

The disposal of effluent and sludge involves either the use of natural ecosystems or the irrigation of a crop (e.g., pasture or trees). Application of sludge to arid or semi-arid vegetation typically enriches the soil with additional nutrients and moisture. Nutrient retention in the soil depends on the system's ability to capture the nutrients (microbial or vegetation response) or store the nutrients (binding to soil particles). Fresquez and colleagues (1990a, 1990b) and Harris-Pierce (1994) reported increasing levels of nitrogen, phosphorus, and potassium in plant tissue with sewage application on semi-arid and arid land vegetation. Biomass removal may be necessary to keep the vegetation in a rapidly growing state. In the case of agricultural or tree plantations, effluent may remove the need for some nutrient fertilizer applications. The soil-vegetation relationship must be used to determine the appropriate application procedures so that effluent stays on site.

Military Training Exercises

Military training is relatively widespread in the arid lands of Israel, Australia, and the United States. Training may include exercises aimed at proficiency at using heavy ground equipment or aerial exercises on bombing targets. In some areas, limited military use of large areas can provide substantial conservation protection because other land uses are then restricted. The Israeli government has set up a unique system whereby nature reserve lands are shared with the military, provided the military uses them only for certain purposes and at stated times.

Ground maneuvers typically do not removal large amounts of biomass. These activities may damage plants, expediting the transfer of live to dead biomass. Removal of the plant canopy may alter nutrient cycling and soil formation processes by altering soil surface characteristics. In Israel, the dust generation caused by tank traffic is carried by wind for long distances, depositing fine layers of dust on desert vegetation.

Problematic Consequences

Disruption of Energy Flow

The disruption of energy flow occurs when removal exceeds the growth of plant biomass such that the ecosystem's ability to capture energy from the sun is impaired. In addition, the ability of ecosystems to produce plant biomass may be altered when conditions are such that an unusual buildup of vegetation occurs. These disruptions can result from disturbances such as fire or insects, through improper management, or through the interactions of management and disturbances. It has been postulated that the system's capability to respond to generalized stress, as well as its overall level of activity and organization, is related to the amount of energy available beyond that used for maintenance (Costanza 1992).

Sustaining populations of wildlife and domestic livestock requires an adequate flow of energy to the herbivores. This flow of energy may become limited at certain times of the year. In the United States, competition between wild and domestic herbivores for spring vegetation has led to some degradation. This problem is solved in the semi-arid region by planting a cool-season plant available for grazing earlier than the native vegetation. Concentration of livestock and wild herbivores on stream- and riverbanks may lead to decreased plant production, loss of plant species, and the invasion of undesirable plant species. Shifting grazing from spring/summer to winter allowed the restoration of riparian zones in the western United States (Chaney et al. 1993) by reducing the grazing pressure on the ecosystem when it is most vulnerable. Similarly, the timing of recreation use or military maneuvers can be managed to reduce the impact on the ecosystem. For example, in the Pukapunyal training area in Australia, critical times to avoid tank traffic are when there is significant rainfall at which time the land is highly susceptible to compaction and long-term damage.

Energy flow through an ecosystem can be altered when grazing pressure or selective harvesting shifts the composition of vegetation. Loss of cool-season plants reduces the amount of solar energy captured in the spring. In the southwestern United States, shrubs gained the competitive advantage when livestock type was shifted from goats and cattle to solely cattle (Heady 1975).

Underutilization of biomass can occur in conservation management situations where protecting the system is confused with the practice of removing all disturbances. Underutilization also occurs in the urban-wildland interface when fire is increasingly suppressed because of higher densities of built-up areas and human life. Restriction of fire in arid and semi-arid systems often results in an increase of plant biomass, at least in the short term (Seligman and Perevolotsky 1994). Plant species composition may shift to species that cannot tolerate fire, or the fuel buildup of existing species may result in fires hotter than the historical pattern. In California, the encroachment of urban areas into the Mediterranean shrub types has led to increased fire suppression, which has led to increased plant biomass production and fuel load as the plant biomass is highly susceptible to fire. Fire inevitably comes, started by either humans or lightning, and moves rapidly through the landscape. The intensity of these California fires is much greater (hotter, more soil degradation) than the Mexican Mediterranean shrub fires, whereas the frequency of fires in northern Mexico is greater than in California (Minnich 1983). Reserve design also faces the complexities of managing land interfaced with urban or other land uses. Unintentional effects arise from management objectives, such as altered fire frequency, differing between reserve and adjoining lands.

Studies in the arid and semi-arid regions suggest that these systems are subject not only to the variability of climate but also to episodic or extreme events such as insect or disease outbreaks. Few cases have documented grazing alone as the cause of desertification (Dodd 1994) but rather cite the variability of plant production in response to highly erratic precipitation. Managing by the "average" is now recognized as inappropriate for areas receiving highly erratic rainfall; the traditional strategy was to migrate to follow the favorable condition (Behnke and Scoones 1992, Dodd 1994, Ellis and Swift 1988).

The interaction of continual overgrazing and episodic events, such as pest outbreaks, drought, or fire, jeopardizes the ecosystem's ability to produce plant biomass. As a result of overgrazing by livestock, Skarpe (1986) showed that Kalahari vegetation is getting increasingly shrubby and often remains so even 30–40 years after abandonment by herdsmen. Stein and Ludwig (1979), citing Buffington and Herbel (1965), described the consequence of overgrazing on the Jornada Experimental Range (New Mexico) when, in 1958, grass cover was over 90% but declined to less than 25% in 1963. Areas protected from any grazing have not returned to grassland. The control of invasive, shrubby, and unpalatable species such as mesquite, junipers, and cacti is a formidable problem in much of the southwestern United States, and restoration of degraded land has often proven impossible or too expensive. Pest and disease outbreaks can alter the system's ability to respond to grazing pressure. Three years after a white grub infestation, graminoids in an ungrazed northern mixed-grass prairie were not as productive as uninfested sites and the prairie was dominated by forbs rather than grasses (Lura and Nyren 1992).

In some cases, these vegetation shifts can be reversed through management techniques such as shifting the season of grazing in riparian areas or intensive manipulation of the vegetation using mechanical, biological, and chemical controls. These attempts shift the ecological balance but rarely exterminate the invading species, plant, insect, or disease. As populations of two species of native prickly pear on Santa Cruz Island rose, the ecosystem effect was a shift from cool-season grasses to plants with the CAM photosynthetic pathway, which shifted transpiration and carbon dioxide uptake from night to day, influencing how water, carbon, and other growth resources were processed (Johnson and Mayeux 1992). While mainland insects were introduced to feed on the prickly pear, the initial decline in cacti was followed by a slight rise as predators of the introduced insects arrived and as the cacti plant population shifted to the species less susceptible to insect predation.

Reversibility requires an understanding of the dynamics of the ecosystem. Unfortunately, plant community interactions with ecosystems are not always known, nor is the relationship between ecosystem structure (species composition) and function (nutrient cycle, energy flow) understood sufficiently to restore degraded ecosystems.††

Soil Degradation

Soil degradation is a general term for disruption of the physical, chemical, and biological processes in the soil. Soil degradation through impacts affecting the physical and chemical processes are discussed here; biological processes are covered in a later section.

The physical nature of the soil can be altered by compaction and by breaking the physical and the cryptogamic crusts. In the arid and semi-arid regions, soil can be compacted by grazing animals, recreationists walking or driving off-road vehicles, and military tanks. Campgrounds and stock watering ponds concentrate the impacts of humans and animals. Many grazing studies in temperate rangelands have supported the basic predictions of range hydrology: increase in grazing pressure reduces vegetation cover and consequently increases bulk density and compaction, thus lowering infiltration rates (e.g., Willatt and Pullar 1983). However, very little research effort has been devoted to semi-arid ecosystems or to rugged terrain (Wilcox et al. 1988). Moreover, it has recently been claimed that, in general, semi-arid regions are less susceptible to waterborne soil erosion because of low precipitation intensity (Finkel 1986).

Studies conducted on arid and semi-arid rangelands reveal a complex picture. For example, three levels of goat browsing on blackbrush shrubs in southern Utah had no significant effect on infiltration rates (Gifford et al. 1983). Four levels of grazing intensity had no effect on runoff following storms in eastern Oregon (Higgins et al. 1989). In another example, heavy grazing reduced shrub cover on semi-arid Somalian rangelands from 100% to 5%, creating potential conditions for excessive erosion (Takar et al. 1990). However, livestock exclusion for three growing seasons did not significantly increase vegetation cover or elevate infiltration rates, principally because of herbivory by other organisms (termites and microorganisms). Shrub adaptations to grazing (thorns) minimized grazing impact, provided a refuge for herbaceous vegetation, and maintained conditions for higher infiltration rates. The overall outcome was similar infiltration rates and erosion in grazed and ungrazed plots. Intensity of grazing pressure also influences the impact on soils. High grazing pressure by cattle, for a short duration, reduced infiltration rates 50% and significantly increased sediment yield and bulk density (Weltz et al. 1989). Infiltration rates vary naturally in time and space as a result of variations in climatic conditions and vegetation (Blackburn et al. 1982). However, only a few studies have dealt with these changes or attempted to evaluate the separate contribution of natural and human-induced (e.g., domestic livestock grazing) processes to infiltration.

Removal of vegetation cover or shifts in the type of vegetation cover can affect the soil surface susceptibility to wind and water by increasing the rainfall impact, decreasing water infiltration, and enhancing waterborne soil erosion through increased overland flow. In situations where woody biomass increases as grass production decreases, soil erosion also increases (Renard 1987). Long-term adaptation to grazing pressure may shift the plant community. In a subhumid to semi-arid habitat in eastern Kenya, soil losses become a serious problem only if vegetation cover drops below 40% (Zobisch 1993). However, in many semi-arid regions where pastoralism has been a traditional mode of subsistence for millennia, the shrub cover of unpalatable species alone is approximately 40%.

The horizontal and vertical biochemical patterns in soils can be altered by breaking the physical or the cryptogamic soil crust by grazing, removing selective species, and, most directly, plowing. Such impacts can lead to the chemical degradation, as in nutrient depletion, shifts toward extremes in the pH of the soil, salinization, and/or contamination by toxic substances when, for example, subsurface contaminants from mining spoils are drawn to the surface. Visitation points for tourists can alter the chemical properties of soils and water when development for habitation ignores the potential effects of local water depletion, effluent and sewage disposal, and garbage disposal.

Under dryland farming, plowing exposes the upper soil profile to wind and water erosion, increases mineralization, and is accompanied by the introduction of chemicals such as herbicides and insecticides. In the Great Plains in the United States, 42% of the organic carbon and 36% of the soil nitrogen in grassland soils at eleven sites was lost after thirty-seven years of cropping (Haas and Evans 1957). The loss of soil carbon and soil may be more significant in aridisols than in the more mesic mollisols; Barrow (1991) reported that soils with less than 2% organic matter content are particularly vulnerable to soil erosion. The problem for management is to achieve an optimum balance between productivity and sustainability, which will be determined by the extent to which ecosystem properties such as soil organic matter, nutrient levels, soil structure, and hydraulic properties can be maintained (NRC 1989, Doran and Werner 1990, King 1990).

The addition of large quantities of waste to arid lands, either nutrients and water in the disposal of effluent and sewage or nuclear and medical waste, also poses potential problems of nutrient seepage into deep water tables (aquifers) with contamination emerging into surface waters, affecting humans, other animals, and plants. If effluent is used for irrigation, the soil-plant system must be able to accept long-term additions of nutrient-rich water. In the case of nitrates, application of effluent depends on soil processes to denitrify the nitrates before reaching groundwater. Surface runoff may also transport nutrients and undesirable chemicals from the disposal sites. The management of effluent disposal requires considerable information about soil buffering and cation exchange capacity, drainage characteristics and water use, nutrient uptake capacity, and growth patterns of the plants.

The disposal of nuclear and other toxic waste in arid and semi-arid lands has also been proposed (e.g., nuclear waste in New Mexico [United States], and European waste shipped to African countries). The management challenge of nuclear waste has been a primary focus for environmental research of many U.S. Department of Energy labs that have generated such waste. The long-term toxicity of these wastes prompts many concerns about seepage into groundwater, and much effort continues to focus on the development of technology for the long-term stabilization of burial sites in semi-arid environments (Nyhan 1989, Nyhan and Barnes 1989).

The extent to which the imposition of management practices on erosive soils and fragile plant ecosystems exacerbates wind and water erosion is difficult to determine. Rangelands, particularly in dry climates, are naturally large contributors of sediments due to their topography and limited vegetation cover (Branson et al. 1981). Soil degradation can result in off-site impacts (e.g., the movement of soil through dust storms and the movement of suspended particulates into waterways). However, this movement of soil and nutrients from one area of the rangeland landscape to another area can be a critical aspect of site regeneration. Research studies have focused on the fine scale (on small plots), but management is often at much larger scales, such as watersheds or larger landscapes.

Landscapes may contain landforms that reflect different sedimentation processes. For example, Pickup (1991) described depositions stable over a thousand years and narrow threads of sediment that are disturbed yearly on the floodplains of central Australia. Soil moisture-capturing ability, as well as water supply, influences plant behavior, which differs greatly on these deposits (Friedel et al. 1993). Use by grazing animals would also reflect these differences. The impact of management on processes such as the distribution of water and nutrients across landscapes may cross the thresholds at which changes are irreversible. Hence, understanding how small-scale impacts affect the landscape or watershed is critical (Pickup 1985, Pickup et al. 1994).

Salinization

Soil salinization is caused by the alteration of the hydrologic cycles, either by the removal of vegetation or the addition of water, and by the accumulation of salt in the soil surface layers. Salinization has been taking place since the beginning of humankind's attempts to increase plant yield through irrigation. Essentially, the application of water in amounts greater than the plants can use results in the accumulation of soluble salts in the root zones or brought to the surface by capillary action. Land clearing, particularly in areas of saline soils, alters water table levels with potential for dryland salinization. Sewage effluents often have very high salinity because of the large amount of salts used for domestic and industrial purposes (Guldin 1989) and, when applied to cropland, may result in soil salinization.

When trees and other deep-rooted vegetation have been removed, rainwater drainage through root zones either causes rising local water tables or flows laterally across impermeable layers of rock or subsoil and emerges in discharge areas. When the chemical characteristics of soil are conducive to leaching salts into free water, the seeps will be saline. For example, large areas of open woodland and semi-arid savanna in southeastern Australia were replaced with either native or seeded pastures. The cleared trees, mainly eucalyptus species, were evergreen and their deep roots extracted water from the subsoil and watershed/jointed bedrock, preventing water tables from rising. The pastures die off in the hot summers so that neither all rainfall infiltrating the profile nor the demand on the subsurface water is as great as it was under tree cover (Ive et al. 1992). Greenwood and colleagues (1985, as cited in Ive et al. 1992) reported that pasture use was 60% of the annual rainfall compared to native tree use of 240–400%. Water tables rose, dissolving salts in profiles that were only intermittently saturated. As the salts moved into the rooting zone, productivity declined. At the soil surface, salts may inhibit vegetation reestablishment.

Salinization can also result from seawater intrusion as water is drawn for irrigation and other uses near coastal areas. In Israel, total water consumption in coastal semi-arid regions is greater than the natural groundwater recharge. The resulting decline of the groundwater table leads to the drying of shallow but active wells and the beginning of seawater intrusion. The remaining active wells then develop higher salt concentrations that are followed by vegetation loss and reduced agroproductivity (Ben-Asher 1994). In the region of northwest Mexico bounded by the Sea of Cortez, large areas were opened for farming crops (wheat, soybeans, vegetables, etc.) with irrigation provided by groundwater pumping. Two decades later, coastal areas were saline from seawater intrusion. As a result of this irreversible process, farmers have either abandoned many lands or have attempted to grow low productivity saline-tolerant crops.

The problem of salinization has been studied extensively, and the solutions are widely known: the difficulties lie not in lack of technical or scientific knowledge but in the fact that management and controls are usually not good enough, irrigation water is not costed at its full value, and irrigation (almost invariably) proceeds until serious problems are apparent in the irrigated areas. Action then tends to take the form of engineering (drainage) works to control the water tables and provide the opportunity for heavy applications of water to flush the salts from the system. This flushing may, in turn, lead to a problem of disposal. Unless rivers and the natural drainage systems of the land carry sufficient flows to allow acceptance of water with a high salt concentration from irrigation areas without causing significant increases in downstream water salinity, the engineering "solutions" may turn out to be short-term approaches, not long-term solutions. Evaporation basins—sacrificial areas with artificial shallow lakes, with high rates of evaporation, into which drainage water is pumped—have been used in Australia where the problems outlined above are urgent: salinity in the Murray-Darling River system is increasing and water tables in the (very large) irrigation areas along the river are rising rapidly.†

One of the first prerequisites for dealing with the problems of soil salinity is a solid information base. Whether or not disruption of hydrological cycles results in dryland salinity depends on geology, topography, vegetation cover, and climate. Whether irrigation leads to salinity depends on the underground drainage systems and aquifers as well as management. The other obvious and important requirement is a sound scientific understanding of the biophysical processes involved.

Disruption of Biogeochemical Cycles

The disruption of biogeochemical cycles occurs when nutrients become tied up in one part of the cycle (plant, soil) or are lost from it (wind and water erosion, leaching). Large pulses of biomass or nutrients (sewage waste), or toxic compounds, can disrupt soil processes. When biomass is not removed or vegetation shifts from herbaceous to woody plants, nutrients can become tied up in the plant component. Ecosystems have typically adapted to some type of disturbance, such as fire or insects, that releases nutrients for regrowth. However, shifts in the type or frequency of disturbance may be catastrophic.

The removal or harvest of biomass, the removal of particular species that fix nitrogen or accumulate particular chemical elements, or a species shift that changes rooting depths will reduce or alter the chemical activity in the soil and nutrient availability. Removal of biomass, such as in grazing, can reduce the organic matter (OM) content of soils by increasing the turnover of organic material. Mineralization supplies even pulses of nutrients to growing vegetation, and OM is the largest pool of nitrogen in soils (Lajtha and Schlesinger 1986). Low OM in soils reduces water holding capacity and soil fertility. Without adequate grassland cover, runoff of NO_3 to downstream ecosystems is quite high

(Schlesinger et al. 1990). Without soil OM and plant litter, seed germination is difficult. Erosion of topsoil in arid regions is probably the most important factor controlling the lack of reestablishment of grasses in areas now protected from grazing.

Nutrient depletion may be severe in areas where wood is scarce and collections are very intensive (e.g., dead and live including leaves and small stems). Some fuelwood species like mesquite and ironwood in Mexico have been locally overharvested causing soil loss and weed encroachment. Harvest can increase nitrogen mineralization in general (Matson and Vitousek 1981) and in pinyon-juniper ecosystems specifically (Thran and Everett 1987), potentially accelerating the loss of nutrients. The pinyon-juniper woodlands of the southwestern United States have been cut repeatedly for fuelwood. Grazing pressure may shift plant composition of an ecosystem toward shrubs, which alters the root distribution through the soil profile. Similarly, deep-rooted trees in agroforestry systems extract nutrients from greater depth than grasses, effectively bringing nutrients to the surface. The introduction of nonnative plants, such as pines in Israel, into areas where the soil is basic alters the soil surface pH as the needles drop onto the surface. *Tamarix aphylla,* on the other hand, can create salinity problems (Griffin et al. 1989). In extensive grazing systems where dung and urine are returned directly to the soil, the net loss of nutrients may be negligible but the redistribution may be very significant (Taylor et al. 1985). Sewage applications on short-grass prairie were associated with increases in forbs, suggesting a decline in range condition (Harris-Pierce 1994), but how long this trend might continue was unknown.

Trampling may be a serious problem in habitats where the cryptogamic crust is common and functions to fix nitrogen, protect the soil from wind erosion, and maintain the hydraulic properties of surface soils (Anderson et al. 1982, Belnap and Harper 1995, Cook 1994, Evans and Ehleringer 1993, Graetz and Tongway 1986, Johansen 1986, Tolsma et al. 1987, Tongway 1990, Walker and Knoop 1987). Crusts can take many decades to recover (Belnap 1993). Grazing and crust disruption can cause lower nitrogen fixation rates in crusts (Terry and Burns 1987). Loss of litter and lichen crust, although taking place on the surface level, is very significant to plant growth and vegetation composition (Graetz and Tongway 1986). Severson and DeBano (1991) reported that heavy grazing by Spanish goats in Arizona chaparral could affect nitrogen accumulation in the soil because of trampling disturbance. The scale at which trampling (as well as other effects of herbivory) is considered is very important: School (1989) pointed out that the proportion of an area actually compacted by hoof prints during normal grazing was relatively small.

Changes in fire intensity and frequency alter the availability of nutrients at the soil surface: more frequent or hotter fires result in increased volatilization of nutrients and eventually change the concentrations through the soil profile. Fire intensity and frequency can be altered by tourism, causing an increased number of fires, or by conservation management where attempts to protect the arid and semi-arid ecosystems have led to fuel buildup. Fire not only has a direct effect on vegetation but also affects the cryptogamic cover on the soil.

Nutrient losses from dryland agriculture are likely to be severe if the practice is poorly managed and leads to erosion (see "Soil Degradation" section). The constant removal of any crop results in the depletion of critical elements contained in the exported plant parts, and these must be replaced from outside (i.e., crops must be subsidized by nutrient replacement) (Buol 1995). Loss of soil organic matter and increased mineralization may be the most significant effect of dryland agriculture. For example, plowing and cropping resulted in losses of more than 36% of the soil nitrogen in Great Plains soils (Haas and Evans 1957).

The interactions of management and extreme events, as in heavy grazing and moderate drought, can result in unplanned changes in the landscape, such as the redistribution of nutrients in the shift from semi-arid grassland to desert shrubland in southwestern New Mexico (United States) (Schlesinger et al. 1990). The black grama canopy enhances water infiltration by lowering the effective energy of raindrops, thereby reducing water and nutrient transport in surface runoff. During moderate drought, heavy grazing decreases black grama cover and increases bare patches. Loss of grass cover and compaction of soil by trampling results in less infiltration of the rainfall, acceleration of soil erosion by wind and rain, and the movement of nutrients from the bare spots to depressions on the landscape. This redistribution of nutrients on the landscape potentially offers germination spots for shrubs such as creosote and mesquite; the shift to a shrubland type further exacerbates the heterogeneity of soil properties because infiltration of rainfall is greater under the shrub canopies. Wind erosion may further increase the transport of nutrients outside the system, altering the system's ability to return to the black grama grassland.

Disruption of Riparian Zones and Wetlands

The disruption of riparian zones and wetlands in arid and semi-arid ecosystems can be initiated by the loss of vegetation cover or reduced biomass production in the riparian zones or the catchment area surrounding the riparian zones, draining of wetlands to channelize streams, or the removal of large quantities of water for irrigation. Loss of riparian vegetation or catchment area vegetation results in streambank instability, increased soil erosion, increased silt deposition in rivers, changing stream and river flow patterns, and possibly altered river courses. In the Macquarie marshes of Australia, grazing cattle destroy the small flow channels through which water spreads through the marshes. An eroded streambed may effectively lower the water table, leaving large areas of meadow in the catchment a dry upland. When natural inland wetlands are destroyed or badly damaged, river flows will become much more irregular with higher flood peaks and longer periods of low flow. Reduced vegetation cover along the streambank reduces the shade on the stream and raises the temperature, jeopardizing fish habitat.

Upland activities can disrupt riparian zones and wetlands. Vegetation loss in the catchment area changes runoff patterns, favoring flash flooding because the runoff from a high intensity rainstorm from a denuded catchment will be greater than from one with good vegetative cover. The resultant floods may exceed the capacity of streambeds, leading to bank erosion and overflow. The east and west Sierra Madre mountains, the most important Mexican mountain range, provide watershed systems to irrigate farmlands in coastal areas and the high plateau. Continual overgrazing and land clearing in riparian areas for farming have resulted in losses of vegetation cover, erosion that accelerates sediment movement and accumulation in dams and reservoirs, and more intense floods on lowlands.

Overland flow from areas where the surface has been modified by grazing or arable agriculture carries large amounts of sediment and nutrients. This overland flow can cause water quality problems, particularly in relation to the development of algal blooms. Toxic blue-green cyanobacteria are becoming increasingly widespread throughout the world, creating particularly serious problems in areas where reasonably high temperatures and ample sunlight are conducive to the growth of these organisms.

Much research has focused on grazing influences on riparian zones, and sufficient technology is available to address many of these problems. Any successful riparian grazing system will include strategies to ensure the following: (1) grazing intensity and season of use are limited to provide sufficient rest to encourage plant vigor, regrowth, and energy storage, (2) sufficient vegetation during periods of high flow to protect streambanks, dissipate stream energy, and trap sediments, and (3) the timing of grazing is controlled to prevent damage to streambanks when they are most vulnerable to trampling (Chaney et al. 1993). Addressing these aspects for a grazing plan necessitates an understanding of livestock use of the riparian zone, the plant community, and the soil types in the riparian and wetland areas.

Changes in Species Diversity

The species diversity of arid and semi-arid lands influences the total plant biomass produced, the rate of biomass production (productivity), nutrient availability, and the cycling of nutrients, including decomposition and other soil processes. While changes in the species present can alter the production of a desired output (e.g., forage production), the effects of biodiversity on ecosystem function (e.g., total plant production) appear to be weak (Schulze and Mooney 1994). Production of biomass and the cycling of nutrients and water continue with reduced numbers of species in most ecosystems, but the occurrence of a diverse species pool may be important to ecosystems facing great environmental shifts (Schulze and Mooney 1994). The relatively rare studies on the relationships between species diversity and ecosystem function have focused on the ecology of more mesic ecosystems. Thus, few general conclusions can be reached about the consequences of species loss for ecosystem function in arid and semi-arid ecosystems.

Improper grazing or harvesting of plant species by humans can result in community composition shifts to species less preferred for grazing or harvest. Total plant biomass production may be similar before and after the community shift. McNaughton (1994) demonstrated that a grassland community was able to maintain plant biomass when grazed by mixed herbivore herds in Africa and that the plant community was able to compensate for species extinction. The rate of biomass production may change as a consequence of a reduction in species diversity. McNaughton (1994) reported no relationship between diversity and productivity when phenological patterns of growth were taken into account in a California grassland study. Examples can be found on the Great Plains of the United States where cool-season grasses take advantage of early spring moisture and warm-season grasses can co-occur and survive the warmer dry periods of the growing season.

Trampling of soil and vegetation by grazing animals may, under special conditions, help to maintain community diversity. The mosaic structure of the grassland in the Serengeti Park in eastern Africa is stabilized by grazing mammals that reduce soil porosity and water infiltration by compacting the soil (Belsky 1986).

Nitrogen availability may be reduced if nitrogen-fixing species are removed or enhanced if nitrogen-fixing species increase. The reduction or eradication of species as a consequence of fuelwood or selective harvesting may prompt changes in the microclimate or albedo, increased wind erosion, alteration inbiogeochemical flux and cycling, and, in some areas, increased salinity (Schulze and Mooney 1994). In some cases, the organism could be a *key species,* the removal of which will disrupt other ecosystem process (e.g., the loss of nest hollows for large birds [Reid and Fleming 1992], the reduced dispersal of important soil fungi [Claridge et al. 1992, Noble et al. 1994], or the direct structural effects of larger species in providing patches of mesic microhabitats). Perhaps even more critical and less well known are the effects of a loss of species diversity on the microbial community. Application of the rhizobacterium *Azospirillum brasilence* generated an increase of nearly 300% in the biomass of herbaceous vegetation in the northern Negev, quite similar to the effect of phosphate fertilization (Zaady et al.

1994). The diversity of soil organisms is great, reflecting the multitude of compounds involved in the decomposition of plants and animals (Meyer 1994). It has been suggested that the lack of appropriate microorganisms is responsible for the delay in revegetating pinyon pine areas chained for forage production in the southwestern United States.

Though little is understood about the relationship between species diversity and ecosystem function, species composition has been the basis for past management monitoring systems in arid and semi-arid lands (Joyce 1993, Westoby et al. 1989). In these systems, degradation is measured as the difference between the species composition of a site at present and the species composition at "climax" or the late successional stages of the ecological type. This system of monitoring arid and semi-arid lands has received much criticism, primarily because of the question of a climax vegetation type for arid or semi-arid lands, the role of disturbance in these systems, and the suggestion that the ratings may be biased toward livestock use. Frost and Smith (1991) concluded that range condition, a measure based on species composition, was not a reliable predictor of the total plant production of the community. They also concluded that, though species shifts were evident as the community moved from high to low condition, this did not necessarily imply a reduction in biomass production. In a study comparing ecosystems with different evolutionary histories of grazing, species composition changes were not tied to changes in aboveground productivity or soil organic matter changes (Milchunas and Lauenroth 1992, 1993).

Monitoring species composition alone may ignore critical changes in the nutrient balance or soil organic matter that may alter the functioning of the system. Arid and semi-arid lands are diverse in their species composition, and questions about the functional significance of individual species remain unanswered. It is not clear that certain species are more critical than others in the nutrient and energy flows of these semi-arid and arid systems.

Arid and semi-arid lands are not only diverse in their species composition but are also diverse in the type of ecosystems distributed across the landscape, reflecting substantial environmental differences. Protecting or conserving ecosystem types on these large and highly variable landscapes is a difficult task, given the variability of climate and soils. Riparian areas may be ephemeral with plant populations somehow collapsing and regenerating on the landscape. Many research questions are at the population or community level, but the resulting design of land use must recognize landscape patch dynamics (cf. Morton et al. 1995). Reserve size may be insufficient to incorporate these dynamics. The implementation of off-reserve conservation also calls for agreements between conservation managers and land users to ensure the integrity of the intervening lands for migration. To make these agreements effective and justifiable for a wide range of different trophic groups may require considerably more knowledge about these landscape processes than is currently available.

Invasion of Undesirable Species

Disruption in the energy flow or nutrient cycling of an ecosystem may provide an opportunity for a species considered undesirable to establish itself. Its role in energy flow and nutrient cycling and its effect on species diversity may be quite different from the plant species commonly associated with the ecosystem. Spread of such species may be facilitated by livestock animals (e.g., cattle spreading mesquite seeds) (Brown and Archer 1987) or by tourists bringing seeds of nonnative plants and disease organisms. Concerns about the spread of nonnatives in U.S. national parks has prompted a policy restricting the type of feed pack animals may use once within the park boundary.

In extensive rangeland areas in western New South Wales and Queensland (Australia), reduction in fire frequency and heavy grazing has led to increasing amounts of wood scrub (Hodgkinson and Harrington 1985, Humphries et al. 1991). There is a dangerous positive feedback: the more scrub, the less grass, and the lower the probability of a fire sufficient enough to destroy the woody plants. Prescribed fire has been successfully used following mechanical control (such as chaining or cutting) of pinyon and juniper in the southwestern United States (Bunting 1987). However, as trees become dominant and large, control by fire becomes more difficult. The earlier in the invasion process that fire occurs or is prescribed, the easier it is to control (Bunting 1987, Everett 1987a).

Conservation management may involve the protection of land considered to have special scenic or biological characteristics by preventing its use for grazing, agriculture, fuelwood collection, or recreation, except in restricted areas. If such areas are left completely alone, then their population dynamics can reflect the introduction of human-dispersed invaders. The policy of no active management has led to the proliferation of plants that are good competitors under the conditions of the management area, such as a lack of grazing pressure from herbivores (Perevolotsky, chap. 15, this volume). Examples are to be found in Australia where rubber vine, buffelgrass (Griffin 1993), and a range of annuals have proliferated in many protected areas.

Scientific Issues and Research Priorities

The essence of scientific understanding for management purposes must be the capacity to predict quantitatively the behavior of an ecosystem and its response to disturbance, environmental change, or management action. Using this quantitative requirement, the type of research needed can be identified in terms of the extent to which we understand a particular ecosystem and the way it functions, the extent to which we can predict responses to change at the scale of interest, the applicability of this understanding to other ecosystems, and our ability to model ecosystem function across a variety of ecosystems at different spatial scales. In many cases, our scientific understanding of the problem is good although the data to make quantitative predictions about the behavior of a particular system is limited. In other cases, we may be unclear about the mechanisms and processes involved and thus have a poor theoretical basis for understanding the system and predicting its behavior. The issues may be further complicated by questions of scale: we may understand the questions at the process level but have a poor basis for scaling up to larger spatial scales (e.g., Ehleringer and Field 1993).

Several scientific questions arise from consideration of management options, ecosystem dynamics, and the consequences of improper management. Research needs specific to particular problems have been identified in the preceding section discussing problems. Here, we have synthesized the research needs that have surfaced in our exploration of the interactions between management options and ecosystem processes. We lack an understanding of: (1) the determination and implementation of optimal harvesting regimes, (2) the significance of species in ecosystem processes such as the flow of material and energy, the production of biomass, and ecosystem states, and (3) the dynamics of landscape processes and ecosystem thresholds.

Optimal Biomass Removal

A major generic question that must be asked in relation to any managed ecosystem is: What is the optimal level of use? Population pressure and resource desires often place heavy demands on landscapes without consideration of the interactions among different uses and objectives, such as livestock production, tourism, agroforestry, and subsistence agriculture. The relationships between rates of removal and growth will influence the long-term sustainability of arid and semi-arid lands. The same stocking rate or harvest of species—in terms of number of stock or biomass per unit area—will have different effects depending on the weather and growth patterns of the vegetation. Although grasses are adapted to moderate herbivory and show compensatory growth after grazing, finding the optimum grazing intensity is difficult (Bartolome 1993). The problem for management is to determine when biomass removal or harvest has altered the ecosystem's ability to respond to changes in soil moisture and nutrient availability in order to produce sufficient biomass to reproduce.

It is critical to evaluate reversibility and the actions that will be required to shift an ecosystem to a desired state reflecting the desired resource outputs. Some ecosystems undergo episodic events that result in dramatic shifts in species composition, and it is crucial to increase our understanding of these shifts and their interactions with management techniques.

It is also important to develop both the capacity to determine the way vegetation management practices interact with historic systems and a basis for decisions on the level and type of restoration or reclamation that may be desired. This is a problem of particular interest in the Old World, where the long history of landscape utilization has resulted in conditions that do not reflect the inherent capacity of the land and climate.

In arid and semi-arid areas, this optimal utilization must consider the spatial and temporal environmental variability and heterogeneity. More will be said about this under landscape processes (below).

What measures can we use to define the carrying capacity for grazing or harvesting? Should these measures reflect changes in the energy flow and nutrient cycling aspects of the systems? Previous monitoring metrics were couched in terms of species composition and often focused on forage plant species. These concepts from community ecology ignore changes in the biogeochemistry or energy flow that may be critical to the system's ability to withstand stresses in arid and semi-arid lands (Milchunas and Lauenroth 1992, 1993). Although there has been much research in the area of soil erosion and degradation, there is, as yet, no agreement about the way soil characteristics could be used as indicators of soil degradation. A multiple attribute approach was recommended where attributes describing soil protection, energy flow, nutrient cycling, and the system's ability to respond to stress would be components of the monitoring system (NRC 1994). Such an approach would focus the assessment of arid and semi-arid ecosystems on the preponderance of evidence where not just one ecological factor but several are used to determine the ecosystems' condition. Managers also need a measure to use in the course of providing feedback for "adaptive management" (cf. Walters 1986), such as those developed by Tongway (1990).

Management at government levels is now moving to assess outcomes rather than proscribe inputs (e.g., to measure erosion levels rather than legislate stocking levels in new legislation in Queensland, Australia); this increases greatly the importance of these types of assessment and monitoring questions.

Functional Significance of Species

Species composition measures have long played a role in monitoring these systems. Though recent questions about their ability to reflect ecosystem-level indicators have been raised, it is perhaps more critical to understand the role of species in ecosystem functions such as energy flow, nutrient cycling, and the system's response to environmental stress. What is the significance of the loss or addition of a particular species? What definitions or metrics of species diversity reflect ecosystem sustainability? What is the role of species in the nutrient saturation potential of arid and semi-arid systems under treatments such as waste disposal? We need to know why particular weeds (undesired invaders) are important in particular areas and under some disturbance regimes. Do they play a critical role in ecosystem function? We need to be able to define the disturbance regimes that will result in elimination of the weed species (i.e., those regimes that will favor desirable species). Is biocontrol of an invading species sufficient if the role of that species in ecosystem function is not understood? What is the role of species such as trees in the movement of nutrients from deep in the soil to the surface layers? Satisfactory definitions of biodiversity must be found to provide a basis for assessing the impact of the loss of species (West 1993).

What is the interaction among soil erosion, nutrient depletion, loss of species, and reversibility? There is an adequate general theoretical understanding of soil processes but, in most cases, we do not have enough information about particular soils and the effects of management practices to make quantitative predictions. There is, in almost all cases, a need for research to determine rates of compaction, organic matter decomposition relative to accumulation, and erosion, relative to the management practices being used in particular cases, and to understand the environmental conditions (rainfall regimes, temperatures). In all cases, it is important to determine thresholds for reversibility: At what point does a process become irreversible so that the soil has suffered a permanent change in condition (NRC 1994, TGU 1995)? This is probably the most important basic question.

The questions for salinity research relate more to rates of salinity development in landscapes under particular conditions than to basic soil physics, so the requirement is for careful measurement and documentation in a range of conditions. Dryland salinization is the complex interaction of hydrologic regimes, variable precipitation, and plant growth. The length of time in which these processes occur varies. There is a need for an early warning of the extent of the dryland salinization process ongoing (Ive et al. 1992). It is possible to construct water balance and water table rise models for both irrigated and dryland situations. The models would describe the development of a salinity profile in the soil based on information about salt concentrations at the water table; they will always need to be parameterized and tested. New knowledge will be acquired in the course of developing such models. It is important that we have the ability to describe salinity profiles in soils—in terms of salt concentrations and the way they change with time—as a basis for defining saline soils in a way that will be useful for predicting salinity's effects on plant growth and ecosystem stability.

Although the effects of salinity on plant growth have been studied, our capacity to predict the effects of saline soils on growth and yield is limited. It should be possible to predict the effects of salt concentrations, integrated through root zones, on plants in a particular ecosystem, their growth rates, probability of survival, and, hence, on plant communities and biodiversity. The question of patchiness also arises here. The reversibility of the soil salinization may not be economically feasible in all cases, which indicates the need for research on salt-tolerant crops (Umali 1993).

Landscape Processes

Processes such as the distribution of water and nutrients across landscapes may cross the thresholds at which changes are irreversible. Erosion, the physical movement of soil, has been widely studied. In most cases, the problem will be to define the erosivity of the soils in a particular ecosystem and the management intensity appropriate to avoid either the loss of protective plant cover or the compaction and reduction of infiltration rates. However, the major problem here is that research is usually conducted in small plots (a few square meters) while the ecological processes occur on larger and more varied units (slope, watershed).

Depletion of nutrients may be a consequence of soil erosion, excessive biomass removal, or fire. While the mechanisms involved in each case are somewhat different though relatively well understood, the most interesting and important research questions address the redistribution of nutrients around the landscape. To quantify this, we need to describe the mechanisms (e.g., rate of dung deposition by stock; location—is it random or concentrated in particular areas; entrainment of nutrients in overland flows, etc.) and to relate these to (micro)topography and climate, taking into account spatial and temporal variations. Is there a point at which the distribution of water and nutrients across the landscape is sufficiently altered such that the ecosystem cannot recover from abiotic stresses such as drought? How does management affect this spatial and temporal distribution of nutrients and water given abiotic influences on that movement?

The causes of riparian zone degradation are often well understood, and determining the management actions needed to maintain these riparian systems at their optimum may simply require clear descriptions of the systems and levels of resource use. Prediction of the flow of waterborne nutrients requires applicable quantitative models of erosion, knowledge about the erodability of the soils over which the water is flowing, and understanding the amount of nutrient entrained in flows. While riparian systems can be studied independently of uplands, their management must nonetheless be in concert with that of uplands.

It is important to determine management's impact on processes (e.g., the distribution of water and nutrients across landscapes) and the thresholds at which changes are irreversible. Understanding how small-scale impacts affect the landscape or watershed scale is critical (Pickup 1985, Pickup et al. 1994).

Arid zone landscapes are characterized by variability. Whatever the geological origin of the soils, the high spatial and temporal variability of rainfall (Prince et al. 1990, Robertson et al. 1987, Stafford Smith and Pickup 1990) results in a mosaic of vegetation patches at a range of scales. At the scale of meters, small depressions accumulate water and provide better environments for plant growth than adjoining higher areas (Shachak and Boken 1994). The plants in small depressions act as miniature windbreaks and trap windblown dust particles that accumulate, accentuating differences in the soil surface. At larger scales, topographical irregularities cause run-on and runoff areas that result in soil moisture differences, leading to differences in plant growth. Also at larger scales, variation in the timing, amount, and intensity of rainfall causes differences in the growth of vegetation, which is largely driven by the availability of water (Le Houerou 1984). Impose on this preferential animal grazing patterns and pressures and it is clear that variability is a fundamental property of arid and semi-arid regions and that research is needed to characterize it in different regions and to understand the relationships among patchiness and vegetation productivity, biodiversity, and animal production at a range of scales (Pickup 1989, Prince et al. 1990, Stafford Smith and Pickup 1990).

Ecosystem stability is related to the issue of use and response to change, prompting questions about how to define and monitor stability and change, particularly against the variability of these arid and semi-arid environments. In this respect, remote sensing, using both satellite and aircraft-mounted sensors (Johnston and Barson 1990, Pickup 1989, Pickup et al. 1994, Stafford Smith and Pickup 1993), could be a potentially valuable tool. Remote sensing offers a means of monitoring large areas, analyzing their spatial heterogeneity, and repeating the evaluation at intervals. Remote sensing does not, as is sometimes thought, yield simple and easily interpreted results; it requires careful ground-based programs to develop its potential. It should also be combined with the application of models in geographic information systems. As part of the monitoring question, we need a clear understanding of the processes that must be monitored on arid and semi-arid lands. For example, while remote sensing can identify changes in cover (Pickup et al. 1994), can it capture soil organic losses and crust changes that are critical in the nutrient cycling of the system? There is a vital need for databases to form the basis for many landscape and regional scale studies.✝

Conclusions

The long history of use of arid and semi-arid lands suggests that the future use will continue to intensify, given available capital, technology, and knowledge. The main points that emerge are that, for ecosystems, we need to understand the interactions among the various components of the ecosystem and the influence of scale. We have relatively sound scientific knowledge of many of the ecosystem processes involved but little ability to incorporate it into an integrated context for predicting the effects of management practices on ecosystem change.

We must develop methods to determine optimum biomass harvesting regimes for specific resource production. That information will in turn lead to a range of questions about the ecosystem and the regeneration and recovery rate of its plant components, as well as about the specific biomass being harvested (e.g., herbage intake rates and preferences of the animals concerned, plants' response to weather and direct influence on the ecosystem). Land use will likely intensify globally, requiring multiple use of the landscape, testing our ability to design optimal harvesting regimes for multiple uses.

The value and importance of biodiversity must be examined in light of the role of species in ecosystem function. Metrics are needed to assess the impact of species loss or reduction on energy flow, nutrient cycling, biomass production, and the ability of the system to respond to stress.

The impact of management on processes such as the distribution of water and nutrients across landscapes needs to be determined. Understanding how small-scale impacts affect larger-scale landscapes or watersheds and whether thresholds exist at which changes are irreversible must be established. How much disturbance can an ecosystem tolerate before the effects become significant and irreversible? Arid and semi-arid ecosystems are highly variable, and this variability affects every question relating to them. The level of each research and management question must be clearly recognized in that context: Is the problem concerned with areas of hectares or square kilometers? Is it concerned with the microsites that may be of paramount importance for small organisms or with the accumulation of nutrients and water and, hence, the distribution of plant species?

With regard to management, we repeat the point made at the beginning of this chapter: management is the manipulation of a system with the objective of achieving a specified end result. This implies that managers (or policy makers) have reasonable knowledge of the way the system will respond to the changes to be imposed on it so that the results will be as desired. In some cases, management is not based solely on such scientific knowledge. Management decisions are driven by economic, sociological, and political considerations, all of which may take precedence over scientific considerations. The need is for managers to acquire, as far as possible, the scientific knowledge they must have to balance the importance of the various factors in any management decision. Scientists have the obligation to provide that knowledge in useful and usable forms, recognizing that it is incomplete and inadequate but accepting that it is the best we currently have and that ongoing and future research will improve the knowledge base.

Note

The last six authors for this chapter are listed alphabetically.

References

Anderson, D. C., K. T. Harper, and R. C. Holmgren. 1982. Factors influencing development of cryptogamic soil crust in Utah deserts. Journal of Range Management 35:180–85.

Barrow, C. J. 1991. Land degradation. Cambridge: Cambridge University Press.

Bartolome, J. W. 1993. Application of herbivore optimization theory to rangelands of the western United States. Ecological Applications 3:27–29.

Behnke, R. H., and I. Scoones. 1992. Rethinking range ecology: Implications for rangeland management in Africa. Paper No. 33. London: International Institute for Environment and Development.

Belnap, J. 1993. Recovery rates of cryptobiotic crusts: Inoculant use and assessment methods. Great Basin Naturalist 53:89–95.

Belnap, J., and K. T. Harper. 1995. Influence of cryptobiotic soil crusts on elemental content of tissue of two desert seed plants. Arid Soil Research and Rehabilitation 9:107–15.

Belsky, A. J. 1986. Population and community process in a mosaic grassland in the Serengeti, Tanzania. Journal of Ecology 74:841–56.

Ben-Asher, J. 1994. Sustainability of water resources utilization in semi-arid zones. White paper presented at the International Arid Lands Consortium workshop, Jerusalem, June.

Blackburn, W. H., R. W. Knight, and M. K. Wood. 1982. Impact of grazing on watershed: A state of knowledge. Texas Agricultural Experiment Station Publication MP-1496. College Station, Texas.

Bormann, F. H., K. R. Smith, and B. T. Bormann. 1991. Earth to hearth: A microcomputer model for comparing biofuel systems. Biomass and Energy 1:17–34.

Branson, F. A., G. F. Gifford, K. G. Renard, and R. F. Hadley. 1981. Range hydrology. Dubuque, Iowa: Kendall/Hunt.

Briske, D. D., and R. K. Heitschmidt. 1991. An ecological perspective. *In* Heitschmidt and Stuth, eds. Grazing management. 11–26.

Brown, J. R., and S. Archer. 1987. Woody plant invasion of grasslands: Establishment of honey mesquite (*Prosopis glandulosa* var. *glandulosa*) on sites differing in herbaceous biomass and grazing history. Oecologia 80:19–26.

Buffington, L. C., and C. H. Herbel. 1965. Vegetational changes on a semi-desert grassland range from 1958 to 1963. Ecological Monographs 35:139–64.

Bunting, S. C. 1987. Use of prescribed burning in juniper and pinyon-juniper woodlands. *In* Everett, ed. Proceedings, pinyon-juniper conference. 141–44.

Buol, S. W. 1995. Sustainability of soil use. Annual Review of Ecology and Systems 26:25–44.

Carter, J. O., M. P. Bolton, and D. C. Cowan. 1989. Prickly acacia: Save dollars by early control measures. Queensland Agriculture Journal 115 (Mar.–Apr.): 121–26.

Chaney, E., W. Elmore, and W. S. Platts. 1993. Livestock grazing on western riparian areas. Denver, Colo.: Environmental Protection Agency.

Charley, J. L., and N. E. West. 1975. Plant-induced soil chemical patterns in some shrub-dominated semi-desert ecosystems of Utah. Journal of Ecology 63:945–63.

Claridge, A. W., M. T. Tanton, J. H. Seebeck, S. J. Cork, and R. B. Cunningham. 1992. Establishment of ectomycorrhizae on the roots of two species of *Eucalyptus* from fungal spores contained in the faeces of the long-nosed potoroo (*Potorous tridactylis*). Australian Journal of Ecology 17:207–17.

Cook, G. D. 1994. The fate of nutrients during fires in a tropical savanna. Australian Journal of Ecology 19:359–65.

Costanza, R. 1992. Toward an operational definition of ecosystem health. *In* R. Costanza, B. G. Norton, B. D. Haskell, eds. Ecosystem health. Washington, D.C.: Island Press.

Dodd, J. L. 1994. Desertification and degradation of Africa's rangelands. Rangelands 16:180–83.

Doran, J. W., and M. R. Werner. 1990. Management and soil biology. *In* C. A. Francis, C. B. Flora, and L. D. King, eds. Sustainable agriculture in temperate zones. New York: John Wiley and Sons. 205–30.

Ehleringer, J. R., and C. B. Field, eds. 1993. Scaling physiological processes. San Diego, Calif.: Academic Press.

Ellis, J. E., and D. M. Swift. 1988. Stability of African pastoral ecosystems: Alternate paradigms and implications for development. Journal of Range Management 41:450–59.

Evans, R. D., and J. R. Ehleringer. 1993. A break in the nitrogen cycle in arid lands? Evidence from delta15N of soils. Oecologia 94:314–17.

Evenari, M., L. Shanan, and N. H. Tadmor. 1971. The Negev: The challenge of a desert. Cambridge, Mass.: Harvard University Press.

Everett, R. L. 1987a. Plant response to fire in the pinyon-juniper zone. *In* Everett, ed. Proceedings, pinyon-juniper conference. 152–57.

———, ed. 1987b. Proceedings, pinyon-juniper conference. Gen. Tech. Rep. INT-215. Ogden, Utah: USDA Forest Service, Intermountain Research Station.

Finkel, H. J. 1986. The soil erosion process. *In* H. J. Finkel, ed. Semiarid soil and water conservation. Boca Raton, Fla.: CRC Press.

Fresquez, P. R., R. E. Francis, and G. L. Dennis. 1990a. Effects of sewage sludge on soil and plant quality in a degraded semiarid grassland. Journal of Environmental Quality 19:324–59.

———. 1990b. Soil and vegetation responses to sewage sludge on a degraded semiarid broom snake weed/blue grama plant community. Journal of Range Management 43:325–31.

Friedel, M. H., G. Pickup, and D. J. Nelson. 1993. The interpretation of vegetation change in a spatially and temporally diverse arid Australian landscape. Journal of Arid Environments 24:241–60.

Frost, W. E., and E. L. Smith. 1991. Biomass productivity and range condition on range sites in southern Arizona. Journal of Range Management 44:65–67.

Gifford, G. F., F. D. Provenza, and J. C. Malechek. 1983. Impact of range goats on infiltration rates in southwestern Utah. Journal of Range Management 36:152–53.

Graetz, D. 1994. Grasslands. *In* W. B. Meyer and B. L. Turner III, eds. Changes in land use and land cover: Global perspectives. Cambridge: Cambridge University Press. 125–47.

Graetz, R. D., and D. J. Tongway. 1986. Influence of grazing management on vegetation, soil structure, and nutrient distribution and the infiltration of applied rainfall in a semi-arid chenopod shrubland. Australian Journal of Ecology 11:347–60.

Greenwood, E. A. N., L. Klein, J. D. Beresford, and G. D. Watson. 1985. Differences in annual evaporation between grazed pasture and *Eucalyptus* species in plantations on a saline farm catchment. Journal of Hydrology 78:261–78.

Griffin, G. F. 1993. The spread of buffelgrass in inland Australia: Land use conflicts. *In* Proceedings of the tenth Australian weeds conference, Brisbane, Australia, September. 501–4.

Griffin, G. F., S. R. Morton, D. M. Stafford Smith, G. E. Allan, K. Masters, and N. Preece. 1989. Status and implications of the invasion of Tamarisk (*Tamarix aphylla*) on the Finke River, Northern Territory, Australia. Journal of Environmental Management 29:297–315.

Guidin, R. 1989. An analysis of the water situation in the United States, 1989-2040. Gen. Tech. Rep. RM-177. Fort Collins, Colo.: USDA Forest Service, Rocky Mountain Forest and Range Experiment Station.

Haas, H. J., and C. E. Evans. 1957. Nitrogen and carbon changes in Great Plains soils as influenced by cropping and soil treatments. Technical Bulletin No. 1164. Washington, D.C.: Government Printing Office.

Harrison, R. B. 1986. The role of improved pastures in commercial production in the tropics and sub-tropics. Tropical Grasslands 20:3–17.

Harris-Pierce, R. L. 1994. The effect of sewage sludge application on native rangeland soils and vegetation. Master's thesis. Fort Collins: Colorado State University.

Heady, H. 1975. Range management. New York: McGraw-Hill.

Heitschmidt, R. K., and J. W. Stuth, eds. 1991. Grazing management: An ecological perspective. Portland, Oreg.: Timber Press.

Higgins, D. A., S. B. Maloney, A. R. Tiedemann, and T. M. Quigley. 1989. Storm runoff characteristics of grazed watersheds in eastern Oregon. Water Research Bulletin 25:87–100.

Hodgkinson, K. C., and G. N. Harrington. 1985. The case for prescribed burning to control shrubs in eastern semiaraid woodlands. Australian Rangelands Journal 7:64–74.

Holechek, J. L., R. D. Piepes, and C. H. Berbel. 1995. Range management. 2d ed. Englewood Cliffs, N.J.: Prentice-Hall.

Humphries, S. E., R. H. Groves, and D. S. Mitchell. 1991. Plant invasions of Australian ecosystems. ANPWS Endangered Species Program Project Report No. 58. Kowari Vol. 2. Canberra, Australia: Australia National Parks and Wildlife Service.

Ive, J. R., P. A. Walker, and K. D. Cocks. 1992. Spatial modelling of dryland salinization potential in Victoria, Australia. Land Degradation and Rehabilitation 3:27–36.

Johansen, J. R. 1986. Importance of cryptogamic soil crusts to arid rangelands: Implications for short duration grazing. *In* J. A. Tidman, ed. Short duration grazing. Pullman: Washington State University Press.

Johnson, H. B., and H. S. Mayeux. 1992. Viewpoint: A view on species additions and deletions and the balance of nature. Journal of Range Management 45:322–33.

Johnston, R. M., and M. M. Barson. 1990. An assessment of the use of remote sensing techniques in land degradation studies. Bureau of Rural Resources Bulletin No. 5. Canberra, Australia: Australia Government Publishing Service.

Joyce, L. A. 1993. The life cycle of the range condition concept. Journal of Range Management 46:132–38.

King, L. D. 1990. Sustainable soil fertility practices. *In* C. A. Francis, C. B. Flora, and L. D. King, eds. Sustainable agriculture in temperate zones. New York: John Wiley and Sons. 144–77.

Lajtha, K., and W. H. Schlesinger. 1986. Plant response to variations in nitrogen availability in a desert shrubland community. Biogeochemistry 2:29–37.

Le Houerou, H. N. 1984. Rain use efficiency: A unifying concept in arid land ecology. Journal of Arid Environments 7:213–34.

Lura, C. L., and P. E. Nyren. 1992. Some effects of a white grub infestation on northern mixed-grass prairie. Journal of Range Management 45:352–54.

Maldonado, A. L. J. 1985. Manejo de la cubierta vegetal en las zonas aridas del noreste de Mexico. Primera reunion regional de estados fronterizos del Rio Bravo sobre parques y vida silvestre. El Paso, Texas. Mimeo.

Manzanilla, H. 1994. Toward the sustainable management of forest species from arid and semiarid zones of Mexico. White paper presented at International Arid Lands Consortium workshop, Jerusalem, June.

Margules, C. R. 1989. Introduction to some Australian developments in conservation evaluation. Biological Conservation 50:1–11.

Marx, E. 1967. The Bedouin of the Negev. Manchester: Manchester University Press.

Matson, P., and P. Vitousek. 1981. Nitrogen mineralization and nitrification potentials following clearcutting in the Hoosier National Forest, Indiana. Forest Science 4:781–91.

McNaughton, S. J. 1994. Biodiversity and function of grazing in ecosystems. *In* Schulze and Mooney, eds. Biodiversity and ecosystem function. 361–84.

Mercer, D. E., and J. Soussan. 1992. Fuelwood problems and solutions. *In* N. P. Sharma, ed. Managing the world's forests. Dubuque, Iowa: Kendall/Hunt. 117–214.

Meyer, O. 1994. Functional groups of microorganisms. *In* Schulze and Mooney, eds. Biodiversity and ecosystem function. 67–96.

Milchunas, D. G., and W. K. Lauenroth. 1992. Carbon dynamics and estimates of primary production by harvest, 14C, and 14C turnover. Ecology 73:593–607.

———. 1993. Quantitative effects of grazing on vegetation and soils over a global range of environments. Ecological Monographs 63:327–66.

Minnich, R. A. 1983. Fire mosaics in southern California and northern Baja California. Science 219:1287–94.

Morton, S. R. 1990. The impact of European settlement on the vertebrate animals of arid Australia: A conceptual model. Proceedings of the Ecological Society of Australia 16:201–13.

Morton, S. R., D. M. Stafford Smith, M. H. Friedel, G. F. Griffin, and G. Pickup. 1995. The stewardship of arid Australia: Ecology and landscape management. Journal of Environmental Management 43:195–217.

National Research Council, Board on Agriculture, Committee on Rangeland Classification (NRC). 1994. Rangeland health. Washington, D.C.: National Academy Press.

National Research Council, Board on Agriculture, Committee on the Role of Alternative Farming Methods in Modern Production Agriculture (NRC). 1989. Alternative agriculture. Washington, D.C.: National Academy Press.

Noble, J. C., P. Clark, and S. J. Cork. 1994. Mesoscale patterning induced by past activities of the burrowing bettong *(Bettongia lesueur)* across arid Australia. *In* Proceedings of the Australian Range Society, eighth biennial conference. 159–60.

Noy-Meir, I. 1985. Desert ecosystem structure and function. *In* M. Evenari, I. Noy-Meir, and D. W. Goodall, eds. Ecosystems of the world. Vol. 12A: Hot deserts and arid shrublands. Amsterdam: Elsevier. 93–103.

Nyhan, J. W. 1989. Development of technology for the long-term stabilization and closure of shallow land burial sites in semi-arid environments. Los Alamos National Laboratory Report LA-11283-MS. Los Alamos, N.Mex.

Nyhan, J. W., and F. Barnes. 1989. Development of a prototype plan for the effective closure of a waste disposal site in Los Alamos, New Mexico. Los Alamos National Lab Report LA-11282-MS. Los Alamos, N.Mex.

Perevolotsky, A. 1987. Territoriality and resource sharing among Bedouin of southern Jinai: A socio-ecological interpretation. Journal of Arid Environments 13:153–61.

Perevolotsky, A., and D. Baharav. 1991. The distribution of "desert kits" in eastern Sinai and sub-regional carrying capacity. Journal of Arid Environments 20:239–49.

Pickup, G. 1985. The erosion cell: The geomorphic approach to landscape classification in range assessment. Australian Rangelands Journal 7:114–21.

———. 1989. New land degradation survey techniques for arid Australia: Problems and prospects. Australian Rangelands Journal 11:74–82.

———. 1991. Event frequency and landscape stability on the floodplain systems of arid central Australia. Quantitative Science Review 10:463–73.

Pickup, G., G. N. Bastin, and V. H. Chewings. 1994. Remote-sensing-based condition assessment for nonequilibrium rangelands under large-scale commercial grazing. Ecological Applications 4:497–517.

Prince, S. D., C. O. Justice, and S. O. Los. 1990. Remote sensing of the Sahelian environment. Commission of the European Communities. Technical Centre for Agricultural and Rural Cooperation. Brussels, Belgium.

Purvis, J. R. 1986. Nurture the land: My philosophies of pastoral management in central Australia. Australian Rangelands Journal 8:110–17.

Reid, J., and M. Fleming. 1992. The conservation status of birds in arid Australia. Australian Rangelands Journal 14:65–91.

Renard, K. G. 1987. Present and future erosion prediction tools for use in pinyon-juniper communities. *In* Everett, ed. Proceedings, pinyon-juniper conference. 502–12.

Robertson, G., J. Short, and G. Wellard. 1987. The environment of the Australian sheep rangelands. *In* G. Caughley, N. Shepherd, and J. Short, eds. Kangaroos: Their ecology and management in the sheep rangelands of Australia. Cambridge: Cambridge Studies in Applied Ecology and Resource Management. 14–34.

Scanlan, J. C. 1992. A model of woody-herbaceous biomass relationships in eucalypt and mesquite communities. Journal of Range Management 45:75–80.

Scanlan, J. C., and W. H. Burrows. 1990. Woody overstory impact on herbaceous understory in *Eucalyptus* spp. communities in central Queensland. Australian Journal of Ecology 15:191–97.

Schlesinger, W. H., J. F. Reynolds, G. L. Cunningham, L. F. Huenneke, W. M. Jarrell, R. A. Virginia, and W. G. Whitford. 1990. Biological feedbacks in global desertification. Science 247:1043–48.

School, D. G. 1989. Soil compaction from cattle trampling on a semi-arid watershed in northwest New Mexico. New Mexico Journal of Science 29:105–12.

Schulze, E. D., and H. A. Mooney, eds. 1994. Biodiversity and ecosystem function. Berlin: Springer-Verlag.

Seligman, N. G., and A. Perevolotsky. 1994. Has intensive grazing by domestic livestock degraded Mediterranean Basin rangelands? *In* M. Arianoutsou and R. H. Groves, eds. Plant-animal interactions in Mediterranean-type ecosystems. Dordrecht, The Netherlands: Kluwer Academic Publishers.

Senft, R. L., M. B. Coughenour, D. W. Bailey, L. R. Rittenhouse, O. E. Sala, and D. M. Swift. 1987. Large herbivore foraging and ecological hierarchies: Landscape ecology can enhance traditional foraging theory. BioScience 37:789–99.

Severson, K. E., and L. F. DeBano. 1991. Influence of Spanish goats on vegetation and soils in Arizona chaparral. Journal of Range Management 44:111–17.

Shachak, M., and B. Boken. 1994. Desert plant communities in human-made patches: Implications for management. Ecological Applications 4:702–16.

Skarpe, C. 1986. Plant community structure in relation to grazing and environmental changes along a north-south transect in the western Kalahari. Vegetatio 68:3–18.

Sopper, W. E. 1993. Municipal sludge use in land reclamation. Boca Raton, Fla.: Lewis Publishers.

Stafford Smith, D. M. 1994. A regional framework for managing the variability of production in the rangelands of Australia. CSIRO/RIRDC Project Report No. 1. Alice Springs, Australia: Commonwealth Scientific and Industrial Research Organization.

Stafford Smith, D. M., and G. Pickup. 1990. Pattern and production in arid lands. Proceedings of the Ecological Society of Australia 16:195–200.

———. 1993. Out of Africa, looking in: Understanding vegetation change and its implications for management in Australian rangelands. *In* R. H. Behnke and I. Scoones, eds. Rethinking range ecology: Implications for rangeland management in Africa. London: Commonwealth Secretariat, Overseas Development Instititue and International Institute for Environment and Development. 196–226.

Stein, R. A., and J. A. Ludwig. 1979. Vegetation and soil patterns on a Chihuahuan desert bajada. American Midland Naturalist 101:28–37.

Stuth, J. W. 1991. Foraging behavior. *In* Heitschmidt and Stuth, eds. Grazing management. 65–83.

Stuth, J. W., J. R. Conner, and R. K. Heitschmidt. 1991. The decision-making environment and planning paradigm. *In* Heitschmidt and Stuth, eds. Grazing management. 201–24.

Takar, A. A., J. P. Dobrowolski, and T. L. Thurow. 1990. Influence of grazing, vegetation life-form, and soil type on infiltration rates and interrill erosion on a Somalion rangeland. Journal of Range Management 43:486–90.

Task Group on Unity (TGU) in Concepts and Terminology Committee Members. 1995. New concepts for assessment of rangeland condition. Journal of Range Management 48:271–82.

Taylor, J. A., D. A. Hedges, and R. D. B. Whalley. 1985. Effects of fertilizer and grazing sheep on pasture heterogeneity in a small-scale grazing experiment. Australian Journal of Agricultural Research 36:316–25.

Terry, R. E., and S. J. Burns. 1987. Nitrogen fixation in cryptogamic soil crusts as affected by disturbance. *In* Everett, ed. Proceedings, pinyon-juniper conference. 387–90.

Thran, D. F., and R. L. Everett. 1987. Impact of tree harvest on soil nutrients accumulated under single-leaf pinyon. *In* Everett, ed. Proceedings, pinyon-juniper conference. 387–90.

Tolsma, D. J., W. H. O. Ernst, and R. A. Verwey. 1987. Nutrients in soil and vegetation around two artificial waterpoints in eastern Botswana. Journal of Applied Ecology 24:991–1000.

Tongway, D. J. 1990. Soil and landscape processes in the restoration of rangelands. Australian Rangelands Journal 12:54–57.

Umali, Dina L. 1993. Irrigation-induced salinity. World Bank Technical Paper No. 215. Washington, D.C.: World Bank.

Vallentine, John F. 1990. Grazing management. San Diego, Calif.: Academic Press.

Vitousek, P. M. 1990. Biological invasions and ecosystem processes: Towards an integration of population biology and ecosystem studies. Oikos 57:7–13.

Walker, B. H. 1985. Structure and function of savannas: An overview. *In* C. Tothill and J. J. Mott, eds. Ecology and management of the world's savannas. Canberra: Australian Academy of Sciences. 83–92.

Walker, B. H., and W. T. Knoop. 1987. The response of the herbaceous layer in a dystrophic *Burkea africana* savanna to increased levels of nitrogen, phosphate, and potassium. Journal of Grassland Society of Africa 4:31–34.

Walker, J., J. A. Robertson, L. K. Penridge, and P. J. H. Sharpe. 1986. Herbage response to tree thinning in a *Eucalyptus crebra* woodland. Australian Journal of Ecology 1:135–40.

Walker, J., P. J. H. Sharpe, K. K. Penridge, and H. Wu. 1989. Ecological field theory: The concept and field tests. Vegetatio 83:81–95.

Walters, C. 1986. Adaptive management of renewable resources. New York: Macmillan.

Weltz, M., K. M. Wood, and E. E. Parker. 1989. Flash grazing and trampling: Effects on infiltration rates and sediment yield on a selected New Mexico range site. Journal of Agricultural Environment 16:95–100.

Werner, P. A. 1991. Savanna ecology and management. Oxford: Basil Blackwell.

West, N. 1993. Biodiversity of rangelands. Journal of Rangeland Management 46:2–13.

Westoby, M., B. Walker, and I. Noy-Meir. 1989. Opportunistic management for rangelands not at equilibrium. Journal of Range Management 42:266–74.

Wilcox, B. P., M. K. Wood, and J. M. Tromble. 1988. Factors influencing infiltrability of semi-arid mountain slopes. Journal of Range Management 41:197–206.

Willatt, S. T., and D. M. Pullar. 1983. Changes in soil physical properties under grazed pastures. Australian Journal of Soil Research 22:343–48.

Williams, M. 1994. Forest and tree cover. *In* W. B. Meyer and B. L. Turner II, eds. Changes in land use and land cover: A global perspective. Cambridge: Cambridge University Press. 97–124.

Yair, A., and M. Shachak. 1987. Studies in watershed ecology of an arid area. *In* L. Berkofsky and M. G. Wurtele, eds. Progress in desert research. Totawa, N.J.: Rowman and Littlefield.

Zaady, E., Y. Okon, and A. Perevolotsky. 1994. Growth response of Mediterranean herbaceous swards to inoculation with *Azospicillum brasilense.* Journal of Range Management 47:12–15.

Zaban, H. 1981. A study to determine the optimal rainfed land-use systems in a semi-arid region of Israel. Ph.D. diss. University of Reading, United Kingdom.

Zobisch, M. A. 1993. Erosion susceptibility and soil loss on grazing lands in some semiarid and subhumid locations of eastern Kenya. Journal of Soil and Water Conservation 48:445–48.

8 The Concept of Sustainability in Dryland Ecosystems

Uriel N. Safriel

This chapter is structured as follows. First, the treatment of the concept of sustainability in dryland ecosystems requires definitions of *ecosystem, drylands,* and *sustainability* that are relevant to the context in which these three terms are joined together in the title of the chapter. These definitions then serve to develop a generalized and rather speculative model—or working hypothesis—of the structure and function of dryland ecosystems. Next, the chapter describes the current, apparently nonsustainable development of these ecosystems. Finally, it discusses whether lessons from the proposed model of the structure and function of dryland ecosystems can serve as guidelines for management for sustainability of these ecosystems when they come under development.

The Structure and Function of Dryland Ecosystems

What Are Ecosystems and What Are Drylands?

Ecosystems As an entity, an ecosystem is viewed by biologists as a community with its environment and by nonbiologists as the environment with its biota. But an ecosystem encompasses more than its physical entity. It includes the concept of rates and reservoirs that affect the biota and, to a larger or smaller extent, are controlled by the biota. This encompasses the rates of mineral recycling, as driven by the rates of energy flows, and the sizes of the mineral reservoir in the environment and in the biota.

For terrestrial ecosystems, the most relevant reservoirs are in the soil. The terrestrial biota is involved in regulating the soil energy transfer rates. For dryland ecosystems, additional flow and reservoir are significant: those of water originating as precipitation, stored rather temporarily in a soil reservoir, and eventually flowing back to the atmosphere. The flows of water to and from the soil are often regulated by the biota. A significant amount of the soil water reservoir leaves it indirectly through the biota itself (fig. 8.1).

From the biological viewpoint, an ecosystem is a production enterprise whereby the biota is the machine and the product is used to maintain, replace, and often increase the machine. This machine appropriates from the environment raw materials and energy for the production process and, in doing so, regulates the sizes of the material reservoirs and the flow rates of energy and matter.

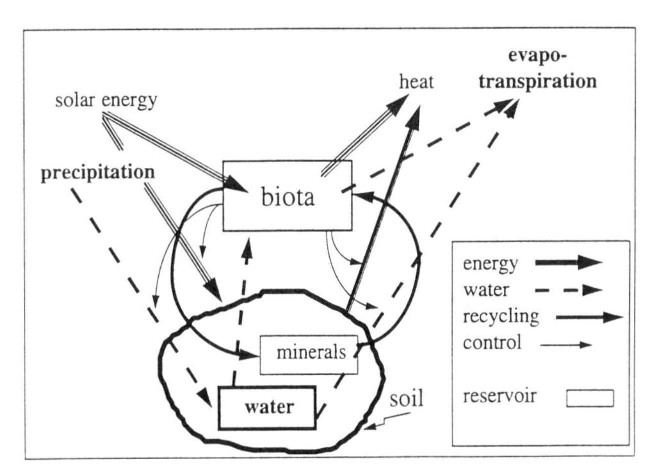

Figure 8.1. The concept of a terrestrial ecosystem—biota and soil reservoirs of minerals and water, energy and water flows, mineral recycling, and the control by the biota of rates of flow and mineral recycling.

Drylands Using production terms, drylands can be defined as ecosystems whose production is limited when water is lost to the atmosphere in the production process at rates that are faster than rates of replacement. Though water constitutes raw material that is chemically transformed in the production process of organic matter, most of the water required by the biota serves for transport and also constitutes a significant building block of biomass. The water contents of the biomass are at high risk of being lost during the process of producing the organic component. Primarily, therefore, dryland production is not limited by a production resource. In contrast, nondryland ecosystems' production is primarily limited by raw materials (i.e., soil minerals) and/or by scramble competition for the energy resource (light). Under some circumstances, dryland production is limited by nutrients, but this limitation, as will be discussed below, is also controlled by water scarcity.

The Range of Dryland Ecosystems

The Aridity Index Water for use of the primary producers is drawn from the soil reservoir. The input to this reservoir is due to precipitation and the output to evaporation. Both output and input are controlled by several other ecosystem mechanisms. Drylands feature low precipitation and high evaporation, and the latter also accelerates the rate of water loss associated with the appropriation of atmospheric carbon in the production process. Devising indices for aridity is a traditional preoccupation of earth scientists. One of these indices has recently gained political significance and a legal stand. This is the ratio of mean annual precipitation to potential evapotranspiration, calculated by a modified Thornthwaite method, using mean monthly temperatures and the average monthly number of daylight hours. The U.N. Environment Program defined areas with an aridity index of < 0.65 as drylands (UNEP 1992); this definition guides legally binding international documents such as the U.N. Convention to Combat Desertification (UNEP 1995).

Categories and Dimensions of Dryland Ecosystems Based on data collected during thirty years (1951–80) on precipitation at 3,758 stations and temperatures at 1,834 stations around the globe, the aridity index was calculated for cells of 0.5° across the globe (UNEP 1992). Dryland ecosystems under the above definition comprise 47.1% of the global land area. These are split into four categories of ranges of the aridity index.

Hyperarid drylands receive precipitation that is less than 5% of potential evapotranspiration. They comprise 7.5% of global land area (mostly the Saharo-Arabian Desert) and often have no seasonal rains, even experiencing years with no rain. *Arid* drylands are those with an aridity index between 5–20%, which is up to 200 mm in winter rainfall areas and 300 mm in summer rainfall areas, and 50–100% variability. They comprise 12.1% of global land area, mostly the Asian and Australian drylands. *Semi-arid* drylands are an aridity index of 20–50%, which is up to 800 mm in summer rainfall areas and 500 mm in winter rainfall areas, and variability of 25–50%. They comprise 17.7% of global land area, mostly in North America and Central Asia. Finally, there is a transitional type of land, the *dry-subhumid* category, with an aridity index of 50–65% and less than 25% variability in rainfall. Dry-subhumid drylands comprise 9.9% of the land and occur usually as a narrow belt between the semi-arid and the humid lands either in the boreal or the tropic regions.

No attempt has been made to relate this "physical" definition of dryland ecosystems to the above proposed production-related "ecological" definition. The question of whether the shift in the factor limiting production from water to mineral or light occurs above the 0.65 precipitation to potential evapotranspiration ratio or below remains open. This is especially valid given the ongoing debate on the measurability and, hence, the utility of the Thornthwaite potential evapotranspiration evaluation (Nicholson chap. 4, this volume).

The Dimensions of Drylands Production

The Low Overall Production Dryland ecosystems have relatively low production as compared to nondryland ecosystems. This does not derive logically from the fact that what limits production of dryland ecosystems differs from what limits production in other ecosystems. It is because dryland ecosystems are not limited by a production resource, either raw material or energy, but by the fact that, when capturing the major raw material for production (carbon dioxide), the producers lose water that cannot be amply replaced. Therefore, resources for production remain unutilized and are not converted to organic mass. The relations between production and the aridity index have not been explicitly demonstrated, but, though the shape of the function is not known, it is safe to state that production falls rather concomitantly with aridity.

The Low- and High-Cover Dryland Ecosystems Note that primary production can be expressed by the dimensions of the individual plant, best described by its height and overall surface cover. These two distinct expressions of plant production seem to differ in their response to the aridity gradient. The average height of the vegetation in the dry-subhumid dryland category is higher than in the semi-arid category. But the drop in cover is much smaller. A sharp drop in cover occurs in the transition from the semi-arid to the arid category. This is probably because, at that threshold of low soil water, further reduction in height will result in aboveground biomass that cannot produce the large amount of belowground biomass necessary for the appropriation of the low amount of soil water (Shmida et al. 1986). The belowground biomass increases, but this results in spacing out of the individual plants such that cover is dramatically reduced. Interestingly, the average height of the vegetation increases with a further increase in aridity for reasons that will be elaborated later. Finally, in the transition to the hyperarid dryland, cover further declines steeply though height in the arid category may be maintained (fig. 8.2).

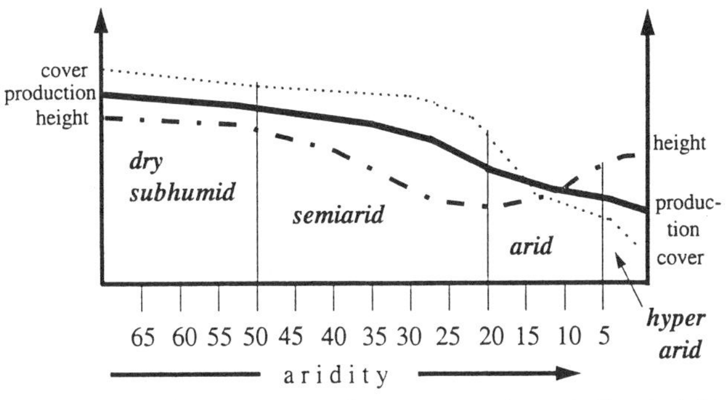

Figure 8.2. Production, vegetation cover, and vegetation height across the aridity gradient (schematic based on observations from Israel; numbers are values of aridity index in percentages).

A Conceptual Model of Low-Cover Dryland Ecosystems

Low Soil Water and Low Production: A Negative Feedback? The significant transition with respect to plant cover generates a distinction between the high-cover dry-subhumid and semi-arid drylands and the low-cover arid and hyperarid drylands. This difference in cover dominates the very different ecosystem functioning of these two types of drylands. In the following, a conceptual model (i.e., a working hypothesis) for the low-cover dryland ecosystems will be presented. The low plant cover increases evaporation from the exposed soil surface. It also increases exposure to wind and reduces rainfall infiltration rates, generating surface runoff. The dry topsoil is then readily eroded by wind and water. Low production also reduces litter cover on the surface, further increasing evaporation and reducing infiltration. More importantly, low production is also expressed in low soil organic matter. This reduces the soil water storage capacity and, especially, the aggregate stability of the soil. This often contributes most to the erodibility of the soil by wind and water forces (Williams and Balling 1994). Due to the low infiltration, most organic matter and nutrient minerals concentrate in the topsoil. Loss of topsoil, therefore, further reduces not only the availability of water but also organic matter and mineral contents. Thus, in this model, low production reinforces further reduction of soil water and even the availability of raw materials for production such as soil minerals (fig. 8.3).

Low Rainfall and Low Soil Cover: A Reinforcing Negative Feedback? The low plant cover results in higher albedo, which generates low surface and near-surface temperatures. This stabilizes the atmospheric column resulting in cooler atmosphere and lower subsiding atmospheric column, which ultimately reduces local and even regional precipitation (Charney et al. 1975, Otterman 1974), provided there is sufficient air humidity (Schlesinger et al. 1990). Thus, two nested negative feedback loops are envisioned by this model of low-cover dryland ecosystems (fig. 8.4). Obviously, under the circumstances described in this model, low-cover dry-

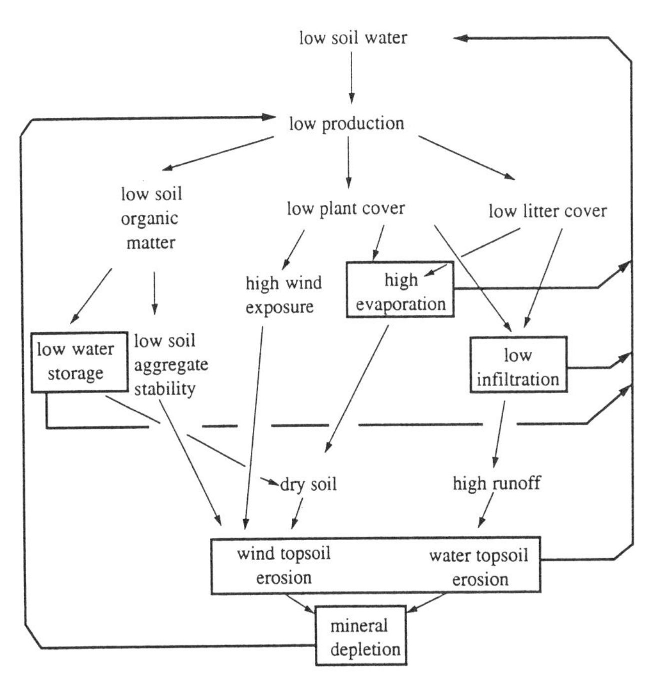

Figure 8.3. A conceptual model of arid ecosystems suffering from soil erosion due to a negative feedback generated by low soil water. Boxes and bold arrows describe the feedback loop.

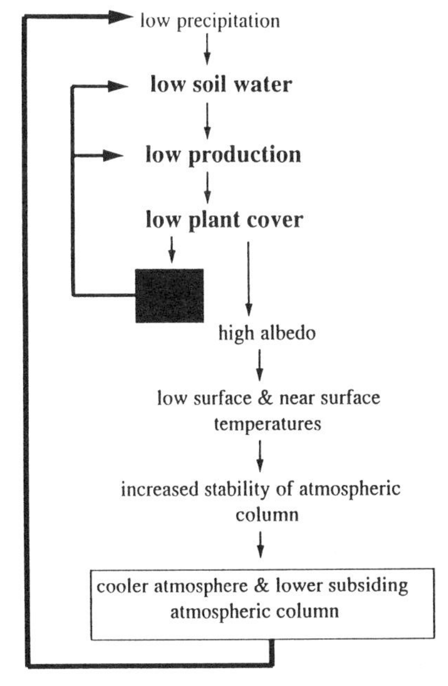

Figure 8.4. The model of fig. 8.3 (bold, box represents all processes of fig. 8.3), reinforced by an additional negative feedback loop generated by the feedback effect of albedo on low precipitation.

land ecosystems cannot be sustained for long. Thus, at this point it is necessary to define the third term in the title of this chapter, sustainability.

Sustainability

Use and Sustainability Sustainability is defined in terms of use. Use is an appropriation of ecosystem components by organisms for an ultimate production of biomass. Sustainability is the long-term continuation of use. Sustainable use is, therefore, use in ways that allow natural processes to replace what is used (Mangel et al. 1993). Ecosystems are biomass production systems made of organisms, materials, and energy. They can function and, hence, exist as long as what is used by their organisms is indeed replaced—materials by recycling, energy and water by flow. Given an appropriate space and time scale, ecosystems are, therefore, everywhere and always sustainable by definition: when materials are not recycled, there is no use for the energy; when energy flow ceases, recycling cannot be driven. Either way, what is used is not replaced, in which case use is not sustainable, production terminates, and the ecosystem eventually ceases to function and hence to exist.

The Nonsustainability of the Proposed Model for Low-Cover Drylands In dryland ecosystems, sustainability also, or even mostly, means replacement of soil water. Unlike light energy, the water flow from clouds to soil surface does not necessarily guarantee replacement of soil water. Due to its negative feedbacks, the dryland ecosystem model is one of ecological nonsustainability leading to ecosystem collapse. Since dryland ecosystems persist and are sustainable, the model is either wrong or incomplete.

Restoring Sustainability at the Level of Rainfall: Soil Cover Feedback? An alternative model that generates a positive feedback between low-cover and precipitation can be proposed (fig. 8.5). Low soil water and low plant cover decrease the radiant energy used otherwise to evaporate and transpire water. This increases surface and near-surface temperatures, which is what reduces the stability of the atmospheric column and creates convection and thermals, bringing more rain (Jackson and Idso 1975). The two nested feedback loops have different signs and can thus stabilize the ecosystem, resulting in sustainability (fig. 8.5). Both mechanisms can prevail simultaneously, and the end result—namely, whether the external feedback loop is ultimately negative or positive—depends on which of the two mechanisms, soil moisture- or albedo-induced, overshadows the other (Williams and Balling 1994). Thus, the outer loop may not suffice for counteracting the inner negative feedback loop, and sustainability of the low-cover dryland ecosystems is not necessarily guaranteed.

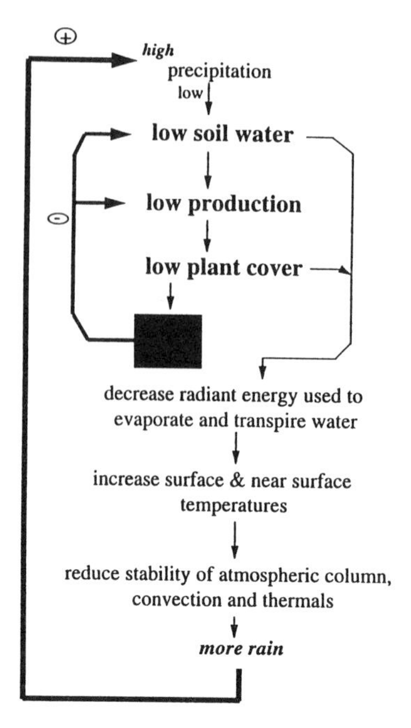

Figure 8.5. The model of fig. 8.4 (bold, box represents all processes of fig. 8.4), with positive feedback loop between low cover and precipitation, restoring sustainability to the dryland ecosystem.

A Nitrogen Submodel of Low-Cover Dryland Ecosystems

The Loss of Nitrogen in Reduced Form The issue of the sustainability of low-cover dryland ecosystems is further complicated by the bottom component of the overall ecosystem model (fig. 8.3): mineral depletion due to topsoil erosion. A specific mineral depletion hypothesis—that is, a nitrogen submodel of the above model of low-cover dryland ecosystems—is proposed in fig. 8.6. To reiterate, the general model assumes the following train of processes: low rainfall, low soil water, low production, low plant and litter cover, low infiltration, dry soil, and high runoff, finally leading to topsoil erosion. In the proposed submodel (fig. 8.6), topsoil erosion leads to high calcium carbonate concentrations exposed at the surface, hence an increase of soil pH. This transforms ammonium ions to volatile ammonia. The volatilization of ammonia to the atmosphere is further accelerated by the dry soil. Ammonia has a short lifetime in the atmosphere, but, due to low rainfall, the local redeposition of ammonia is low and most is lost by long-range transport in the atmosphere (Schlesinger et al. 1990).

The Loss of Nitrogen in Oxidized Form Whereas nitrogen is lost in its reduced form mainly due to abiotic processes, nitrogen in oxidized form is lost due to biotic processes. Due to the low production, the uptake of nitrates by the vegetation is low and can accumulate during the dry period to become readily available to denitrifying bacteria. Runoff

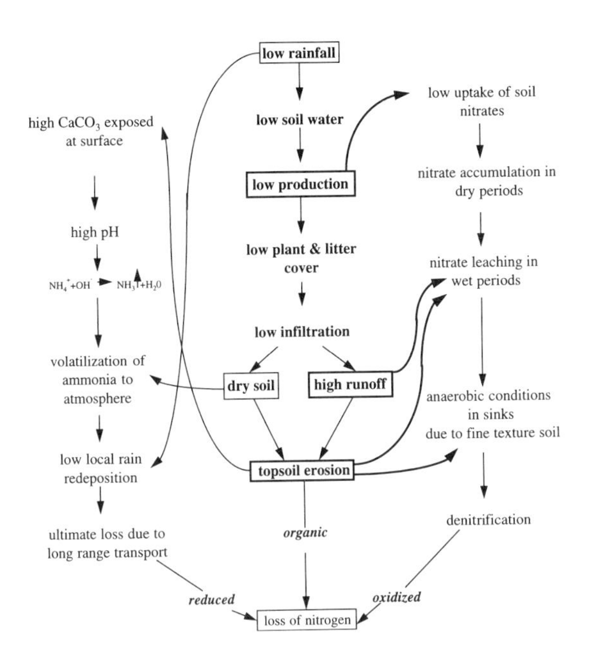

Figure 8.6. The nitrogen submodel of low-cover dryland ecosystems, showing loss of organic matter and of reduced and oxidized nitrogen due to low rainfall.

and soil erosion leach the nitrates to sinks where anaerobic conditions develop mainly due to the accumulation of fine-texture transported soil. Nitrogen is then lost in the biological process of denitrification (Schlesinger et al. 1990). To conclude, nitrogen is constantly being lost in its organic reduced and oxidized forms. Hence, low-cover dryland ecosystems are nonsustainable not only due to the negative feedbacks of soil water, production, precipitation, and plant cover but also due to the constant depletion of nitrogen (fig. 8.6).

Restoring Sustainability to the Nitrogen Submodel The loss of nitrogen can be somewhat counterbalanced, considering the possibility that, due to the high albedo of these drylands, the soil is cool, thus reducing the loss of nitrogen (Schlesinger et al. 1990). If, however, cooling is not obtained due to the prevalence of relatively low evapotranspiration, the soil becomes warmer, exacerbating nitrogen loss. However, the sustained loss of nitrogen need not make the low-cover dryland ecosystems nonsustainable as long as the fraction appropriated by the biota is replenished. For example, the biota may be limited just by water availability, as implicit in the proposed ecological definition of dryland ecosystems, and thus unutilized nitrogen is always left over. This is, however, unlikely given two biological observations: (1) the extremely high nutrient utilization efficiency demonstrated by some arid drylands plants and (2) the high incidence of nitrogen-fixing organisms in these ecosystems (Schlesinger et al. 1990). To conclude, the overall ecosystem model just presented, and its nitrogen submodel, fail to resolve the issue of sustainability of low-cover dryland ecosystems.

Restoring Sustainability: The Source-Sink Spatiotemporal Dynamics Model

The Spatial Dimension

The Source-Sink Mosaic As for most ecosystems, the appropriate description of dryland ecosystems by models depends on scale (O'Neill et al. 1986). The negligence of scale in constructing the model just described may account for its inherent nonsustainability. To include scale, a model that depicts low-cover dryland ecosystems as spatially heterogeneous will be now described. In the proposed model, the spatial heterogeneity is expressed as a mosaic of patches of two types: patches that serve as source and patches that function as sinks, for water, soil, and minerals. Furthermore, this mosaic is structured in a hierarchical scale (fig. 8.7) translated into channels of hierarchical order within a watershed. Thus, the dryland ecosystem model that describes a spatially uniform ecosystem at a large scale may, in fact, describe a spatially heterogeneous ecosystem at the smallest scale, and there we find only the source patch. For the sink patch in this mosaic, a different model is proposed.

The Sink Model This model is initialized by high soil water in the sink patch, because direct precipitation is generously

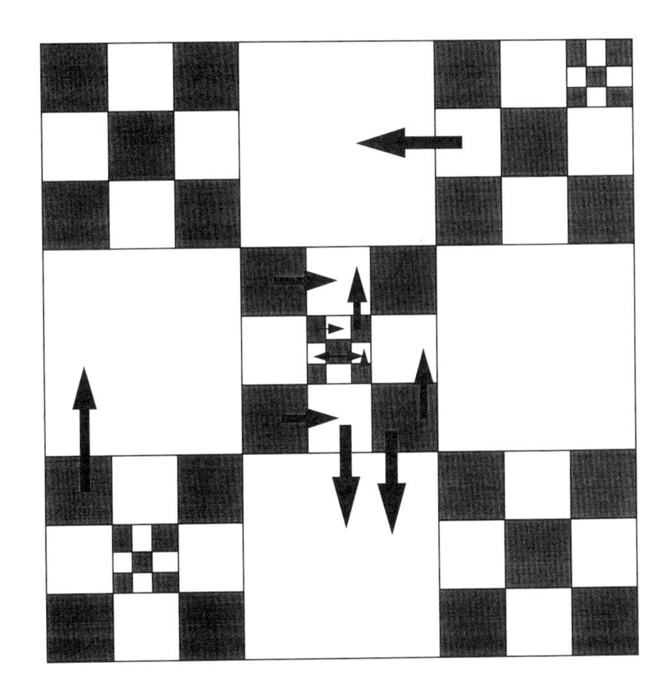

Figure 8.7. A schematic representation of the hierarchical mosaic of a dryland ecosystem landscape. Arrows stand for water flows in a watershed, from sources to sinks (see text).

augmented by the contribution of the source-generated runoff. The resulting high soil water contributes to high production. This promotes plant cover that protects the soil from wind. It also generates high litter cover and high soil organic matter that jointly increase infiltration and reduce runoff, resulting in soil conservation. Furthermore, plant cover and litter reduce evaporation, so reinforcement of soil water contents may occur—a positive feedback. The source-generated runoff also carries some soil and litter. This litter is later protected from wind when the sink's plant cover acts as a windbreak. It is likely that the high production appropriates nitrates quickly, thus outcompeting denitrifying bacteria even if at times anaerobic conditions may prevail in the sink. Finally, for many small animals, the vegetation of the sink provides refuge from predation and shelter from heat. But these animals also forage in source patches, some carry food items from source to sink patches (e.g., tenebrionid beetles; E. Gruner and Y. Ayal personal communication), and all deposit feces while sheltering in the sinks. Thus, there is a significant nutrient enrichment of the sink that may enhance production should water availability in the patch be so high that minerals become the limiting factor (fig. 8.8).

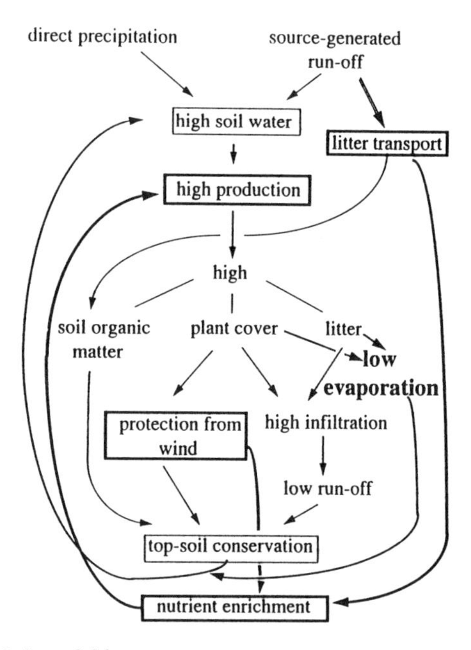

Figure 8.8. A model for processes in the sink patch. Soil water is high because the sink patch receives runoff water generated by the source patch.

The Replenishment of the Source At the scale of the whole ecosystem, energy and water flow through and nutrients are recycled within. At the scale of patches, the source contributes water and materials to the sink. Rainfall replenishes the source. Carbon losses to the sink in the form of litter and organic matter are replenished by photosynthesis of the source's primary producers. But the outflow of nitrogen must be replenished by an inflow in order to attain sustainability, so that at least what is used by the source's biota is replenished. Two mechanisms for replenishing minerals in the source are envisaged (fig. 8.9). One is atmospheric dust, transported by wind and deposited by wind

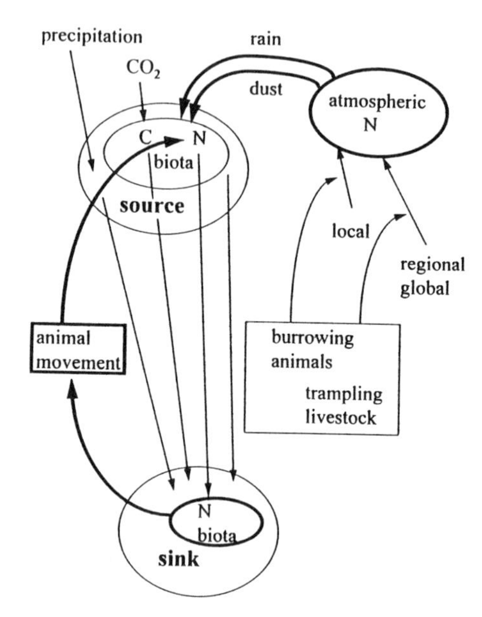

Figure 8.9. The joint source-sink patch model. The nitrogen replenishment of the source patch is in bold lines.

and rain (Jones and Shachak 1990). The source of this dust can be local, regional, or global. Its generation can be prompted by animals, either through trampling by large ones or through burrowing by small ones. (Burrowing organisms are common in dryland ecosystems for reasons that will be discussed later.) The second mechanism is provided by animals that perceive the environment at a larger scale (or coarser grain) and move among patches, so they may feed in sinks and deposit droppings in the source patches. It should be noted that whereas the movement of nitrogen from source to sink is downslope, the replenishment can be effected by animals moving the nitrogen upslope.

The Generation of Patchiness: Dynamics of the Source-Sink Patches To answer the question of what generates the proposed mosaic of patches, it is necessary to address the issue of what makes the source a source. In other words, what contributes to the low soil infiltration of the source patch? Two different factors contribute to the low infiltration: hard surfaces, such as rocks, and compacted or crusted soils (fig. 8.10). The crust can also be biogenic (Boeken and Shachak 1994), thus fixing carbon and possibly also nitrogen (Jones and Shachak 1990).

The spatial distribution of nonbiological hard surfaces is mostly determined by geomorphology. Where soil cover on

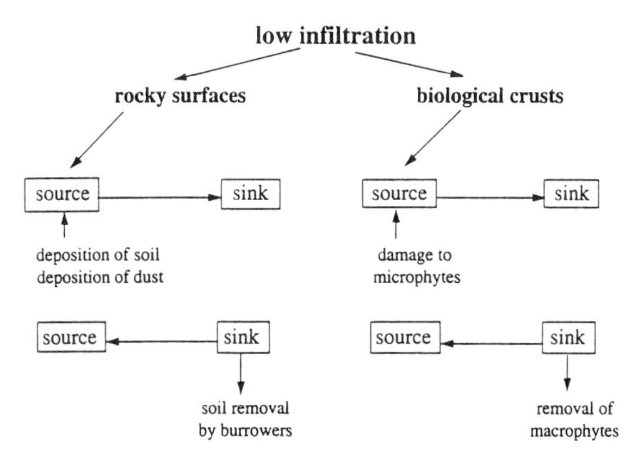

Figure 8.10. The two causes of low infiltration that result in a source patch, the processes that transform a source patch into a sink patch, and vice versa.

rocky surfaces is relatively shallow, the spatial position of source patches depends on local mechanisms that deposit or remove soil. Where soil is deposited on a rocky surface after being transported by wind or water, the deposition site becomes a source. Soil can also be removed by burrowers, until the underlying rock becomes exposed so that a sink patch becomes a source. The patchiness is thus spatially dynamic, and its changes in time are controlled by ecosystem functions.

Burrowers are common in dryland ecosystems probably due to (1) the biophysical advantages of burrows with respect to air temperature and humidity (e.g., desert woodlice), (2) the avoidance of predation risk during daytime where hiding opportunities are scarce due to the low vegetation cover (e.g., small rodents), (3) the need to store food during the nonproductive season (e.g., ants), and (4) the attraction of hidden, belowground plant reserves such as bulbs (e.g., porcupines).

The dynamics of the biological crusts and, hence, their spatial positioning are less well understood. One hypothesis is that a source patch becomes a sink when the microphytes that form the crust (cyanobacteria, algae, lichens, mosses) are overshadowed by higher plants (macrophytes). The sink becomes a source when higher plants are removed or cannot become established for as yet unknown reasons. The biological soil crust may be significant only in semi-arid ecosystems. In drylands of low aridity, the microphytes may be outcompeted by higher plants. In those with high aridity, there is not enough moisture to sustain the microphytes.

The Temporal Dimension

The Larger-Scale Patchiness The source-sink mosaic model generates sustainability at the smallest scale. Sustainability is due to a mosaic of patches in which source patches contribute to sink patches so that the latter are sustainable. Also, there are mechanisms to replenish the source patches so that the whole system is sustainable at that scale too. How then does this system of sources and sinks at the smallest scale act as source for a sink at larger scales? A mechanism for this can be proposed by invoking temporal variability and superimposing it on the spatial heterogeneity. With increased aridity, there is increased temporal variability of rainfall, expressed in both intensity and frequency of rainstorms. With increased aridity, rainfall frequency decreases but intensity of some rainstorms increases, and precipitation is best described as occurring in pulses. High pulses of precipitation in the arid dryland ecosystems saturate the sinks at the smallest scale so that they literally overflow, such that the smallest-scale ecosystem serves as a source to the larger-scale one.

Quantity and Frequency of Water Recharge: Depth and Scale High-intensity pulse results in more runoff reaching the sink, hence in deeper penetration. Furthermore, increase in scale reinforces intensity and further increases depth of penetration. Depth of storage reduces evaporation; thus, the higher the intensity of pulses and larger the scale, the more water stored, and the more stored water persists. Since deep levels are recharged only by high-intensity pulses, the frequency of recharging deep levels is low. For similar reasons, the frequency of recharging declines with scale (fig. 8.11). This decline in depth of penetration and recharge of sinks has probably generated an evolutionary response by plants, expressed in a dichotomy of growth forms. There are plants with shallow root systems and low height, such as perennial grasses and annuals, and others with deep roots and high height, such as shrubs and trees. It is, therefore, also convenient to categorize depth of the soil water reservoir in low-cover dryland ecosystems into "shallow"

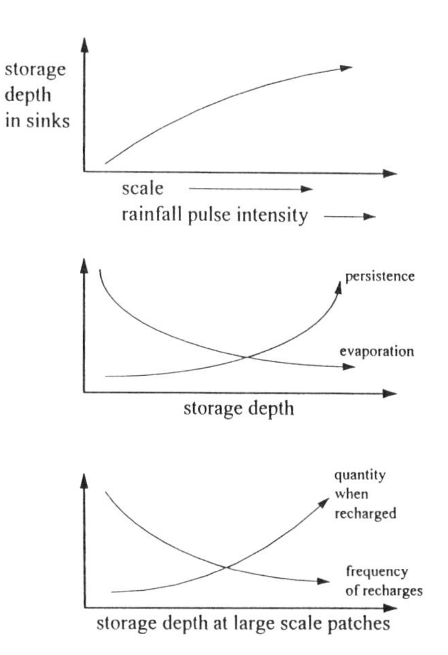

Figure 8.11. Storage of water in the soil reservoirs, including relations among depth, evaporation, persistence, frequency of recharges, and scale.

and "deep" categories. Many perennials can use both shallow and deep reservoirs (Adar et al. 1995), whereas most ephemerals can only utilize the shallow one.

Depth of Water Reservoir and Plant Growth Forms across Scale In the deep reservoir, water amount steeply increases and the frequency of recharges steeply declines with scale. In the shallow reservoir, the amount stored is always less than in the deep reservoir, but it increases slowly with scale. Frequency of recharge is always higher, and it declines more slowly with scale in the shallow than in the deep reservoirs. Thus, the difference of water storage in the shallow and deep soil reservoirs increases with scale. In the shallow reservoir, water constitutes a "reliable" but very "poor" resource (following the terminology of Shmida et al. 1986). Therefore, it supports small annuals recurrent every year. In the intermediate scale, say a first-order gully, the shallow reservoir is "very rich" but "less reliable." This may support larger annuals that germinate and reproduce only in years of high-intensity pulses. At the largest scale, water constitutes a richer but less reliable resource, perhaps supporting only a few annuals adapted for such a degree of unreliability. Thus, the production of ephemeral vegetation will have a hump-shaped curve across scale (fig. 8.12). In the deep reservoir and at the smallest scale, water is a relatively "poor" and "unreliable" resource, though more so in both respects as compared to the shallow reservoir. This may support perennial grasses. In the intermediate scale, the water resource is richer but less reliable, and shrubs are supported. In the largest scale with much richer but highly unreliable water resource, trees can be maintained. Thus, in the deep reservoir, the large quantity of stored water cancels out the effect of low frequency of rainfall pulses. Probably before the water is exhausted by the plants (almost nothing evaporates directly from the soil), they are replenished even though the frequency between replenishment events is low.

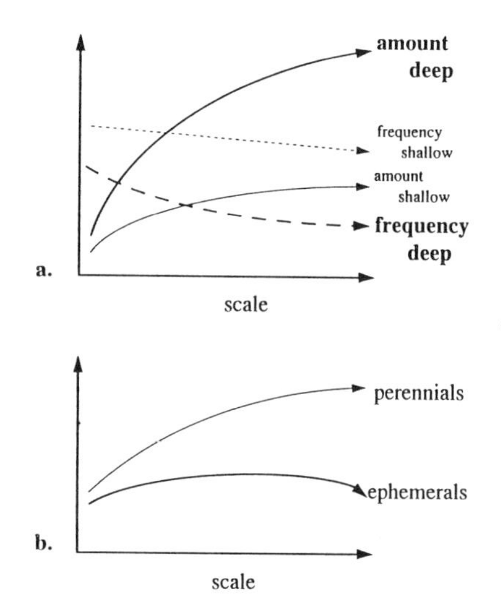

Figure 8.12. Shallow- and deep-water storages and scale—relations between amounts stored and frequencies of recharge (a) and the response in plant lifeforms (b).

Pulses and Reserves The precipitation pulses not only determine the spatial distribution of production and lifeforms but also the timing of production. Rainfall pulse constitutes a trigger that sets off a pulse of production. This pulse of production generates reserves that enable circumvention of the low and variable frequency of rainfall pulses (Noy-Meir 1973). The reserves are either in the form of resistant plant parts, such as apparently dry but live shoots and branches, or storage units and other propagules that constitute *memory.* These reserves contribute to sustainability, especially with respect to plants that depend on deep water storage and those in the larger-scale patches: their production reserves guarantee persistence even when the water reservoir is depleted prior to the occurrence of the rainfall pulse (fig. 8.13). The production of plants induces a pulse in the production of the consumers. Decomposition and mineralization, however, are decoupled from the pulse and can be carried out during the period between successive pulses (Whitford et al. 1983).

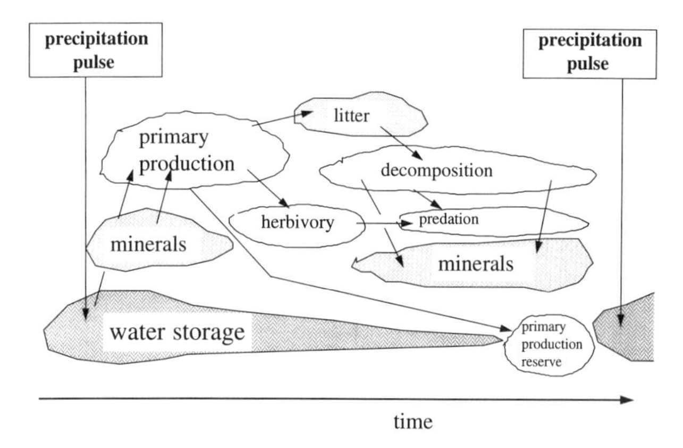

Figure 8.13. The effect of precipitation pulses on timing of processes (blanks) and storages (shaded) in a dryland ecosystem.

The Sustainability of the Source-Sink Spatiotemporal Model Earlier it was proposed how sources at the smallest scale can be replenished to attain sustainability. Not only nutrients but also memory in the form of seeds are transported from source to sink. The seeds can be redistributed by animals to replenish the source. It is likely that similar mechanisms also operate at the larger scale with respect to both nitrogen and seeds. Consideration should be given to the processes that leach salts from sources to sinks since salts readily accumulate where infiltration is shallow and evaporation is high. Runoff should flush salts from source to sinks, and intense pulses should flush salts from small- to large-scale sinks. Salts may eventually accumulate where drainage is poor, as in flat plains, or due to particular soil properties.

Ecosystems will then be replaced by other ecosystems whose communities are made of salt-tolerant plants. Note that in the proposed model, large-scale sinks should be highly productive but salinity may counteract this trend.

A Model for the High-Cover Dryland Ecosystems

The spatial heterogeneity–temporal variability source-sink model can explain the sustainability of low-cover dryland ecosystems, namely, arid and hyperarid ones. High-cover dryland ecosystems—the semi-arid and the dry-subhumid ones—are spatially less heterogeneous and temporally less variable. In the following, a conceptual model for these dryland ecosystems is developed (fig. 8.14). The relatively high precipitation and low evapotranspiration result in relatively high water content in the topsoil. This generates high production but of low, shallow-rooted plants that produce high plant cover, high litter cover, and high soil organic matter. These reduce evaporation, increase infiltration, and reduce runoff, leading to soil conservation. Due to topsoil conservation, calcium carbonate is not exposed and pH is low; this together with the wet soil prevents loss of reduced nitrogen. Soil conservation and low runoff prevent nitrate leaching. Thus, the likelihood of denitrification is low, both due to low incidence of anaerobic conditions and to fast uptake by plants. Thus, there is a complete recycling of nitrogen that promotes sustainable production.

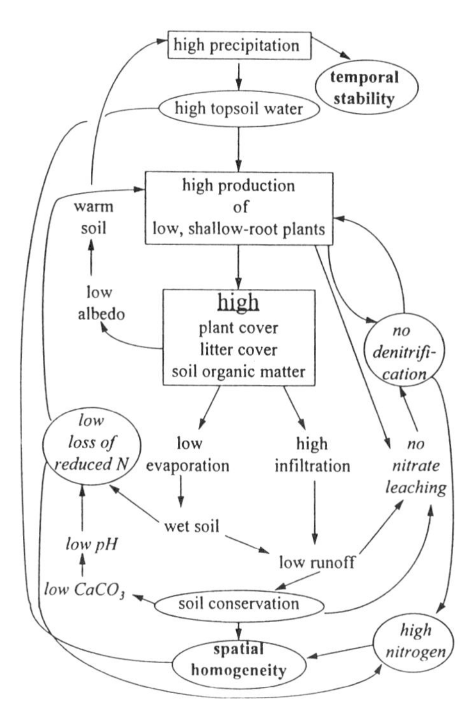

Figure 8.14. A conceptual model of high-cover dryland ecosystems. Nitrogen dynamics are italicized.

There is also the possibility that the high plant cover and, especially, the high incidence of litter reduce the albedo. This warms the soil, possibly resulting in more precipitation. The critical feature of this model is the promotion of spatial homogeneity in soil and, hence, landscape features but also in topsoil water and nitrogen. Finally, the high precipitation is also associated with lower variability in rainfall properties resulting in an increase in temporal stability.

Toward a Unified Dryland Ecosystem Model

The model for the high-cover drylands describes mainly semi-arid grassland ecosystems but can also be applied to the dry-subhumid ecosystem with its even higher precipitation infiltrating deeper than in semi-arid ecosystems. This sustains taller plants that coexist with those of low stature that utilize the shallow reservoir only. The resulting landscapes are savannas and scrublands. In both the dry-subhumid and the semi-arid drylands, spatial homogeneity is maintained by ecosystem functions; temporal stability may be facilitated as well. Ecosystem functions of arid drylands generate spatial heterogeneity, and this, together with the high temporal variability that elicits an adaptive evolutionary response of the individual species, contributes to sustainability. Arid ecosystems are often shrublands in which shrubs and trees utilize the deep reservoir; the shallow reservoir is utilized mostly opportunistically, usually by ephemerals. The model for the hyperarid drylands is similar, though the spatial heterogeneity is at the large scale only. Source patches are large and generally devoid of vegetation; the shallow reservoir rarely stores water; thus occurrence of low vegetation and ephemerals is rare in time and space; and large bushes and trees may predominate though the overall cover is low. Also, saline ecosystems with their typical vegetation are common in hyperarid drylands. Thus, dryland ecosystems can be described by two types of spatial heterogeneity: (1) vertical, the stratification of shallow and deep water (and mineral) reservoirs, and (2) horizontal, the transition from relative homogeneity to a landscape mosaic of source and sink patches at various scales (fig. 8.15).

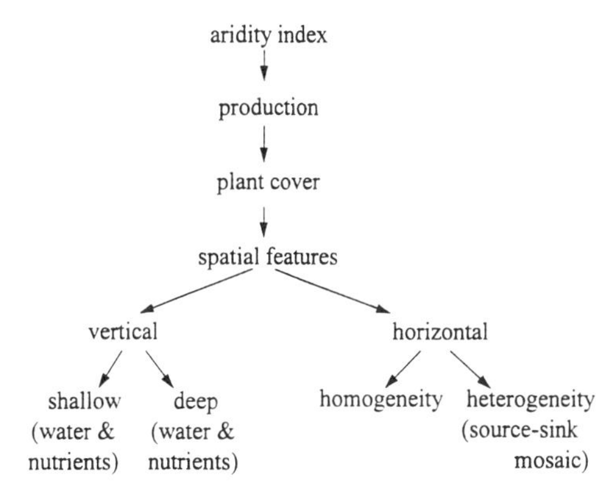

Figure 8.15. Determinants of the vertical and horizontal spatial heterogeneity of dryland ecosystems.

Incorporating Consumers in the Dryland Ecosystem Model

Herbivory Interactions

Plant Defenses The proposed dryland ecosystem model addresses primary producers explicitly and decomposers implicitly. In the following, the ecosystem functioning of consumers is addressed within the context of the proposed model. To reiterate, primary production declines with aridity and the stature of the perennial vegetation first declines and then increases across the transition between semi-arid and arid drylands. Also, the incidence of ephemerals is lowest at the semi-arid and often peaks in the arid dryland. It is hypothesized that the incidence of defense against herbivory among plants follows the trend of perennials' stature, but the incidence of defense is greater in the arid than in the dry-subhumid ecosystems. These combined features result in an interesting secondary production curve that still requires verification (fig. 8.16): it is high in dry-subhumid, highest in semi-arid, and sharply declines in arid and hyperarid ecosystems.

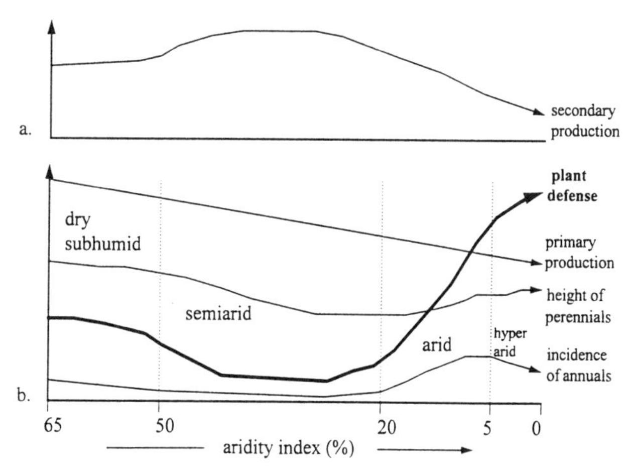

Figure 8.16. The hypothesized relations between plant defense (b) and secondary production (a) in dryland ecosystems, across the aridity gradient.

The Low Herbivory The shape of the secondary production curve is determined by several factors. Obviously, low primary productivity supports only low secondary productivity. But either mechanical or chemical (or both) defenses reduce herbivory. The investment of perennials in defense is far greater than that of annuals, as predicted by the difference in their life history strategies. Also, there is a greater evolutionary incentive for investment in defense where the defended production is expensive. Ephemerals do not invest in protection but rather in rapid growth and reproduction. Yet, their production often does not support secondary production because it is hard for herbivores to adapt to the unpredictability, rarity, and short duration of the ephemerals' production pulses except by migration. The low herbivory results in short trophic chains, low overall species richness, and hence low redundancy in the execution of ecosystem functions (fig. 8.17). Disturbances that lead to drastic reduction in population sizes and extinction of species may, therefore, disrupt ecosystem functions and reduce

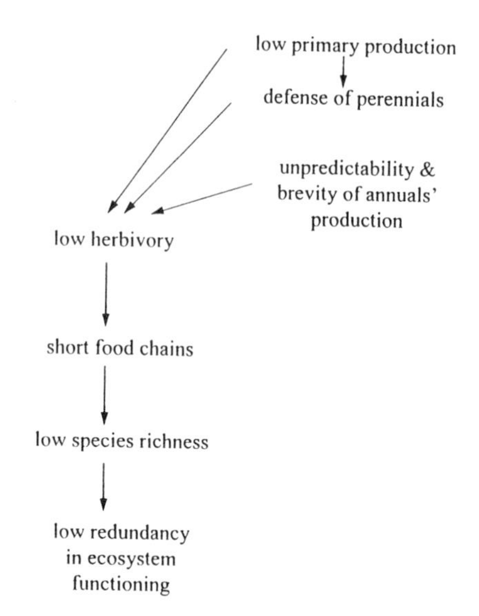

Figure 8.17. Causes (right) and effects (left) of low herbivory in arid ecosystems. In semi-arid dryland ecosystems, herbivory is stronger, and there are longer and more complex trophic webs.

sustainability. This scenario is typical in arid ecosystems. Semi-arid systems have higher herbivory, more complex trophic structure, and higher species richness.

Interactions of Decomposers

Low herbivory results in a high generation of plant litter, hence in high detritivory. Furthermore, due to short durations of surface moisture and overall low topsoil moisture, conditions for microdecomposers are unfavorable, and the role of macrodecomposers is highly significant. Thus, unlike the live plant food chain, the detritivorous food chain is long and complex (Whitford et al. 1983), species richness is high, and redundancy is high (fig. 8.18). Therefore, it is hypothesized that loss of species due to disturbances will not disrupt the rates of nutrient cycling or their reservoir dynamics. Some predators of the detritus food chain also prey on herbivores and, thus, may stabilize the herbivory food chain (fig. 8.19).

Dryland Ecosystems Sustainability: Ecology and Economy

The Low-Cover–High-Cover Dryland Ecosystem Transitions

The "Desertification" Transition From the models above, it is clear that reduction of cover in high-cover dryland ecosys-

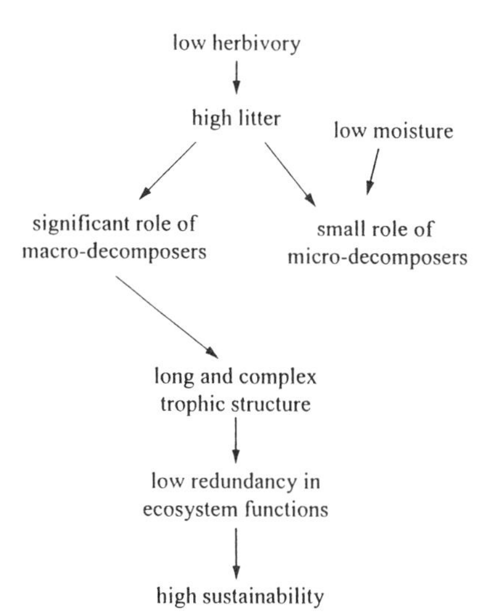

Figure 8.18. The role of decomposers in the functioning of dryland ecosystems.

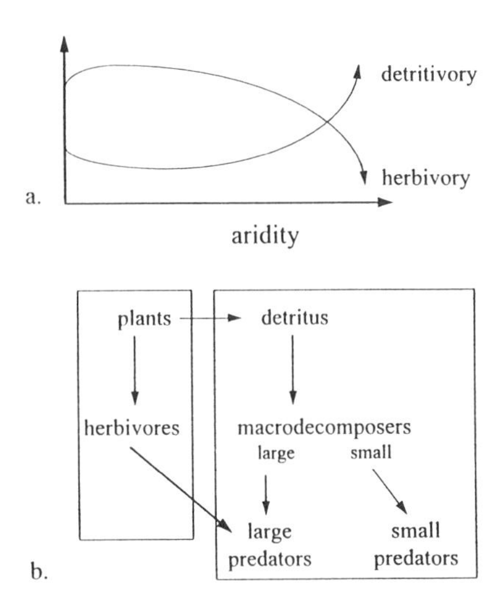

Figure 8.19. Changes of the relative role of herbivory and detritivory related to increase in aridity (a), the herbivory food chain (b [left]), and the detritivory food chain (b [right]) of dryland ecosystems. Predation links the two chains and generates stability (thick arrow).

tems can induce spatial heterogeneity resulting in low-cover dryland features. In other words, a semi-arid dryland ecosystem can transform into an arid ecosystem. This process is *desertification,* as defined by ecologists (e.g., Krebs and Coe 1985, Schlesinger et al. 1990, Sinclair and Fryxell 1985). Climate change bringing about lower precipitation can also induce this transition. But, in the model of the low-cover dryland ecosystem, it is low production that generates all other ecosystem processes related to water and topsoil distribution and redistribution. Only in the depletion of nitrogen in its reduced form is precipitation directly involved in augmenting other major processes. Thus, reduction of vegetation induced by causes other than low precipitation can easily generate the desertification transition. Furthermore, reduction of vegetation coupled with the higher precipitation of semi-arid drylands (as compared to arid drylands) can further increase runoff and accelerate the desertification transition (fig. 8.20). Thus, desertification due to climate change will be slower and possibly less severe than desertification due to direct reduction of cover that is usually induced anthropogenically.

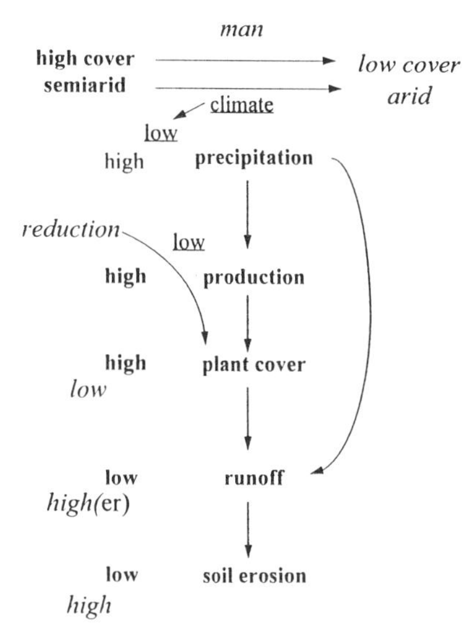

Figure 8.20. Soil conservation in semi-arid dryland (bold) and the desertification transition from semi-arid to arid conditions: anthropogenically induced (italics), and climatically induced (underlined). The relatively high precipitation of semi-arid dryland, coupled with low plant cover due to humans, makes runoff even higher.

"Making the Desert Bloom" Transition Can the transition go the other way, from arid dryland to semi-arid dryland ecosystem (fig. 8.21)? This question applies both to restoration of anthropogenically desertified ecosystems and to

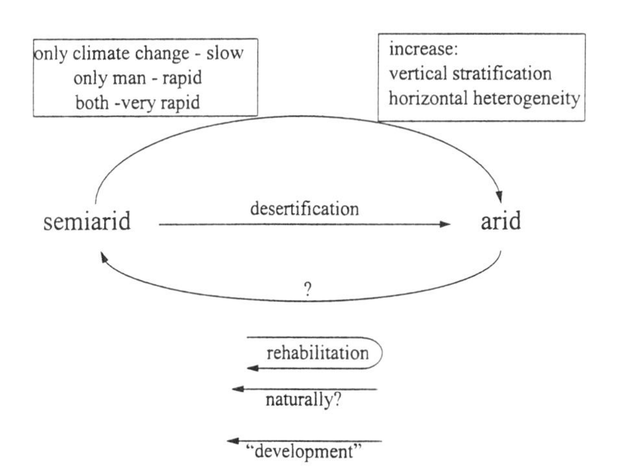

Figure 8.21. The desertification transition (top left box—rates of desertification; right box—expressions of desertification) and rehabilitation options (bottom).

what is called "desert development," which may imply under some circumstances the transformation of arid to semi-arid ecosystems. To approach these issues, it is first examined whether a reduction in aridity due, for example, to climate change can generate this transition. In the source-sink spatiotemporal heterogeneity of the low-cover dryland ecosystem model, a high-intensity rainfall pulse makes the small-scale sink patches overflow, thus acting as sources for the higher-scale sinks. It may, therefore, be assumed that increased precipitation will simply intensify the scale effect in the low-cover dryland ecosystem but not transform it to a high-cover, semi-arid ecosystem. However, due to the inverse relations between magnitude and variability of rainfall in dryland ecosystems, an increase in precipitation will be expressed in an increase in the frequency of pulses coupled with a reduction of their intensity.

The Effect of Climate Change on Transitions

The Source Formation Model To address this issue, it is necessary to propose a model for the formation of a source patch. This model is highly speculative (fig. 8.22). It initiates with bare soil surface. The high-intensity rain pulses impact

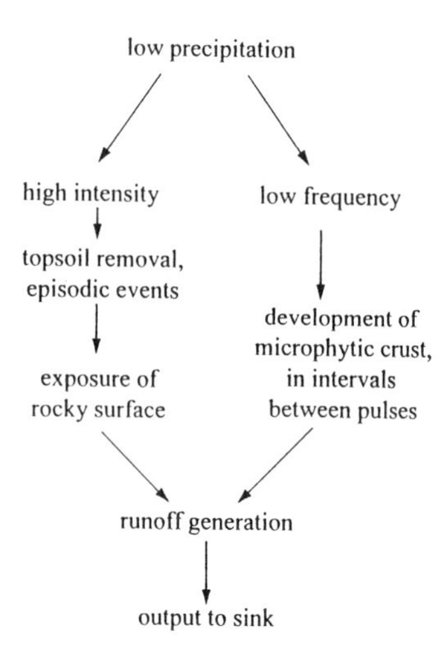

Figure 8.22. A proposed model for the formation of a source patch.

the soil and cause episodic events of topsoil removal. If there is a shallow underlying rock, it soon becomes exposed and can generate stronger runoff. Alternatively, the low frequency is expressed in long intervals between episodes of topsoil removal. During these intervals, microphytes can utilize the little water there is and form a biogenic crust. Either this crust or the bare rocky surface, or both, generate runoff to sink patches.

Source Dynamics under Climate Change It is now possible to examine what happens when precipitation increases due, for example, to climate change (fig. 8.23). Intensity becomes

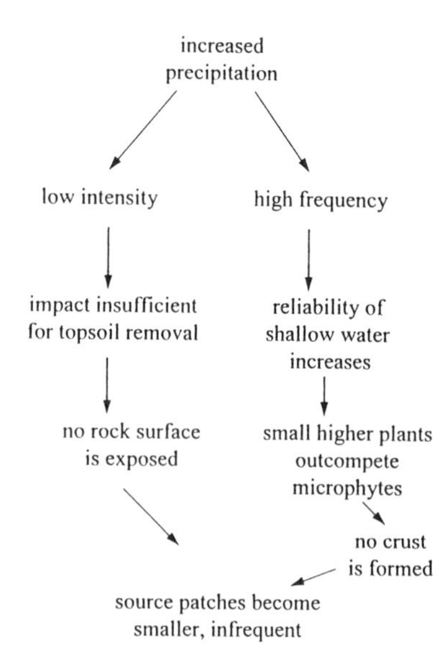

Figure 8.23. A proposed model for the formation of a source patch due to increased precipitation induced by climate change.

lower, and the impact of rainstorm is insufficient for topsoil removal, hence no rock surface becomes exposed. The increased frequency of rain pulses increases the reliability of water in the shallow reservoir, and relatively small plants can develop and outcompete the microphytes. Thus, biogenic crust is not formed. Source patches become small and rare.

Sink Dynamics under Climate Change How do the source's dynamics under increased precipitation affect the sink's dynamics? Since the sources become smaller and rarer, the input to sinks becomes smaller. This reduction can be counterbalanced by the overall higher precipitation. Thus, the amount of water to sinks need not necessarily decline. Yet, the intensity of the flows from sources to sinks is reduced, hence less water is stored in the deep reservoir. On the other hand, the frequency of flows from source to sink increases. This results in a greater amount and, more importantly, a persistent water supply in the shallow reservoir. The scarcity of water in the deep reservoir precludes tall vegetation, and the more reliable shallow-water reservoir promotes low-stature vegetation but high cover of this vegetation (fig. 8.24). This vegetation may even encroach upon the adjacent sources. Thus, there is an overall reduction of spatial heterogeneity that also affects the nitrogen budget.

Natural and Human-Effected Transition from Arid to Semi-Arid Ecosystem The model just presented allows for natural transformation of low-cover arid to high-cover semi-arid ecosystem as a result of increased precipitation duc to climate change. Since the transformation in the opposite direction can be induced by climate change as well as by anthropogenic reduction of plant cover, the question is whether the

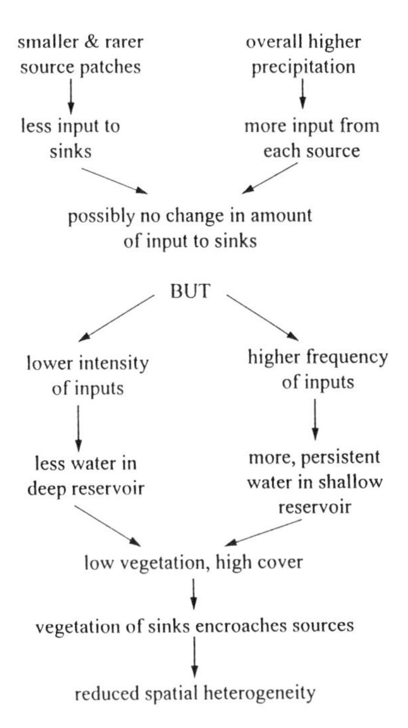

Figure 8.24. Processes affecting sink patches when precipitation increases.

transformation from arid to semi-arid ecosystem without climate change but by human intervention, such as an increase in plant cover or other means, can affect this transformation and result in an ecologically sustainable semi-arid ecosystem. Obviously, the interest to invest in such a transition is economic. It is, therefore, necessary to explore the coupling between ecological and economic sustainability dealt with in the next section of this chapter.

Ecology and the Sustainable Development of Dryland Ecosystems

The Concept of Economic Sustainability

Terminology The motivation of investing in a transition from arid to semi-arid ecosystem is economic benefit. This type of investment is termed *development.* The term *sustainable development* was introduced in 1987 via the Brundtland Report (WCED 1987), coupled with the term *environment* (the Brundtland Report was prepared by the World Commission on Environment and Development). *Ecological sustainability* was highlighted by the Sustainable Biosphere Initiative (Lubchenco et al. 1991) of the Ecological Society of America and is expressed in the title of the present volume: Toward Ecological Sustainability. The relations between ecological and economic sustainability will be now explored.

Ecological and Economic Sustainability If an ecosystem can only be sustainable—otherwise it does not persist—what is the usefulness of the term sustainability? It is valid only in the context of development, which is a term confined to human activity. Whereas species vary with respect to their rates of production in time and space and often exhibit variable trends, production by the human species is unique in that it is ever accelerating, at least ever since the "agricultural revolution." Furthermore, while the mechanisms of change in species' production are not well understood, those that govern the steadily accelerated human production are clearly identified: manipulation of the production of some species other than human and transformation of habitats and ecosystems to cater for these particular species. Thus, development is an acceleration of human use expressed in increased per capita resource consumption. Though not obligatory, this consumption usually results in population growth. Population growth, or even just the increase in per capita resource consumption, generates more development. This is *economic growth.* It is evident that ecological sustainability and economic sustainability may be quite at odds. An agricultural field that becomes salinized is a nonsustainable development; a grassland valley that has become salinized is a transformation of sustainable grassland ecosystems to a sustainable salt-pan ecosystem.

Global Ecosystem and Its Subsystems Whereas ecosystem is a functional concept, its spatial delineation is artificial, often determined by research convenience. The larger the space of reference, the less dependent it is on the functions of other adjacent ecosystems. The largest ecosystem is, therefore, the global ecosystem: the sum of all ecosystems regardless of delineation. As of the time of emergence of development, human effects on the global ecosystem have become prominent to the extent that it is now useful to identify human effects as the "economic subsystem" within the global ecosystem (fig. 8.25). Thus, as development proceeds, the economic subsystem increases at the expense of the "natural" subsystem of the global ecosystem (Goodland 1991).

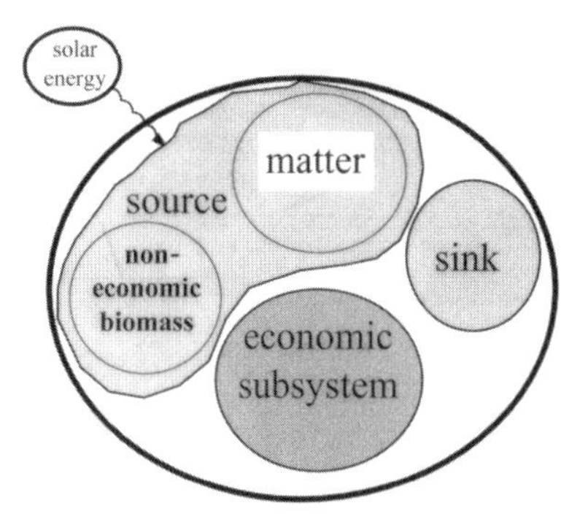

Figure 8.25. The global ecosystem. Assuming solar energy, matter, and space are constant and finite, the economic subsystem is sustainable when its size is optimized with respect to the size of the source (matter and noneconomic biomass) and the sink, so that it is not limited by source or sink.

Development and Sustainability Given the above definition of use and sustainability, sustainable development is human use carried out in ways that allow natural processes to replace what is used. The quest for sustainable development emanates from the concern for future generations; if what is used is replaced, the current use of resources does not affect the options of future generations (WCED 1987). But, since development is defined as accelerated rate of use, it follows that sustainability of development is incumbent upon an accelerated replacement rate. With respect to energy replacement, the rate of fixation of solar energy is ultimately limited by space, and the rate of storage as fossil fuel is practically nil when compared with rate of use. With respect to material recycling, there is an upper ceiling on the rates of the physicochemical and biochemical processes involved. Furthermore, many products of the economic subsystem have very slow recycling rates that require appropriation of space for "sinks" at the expense of the natural global ecosystem (fig. 8.25). There should come a time when rate of replacement cannot increase to match rate of use—that is, development becomes nonsustainable. If nonsustainability leads to an eventual cessation of development, then the only option for maintaining sustainability is to increase the economic subsystem at the expense of the global natural ecosystem.

Is Sustainable Development Feasible? Since the global ecosystem is finite in space, the economic subsystem cannot expand indefinitely. It will become nonsustainable when it completely replaces all natural ecosystems and occupies the whole global ecosystem. It is hypothesized that nonsustainability will happen more quickly, and the major premise of sustainable development is that a certain portion of the global ecosystem is required for maintenance of the economic subsystem sustainably (Lubchenco et al. 1991). Beyond that "threshold," neither the natural nor the economic subsystems remain sustainable. The economic subsystem will react to the demise of the natural subsystem—with no sustainability it will cease to function and, hence, to exist. It follows that development as an accelerated use is only temporary. Making it sustainable may defer the collapse of the economic subsystem but not prevent it. The accelerated use should be eventually replaced by sustainable use—that is, growth, expressed in per capita increase in either use and/or population, should eventually cease.

Sustainable Development for Cessation of Growth Sustainable development is, therefore, growth leading to eventual sustainable economic use where neither production per capita nor population increase. Sustainable development is defined as development that enables and makes socially acceptable an eventual cessation of growth. This development is bound to affect options of future generations, but if sufficient restrictions on current development are implemented, as many future options as possible will be maintained (Fuentes 1993). Furthermore, this kind of sustainable development, which leads to eventual cessation of acceleration of use, may never be reached within conventional time horizons. Sustainability is thus a goal; it is not a fixed endpoint to be reached but a direction to be taken for alleviating prospective hardships (Lee 1993). Moreover, if cessation of growth does not become socially acceptable, practices for sustainable development will defer the unavoidable collapse as far as possible beyond conventional planning horizons.

Development of Dryland Ecosystems

The Role of Dryland Ecosystems in Global Sustainability Development may become sustainable now, or temporarily current sustainability can be traded off with expansion of the economic subsystem at the expense of natural ecosystems. Since at least the industrial revolution, development has been by and large unsustainable, as manifested by the rapid expansion of the economic subsystem. This expansion has encompassed first the nondryland ecosystems of the biosphere. This is because economic production is high where natural production is high, hence a highly productive natural ecosystem is likely to become an economic ecosystem before less productive ecosystems. Instead of developing productive ecosystems sustainably, humans have initiated nonsustainable development in less productive ecosystems, that is, in the dryland systems. This nonsustainable development manifests itself in what is termed by land managers, policy makers, and developers *desertification,* that is, a reduction of production as compared to the one prior to the recent development. Within the drylands, development proceeds along the aridity gradient from the less arid to the more arid. After the more arid ecosystems enter the vortex of nonsustainable development, the final collapse of the global ecosystem may be in sight. Thus, rather than being the "last frontier" for human development, drylands may serve as the last stronghold in the struggle for global sustainability.

Prospects of Sustainability for Dryland Ecosystems Among several factors responsible for failure to attain sustainability in general, Ludwig and colleagues (1993) pointed at (1) the complexity of the underlying natural systems involved, which precludes a reductionist approach to research and management, and (2) the substantial levels of natural variability that mask the effects of overexploitation. With respect to complexity, drylands have been customarily regarded as ecologically simple. This notion has been recently challenged: even though drylands may be structurally simple, they are functionally very complex (Safriel et al. 1989). The levels of natural variability are particularly large in dryland ecosystems as compared to all other ecosystems. Therefore, it is in dryland ecosystems, more than in others, that for achieving sustainability use must be reduced to far below the maximum yield, in order to compensate for expected and unexpected variations in production. An even greater reduction may be prescribed as a safety margin as long as knowledge of the underlying ecosystem processes remains insufficient (Ludwig 1993).

Dryland Degradation and Desertification The 6.1 billion hectares of drylands comprise 47% of the 13,013 billion hectares of the earth's land. Nearly half of the earth's land area, drylands are inhabited by 900 million people—17% of the global population. Thus, as expected, the drylands are underpopulated by area, though obviously overpopulated with respect to their productivity. That this constitutes overpopulation is shown by the high degree of nonsustainability expressed by *land degradation,* which is often synonymous with desertification. Land degradation is an economic term that implies a reduction in soil fertility expressed in reduced production of economically valuable biomass in dryland ecosystems transformed into either agricultural land or pasture. The U.N. definition of desertification, which is at the heart of the Convention to Combat Desertification (adopted in 1994 with expected implementation in 1997), follows: "land degradation in arid, semi-arid and dry sub-humid areas resulting from various factors, including climatic variations and human activities."

Stability and Desertification This degradation can be measured by the economic subsystem's resistance and/or its resilience, two measures of ecosystem stability. Resistance in the economic context may be the degree of reduction in the production of economic biomass following natural disturbance (e.g., drought) and/or human "mismanagement" (Puigdefabregas 1995): the higher the reduction, the lower the resistance. Resilience in the economic context may be the rate of recovery, that is, the full restoration of economic production once the natural disturbance is removed and/or the mismanagement is rectified. Thus, the acid test of degradation is in the land's resilience; severe disturbance and extreme mismanagement are likely to weaken resilience, increasing land degradation. A useful measure of resilience, hence of degradation, is the cost of restoration. Restoration may also mean the return of the original natural ecosystem, and the resilience is then that of the natural ecosystem—the rate of its recovery from development effects. When a natural dryland is transformed to an economic ecosystem, then suffers disturbance and/or mismanagement, and then has the disturbances removed, unit area was calculated according to a combination of the degree of soil degradation and the percentage of the area affected. Light, moderate, strong, and extreme degradations were categorized as infrequent (up to 5% of the unit affected), common, frequent, very frequent, and dominant (more than 50% of the unit affected) to produce twenty combinations lumped to four color categories (low, medium, high, and very high degradation) in the *World Atlas of Desertification* (UNEP 1992).

Global Desertification Statistics GLASOD defined hyperarid drylands as "nonsusceptible" to degradation since they are hardly developed and are regarded nondegradable. Of the 5,169 million hectares of susceptible drylands, about 20% are degraded. It is noteworthy that land degradation or desertification is not confined to drylands, as 929 million hectares of the 6,865 nondryland (13%) is degraded. Of the susceptible degraded drylands, 41% are lightly, 45% moderately, 13% strongly, and 0.7% extremely degraded. By continents, 33% of Europe's, 25% of Africa's, 13% of Australia's, and 11% of North America's susceptible drylands are degraded. But, the majority of Africa's degradation is strong and extreme, while the majority of Australia's degradation is light.

Stability of Natural Ecosystems and Sustainability of Economic Ecosystems By definition, natural ecosystems remain sustainable even when their production fluctuates or declines and stabilizes at a lower level. This is because they continue to function and, hence, to exist. Instability of the economic subsystem and especially a decline in the economic production are signs of nonsustainability. The stability of a natural ecosystem can be assessed by its reaction to a stressor. Development can be treated as a stressor. If the resistance and/or the resilience of a natural ecosystem are low, its production will fluctuate under the impact of a stressor. Such an ecosystem, when transformed to an economic ecosystem, will also fluctuate and then become a nonsustainable economic ecosystem. The hypothesis is, therefore, that development of unstable natural ecosystems yields nonsustainable economic ecosystems. Conversely, natural ecosystems with high stability are more likely to adapt to economic production sustainably than natural ecosystems with low stability.

The Relations between Aridity and Economic Nonsustainability

Stability and Sustainability Two alternative hypotheses can be formulated: (1) Dryland ecosystems track environmental variability. As aridity increases, "natural" production is more variable. Development becomes a stronger stressor with aridity; thus, sustainability, expressed in land degradation/desertification, is reduced. (2) Dryland ecosystems are stable due to the evolutionary response of their species to instability. Adaptations are more common and more effective as aridity increases. The natural ecosystem is, therefore, resistant and more resilient to the stress of development as aridity increases. Development and land degradation/desertification decline as aridity increases. Equating nonsustainability with land degradation, the spatial extent and the degree of land degradation will now be examined in the drylands susceptible to degradation—the dry-subhumid, semi-arid, and arid ecosystems.

Dryland Degradation Statistics An index for degree of degradation may be the ratio between the area that is strongly and extremely degraded and the area that is only low and moderately degraded, in each of the three aridity categories. On the global scale, the ratio is low: 0.19 in semi-arid lands, 0.14 in dry-subhumid lands, and 0.12 in arid lands (fig. 8.26a). By continents, only Australia and South America show a decline in degradation with aridity, and only Asia shows an increase in degradation with aridity. The highest ratios are in Africa—more degraded lands are twice as common than less degraded lands in its dry-subhumid and semi-arid regions. The highest severe degradation of arid drylands relative to a mild one is in North America (fig. 8.26b). Thus, these inconsistencies may mean that the degree of nonsustainability, as expressed in the land degradation index, is affected more by the magnitude of development pressure than by the stability properties of the ecosystems, as implied from their degree of aridity.

Area Relations and Degradation The mere size of the dryland area may determine the degree of land degradation because large sizes attract more development. The spatial extent of degradation should, therefore, increase with the increase of the spatial extent of the dryland category within the continent. Indeed, there is a relation between the absolute size of the continent's dryland type and the absolute size of degradation. By continents (except for the dry-subhumid category [fig. 8.27a–c]) and globally (fig. 8.28), the largest dryland area type is the semi-arid, which has the largest area of degradation in it. Next in both respects is the arid and finally the dry-subhumid drylands. But, the low and

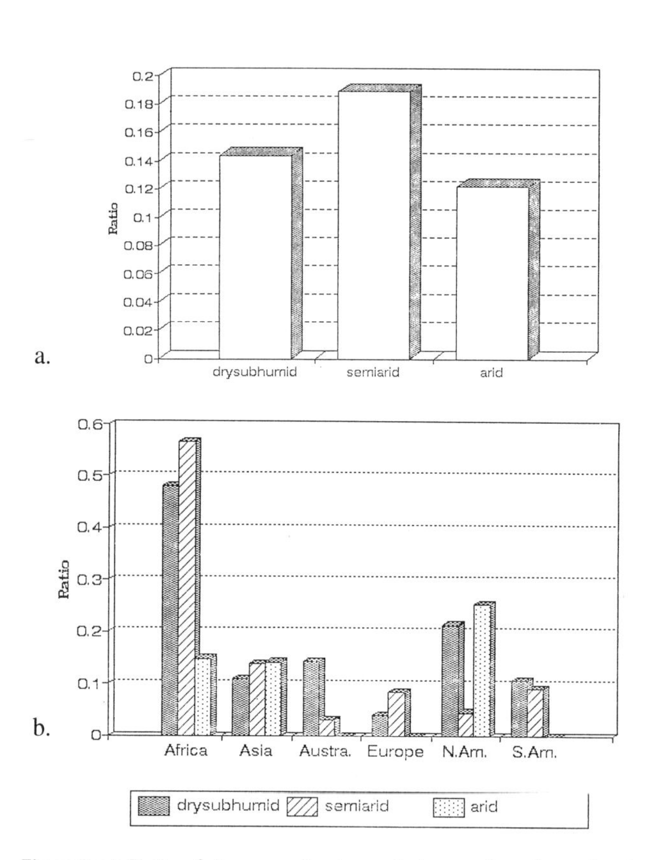

Figure 8.26. Ratio of strong and extreme to low and moderate land degradation: (a) global; (b) by continents (data from UNEP 1992).

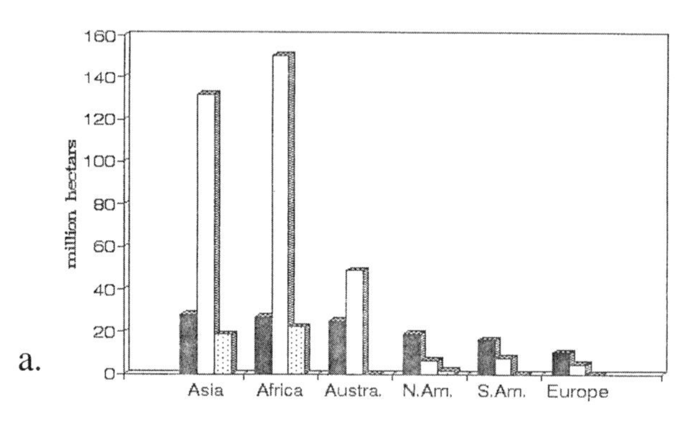

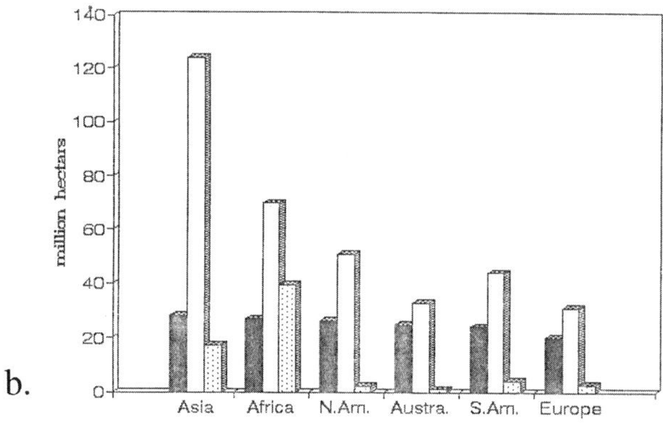

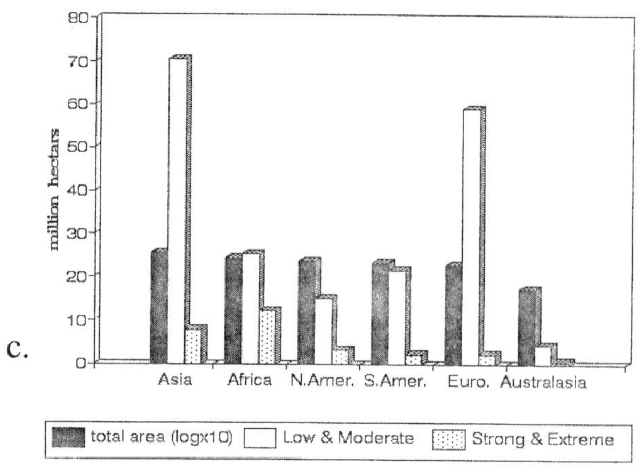

Figure 8.27. Land degradation by continents (compared with continent's total dryland area, expressed in log x 10 and arranged in descending order [data from UNEP 1992]).

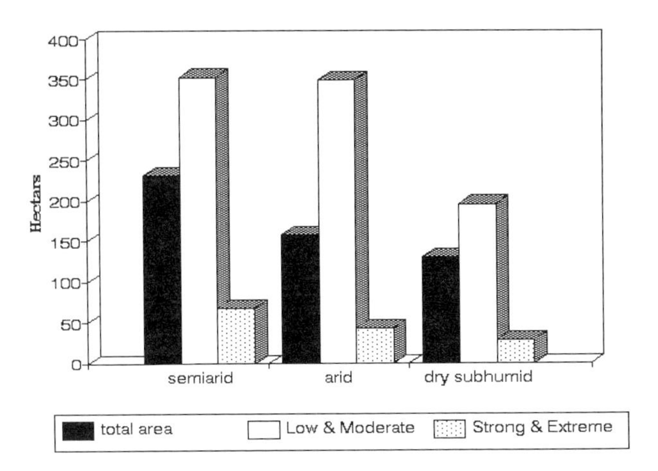

Figure 8.28. Global land degradation. Total area of dryland type expressed by x 10 million ha; degraded areas expressed by x 1 million ha (data from UNEP 1992).

moderate degradation are proportionately greatest in the arid ecosystems, as compared to the two other categories. To summarize, it is not clear whether development pressure per se determines the magnitude of degradation, but it is likely that the degree of nonsustainability may increase with aridity, regardless of cause.

Degradation by Continents Globally, arid drylands suffer the highest degradation expressed in percentage degraded of the global arid dryland area. This is with respect to the low degradation category, whereas, with respect to the high degradation categories, percentages are much smaller and do not substantially vary among the three dryland ecosystem types (fig. 8.29). This combined global picture is, surprisingly, most remarkable in Europe (fig. 8.30a), which has the highest figure—nearly 45% of its arid ecosystem is desertified. This may be associated with the fact that arid ecosystems comprise a very small percentage of the continent. In South America, with a similar percentage, though a higher overall area of arid drylands (fig. 8.30b), the percentage of desertified arid ecosystems is lower, less than 20%. North America, with a similar percentage of continent area as arid drylands, has less than 10% of these lands desertified (fig. 8.30c). Thus, the pressure on land resources in Europe

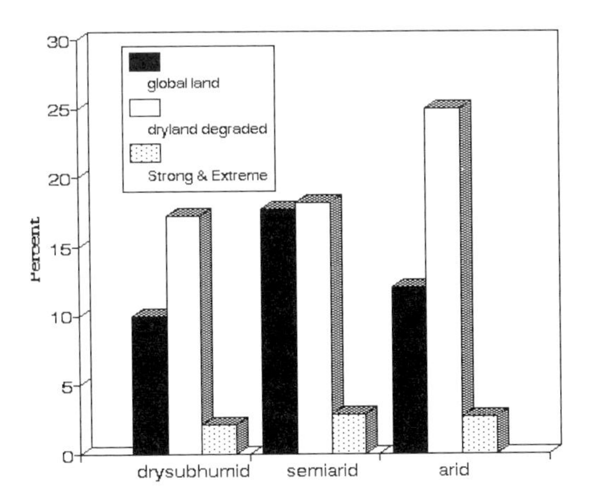

Figure 8.29. Global land degradation, by percentage (data from UNEP 1992).

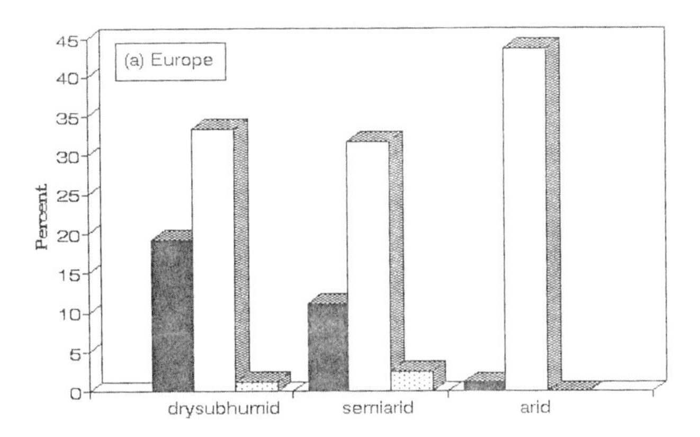

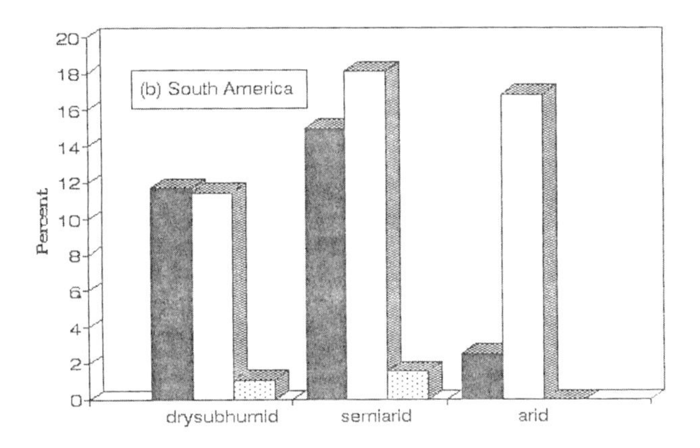

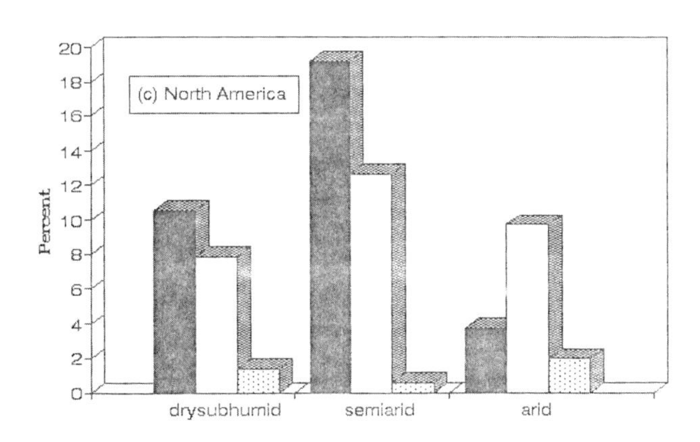

Figure 8.30. Land degradation by continents, by percentage (degraded land of total area of each dryland category [data from UNEP 1992]).

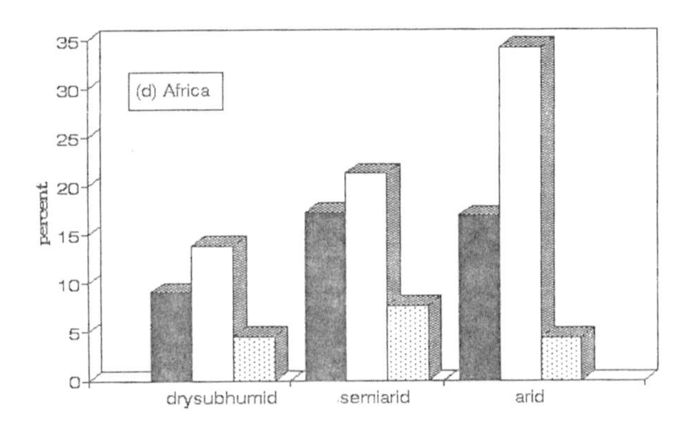

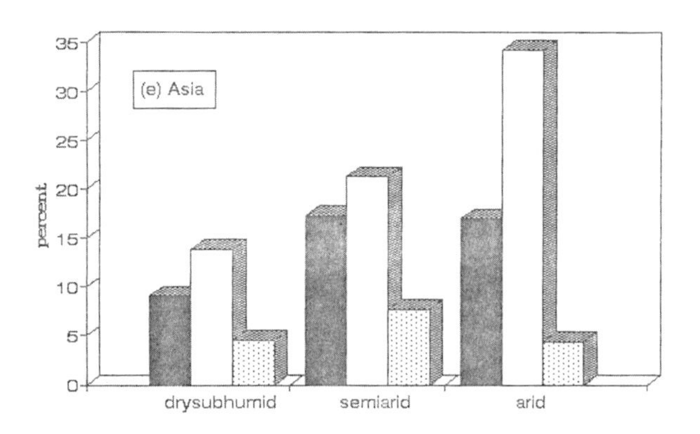

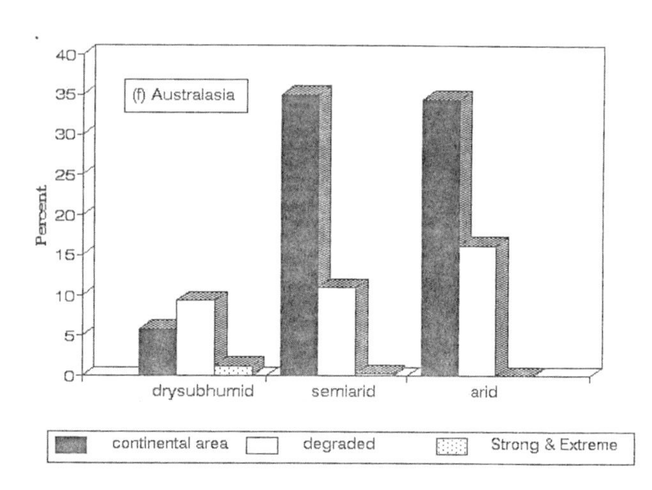

Figure 8.30. (cont.)

seems to be great, resulting in a high extent of development and degradation of its most arid lands. In terms of severity, Europe has not yet degraded its arid drylands to the maximal degree whereas, in Africa, the percentage of moderately degraded arid ecosystems is approaching that of Europe. In Africa, 5% of these drylands are extremely degraded (fig. 8.30d). In Asia, dry-subhumid ecosystems are degraded much more than in Africa (fig. 8.30e). In Australia, moderate desertification increases with aridity, but extreme degradation decreases with aridity (fig. 8.30f).

The Mechanisms of Dryland Economic Nonsustainability

Water Erosion Dryland development practices that cause water erosion include (1) reduction of plant cover, (2) replacement of natural perennial plants with annual plants that do not cover the soil surface all year, and (3) compaction. It is not clear if mechanized machinery and livestock trampling cause compaction and thus increase runoff and eventual erosion or if they damage the biogenic crust and thus reduce runoff and erosion. Degradation as a result of water erosion decreases with aridity. Altogether, 48% of degraded areas in susceptible drylands are degraded due to water erosion. As expected with increased aridity (and reduction in rainfall), the contribution of water erosion to degradation declines (fig. 8.31).

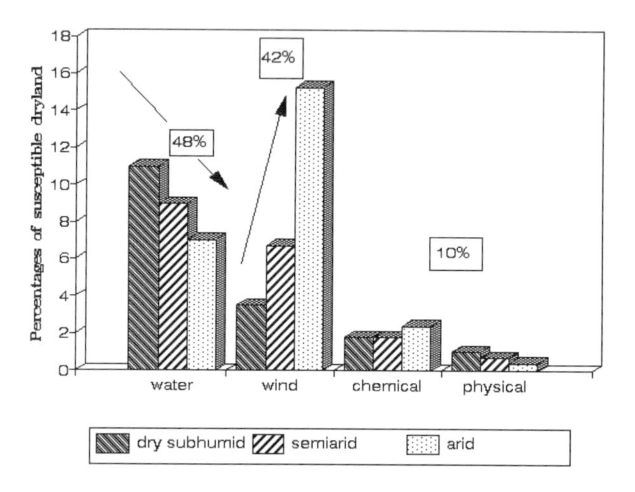

Figure 8.31. Global land degradation by mechanism, by percentage of degraded susceptible drylands. Percentages in boxes stand for degradation due to water erosion, wind erosion, chemical causes, and physical causes in all susceptible dryland types combined. Arrows point at trends along the aridity gradient (data from UNEP 1992).

Wind Erosion Wind erosion exhibits the opposite trend: it increases with aridity. It is more selective than water erosion with respect to the components of soil removed and, therefore, affects the structure of the soil, hence its ability to retain moisture. It also removes particles that are rich in bound nutrients. Apart from damages by deflation, transport and deposition damages are expressed in abrasion of soil and plants and in increased salinity. Development causes wind erosion through removal of vegetation, agricultural practices, use of vehicles, and military uses. In North America and Asia, wind erosion is mainly due to cultivation practices—mainly plowing. In Africa, it is caused primarily by firewood collection and overgrazing.

Chemical and Physical Deterioration Chemical deterioration is caused by intensive agriculture and can occur also in oases within hyperarid drylands. Chemical deterioration encompasses (1) nutrient depletion, either indirectly by wind and water erosion, where fertility is not replenished by floods, by removing vegetation in cases where most nutrients are stored in the vegetation biomass, or by removing economic plants with a subsequent long period with no uptake, which increases vulnerability to removal by other factors, and (2) salinization, including chlorides, sulphates, and carbonates of sodium, calcium, and magnesium. These salts reduce soil pore space and hence the ability of soil to hold air and nutrients. They may also be toxic to the plants. Drylands are susceptible because (1) there is not enough water to leach salts, (2) the low precipitation increases the demand for irrigation and irrigation water carries salts that often accumulate due to the high evaporation, and (3) the tendency to cultivate flat areas that on the one hand require irrigation and on the other are difficult to leach. Poor drainage can also cause waterlogging—that is, the elevation of the water table and salinization of the upper soil horizons. The GLASOD survey has detected an increase in salinization during the last fifty years. Of soil degradation, the 10% in susceptible areas is due to chemical deterioration. The physical deterioration includes compaction, sealing and crusting of the soil surface, and other localized effects. Altogether, loss of soil by water erosion accounts for 48% of land degradation and decreases with aridity. Loss of soil by wind erosion accounts for 42% and increases with aridity. Chemical and physical degradation combined account for 10% and do not show a trend with respect to the aridity gradient (fig. 8.31).

The Causes of Dryland Economic Nonsustainability

These are discussed in the order of their significance (fig. 8.32a):

1. Overgrazing contributes the highest percentage to dryland degradation. It includes the actual removal of production and trampling, both of which lead to soil erosion. Overgrazing affects mainly semi-arid ecosystems when they are overstocked but can also impact other drylands.
2. Cultivation is next in significance. The complete removal of vegetation cover during the fallow periods reduces nutrients and increases dependence on fertilization. In semi-arid drylands, the resulting soil erosion generates spatial heterogeneity. In arid drylands, cultivation includes reduction of the spatial heterogeneity and redistribution of water and increases the dependence on irrigation. Irrigation increases salinity. Pesticide use reduces richness and numbers of macrodecomposers and may lead to loss of nutrients, too.
3. Deforestation is common in the dry-subhumid drylands. It encompasses removal of natural vegetation cover in totality, for new land use. The degradation occurs between the time the vegetation is removed and the new use is effected.
4. Overexploitation, mainly for domestic use, such as firewood, fencing, and building materials is next in significance.
5. Finally is industrial pollution, which causes a reduction of soil fertility (fig. 8.32a).

Only in the Americas is overgrazing not the most important cause. In North America, agriculture has the most impact and in South America overexploitation of forests (fig. 8.32b).

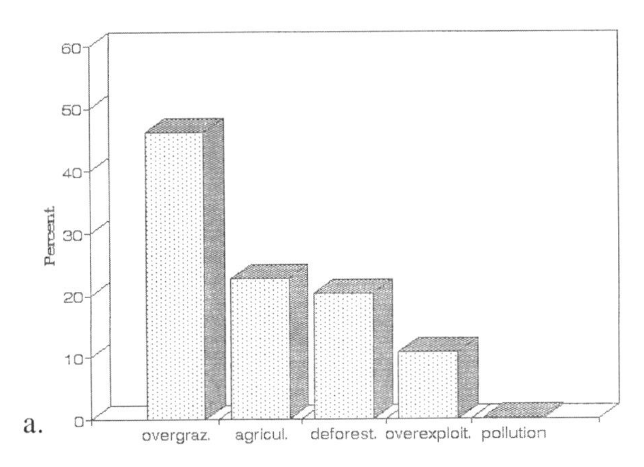

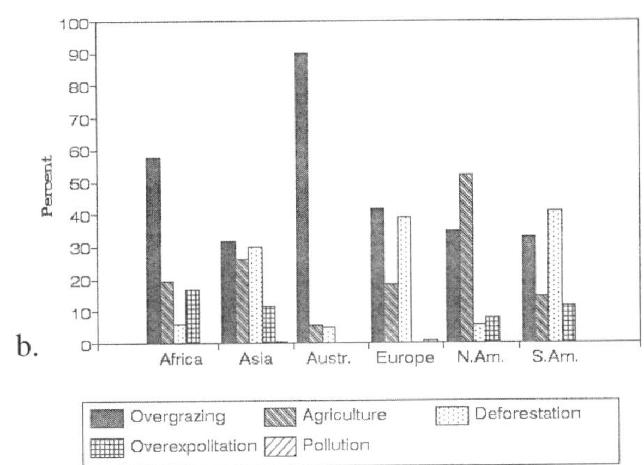

Figure 8.32. Causes of land degradation: (a) percentage of global degraded susceptible drylands, by causes; (b) percentage of degraded susceptible drylands, by continents (data from UNEP 1992).

The Cost of Desertification

Is desertification really damaging, and what is the cost of combating it? The annual income lost to desertification is estimated as U.S.$26 billion in 1980 and U.S.$42.3 in 1991. These costs reflect both the decreases in agricultural production due to land degradation and expenditures required to rehabilitate desertified land. Such expenditures amount to about $5 billion and $10.6 billion per year for 1980 and 1991, respectively (Greijn 1994). These figures illustrate the necessity to find a means of combating desertification.

Combating Desertification

Production of Natural vs. Economic Ecosystems How can we develop and manage drylands for sustainability? Obviously, if humans reduce actions that cause desertification (e.g., reduce overgrazing, cultivation, etc.), the process will be halted. But this is like prescribing reducing emissions to avert global climate change—it's easier said than done. An alternative strategy for sustainable development of drylands could be sought. The statistics on the extent of desertification indicate that development of drylands has so far always been nonsustainable. All attempts to increase production above the natural level have already resulted, or will inevitably result, in an eventual reduction of production. An interesting conclusion can therefore be drawn: it is impossible to increase the production of natural ecosystems, or at least of natural dryland ecosystems. This assumes that the "goal" of natural ecosystems is to maximize production, and this objective has been already met. Implicit in this notion is the assumption that the ecosystem "evolves," resulting in maximization of production. This evolution is the sum of the development of all the ecosystems' species, which have optimized their fitnesses when struggling to maximize the productions of their individuals.

Replacement of Species and Maintenance of Structure and Function The goal of development is to increase economic production, namely, that part of production utilized by humans. Thus, development should strive to replace partially natural production by economic production. This usually entails replacement of the species of the natural ecosystem by species from other ecosystems. This practice can be sustainable, because the ways communities have been assembled is not necessarily the only one for achieving optimization of fitnesses of an ecosystem's species. Historical relationships among species are involved and persist in the current species composition of an ecosystem and their adaptations. But all of the species combined produce the maximal production. So the strategy for sustainable development of drylands may be replacement of species with those that maximize the economic part of production, at the expense of the natural production, while maintaining the structural and functional features of the natural ecosystem. This means that the economic ecosystem should be constructed such that it imitates, as much as feasible, the natural ecosystem.

A Strategy for Sustainability The goals for such a strategy would be to maintain soil cover and to exploit soil water across the soil profile. The method is diversification, whereby development is a spatial and temporal mix of forestry, cultivation, and grazing. The proportions in this mix change with aridity: more afforestation in dry-subhumid drylands, more cultivation in semiarid drylands, and more pastoralism in arid ecosystems. Accordingly, this mix will be expressed in vegetation stratification. The three layers are trees, shrubs and other perennials, and annuals. With respect to trees, the harvestable economic production, usually fruit, is small relative to the rest. Most biomass is retained and thus cover is maintained throughout the year. Water is drawn from the deep reservoir. Another portion of the production can be harvested as firewood. The shade reduces evaporation and transpiration but not photosynthesis. Hence it promotes the growth of shrubs and annuals. Economic shrubs can sustain livestock, and annuals can produce additional economic products. Mineral balance should be maintained; thus subsidizing with fertilizers is not encouraged. An initial subsidy of water investment may be required for the trees but, later, irrigation may be minimized, so that salinization is avoided. The heterogeneity does not encourage local consumers to become pests, and pesticide application is reduced, which improves nutrient cycling, since the macrodecomposers will not be decimated. This scheme is of course much more complicated than conventional agricultural development and requires much research—but it may be sustainable.

Sustainability and Biodiversity In the proposed scheme, much of the local biodiversity is conserved. This is significant in two ways. First, many of the local species may be involved in the generation of "ecosystem services" in ways that are not yet understood. There is some experimental evidence that biodiversity is instrumental in the functioning of ecosystems (Naeem et al. 1994). But the notion that biodiversity is critical for economic sustainability is still a hypothesis that requires research support. Second, some of the species may constitute assets of potential economic significance in the near and far future. Finally, biodiversity can generate income from an economy based on recreation, leisure activities, and ecotourism.

Educational Significance of the Goal of Sustainability Even if dryland development approaches sustainability, as long as it generates population growth, sustainability will not be achieved. The main significance of investing in attempts to achieve sustainability may be the social and educational values of this effort. This in itself may affect decisions of individuals.

References

Adar, E. M., I. Gev, and A. S. Issar. 1995. The effect of forestation on a shallow groundwater reservoir in an arid sand dune terrain. Journal of Arid Land Studies 5:259–62.

Boeken, B., and M. Shachak. 1994. Desert plant communities in human-made patches: Implications for management. Ecological Applications 4:702–16.

Charney, J. G., P. H. Stone, and W. J. Quirk. 1975. Drought in the Sahara: A biogeophysical feedback mechanism. Science 187:434–35.
Evenari, M. 1985. The desert environment. *In* M. Evenari, I. Noy-Meir, and D. W. Goodall, eds. Ecosystems of the world. Vol. 12A: Hot deserts and arid shrublands. Amsterdam: Elsevier. 1–22.
Fuentes, E. R. 1993. Scientific research and sustainable development. Ecological Applications 3:576–77.
Goodland, R. 1991. The case that the world has reached limits. *In* R. Goodland, H. Daly, S. El Serafy, and B. von Droste, eds. Environmentally sustainable economic development: Building on Brundtland. Paris: UNESCO. 15–27.
Greijn, H. 1994. Let the figures speak for themselves. Ecoforum 18(1):9.
Jackson, R. D., and S. B. Idso. 1975. Surface albedo and desertification. Science 189:1012–13.
Jones, C. G., and M. Shachak. 1990. Fertilization of the desert soil by rock-eating snails. Nature 346:839–41.
Krebs, J. R., and J. M. Coe. 1985. Sahel famine: An ecological perspective. Nature 317:13–14.
Lee, K. N. 1993. Greed, scale mismatch, and learning. Ecological Applications 3:560–64.
Lubchenco, J., A. M. Olson, L. B. Brubaker, S. R. Carpenter, M. M. Holland, S. P. Hubbell, S. A. Levin, J. A. MacMahon, P. A. Matson, J. M. Melillo, H. A. Mooney, C. H. Peterson, H. R. Pulliam, L. A. Real, P. J. Regal, and P. G. Risser. 1991. The sustainable biosphere initiative: An ecological research agenda. Ecology 72:371–412.
Ludwig, D. 1993. Environmental sustainability: Magic, science, and religion in natural resource management. Ecological Applications 3:555–58.
Ludwig, D., R. Hilborn, and C. Walters. 1993. Uncertainty, resource exploitation, and conservation: Lessons from history. Science 260:17–36.
Mangel, M., R. J. Hofman, E. A. Norse, and J. R. Twiss. 1993. Sustainability and ecological research. Ecological Applications 3:573–75.
Naeem, S., L. J. Thompson, S. P. Lawler, J. H. Lawton, and R. M. Woodfin. 1994. Declining biodiversity can alter the performance of ecosystems. Nature 386:734–37.
Noy-Meir, I. 1973. Desert ecosystems: Environment and producers. Annual Review of Ecology and Systematics 4:25–41.
O'Neill, R. V., D. L. DeAngelis, J. B. Waide, and T. F. H. Allen. 1986. Hierarchical concepts of ecosystems. Princeton, N.J.: Princeton University Press.
Otterman, J. 1974. Barring high-albedo soils by overgrazing: A hypothesized desertification mechanism. Science 186:531–33.
Puigdefabregas, J. 1995. Desertification, stress beyond resilience: Exploring a unifying process structure. Ambio 24:311–13.
Safriel, U. N., Y. Ayal, B. P. Kotler, Y. Lubin, L. Olsvig-Whittaker, and B. Pinshow. 1989. What's special about desert ecology? Introduction. Journal of Arid Environments 17:125–30.
Schlesinger, W. H., J. F. Reynolds, G. L. Cunningham, L. F. Huenneke, W. M. Jarrell, R. A. Virginia, and W. G. Whitford. 1990. Biological feedbacks in global desertification. Science 247:1043–48.
Shmida, A., M. Evenari, and I. Noy-Meir. 1986. Hot desert ecosystems: An integrated view. *In* M. Evenari, I. Noy-Meir, and D. W. Goodall, eds. Ecosystems of the world. Vol. 12A: Hot deserts and arid shrublands. Amsterdam: Elsevier. 379–87.
Sinclair, A. R. E., and J. M. Fryxell. 1985. The Sahel of Africa: Ecology of a disaster. Canadian Journal of Zoology 63:487.
U.N. Environment Program (UNEP). 1992. World atlas of desertification. Editorial commentary by N. J. Middleton and D. S. G. Thomas. London: Edward Arnold.
———. 1995. U.N. Convention to Combat Desertification. Geneva: UNEP.
Whitford, W. G., D. W. Freckman, L. W. Parker, D. Schaefer, P. F. Santos, and Y. Steinberger. 1983. The contributions of soil fauna to nutrient cycles in desert systems. *In* P. Lebrun, H. M. Andre, A. De Medts, C. Gregoire-Wibo, and G. Wauthy, eds. New trends in soil biology. Proceedings, eighth international colloquium of soil zoology. Ottingnies-Louvain-La-Neuve: Dieu-Brichart. 49–59.
Williams, M. A. J., and R. C. Balling Jr. 1994. Interactions of desertification and climate. Geneva: UNEP and WMO.
World Committee on Environment and Development (WCED). 1987. Our common future. The Brundtland Report. Oxford: Oxford University Press.

PART 3 *Arid Lands Management Principles and Analyses*

Introduction

The previous section began to define the management situation for the various biophysical arid and semi-arid lands of the world. This section addresses the management situation directly in the initial chapter, which provides an overview of the arid and semi-arid land situation in Israel, North America, and Australia. Management involves planning for action. The next three chapters of this section define a planning process and describe a simulation analysis approach and an operations research approach to management planning. The latter two methods of planning analyses are alternative methods, depending on the questions and objectives of the management planning.

The first chapter of this section, by James T. Fisher, Mark Stafford Smith, Rafael Cavazos, Hugo Manzanilla, Peter Ffolliott, David Saltz, Mike Irwin, Theodore W. Sammis, Darius Swietlik, Itshack Moshe, and Menachem Sachs, begins with an overview of the biophysical resources, land use and management practices, and land management issues for the arid and semi-arid lands in Israel, North America, and Australia that were the focus of the workshop on which this book is based. The second section of the their chapter examines the similarities and differences among the three arid lands regions, each influenced by their biophysical resources, management actions and policies, and the social and political human dimensions of these areas.

In chapter 10, Peter F. Ffolliott, Itshack Moshe, and Theodore W. Sammis describe the essentials of the planning process used for land management agencies: the development of objectives, understanding of constraints, and selection of analysis techniques for evaluation of alternative actions. They suggest that planning occurs in the context of both anthropocentric and ecocentric factors. The planning process is iterative and the steps that are included typically and described here include monitoring, issue identification, objective setting, identification and evaluation of alternatives, and a recommendation and decision on alternatives for management actions.

In the management of metaecological systems, as is almost always the case (as opposed to kinds of ecological systems, such as communities), the analysis process must be capable of assisting decision makers in evaluating complex alternatives. Rarely can a decision maker do such an analysis without some assistance from a quantitative algorithm. If the question that the decison maker wishes to answer tends toward "what will happened if I do this" action, simulation models are of assistance. On the other hand, if the decison maker is interested in the answer to a question regarding optimal use of land, labor, and capital, then an operations research approach is appropriate.

Geoffrey Pickup and Mark Stafford Smith have described approaches to simulation that have relevance to arid lands and focus on the importance of ecological characteristics of temporal and spatial heterogeneity. They review a variety of models, discussing the strengths and weaknesses of each in terms of management decision making.

John Hof describes the rationale for considering the use of optimization models for planning purposes and then offers examples for the use of this analysis method. In addition to the traditional use for determining the optimum mix of management actions for the output goals and investment levels, Hof demonstrates the use of optimization models for a species richness or biodiversity objective and for static and dynamic spatial optimization objectives.

9 Land Use and Management: Research Implications from Three Arid and Semi-Arid Regions of the World

James T. Fisher, Mark Stafford Smith, Rafael Cavazos, Hugo Manzanilla, Peter F. Ffolliott, David Saltz, Mike Irwin, Theodore W. Sammis, Darius Swietlik, Itshack Moshe, and Menachem Sachs

The variety of arid and semi-arid systems in the world is immense; they are located in areas with different temperatures, soil fertility, seasonality of rainfall, and elevations. Further diversity is introduced by different human social systems whose impact is affected by remoteness, access to water resources, general level of affluence, political systems, and land tenure, among many other historical and cultural factors. This variety of arid and semi-arid regions precludes too many sweeping generalities about their use. However, in all areas, humans need to manage their biological resources for sustainability and, where possible, to restore the biological integrity of degraded lands. Meeting this challenge on a global or even regional scale is beyond individual scientific or management disciplines. This chapter, therefore, describes a number of different land-use systems from around the world, including their underlying resources and management problems, and aims to stimulate creative interdisciplinary approaches to ecological sustainability.

By necessity, the chapter's geographic coverage is limited by the expertise of the authors. We describe the resources and management of three arid and semi-arid regions in detail: the southwestern United States and northern Mexico (referred to here ecologically as "arid North America"), the Negev in Israel, and Australia. These regions do not represent the world's full diversity of arid and semi-arid lands, but they do include systems with different levels of investment, different levels of aridity, different patterns of resources, and, consequently, different priorities for land use. We highlight the similarities and differences between these regions and then seek to identify and discuss the factors that lead to different types of management problems. It is these factors and problems that must ultimately direct our science toward usefulness. A final section lists some specific research topics and approaches to research.

Regional Examples

The purpose of this section is to describe three arid systems that were of special concern to participants at the workshop and then extract some useful similarities and differences that can help to generalize research problems.

Southwestern United States and Northern New Mexico

Biophysical Resources Extensive and diverse arid and semi-arid ecosystems are found in the southwestern United States and northern Mexico between 24° and 38° north latitude and 103° and 118° west longitude. For conciseness, we refer to this region as arid North America. It incorporates the Chihuahuan, Sonoran, and Mohave deserts (Brown 1982, Brown and Gibson 1983). The Chihuahuan and Sonoran deserts cover 728,000 km² in arid North America. The Sonoran desert lies below 600 m in altitude and is generally hotter than the Chihuahuan, which lies above 1200 m. The semi-arid zones of arid North America are a discontinuous prolongation of both deserts (MacMahon and Wagner 1985, Medellin 1983).

A notable feature of this region is the presence of mountains delimiting deserts and semi-arid high plateau provinces. The eastern edges are defined by the Sierra Madre Oriental, the western edges by Sierra Madre Occidental, and the arid to semi-arid southern plains and high plateaus by the Trans-Mexican Volcanic Belt. A system of scattered mountain islands of temperate subhumid vegetation (i.e., Madrean Archipelago [Peet 1988]) extends into the United States. These "sky island" regions have played a major role in determining the species richness of both Mexico and the United States and continue to provide a critical corridor for floral and faunal migrations as they did during the glacial epochs (Axelrod and Raven 1985). Although these mountains may receive more than 800 mm precipitation yearly, they may also have periods of prolonged aridity. The management of arid and semi-arid regions is interwoven with that of the adjoining mountains.

Precipitation Patterns Rainfall in the eastern portions of the region (characterized by the Chihuahuan desert) depends on summer monsoonal flows. Rains begin earlier in the southernmost areas, providing moisture to central Chihuahua a few weeks before it is received in southern New Mexico, generally in mid-July (Mosino-Aleman and Garcia 1974). The flow begins retreating toward the equator in August. Winters are dry and relatively mild. Two precipitation periods occur in the Sonoran desert of the central region, one in the hot summer and another in the cool winter. While total annual rainfall is generally less than 300 mm, its distribution into two seasons of the year partly explains the relatively dense cover of vegetation that is common to the Sonoran desert (MacMahon and Wagner 1985, Medellin 1983). Precipitation falls in the western portions of the Madrean Archipelago (including the Mohave desert) in the autumn and winter. Summers are hot with no rains while winter temperatures are relatively mild. Because monsoonal rains begin earlier and persist longer in the Sierras west of the Chihuahuan desert, forest growth is more rapid in this region than in Madrean Archipelago and other mountain areas farther north.

Topography and Soils These lands have an extremely varied topography, and rivers and intermittent streams traverse the floodplains at lower elevations. Soil characteristics are also influenced by the distribution of coarse and fine soil fractions from transportation processes, reworking and deposition by wind and water, and periodic inundation of floodplains, all of which contribute to soil nutrients and leaching. Aridity means that the ability of soils to absorb and retain water is of critical importance. Soil depth largely governs the amount of water that can be held in a single profile. However, soil depth and the extent of the rooting zone are often limited by a hardpan layer—carbonate layers are commonly found 5–60 cm below the soil surface. Low plant productivity, high temperatures promoting decomposition, and intensive weathering of minerals all contribute to soils with relatively low organic content and often nitrogen deficiencies.

Vegetation The vegetation types of arid North America are correlated with environmental conditions (Ehleringer 1985). The main vegetation types include desert shrubs and desert grasslands. Chaparral shrublands, pinyon-juniper and oak woodlands, and montane pine forests are found at middle elevations of the mountains in the region.

1. *Desert shrublands:* Warm desert shrubs typified by creosote bush (*Larrea divaricata*) occur at lower elevations in the foothills and valleys. The Sonoran desert is richer in associated species due to its bimodal precipitation pattern, providing a climate favorable for species such as *Cercidium microphyllum, C. floridum, Olneya tesota, Ambrosia dumosa, Cereus gigantea,* and many cool-season annuals. Characteristic plants in the Chihuahuan desert are *Larrea, Fluorensia cernua, Parthenium incanum,* and others. *Prosopis* and *Atriplex* are common throughout.
2. *Desert grasslands:* The desert grasslands occur at a higher altitude than the desert shrublands. *Bouteloua curtipendula, B. gracilis, Eragrostis intermedia,* and *Hilaria belangeri* are characteristic species in the higher altitude grasslands while lower grasslands contain more desert species, including *Bouteloua eriopoda* and *Muhlenbergia porteri.* Shrubs and cacti such as *Prosopis, Acacia* (in northern Mexico), *Fluorensia, Larrea, Fouquieria, Yucca, Agave, Opuntia,* and *Dasylirion* occur in varying densities. Since the 1880s, shrubs have increased relative to grasses in transition areas between desert shrub- and grasslands. Factors responsible for increasing shrubs include heavy selective cattle grazing, livestock dissemination of shrub seed (e.g., *Prosopis glandulosa*), and periodic drought (Buffington and Herbel 1965).
3. *Chaparral shrublands:* Chaparral ranges in elevation between desert shrubs and ponderosa pine vegetations, occupying deeper, coarse-textured soils; juniper woodlands or grasslands are more common on the heavy soils (MacMahon and Wagner 1985). The communities are dominated by evergreen sclerophyllous shrubs with small coriaceous leaves, including various species of *Quercus* and *Arctostaphylos* spp., *Ceanothus greggi, Cercocarpus, Rhus,* and *Adenostoma.*

Riparian Vegetation In the southwestern United States, riparian vegetation is associated with perennial streams and rivers or with seasonally wet drainage systems. Drainage systems that are ephemeral with regard to carrying or holding water include arroyos, closed basins, alkali sinks, swales, and playas (Dick-Peddie 1993). The floodplain portion of this vegetation type typically occurs near meandering river systems such as the Rio Grande. Undisturbed floodplains can support mixed cottonwood (*Populus fremontii*) and willow (*Salix goodding*) forests (Minckley and Brown 1994). However, the redirection of natural water courses, woodland clearing, and livestock grazing have promoted the rapid and expansive spread of alien salt cedar (*Tamarix chinensis*) along floodplains by limiting the reproduction of native trees. Mesquite (*Prosopis* spp.) thickets, or basques, are also found along rivers as are diverse shrub species. Shrub-trees (e.g., *Chilopsis linearis*) and nonsclerophyllous shrubs are commonly associated with arroyos running through desert shrublands (Dick-Peddie 1993). Closed basins are usually broad, flat, or gently sloping areas where water tends to spread out rather than form gullies. Closed basins often become salinized and support salt-tolerant plant species (e.g., *Atriplex canescens*). Normally dry, closed basin lakes called playas are found throughout the deserts of the Southwest (Minckley and Brown 1994).

1. *Pinyon-juniper woodlands:* These woodlands represent transition communities between the colder montane pine forests at higher elevations and desert grasslands at lower elevations in generally more humid areas. Endemic elements are *Juniperus osteosperma,* sometimes associated with *J. monosperma,* and often with *Pinus edulis* or *P. cembroides.* The vegetation type contains few other endemic species with an understory typical of adjacent types.

2. *Oak woodlands:* This woodland type contains open to moderately dense stands most commonly dominated by *Quercus emoryi.* It ranges from lower elevation open savannas with grasses of the upper desert grassland to denser woodlands at higher altitude containing junipers, pines, and chaparral species.

3. *Montane (pine) forests:* These forests are found on the middle elevations of the mountains scattered throughout arid North America. A variety of pines, dominated by *Pinus engelmannii* and *P. arizonica,* in northern Mexico are largely replaced by a single species, *P. ponderosa,* in the southwestern United States. Scattered pinyon, juniper, and oak are associated with the pines, and a variety of grasses, forbs, and half-shrubs form the understory.

Wildlife The Chihuahuan and Sonoran deserts of arid North America contain critical habitats for more than 350 species of mammals, birds, amphibians, fish, and reptiles. Game species include javelina (*Diocotyles tajacu*), bighorn sheep (*Ovis canadensis*), mule deer (*Odocoileus hemionus*), Gambel's quail (*Lophortyx gambelii*), mourning and white-winged dove (*Zenaida macrorura* and *Z. asiatica*). Among the predators are bobcat (*Felis rufus*) and coyote (*Canis latrans*). Threatened and endangered species include the peregrine falcon (*Falco peregrinus*) and others (MacMahon and Wagner 1985, Martin and Patton 1981).

Wildlife is especially dependent on riparian zones that are near surface water with an abundance of mesic tree and shrub genera (e.g., *Populus, Salix,* and *Alnus*) excluded from the surrounding drier areas. Ephemeral, as well as permanent streams, can create these zones that contribute greatly to wildlife diversity. The riparian zones associated with major rivers and their tributaries provide migratory corridors for neotropical birds and are of critical importance to numerous endangered predatory birds. Water diversion, livestock grazing, and recreation can drastically alter the suitability of riparian zones for wildlife.

Land Use and Management Practices There are many land uses in arid North America. Because most of the United States in this region is owned by the public, society plays a major role in determining how federal, state, county, and other public agencies manage land and prioritize uses. Uses on large areas controlled by indigenous American people depend primarily on the decisions of representative tribal or communal councils. In the distant past, priorities set by these groups often reflected long-standing traditions. However, tribes have become more commodity-oriented as they have become more assertive in demanding rights from nineteenth-century treaties. In Mexico, *ejidos* and agricultural *comunidades* have historically used land in accordance with rural community goals. In Mexico and the United States, land uses reflect the values that society places on these lands.

Livestock Production Livestock grazing occurs throughout the region. The principal types of livestock raised under largely unconfined conditions are cattle, sheep, goats, and horses. Nearly 8.5 million cattle, 1.9 million sheep, and 1.7 million goats graze these rangelands annually (Downing and Ffolliott 1983). Estimates of the numbers of horses that graze in the region are unavailable. Herd compositions in the southwestern United States differ markedly from those in northern Mexico, however. Cattle remain most common throughout the region, but an increasing proportion of goats graze Mexican ranges, providing milk as well as meat.

The value of rangelands for sustainable livestock production is often constrained as these lands are fragile, easily damaged by abuse, and subject to droughts (Byington 1990, Herbel 1979, Holechek et al. 1989). Range management practices may be sustainable when properly planned and implemented, although improvements (e.g., watering devices and fencing) are still inadequate in Mexico. There has been widespread clearing of the woody overstory to increase forage biomass, even though forage yield eventually reverts to pretreatment levels. Livestock grazing systems that are common in the region include year-long (normal in Mexico), seasonal, and rotational grazing. Selection of grazing systems depends largely on the type of livestock grazed, forage availability, rangeland condition, and carrying capacity, as well as prevailing weather patterns.

Agricultural Crop Production Both large-scale commercial crop production and small-scale rain-fed farming are practiced in the region, with the system depending largely on land productivity and the ability to invest capital. Commercial crop production is rarely possible without large-scale irrigation systems, however. Small-scale farming to produce seasonal crops is common where local water supplies are sufficient, with crop production mainly for family subsistence or for sale to local markets. Commercial farming generally poses more rigorous technical and economic challenges than small-scale farming. A wide variety of crops are produced in the region, including cereals (wheat, maize, barley, oats, rye, sorghum), legumes (beans, soybeans, peas), root crops (potato), leaf vegetables (cabbage, lettuce), fruit and seed vegetables (tomato, pepper, pumpkins), fruit (grapes, citrus, some apples), and forage crops (alfalfa, clover, and other annual and perennial grasses preferred by livestock).

Mexico's agricultural production has undergone considerable recent change to balance trade, primarily with the United States. An original emphasis on government subsidies to capital-intensive agriculture and an interventionist approach to crop pricing caused major shifts in land uses (Nuccio 1991). With corn prices held down for the sake of the urban population, for example, growers produced sorghum for livestock feed instead. Such distortions eventually created a trade imbalance in basic foods. In recent years, there have been major reforms to encourage production. By lifting restrictions on foreign investments and forms of company ownership, Mexico is providing a source of much needed working capital to its agricultural communities. In essence, reorganized ejidos provide land and labor while foreign companies provide the capital and advanced technology needed for modern production.

The full impact of the recently negotiated North American Free Trade Agreement (NAFTA) on agricultural crop production in the region is as yet unknown. Growers north of the Mexican border will probably have a greater opportunity to respond to market demands created by NAFTA than others. However, U.S. growers must deal with the issue of urban encroachment and reallocation of water. Current trends suggest that the area of cultivated lands in the southwestern United States will continue to decline (on a per capita basis) in response to population growth.

Wildlife Resources The wildlife resources of the arid and semi-arid ecosystems of arid North America have both "consumptive" and "nonconsumptive" values. In terms of consumptive use, management activities are centered on big game and, to a lesser extent, on small game species that are widely hunted in the region. Game animals (including elk, deer, squirrels, dove, and quail) furnish hunting opportunities for many people. Hunting success for big game ranges is 15–25% for big game species. Hunting success for small game is more difficult to quantify. The hunting seasons for big and small game frequently overlap, and, as a consequence, other small mammal species are often hunted as "buffer species" when big game is scarce.

In recent years, wildlife management agencies in the region have devoted more attention to the nonconsumptive uses of wildlife. Nongame species of birds, rodents, and reptiles play important roles in ecosystem functioning. They also possess aesthetic values (e.g., for bird-watching) that are unfortunately difficult to measure. Accordingly, species and habitat management practices aimed at nongame species continue to be developed, as well as efforts to conserve wildlife species categorized as rare or endangered. To maintain stable, diverse wildlife populations, it is necessary to foster favorable habitat conditions while achieving a balance among factors regulating population size. Maintaining this balance is largely the responsibility of wildlife management agencies, although nongovernmental organizations (NGOs) are increasingly involved.

Forestry and Other Woody Production Systems Forestry is diverse and widespread, including amenity planting and soil conservation practices, as well as conventional timber production and wood processing. Policies governing harvesting practices and quantities are becoming more restrictive in response to public concerns about the environment. Windbreak and streambank plantings help to control the erosive effects of wind and water on soils (Dregne 1983). Salinized agricultural lands have been reclaimed through the planting of salt-tolerant plants (halophytes) such as *Atriplex* and *Salicornia* (Goodin et al. 1990). The production and protective functions of forestry are often linked to provide multiple products in addition to more indirect benefits. The term *agroforestry* collectively represents those land-use systems and practices in which trees and shrubs are combined deliberately with agricultural crops or livestock (Raintree 1987). In areas exposed to strong winds, the dominant form of agroforestry is the establishment of tree and shrub windbreaks (Byington 1990); these stands protect crops and livestock, as well as provide habitat for game. *Silvipastoralism* represents another major type of agroforestry in the region, mixing forestry with grazing practices.

In many rural areas, wood (especially mesquite) continues to be a principal cooking and heating fuel and source of fence posts. Pinyon is a valuable source of edible nuts and Christmas trees. Shrubs and trees are planted for amenity purposes in parks, buffer strips along streets and sidewalks, roadside plantations, and greenbelts around towns and villages throughout the region, largely for beautification and protective purposes. In many cases, the species planted are native to the area and only require irrigation for establishment, thus conserving the limited water resources. Planting of home gardens is also a long tradition of land use in the region.

Use of Xerophytic Plants A number of xerophytic plants are used commercially in arid North America (Carr et al. 1986, Foster and Vardey 1987, Nabham and Felger 1985), including guayule (*Parthenium argentatum*), a rubber-producing perennial shrub native to the Chihuahuan desert, and jojoba (*Simmondsia chinensis*), which grows in the Sonoran desert; jojoba seed oil can substitute for sperm whale oil in industrial products. Species of the genus *Lesquerella* are also oilseed producers, homologous to castor oil; one species, bladderpod (*L. fendleri*), can be grown with less water and on sites too poor for castor-oil plant (*Ricinus communis*). Numerous other xerophytes have economic potential and are now receiving attention, including *Cucurbita* spp. (e.g., buffalo gourd, *C. foetidissima,* with roots high in starch that can be hydrolyzed to sweeteners) and *Grindelia* spp. (gumweed: e.g., *G. camporum* has high yields of crude resin). Other products are wax from candelilla (*Euphorbia antisiphillitica*), fiber from lechuguilla (*Agave lecheguilla*) and *Nolina* spp., and chemicals from *Yucca* spp.

Water Resources Much of the region's water is in the large rivers that originate in the mountains; however, groundwater is recharged locally by deep drainage through the soil. Surface runoff events, soil moisture storage, and groundwater recharge are more variable and less reliable than in wetter climates, but the large rivers provide more reliable water than in other arid regions. The critical importance of watersheds in arid and semi-arid ecosystems is sometimes neglected, resulting in severe erosion, reduced or redirected stream flow, and reduced water quality because of increased stream turbidity. The many watershed management programs include promotion or enforcement of best management practices in silviculture and agriculture. In arid North America, programs can be grouped into activities that (1) minimize adverse impacts to soil and water resources, (2) increase the availability of high-quality water supplies, and (3) restore watershed productivity:

1. The degradation of soil and water resources is usually associated only with intense rainfall events. Management programs focus on protecting watersheds against this. Many of the actions are similar to those employed to control erosion.

2. Most high-quality water in the region is already appropriated for human use so that most additional water for the expansion of population and other developments is pumped from underground basins, often at rates exceeding recharge. Water resources are being augmented by water harvesting, the reuse of wastewater, water salvage by removing riparian vegetation, and vegetation modification on upland watersheds.

3. Management to rehabilitate watersheds includes the control of gullies and mass wasting with upstream check dams, establishment of tree or shrub cover on degraded sites, and curtailment of wood harvesting, livestock grazing, and other exploitation. Degraded watersheds are difficult to rehabilitate, so emphasis is placed on avoiding damage.

Urbanization and Environmental Degradation Urban areas immediately north of the U.S.-Mexican border have witnessed the highest population growth in the United States over recent decades. Mexico's border town populations reflect its rapid national growth (from 85 million in 1990 to 200 million by 2010) (Nuccio et al. 1990) and resettlement. Within the United States, people have moved to the Southwest in pursuit of a milder climate, less congested living, and work. The expansion of urban areas puts pressure on habitats for wildlife species, increases uncontrolled recreation, withdraws water from nonrenewable supplies, and threatens regional water quality. More than 60% of Mexico's streams and rivers are already contaminated (Mumme 1992). Air pollution, a minor concern in the past, is becoming more severe, and the public increasingly objects to the impacts of agriculture, maquilidora industries, and waste disposal. In Mexico, fire remains a major form of disturbance on both rangelands and forested lands and is often related to agricultural encroachment into natural plant communities. Resolution of these conflicting land uses is a major problem confronting planners in the region.

A key issue in the public debates concerning NAFTA was the environmental risks created by industrial relocations, the international transport of potentially hazardous products, and specific effects on water supplies (Manzanilla 1992). Before NAFTA, the United States and Mexico agreed on several measures to reduce the risk of soil and water contamination (Herrera-Toledano 1992), and Mexico reorganized its environmental protection agency (SEDESOL, Secretariat of Ecology and Social Development) to meet demands for more rigorous enforcement of industrial regulations on both sides of the border. Monitoring for environmental compliance has increased sharply (Olivier 1993), and Mexico has begun closing or suspending the operations of offending companies. U.S. companies operating in Mexico believe economic opportunities arising from NAFTA will lead to improved environmental protection in Mexico (Kelso 1992).

Tourism and Recreation Recreation and tourism are growing rapidly in the Madrean Archipelago and represent one of the largest income generators in arid North America. The values of many of these activities (e.g., hunting and fishing) can be partially quantified in an economic sense through the expenditures for licenses, permits, and purchase of equipment. In contrast, the value of activities such as camping, hiking, and picnicking are harder to determine but are certainly a major part of tourism. Unfortunately, there are no comprehensive statistics on the levels of these nonconsumptive uses. For whatever reason, however, a growing number of both local people and other visitors are enjoying the unique recreational and aesthetic resources of the region. The pressures of increasing recreation and tourism are causing planners and managers to furnish amenities more explicitly.

Conservation Reserves In the southwestern United States, the National Park Service administers sixteen national parks, national monuments, and recreational areas, many of which contain unique components of arid and semi-arid ecosystems. The forest, woodland, and grassland resources on fourteen of these sites have been protected from exploitation for fifty years or more. Importantly, relatively high concentrations of archaeological sites are also found in these areas. Wilderness areas have been and continue to be designated in the region in response to the public's interests in protection of these areas. In northern Mexico, there are five biosphere reserves covering 42,500 km^2 (components of UNESCO's Man and the Biosphere Program), as well as eight national parks and five forest reserves, all under the control of federal and state agencies.

Mining Mining has been important historically to the economy of people throughout arid North America. Surface mining of copper, molybdenum, sand and gravel, silver, gold, gem stones, and other minerals is still widespread. There are also large coal deposits, with the area's largest deposit located on the Black Mesa of northern Arizona. Much of the world's copper comes from the region, although reduced prices have curtailed copper mining in recent years.

Active and abandoned mining sites frequently scar the landscape and sometimes leak toxic chemicals into stream systems. Reclamation efforts are determined by regulations, but existing laws do not adequately address either the environmental or economic consequences of changing the level of mining. Reform advocates are attempting to change these laws, resulting in conflicts between environmentalists and representatives of the mining industry.

Military Military units are garrisoned throughout arid North America, although military activities are more intensive in the southwestern United States than in northern Mexico. Air force and army units conduct extensive training activities in desert terrain in Arizona, New Mexico, and California. Environmental impact statements are frequently required before exercises are allowed, and considerable resources are now expended on environmental research by the army's Construction Engineering Research Laboratory (CERL). Conflicts about the use of limited water resources in southeastern Arizona have placed the military at odds with environmentalists and water resource managers. A potentially contentious relationship also exists in El Paso, Texas, where 40–45% of the underground formation that supplies the city's groundwater is located beneath Fort Bliss Military Base (Holland 1989).

Current Management Issues Arid North America will continue to be used for livestock production. However, the ecosystems also have values for wood production, wildlife resources, water management and recreation, and tourism, and they are under pressure from urbanization. As these demands continue to increase, there is a growing awareness of the need to plan and implement comprehensive management practices that integrate the multiple values on a long-term, sustainable basis.

The multiple-use philosophy of land management served the people of arid North America reasonably well into the 1980s. Lands were characterized largely in terms of their capacities to provide commodities and amenities. Research aimed to identify the factors that limited the realization of these capacities, and a key objective of management was to reduce or remove these limitations. Decisions about the resources required the determination of optimum yields among the desired, and often competing, uses (Kessler et al. 1992). However, multiple use is not necessarily the best way to approach management when stakeholders begin to ask how to balance a wide range of potential uses and values, which is the situation in the region today. An alternative paradigm, the ecosystem approach, has been put forward with the intent of identifying natural resource management with ecology, a field that Odum (1977) called a holistic science. The ecosystem approach broadens the multiple-use paradigm to one viewing the comprehensive context of living systems (including soils, plants, animals, minerals, climate, water, topography, and all of the ecological processes that link them together). In this view, the functional properties of the system are not compromised to optimize the use of a few resources, and system sustainability is not sacrificed for short-term gain. Terms frequently associated with the ecosystem approach are sustainable development, sustainable resource use, and sustainable economic growth. These terms have been defined in various ways but, in one way or another, usually express concern about the effects of present-day activities on the future ecosystem, the importance of maintaining ecological processes, and the benefits of improving the quality of life now without denying future generations a similar opportunity (Young 1992).

Regardless of the management paradigm, the roles of public participation in endorsing management practices are changing in the region. In the past, stakeholders' roles have largely been to respond to professionally prescribed alternatives that estimate outputs of management in terms of cords of fuelwood, animal-unit months of grazing, recreational user days, or breeding pairs of Mexican spotted owls. The public now frequently expresses the need for management to operate harmoniously with nature rather than exert more control and unbridled dominance (Norton 1988). The public is particularly concerned about decisions regarding water and is demanding a higher level of accountability for the consequences of management decisions (Getches 1993). Researchers and managers are, therefore, challenged to provide new approaches that are more congruent with the current views of the public. Some of these issues are summarized in tables 9.1, 9.2, and 9.3.†

Table 9.1. Some Trends in Management Problems and Responses in Relation to Different Land Uses in Australian and Mexican Arid and Semi-Arid Lands, at the Individual Enterprise and the Regional Scales.

Land-Use Issue	Local/Enterprise Scale	Regional/Societal Scale
Grazing	Economic viability problems Intensification of management Diversification Rehabilitation and control of feral animals Use of native and exotic species Increased use of formal adaptive management	Increased concerns about sustainability Abandonment due to losses of viability Pressure to fit with regional conservation planning Formal assessment under international conventions
Recreation/tourism	Continuing rapid growth Greater control of local human impact Management of private off-road vehicles	Stratification of access More regulation of land use Greater role in resource conservation
Traditional uses by indigenous people	Health, security, and education concerns Resolution of internal land-use conflicts Understanding of acceptable harvesting levels	Increased community concerns about sustainability More incorporation of community concerns into regional conservation planning
Conservation management	More weed/pest control Conservation agreements with landholders New approaches to reintroduction of endangered species	Better understanding of ecosystem conservation Better import controls on weeds Control of artesian water use Better regional reserve/off-reserve coordination
Marginal dryland/irrigated agriculture	Economic viability problems Increased use of agroforestry More management of salinity Better control of water resources and efficiency of water use Better agricultural techniques	"Regional adjustment" (regional planning for optimal land use) Rural community decline Community concerns about sustainability Diversification in agriculture
Mining	Continued reclamation and restoration	Exploration increasingly remotely sensed
Information transfer	"Participatory problem solving"	Regional objective setting

Israel: Land Use in the Negev Desert

Biophysical Resources The Negev desert covers about 11,000 km^2 south of the latitude 31°37'N, forming a triangle bounded by the Mediterranean Sea and Sinai desert in the west, the Arava and Dead Sea Valleys in the east, and Elat on the Gulf of Aqaba at its southern tip. Mean annual rainfall ranges from 400 mm in the northwest to 30 mm in the southern Arava Valley. Precipitation occurs in winter, mainly between December and March, and is highly variable over time (monthly and annual amounts) and space.

Regional Climates and Soils The Negev can be divided into five major geographic regions that are associated with a north-to-south rainfall gradient and have distinctive soil types:

1. the northern Negev, including the southern and western inclines of the Hebron Mountains and the southern region of the coastal plain (250–400 mm annual rainfall); soils are light rendzina and dark brown loess;
2. the western Negev, characterized by light, sandy soils (250–300 mm annual rainfall);
3. the central Negev, characterized by serozeum soil with poor water penetration due to surface crusting (150 mm annual rainfall);
4. the Negev highlands, 400–900 m above sea level, characterized by relatively extreme temperatures, lithosoils on the slopes, and colluvial soils in the valleys (100–150 mm annual rainfall); and
5. the Arava, Dead Sea Valley, and southern Negev, with saline soils (30–80 mm annual rainfall).

There has been no significant change in the Negev's climate for the past 3,000 years. Despite the harsh conditions, the Negev desert and its land resources have supported a human population for the past 4,000 years. Between the fourth century B.C. and seventh century A.D., urban and rural settlements survived by collecting runoff water in cisterns and developing runoff agriculture. In the seventh century, migrating tribes invaded the area and the settlements were abandoned. The runoff agricultural systems that cov-

Table 9.2. Specific Research and Development Issues Critical for Analyzing the Sustainability of Pastoral Land Use in Australia

a. Collate appropriate socioeconomic survey work to properly define what different categories of managers perceive as adequate profitability to be viable.
b. Collate information on biological rates for existing enterprise types, initially using existing records but also developing plant and animal production models to allow prediction.
*c. Extend predictive point-based models of forage production and forage utilization to more rangeland vegetation types; priorities can be set, see j (below).
*d. Develop a more comprehensive database of the appropriateness of different breeds, herd structures, and management technologies for forage use and forage-use control.
*e. Develop a better understanding of the resilience of soils and vegetation to stocking pressure, focusing on the thresholds at which change takes place and on recovery.
*f. Develop a better understanding of management techniques (other than fire, which is well understood) that use the knowledge of thresholds and landscape resilience for maintenance and rehabilitation.
*g. Apply point-based forage production models at the landscape scale using points to represent conditions in whole areas.
h. Extend grazing distribution models to more landscape types and animal species.
*i. Further develop landscape models that represent spatial processes in erosion, vegetation change, and grazing distribution to identify resilience at a landscape rather than point scale.
j. Identify critical areas for further ground-based research, monitoring, and management as a result of i (above); in particular, develop an increased understanding of the use of early warning or "sentinel" areas in the landscape.
*k. Determine methods of property planning and paddock management that reduce effects of spatial variability and can be used as alternative strategies in the assessment of sustainability.
l. Define socioecologically what scale of "sacrifice zone" is necessary and acceptable.
*m. Determine probability distributions for safe levels of forage utilization (i.e., develop the probability distribution through time of h and i [above]).
*n. Determine the existence of critical decision points when required stocking level changes could be forecasted.
o. Determine what the impact of errors would be on critical parts of the landscape (usually in drought conditions) and the cost of the consequences so an appropriate form of cost/benefit analysis can be made.
*p. Determine whether and what stock management strategies exist to handle variability in time.
q. Further develop and test an appropriate, multilayered, regional degradation monitoring capacity for government agencies.
r. Further develop and test appropriate on-property monitoring systems for managers (cf. Hacker et al. 1990).
*s. Further develop appropriate decision support systems with the involvement of managers to address matters that are problematic in management decisions.

Source: According to Pickup and Stafford Smith 1993.
Note: Asterisked items were seen as important also in arid North America.

Table 9.3. General Research and Development Issues in Relation to Land-Use Objectives Other Than Grazing in Arid North America

Biodiversity

a. Identify biological connnections needed to maintain the genetic flow between mountain ranges.
b. Intensify basic research on vertebrate and invertebrate community structures because of their high diversity.
c. Enhance baseline flora data, including extinctions and colonization, especially in Mexico.
d. Better understand how plant, vertebrate, and invertebrate communities respond to fire frequency.
e. Focus on the interactions between watershed condition and the health of riparian systems.
f. Enhance the scientific basis for ecosystem-level management in the context of a change in management priorities from maximal production in the past to system integrity today.

Wildlife

g. Develop a better understanding of the impacts of livestock grazing and residual vegetation cover on wildlife species such as quail, nongame birds, and deer.
h. Study the effects of fuelwood harvesting and other tree removal on wildlife habitats and species.
i. Develop a better baseline inventory of wildlife species, both endemic and visiting, and their habitat requirements from ecosystem management's viewpoint.

Forestry

j. Enhance basic ecophysiological research on regeneration in relation to different frequencies, cycles, and quantities of harvest.
k. Establish long-term landscape goals for forest and woodland communities, incorporating appropriate levels of diversity.
l. Improve knowledge on regulating forest and woodland communities at the stand level.
m. Develop the knowledge needed to specify rotation cycles for harvested tree species beyond the few species currently studied on a limited range of sites.
n. Understand the effect of different silvicultural treatments on overstory-understory relationships and their influence on plant and animal communities.
o. Study entomological, pathological, grazing, and fire influences on the sustainability of forest and woodland communities.

Water Resources

p. Develop a better understanding of hydrological processes and relationships between hydrological and nutrient cycles under different land management practices.
q. Construct better baseline data on rates of erosion and sedimentation to ascertain the effects of hydrological processes on ecosystem sustainability.
r. Improve understanding of the impacts of land management and natural events (fire, flooding, drought) on soil resources and long-term site productivity.
s. More actively incorporate soil and water conservation into existing land management programs.
t. Identify the water requirements of native plants, especially common herbs, in the context of watershed management.

Recreation and Tourism

u. Incorporate the private sector in the planning of recreation and tourism on publicly owned lands.
v. Develop an understanding of how to zone activities to minimize degradation.
w. Identify the appropriate level of fees to be charged for recreation and tourism activities.

Note: Many of these are also relevant to Australia. See also the list in Morton et al. 1994.

ered tens of thousands of hectares were destroyed. The nomadic tribes overexploited the environment by overgrazing and cutting wood, which accelerated soil erosion and desertification.

There are few permanent water sources. These include two streams along the Dead Sea that flow over a short distance (< 5 km) and less than twenty small springs with too limited a flow to produce a stream. Water flow in the dry riverbeds (wadis) occurs for several hours after rainfall, mostly in the form of flash floods.

Vegetation The Negev desert is a junction between four biogeographic zones, with species richness increasing with latitude (Ward and Olsvig-Whittaker 1993) and few species common to all four zones:

1. The Mediterranean zone in the north and northwest.

2. The Irano-Turanian zone, covering the central Negev. This consists mostly of steppe vegetation dominated by small bushes such as *Artemisia herba-alba, Zygophyllum dumosum, Gymnocarpos decander,* and *Anabasis articulata.* Boyko (1949) argued that the dominance of these species is due to centuries of overgrazing and that, in the past, the area was dominated by species of *Stipa* and *Stipagrostis.*

3. The Saharo-Arabian zone in the southern Negev. This is defined as true desert (Zohary 1962). Much of the dominant vegetation is similar to the Irano-Turanian zone, but it is restricted to the dry riverbeds due to the low annual rainfall (< 80 mm).

4. The Sudanian zone (Zohary 1973), along the Syrian-African crack (the Arava Valley) flanking the eastern end of the Negev. This is typified by savanna vegetation such as acacia and ziziphus trees. Like the Saharo-Arabian zone, Sudanian zone rainfall is below 80 mm and vegetation is restricted to the dry riverbeds.

Wildlife The Negev desert supports a variety of wild animals, many of which are endangered. The mammals include several ungulates: Nubian ibex (*Capra ibex nubiana*), dorcas gazelles (*Gazella dorcas*), and mountain gazelle (*Gazella gazella*), as well as a remnant subpopulation of *Gazella gazella acacia* in the southern Arava Valley consisting of less than twenty individuals, and carnivores such as wolf (*Canis lupus pallipes*), striped hyena (*Hyaena*), caracal (*Lynx caracal*), and leopards (*Panthera pardus*). There is no hunting, and the value of these populations is strictly nonconsumptive.

Several of these species were on the verge of extinction in the Negev in the recent past. In the past twenty years, the enactment of wildlife protection laws and exclusion of domestic grazing south of latitude 30°35'N has enabled many of them to rebound. Although many of the species are well adapted to desert living, the long-term persistence of their populations is strongly dependent on access to water and to the vegetation along riverbeds.

Current Land Use Practices in the Negev The Negev desert covers about 50% of the state of Israel, but only 6% of the country's residents live there. Almost all of the Negev lands are publicly owned. Indicated in table 9.4 are the major land uses, dominated by military and conservation interests (fig. 9.1).

Table 9.4. Major Land Uses and the Proportion of the Negev Involved

	Land Area[a]
Military zones (50% of the area) and nature reserves (36% of the area)[b]	69%
Agriculture	
rainfed	10
irrigated	6
Rangelands	10
Landscape enhancement	2
Mining	1
Urban (including roads)	3

a. Total area about 11,000 km^2.
b. These overlap considerably.

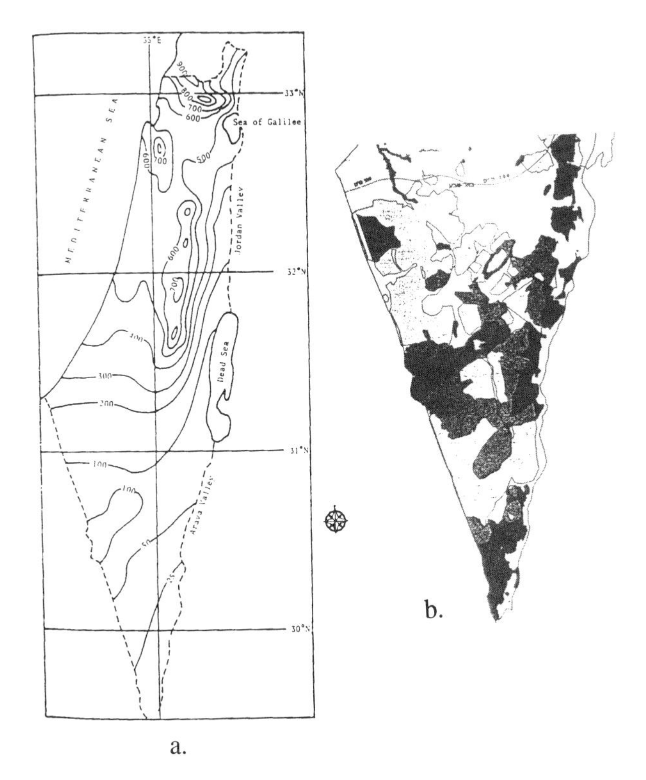

Figure 9.1. Climate and land use in Israel: (a) annual mean rainfall throughout the country (isohyet units are mm); (b) spatial distribution of the primary land uses of the Negev (the solid line is the 200 mm isohyet); black is nature reserves, light gray is military firing zones, and medium gray shows areas where these two uses overlap.

Management Objectives Several categories of factors drive the management of the Negev:

1. *Land uses:* The northern part of Israel is densely populated relative to the Western world. Thus, the Negev desert is the last available open area in Israel, and many activities requiring open space have been squeezed into this region. These include military training zones, polluting industries (in situ), urban waste disposal (imported from the north), and tourism and recreation.

2. *Economics:* Unique characteristics make the Negev desert suitable for producing out-of-season crops and for tourism, hiking, four-wheel driving, and rangeland grazing.

3. *Conservation:* Past overutilization of the land has led to degradation of soil, water, plant, and animal resources. Rehabilitation of degraded lands and conservation practices incorporate concerns such as soil and water resource conservation, afforestation and savannization as land development practices, and conservation measures to increase biodiversity, including designation of nature reserves and reintroduction of species that were locally extinct.

Land Use Problems and Management Approaches

Military The impact of military activities varies considerably depending on the type of exercise. Areas used for training ground forces are subject to significant physical disturbance by heavy off-road vehicles. Troop movements create dirt roads and dust, disturb the soil crust, and remove plant cover. In regions designated for air force practice, damage is localized while the large buffer zones around the targets remain relatively undisturbed.

Conservation

1. *Water and soil:* In the semi-arid regions, management for rehabilitation is at the watershed level. Contour plowing and cultivation on the slopes, and terracing in the valleys, are used to prevent erosion and runoff. In the more arid regions and when resources are limited, efforts are restricted units with high potential within the watershed. Stone and earthen dams are used to catch runoff water to enhance local vegetation growth. Grazing is controlled, trees are planted, and drainage conservation is employed to prevent future damage. Dams and off-stream reservoirs are constructed to collect runoff water for irrigation.

2. *Nature reserves:* Nature conservation is implemented through the declaration of nature reserves and the protection of endangered species. All animal and plant life and the physical environments are protected by law in nature reserves, and rangers in these areas have authority similar to that of police elsewhere. The movement of vehicles in nature reserves (including those overlapping with military zones) is restricted to specified roads. Particular areas may be closed to vehicles altogether. Hunting, grazing, and mining are not permitted within the reserves. Reserves located around habitats that are critical for wildlife are often fenced to exclude human interference, especially during reproductive periods.

In reserves that overlap with military training zones, military activity must be coordinated with the authorities and must comply with restrictions such as keeping to designated roads. Certain areas are allocated for intense activity (aggregation areas) to minimize damage elsewhere. Close cooperation between military authorities and local rangers is encouraged to facilitate open communication and effective coordination.

3. *Restoration:* Efforts are made to restore damaged areas by physical means, such as using volunteers to blot out off-road vehicle tracks, and by careful road planning so that runoff is unobstructed. Wild species that were locally extirpated are being reintroduced. These include Asiatic wild ass (*Equus hemionus*), with a herd of over fifty animals already in the wild, and white oryx (*Oryx leucoryx*), which will be reintroduced within the next few years. The north African ostrich (*Struthio camelus*) is also being considered for reintroduction.

Livestock Production Traditionally, land has been grazed by nomadic herds of mainly sheep and goats raised for subsistence purposes (Ben-David 1990). Today, international borders, nature reserves, military zones, and increased utilization have limited nomadic herd movements. Consequently, there is a shift in grazing strategies to a seasonal pattern. During the summer, the herds are grazed on the stubble fields of agricultural settlements. In winter, animals are stabled and fed supplements. Only in late winter and spring are the herds grazed in open and afforested areas. It is anticipated that there will be fewer large herds in the future and that livestock practices will increasingly conform to modern range management and husbandry techniques. Modern commercial grazing is limited to fenced areas with cultivated range plant communities, using waste water to irrigate fodder. Modern range management and husbandry techniques are used.

Tourism and Recreation Large-scale afforestation projects have created forest belts around urban areas of the semi-arid regions. Plantations made possible on marginal sites by soil and water harvesting improve the quality of life for local residents and increase the potential for recreational use. Artificial parks created by small-scale afforestation projects in the vicinity of natural water sources and settlements are especially attractive to recreational users. Visitor centers, located in various places throughout the Negev desert, offer guidance and help to travelers. Off-road vehicles are restricted to designated roads, thus minimizing disturbance.

Current Management Issues Increased cooperation among different authorities is needed to enhance the development of ecologically sound environmental programs. The Land Development Authority of the Jewish National Fund and the Nature Reserves Authority cooperate on several programs, but further expansion of this cooperation would be beneficial.

An integrated information system about the area is lacking. Geographic information systems (GIS) databases are beginning to be developed. The usefulness of remote sensing in desert environments in Israel is as yet unclear and, although there is a considerable body of research, comparatively little of this has been applied. Various organizations maintain databases, but these are not integrated. For example, the Land Development Authority has a database on rainfall, runoff, and soils in which the soils data were obtained from the Ministry of Agriculture. The Nature Reserves Authority is separately developing a database focused on flora and fauna, while the Society for Nature Protection in Israel has a similar database in cooperation with the Hebrew University of Jerusalem.

The existing overall development plan for the Negev desert defines land objectives for the different region. However, the plan was mostly politically driven with little input from the scientific community. Conflicts between the various interested parties are often left unresolved or are temporarily solved at the lower levels.†

Australia

Australia's arid and semi-arid lands cover almost three-fourths of the continent (fig. 9.2) (Stafford Smith 1995, Williams and Calaby 1985). In the 5.6 million km^2 involved, about 66% is used for sheep and cattle grazing, 14% is Aboriginal land, 4% is in conservation reserves, and 16% remains unallocated; other land uses include mining, tourism, and recreation, as well as minor commercial harvesting of seeds, timber, native flowers, and traditional foods and medicines by Aboriginal people. Mining exploits small areas but is being explored over much of the continent; it leads annual economic outputs at about $10 billion (figures in Australian dollars), followed by tourism, the fastest growing industry dependent primarily on conservation reserves, at $3 billion, and grazing at $0.85 million. Thus, the largest users of land in terms of area are not the most economically productive.

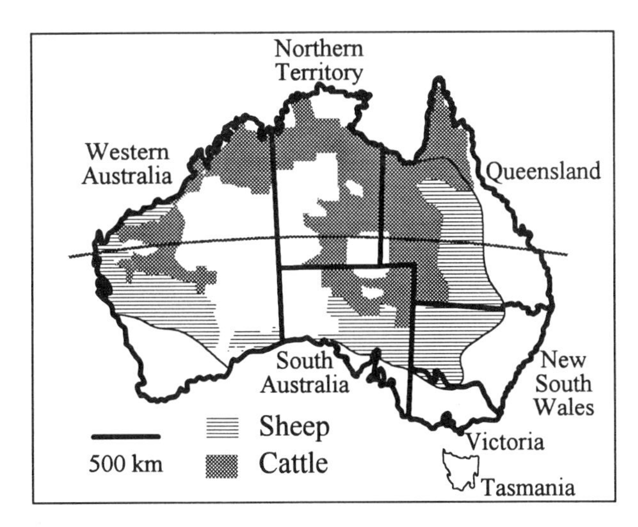

Figure 9.2. Australia's arid and semi-arid zones, state boundaries, and regions of the rangelands used by the pastoral industry primarily for cattle as compared to sheep.

The periphery of the vast interior is bordered to the east, southeast, and southwest by marginal agricultural zones situated on cleared semi-arid lands; these commercial farms range from sheep production to mixed farming and opportunistic cropping. In the southeast, around the Murray-Darling River system, there are several irrigation areas; these are small on the continental scale but produce as much as most of the more arid region.

Biophysical Resources

Geomorphology Inland Australia is generally flat. The highest point in the arid interior rises to 1,500 m from plains that are already at 600 m; the small total area of isolated mountain ranges (about 16% of the arid zone) has important effects on water redistribution. Apart from the uplands and piedmonts, the main physiographic types are shield deserts (mainly in western Australia), stony deserts (mainly southern and eastern), riverine and clay plains (scattered throughout, but large areas in western Queensland), and sand deserts (most of the core area). Geologically, the mountains and shield areas are very old, generally being of Precambrian origin; furthermore, some erosional surfaces have been exposed continuously since the Cambrian, some 500 million years ago. Some plateaus represent gently folded Palaeozoic rocks. The Great Artesian Basin—water-holding strata underlying the eastern half of the continent—was formed as a result of inundation during the Jurassic and Cretaceous. Over the top of most lower-lying areas are Quaternary alluvium and the windblown deposits that form the large sand deserts.

Soils and Vegetation Inland Australia is dominated by infertile soils, notoriously poor in phosphorus (e.g., total P: 100–300 ppm; available P: 3–22 ppm), although there are more productive patches. The cracking clays, calcareous earths, and alluvial soils are usually more fertile. Low soil phosphorus levels can also limit nitrogen accumulation, and, in any case, most nutrients are typically held in the top 2–5 cm of soil where they are highly susceptible to erosion. Most widespread are infertile red siliceous and earthy sands throughout the core of the arid zone; these are dominated by two vegetation communities (table 9.5, fig. 9.3): the acacia shrublands (most commonly mulga, *Acacia aneura,* that occurs over 20% of the continent) and spinifex hummock grasslands (based around the spiky hummock grasses of the genera *Triodia* and *Plectrachne*). At the opposite extreme are very sticky, fertile cracking clays, often with gilgai micro-

Table 9.5. Main Vegetation Types of the Australian Arid Zone

Formation	Area[a] (m km^2)	Dominant Plant Genera
Arid		
Shrub steppe	0.43	*Atriples/Maireana* chenopod shrubs
Acacia shrublands	1.60	*Acacia* shrubs
Hummock grasslands	1.60	*Triodia/Plectrachne* grasses
Arid tussock grasslands	0.50	*Astrebla* grasses
(Salt lakes	0.44	unvegetated)
Periphery, including semi-arid		
Semi-arid shrub woodlands	0.39	*Eucalyptus/Acacia* trees/shrubs
Mallee woodlands	0.42	*Eucalyptus* shrubs
Arid/semi-arid low woodlands	—	varied, *Eucalyptus, Acacia, Casuarina, Callitris*

a. The areas are approximate, due to overlaps between zones.

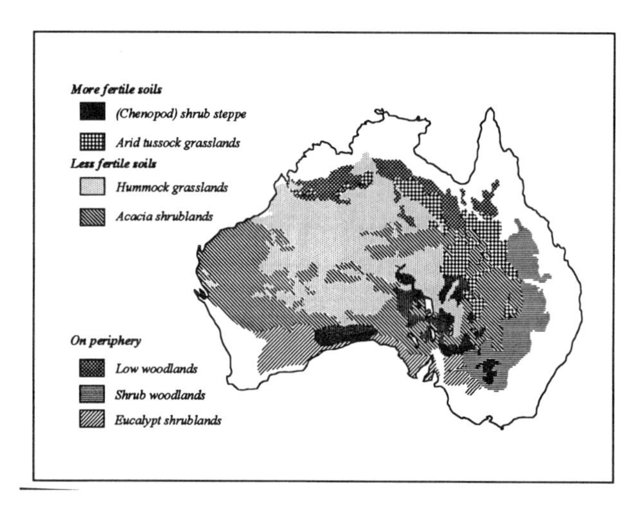

Figure 9.3. Simplified distribution of the major vegetation types of the Australian arid zone and its semi-arid periphery (after Williams and Calaby 1985, and others). Note the limited contiguous areas of fertile soils.

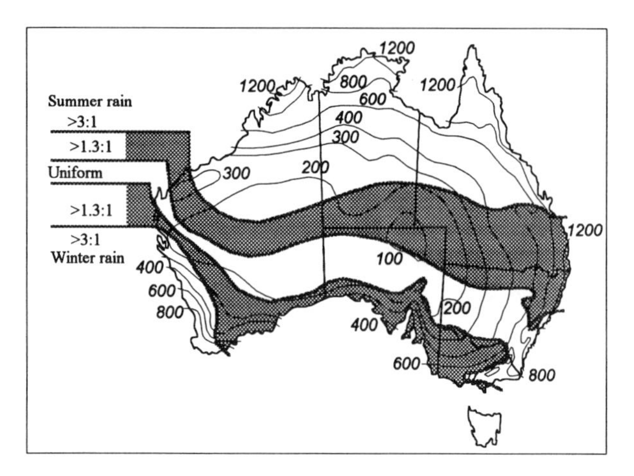

Figure 9.4. Seasonality and median annual total rainfall across Australia (after AUSLIG 1992).

relief, on alluvial plains feeding southwest from the eastern half of the continent; these are characterized by an open grassland dominated by Mitchell grasses (*Astrebla* spp., especially *A. pectinata*). Most of the gradational soils are relatively fertile calcareous earths in the southern half of the continent dominated by low (1–2 m high) chenopod shrubs, most notably saltbush (*Atriplex vesicaria*), and long-lived bluebushes (*Maireana* spp.), sometimes with an open *Acacia* spp. or *Eucalyptus* spp. overstory. Texture contrast, duplex soils occur in smaller areas throughout, especially in alluvial areas where they are susceptible to degradation.

Climate Contemporary weather patterns in northern Australia are dominated by tropical influences and, in southern Australia, by the passage of frontal systems around the South Pole with a mean periodicity of 6.8 days. This leads to an enormous climatic gradient across the continent (fig. 9.4). Rainfall occurs predominantly in the summer and is cyclonic in nature in the north, which experiences distinct wet and dry seasons without frosts. In the south, rain falls more gently and dominantly in winter, although occasionally cyclonic incursions track across the continent to bring heavy rains in summer. The southern half of the continent experiences significant numbers of frost. Throughout the continent, mean monthly summer temperatures (January) are around 30 °C with most arid areas experiencing daily maximums above 37.8 °C on more than a hundred days per year; daily temperatures average 10–17 °C in both summer and winter. Pan evaporation exceeds rainfall tenfold in most areas.

These statistics are relatively consistent year to year, except for rainfall; hourly, monthly, and annual precipitation totals are severely skewed. Rainfall is both variable and unpredictable to the extent that plant biomass production may vary by a factor of ten from one year to the next. Rainfall variability in Australia fits with the general global pattern of increasing toward the equator and as annual totals decline. However, it shares an additional influence with India's Thar desert and southern Africa; the El Niño–Southern Oscillation (ENSO) effect and other climatic syndromes driven by sea-surface temperatures in the Pacific and Indian Oceans cause intermittent periods of severe drought and deluge. For example, the Alice Springs record with a mean of 283 mm/year (s.d. = 144 mm, median = 263 mm) includes periods of over a year with no rainfalls exceeding 12.5 mm and over five years with no rainfalls exceeding 50 mm. The resulting variability is remarkable in global terms and has profound implications for the ecology and management of arid Australia (Stafford Smith and Morton 1990).

Land Uses and Problems

Livestock Production Pastoralism is conducted on a very extensive scale on areas varying between 10,000 ha and 30,000 km² that are held mainly under lease. The development of pastoralism in the rangelands of arid and semi-arid Australia in the nineteenth and twentieth centuries has been followed by widespread land degradation. There are few reliable surveys of the extent of this problem, but a widely cited report (Woods 1984) estimates that of the 3.36 million km² used for grazing, 880,000 km² is experiencing soil erosion and adverse changes in pasture characteristics, and of this, 150,000 km² is severely eroded.

Much of the damage occurred in the early days when explorers' exaggerated reports caused overly optimistic assessments of carrying capacity. Such mistakes were perpetuated by administrators who allowed subdivision of land into holdings that were too small to be economically viable. Animal impact per unit area has now been greatly reduced. Initially, grazing was based around a limited number of natural waters where devastation was high. More recently, the construction of dams and wells (bores) has spread animals more widely across the landscape. At the same time, improved transportion infrastructure has made it much easier to move stock out of drought-affected areas so it is now technically possible to avoid heavy grazing pressure during drought.

Despite these innovations, rapid degradation at a few locations in the landscape has probably been replaced by a slower but more widespread rate of decline in pasture productivity. This is because, first, stock numbers are not consistently synchronized with variations in rainfall (Friedel et al. 1990) so that grazing pressure is often at its highest when the quantity of forage is declining. Second, there is still considerable ignorance about the level of grazing that soils and vegetation can tolerate without compromising future pasture production as rainfall conditions improve. Third, livestock producers in rangelands are price takers rather than price setters and are subject to the vagaries of international markets. Many of them also carry high levels of debt, particularly after a drought. Faced with the need to service debt and with a cost-price squeeze, grazers may be forced to overgraze merely to survive (Mott and Edwards 1992). Few like to do this but the alternative can be financial ruin.

Most of the land degradation occurs in the pastoral area, but there are significant problems elsewhere, including altered fire regimes, extinction of native species, feral animals, and the spread of introduced and unwanted weed species. Furthermore, not all land degradation is the direct result of commercial grazing. Rabbits remain Australia's principal environmental problem, exerting tremendous pressure on pastures and virtually halting tree regeneration in some areas (Morton and Pickup 1991). In the northern rangelands, there are problems with feral horses and donkeys, while in the south there are large numbers of feral goats. Although the kangaroo is a native animal, its population has also greatly increased in response to the greater availability of water in some grazed areas. It now competes directly with domestic grazers, especially during drought (Caughley et al. 1987).

Australian rangelands are mostly leased from various state governments. The spatial and temporal variability that characterizes Australian rangelands has lead to serious problems with land degradation and policy paralysis in the past (Chisholm and Dumsday 1987), partly driven by lack of appropriate understanding and information. Given the complexities described earlier, in retrospect this is hardly surprising; however, today's challenge is to change this situation. The implications of both the temporally and spatially variable natures of rangelands and the quality of models that science has so far devised are far-reaching for managers and policy makers alike (Mentis et al. 1989). Additional details about these issues can be found in Foran and colleagues (1990), Pickup (1989), and Pickup and Stafford Smith (1993); the latter identify a series of research and application questions that need answers in order to define sustainability in the pastoral industries of Australia, given their broad management scale (tables 9.2 and 9.3).

Aboriginal Lands Aboriginal lands are occupied by the indigenous people of Australia who were displaced by the arrival of pastoralists and their stock in the late nineteenth century (Ross et al. 1994, Stafford Smith et al. 1994). They have recently regained ownership of substantial areas of land, but these are dominated by areas with poorer soils that were perceived to have little value for commercial production. Aboriginal people are currently making a greater effort to acquire productive land, and in areas where tourism, mining, and traditional uses collide, they are active members of park management boards (Hill and Press 1994).

Although Aboriginal people may use their lands for any of the other land uses discussed here, especially grazing, tourism, and mining, their land management problems are different:

1. They have overriding concerns of community health and education, which have yet to be fully met, so conservation concerns are often secondary. Most Aboriginal lands are remote with little access to mainstream employment opportunities.

2. They have several additional uses for land, including hunting and collecting of traditional foods, maintaining sites of cultural and religious significance, gathering firewood, and collecting wood and other materials for the creation of commercial artifacts.

3. Much of their land is held under an inalienable communal title (formally vested in a Land Trust on behalf of a community) so that there is considerable scope for conflict between land-use objectives within each management unit (e.g., the impact of cattle grazing on traditional food resources or casual hunting for game on fence maintenance); this compares to pastoral land where the management unit is generally held by one family with a relatively simpler suite of land management objectives.

There is a considerable opportunity for research to clarify land-use conflicts arising, for example, from the impacts of grazing on traditional foods and locations such as waterholes, the social effects of tourism on traditional culture, the effects of traditional food harvesting (with today's priorities) on the maintenance of biodiversity, and so on. In the past, the issue of equitable land ownership rights has tended to obscure the need for responsible stewardship of Aboriginal lands once acquired. However, land management now receives greater attention heightened with an increasing recognition that traditional knowledge held by the current generation of Aboriginal people will be useful in overcoming many problems. Traditional knowledge may be less relevant for the problems related to modern impacts, including the introduction of weeds and feral animals, the loss of many native species, and access to guns and off-road vehicles.

Tourism and Recreation Tourism is the fastest growing industry of the arid lands in Australia; it is dominated by visitors' desire to see the landscapes and experience the relatively pristine, vast, open spaces that are now rare in the world and nowhere else so readily accessible. As a consequence, most commercial tourism is focused on the conservation reserve system, and a large part of the public resources expended on conservation reserves goes toward managing tourists rather than ecosystems. "Hot spots" like Uluru-Kata Tjuta National Park ("Ayers Rock"), the Hammersleys, Kakadu, and Wilandra Lakes (three of these are World Heritage Areas) attract the majority of visitation, and thus can be managed quite intensively. However, some commercial tourism and most private recreation also penetrates the wider landscape more diffusely, leading to problems associated with off-road vehicles that are less easily managed, such as weeds, rubbish, and erosion. More subtly, several million visitors require shelter, water, and waste management facilities each year. Some villages such as Ayers Rock Resort at Uluru have been designed to cope with the numbers, but others, such as the Glen Helen Resort in the West MacDonnell Ranges, may have problems with water and waste disposal. The tourist industry increasingly sees the need to protect its own future by treading lightly on the landscape. Central to this issue is the need for research to determine the best ways of stratifying levels of use to allow some "sacrifice zones" without compromising the landscape-level function.

Conservation Management Conservation management in Australia is a strong commitment under national and international obligations, and a considerable effort has been put into developing the concept of *ecologically sustainable development* as a national government policy. This is implemented through the management of the reserve network that is controlled mostly by the Australian states: through resource inventory, again mainly carried out at the state level but somewhat coordinated nationally, through voluntary programs, such as "LandCare," that seek to encourage sustainable use of land, especially in relation to soil and water conservation in agriculture, and through other specific programs such as Greening Australia's "Billion Tree Program."

Basic conservation of biodiversity has serious problems in Australia (Morton et al. 1994). For example, the arid lands have lost half their complement of native ground-dwelling mammal species in the last two hundred years. In comparison to the arid lands of Eurasia, Africa, and the Americas, this may be part of an inevitable selective reconciliation to new management practices that include grazing impacts, feral animal competition and predation, and changes in fire regimes; however, the populace does not take a sanguine view of such losses. Considerable efforts are placed into reintroduction from remnant populations, often offshore, but these have rarely been successful due to the combination of factors that must be controlled: habitat loss, feral predators, and natural environmental variability, against which large populations may be buffered but that can be catastrophic when numbers are low.

Mining and Exploration Mining is a major part of the economics of the inland, even though most money is exported from the region. In some mineral-rich regions, such as much of western Australia, mining of gold, iron ore, tin, copper, and silver affects large areas. Gas and oil wells occur toward the center of the continent. Overall, the extraction process itself affects modest areas, but exploration is much more extensive. This can facilitate weed invasion, provide access tracks that recreational users subsequently exacerbate, and create demands for rehabilitation. In general, the larger operations have become aware of these problems in recent years. Moreover, although they may be very controversial, mining tracks may cause far fewer problems than tracks created by unskilled pastoralists on their own properties.

Other Land Uses With only 300,000 people in the entire region, encroachment is not a major issue in Australian arid lands. Around townships such as Alice Springs, Mt. Isa, Broken Hill, Kalgoorlie, and some smaller villages, there are certainly land-use conflicts and water-use questions of local significance. Likewise, although arid Australia has been proposed as a toxic or nuclear *waste disposal* area, so far the dangers involved in lengthy transport distances and concerns for impacts in Aboriginal lands have foiled most developments. Negative feelings build on one major such use in the past, that of Maralinga for British nuclear testing after World War II, which caused massive clean-up problems that are still not fully resolved and serious impacts on the local Aboriginal tribes of the time.

Pastoralists do use run-on *water harvesting* to help with rehabilitation efforts (e.g., Purvis 1986) and even to develop small irrigation blocks, but this is on a very small scale in the arid zone proper. There are also some small areas of *irrigation* using groundwater, producing table grapes or melons; these have enormous transport and production costs so they function only where a critical early market can be accessed. Water resources are rarely of adequate widespread quality to allow larger enterprises, and desalination could not be justified in commercial returns given the high costs of labor and transport. Key issues are summarized for different land uses at different scales in table 9.1.†

Management and Policy Issues

From the preceding descriptions of three arid regions of the world, it is evident that there are many similarities but also some fundamental differences. These similarities and differences are driven by at least three categories of issues: the underlying biophysical resources available to each region, the management and land-use systems applied to regional resources, and the social and political forces that human culture has imposed on each region. The following sections explore some of the regional differences to highlight issues that *may* be researched in common among regions and also issues that *cannot* be transferred from one region to another.

Biophysical Resources

Fundamental to land use in any area is the resource base from which it operates. The characteristics of the different regions have been described above; this section compares and contrasts some critical factors from the point of view of land use.

Topography The Israeli Negev, North American, and Australian arid and semi-arid zones differ considerably in size. The Negev covers 11,000 km^2, the North American arid zone covers an area of 1 million km^2, and Australia's arid and semi-arid regions cover 5.6 million km^2. A much larger proportion of the Australian arid area is remote, resulting in high transport and labor costs.

Topography also differs among regions. The Negev desert consists mostly of rolling hills in the north, with a sharper terrain traversed by canyons in the south. Australian deserts are dominated by flatlands with only small areas of very low ranges. The topography in the North American dryland zones differs considerably from that of Australia and Israel because its flat terrain is interspersed by high mountains. This creates some permanently flowing rivers through the arid areas of North America, effectively fed from outside, while in Australia the emphasis is on the dynamics of flat landscapes.

Soils The soil's physical and chemical properties are critical determinants of land use owing to their influences on moisture infiltration and retention, drainage, and nutrient availability. For example, in the Negev, the physical properties of sandy soils are more conducive to irrigated crop culture than are those high in clay or loess (Kovda et al. 1979). Pockets of high salinity constrain crop productivity in each of the regions. In North America and Israel, arid and semi-arid zone soils tend to be neutral to alkaline and fertile. All soil reactions can be found in Australia, with the acidic soils usually exhibiting low phosphorus availability. It is important to separate the very different functional responses of fertile and infertile soils. Although Australia contains some areas of fertile soils, it differs from the other regions by being generally dominated by very infertile soils. Large management units are, therefore, necessary for economic viability.

The potential for water and wind erosion is usually uniformly high in the Negev and North American arid and semi-arid zones. In Australia, however, erosion is less pronounced and drainage is rarely competent due to the flat terrain and denser plant coverage; although the continent has large areas of degradation, including salinization and erosion, on a world scale this degradation is regarded as being of only moderate intensity (Safriel chap. 8, this volume).

Water Rainfall patterns in most arid and semi-arid regions are seasonal. In the Negev, the rainy season is winter. In different areas of North American arid and semi-arid zones, the rainy season is summer or winter, or both, with heavy rains derived from tropical depressions occurring at multiyear intervals. Semi-arid zones in the eastern portion of the deserts are characterized by summer monsoonal rains, while those to the west are typically winter rains. Northern Australian deserts are characterized by summer rains, while in the rest of the continent rains may fall in both winter and summer with winter becoming more common in the south (cf. fig. 9.4).

For a given mean annual rainfall, sites in Australia exhibit greater interannual variability than the other regions (Nicholls and Wong 1990), although more work is needed to understand the ecological significance of these differences (Stafford Smith and Morton 1990). On the other hand, areas in the Negev experience extremely arid conditions, while nowhere in Australia receives a long-term mean annual rainfall of less than about 150 mm.

Underground water is limited throughout the Negev Desert and is found only at depths of approximately 1,000 m. Water supplies closer to the surface are highly saline. By contrast, in both North American and Australian arid and semi-arid regions, underground water is often available, but the degree of salinity varies. In Mexico, underground water is variable and frequently saline. Groundwater-based irrigation is uncommon in Australia. By contrast, water for stock is usually harvested into stock tanks in arid North America while wells are widely used in Australia.

Surface water in the form of streams, rivers, and lakes is absent in the Negev. In Australia, a single permanent river system, the Murray-Darling, runs through the semi-arid region, while in North America, there are several rivers and various streams and lakes fed from the contiguous mountainous regions. Water salinity in these rivers varies with location from 50 ppm to 1,500 ppm total dissolved salts. There is high potential for water harvesting in Israel and arid North America but much less in Australia.

Natural Deposits All three arid and semi-arid zones contain mineral deposits in some areas. These include phosphates in the Negev; copper, silver, uranium, and salt in arid North America; and a very wide variety of minerals in the larger area involved in Australia. Oil and natural gas are only found in Australia, the United States, and Mexico. Significant coal deposits exist in northern Arizona, New Mexico, and South Australia. Mining is, thus, significant in all three regions.

Renewable Energy Sources The three arid and semi-arid zones have high potential for solar energy utilization. In Israel, the use of solar collectors for heating water is required by law and is, therefore, widely used. To a lesser extent, this approach is used in Australia and much less in arid North America. Wind energy is not readily available in the Negev but is plentiful in some areas of arid North America and Australia. It is used in the United States and, to a lesser extent, in Mexico and Australia. Geothermal energy can be harvested at low levels along the Arava Valley in the eastern Negev where hot springs are abundant. These are used occasionally for heating greenhouses. This source of energy is also utilized in some of arid North America but is not available in Australia.

Land Use and Management Systems

Many different management systems are present in all the regions but not all are feasible in each region, and their relative importance differs among regions. Land uses are ultimately driven by investment levels made feasible by regional population and political pressures. Seligman (personal communication) suggests that the relative importance of these factors could be captured by creating a series of indices based on ratios between socioeconomic and primary productivity criteria; this could be a valuable exercise for future research to help develop a better characterization of regional drivers. Criteria might include the population, area of land, and income levels of a region by sectoral land uses, as well as regional and sectoral imports (including fossil fuel use) and exports in terms of production. Relative indices of the importance of particular land uses could then be obtained by land-use criteria by regional totals (e.g., population engaged in grazing industries compared to total population), as could indices of regional and sectoral potential affluence (outputs per person) and efficiency of production (outputs per unit input). Table 9.6 shows a simplified summary of such an approach.

These differences are driven by differences in inherent land productivity, production risk, and remoteness from dense settlement, which create considerable disparities among the regions in management intensity and the characteristic size of management units (ranches, conservation reserves, nomadic ranges, etc.). Australian ranches can exceed 12,000 km^2 (larger than the entire Negev) with individual fenced areas exceeding 500 km^2. Consequently, levels of investment per unit area, costs of fencing and water supply, and even fuel costs within an enterprise differ hugely (Pickup and Stafford Smith 1993).

The next section examines the way the major management systems vary among regions, in order to identify constraints and opportunities to research as well as issues that are widespread as opposed to those of local concern.

Livestock Production Systems Production systems and the types of livestock grazed differ across regions. For example, the proportion of goats is less in the United States and Australia than in the other regions; the goat population is increasing in Mexico as a dual purpose source of meat and milk, even though beef is still the main source of meat. Thus, livestock research has been strongly biased toward sheep and cattle in Australia and the United States, yet information on other herbivores is increasingly needed.

Not surprisingly, areas with commercial ranching tend to be more strongly linked to markets than subsistence grazing systems; a much larger part of production is exported from Australia than the other areas, making it more vulnerable to world market conditions. Although once considered subsistence grazers, Israel's Bedouins market considerable amounts of surplus produce and are shifting from many small to fewer large herds. This will provide them with more opportunities to implement modern ranching practices, but the overall effect that this change will have on sustainability is unknown. Where objectives of market quality come to

Table 9.6. Significance of Different Land-Use Management Objectives in the Different Regions, as Assessed by Workshop Participants

	Israel	Southwestern United States	Mexico	Australia	LDC[a]
Grazing	Low	Medium	High	High	High
Conservation	High	High	High	High	Low
Recreation	Medium	High	Medium	Medium	Low
Agroforestry	Low	High	High	Low	Medium
Forestry	Low	Medium	Low	Low	Low
Military	High	Medium	Low	Low	Medium
Agriculture	High	Medium	Medium	Low	Low
Waste Dump	Low	Medium	Medium	Low	Low
Mining	Low	Medium	Low	Medium	Medium
Land enhancement					
Urban	Medium	Medium	Medium	Low	Low
Rural	High	Low	Medium	Medium	Low

Note: Precise values are debatable, but note that spatially intensive uses tend to become more important in countries with higher population densities and that land uses other than primary production tend to become more important in countries with higher levels of affluence.

a. LDC = less-developed countries. This column was included to compare likely patterns with countries around the edge of the Sahara, for example; it is even more generalized than the region-specific columns since there is a huge diversity in LDCs.

dominate those based on numbers, different levels of stocking become optimal. Mixed objective grazing use is becoming important in Australia, too, as Aboriginal people seek to use newly acquired land for pastoralism (Stafford Smith et al. 1994). Again, research has tended to focus only on commercial land uses, and a more thorough understanding is needed of subsistence systems and how those systems may make the transition to a market base.

Forage composition and the seasonal availability of the forage resources to livestock differ among the regions, controlling to some extent the type of grazing systems (year-long, seasonal, rotational) that are feasible for range management. The intensities of range improvement vary among and within regions: range improvement considers factors such as optimal livestock numbers, range rest periods, water development for livestock, brush control, and seeding of more productive species (Smith 1899). The most critical decision that a manager has to make is that of livestock numbers or *stocking rate* (Van Pollen and Lacey 1979) and how this should vary spatially and through time. Research on risk management issues, such as the net economic benefits of forage cropping, buying and selling livestock, and long-term impacts on pasture production, can help managers to adopt a better reactive stocking rate regime instead of stocking for average forage supply levels. However, the way in which this should be implemented differs among regions because of socioeconomic (e.g., markets, political support, etc.) and physical (e.g., interannual climatic variability) factors.

Agricultural Crop Production The regions are characterized by rain-fed subsistence practices and irrigated commercial agriculture, both at a relatively small scale, and by large-scale dryland commercial agriculture. However, the relative importance of these crop production practices is largely dependent upon the level of capital investments made. Resources and markets determine the kinds and amounts of crops produced. Increasingly, irrigated crops must have high value for human consumption or use (e.g., pharmaceutical). Water-use efficiency relative to costs and return is a critical factor. Levels of water use are probably linked with the likelihood of salinization problems; the link is a high priority for research in most areas.

Crop selection is determined by dietary preference, farming practices, marketing opportunities, and the economic status of the people in the regions. Production incentives (price supports, extension services, low-cost water, new plant varieties developed by public agencies, etc.) vary in kind and amount by region, type of crop, and marketing opportunities. Crop types requiring minimal supplemental water provided by water harvesting techniques are options where markets are near and demands predictable. The most appropriate balance of all these factors is a key research concern in most regions, as well as the need for agroforestry.

Biodiversity Individual countries are viewed as critically important to global biodiversity if they possess unusually large numbers of endemic plant or animal taxa or provide critical habitat or migration routes for endangered species. Countries contribute to global biodiversity in different ways (table 9.7). Australia ranks very high in vertebrate endemism, as does Mexico, while the United States is not ranked among the top twelve countries (table 9.7). The desert areas of North America and Australia possess similar species richness in avifauna (130–40 species), but insectivorous small mammals,

Table 9.7. Known, Endemic, and Threatened Plants and Vertebrates

	Australia	United States	Mexico	Israel
Flowering plants				
Known	15,000 (9)	18,956 (8)	> 20,000 (4)	2,294
Endemic	12,000	3,924	> 2,780	154
Threatened	2,024	2,262	883	3
Mammals				
Known	282	346	438	—
Endemic	210 (1)	93 (4)	136 (3)	2
Threatened	38	27	25	8
Birds				
Known	571	650	961	169
Endemic	349 (2)	69	88 (7)	0
Threatened	39	43 (9)	35	15
Reptiles				
Known	700	—	717	—
Endemic	616 (1)	—	368 (2)	—
Threatened	9	25 (1)	16 (3)	1
Amphibians				
Known	180	—	284	—
Endemic	169 (3)	122	169 (2)	0
Threatened	3	22 (1)	4 (3)	1

Note: Dashes indicate unavailable values. Numbers in parentheses show rank among countries of the world (WCMC 1992). All figures are country totals. A disproportionate number of the threatened Australian mammals occur in the arid regions compared to other countries.

granivorous birds, lizards, termites, and ants are all more regionally and locally diverse in Australian deserts (Morton 1993). Mexico, Australia, and the United States all have rich floras. Owing to Israel's land area and long history of human influence, it contains relatively small numbers of endemic taxa. However, on an areal basis, it is relatively rich in known Mediterranean flora. Mexico's flora probably includes about 30,000 species (Bye 1993). Its arid and semi-arid zones contain about 6,000 plant species with 60% being endemic (Toledo and Ordonez 1993). Explosive speciation has occurred in groups such as cacti, composites, and grasses (Rzedowski 1993).

Israel, the United States, and Mexico also serve as migration corridors between continents. Their wetlands, riparian zones, and coastal areas play a critical role in maintaining flightways. North American forests create corridors for both birds and migratory butterflies. Maintaining the integrity of these routes is often the focus of international policy and multilateral agreements.

All these statistics indicate which groups have received most attention in the research of different regions. Most countries value their biodiversity, and management practices are increasingly required to reflect public pressure to preserve it. On a global basis, most of the causal factors currently threatening mammals are anthropogenic in origin, with habitat loss or fragmentation posing the gravest threat. Cultivation and settlement followed by pastoral development and deforestation are considered the most threatening types of human disturbance. In all regions, there is an urgent need to better understand how much anthropogenic disturbance is compatible with the maintenance of biodiversity (i.e., to go beyond observations that say grazing impacts on biodiversity to specifying the level of impact). This principle is universal, although the anthropogenic disturbance of concern differs among regions.

Wildlife Resources, Hunting, and Tourism Both consumptive and nonconsumptive utilization opportunities differ among the regions. In the United States, wildlife populations are managed with the goal of sustaining the harvest while providing supplementary income to private landowners or income multipliers to communities near the resource. By contrast, hunting in Israel is nonexistent, except as a means of controlling small animal pests, and minor in Australia; significant harvesting of kangaroos takes place in Australia as a controversial pest-control measure in areas with sheep ranches and as a subsistence contribution to diet in areas with Aboriginal occupation. Research on the optimization of hunted populations is similar to that on domestic herbivores and is widespread in the United States. Hunting on Aboriginal lands in Australia is largely laissez faire at present and urgently needs reassessment.

Where hunting is a popular sport, wildlife management produces synergistic effects with other uses—making ecotourism more attractive, for example. For wildlife to attract tourism, it must be visible to visitors, as is the case with large game animals in North America and Africa. The chain of ecological links in Australia, leading to dominance by lizards (Morton and James 1988), means that tourism there depends more on landscapes and birds than on large mammals, with significant consequences for management including a lower pressure to maintain high and viable populations of the remaining large wildlife. Research to help manage (and display) the more cryptic wildlife in Australia will be needed in coming years. In some areas of the United States and Australia, wildlife such as wild horses are neither hunted nor visible to the extent that tourism is attracted. Nevertheless, such wildlife pose major management problems.

Recreation pressures on arid lands are highest in the United States where off-road vehicles, sightseeing, developed camping, and water-based recreation are all extremely popular. Recreation pressures are more moderate in Mexico where they are usually quite concentrated and in Israel where they occur primarily in short trips and day visits. In Australia, commercial tourism and off-road vehicle (ORV) recreation are growing rapidly. ORV use is particularly noteworthy because it occurs primarily in arid lands; it is highly dispersed with significant impacts on those lands.

Recreation and tourism increasingly concentrate pressure on dry coastal lands and riparian areas, especially those with open water bodies. These areas are often significant for landscape function, so even low levels of land degradation created by tourism can cause serious land-use conflicts. Research needs thus relate to the same issues as for biodiversity: In this case, what direct and indirect impacts do different levels of tourism have on the biota?

Although tourism and recreation in arid regions revolve predominantly around the natural ecosystems, dryland systems are also the sites of historically significant artifacts, thanks to their high levels of preservation compared to mesic regions. This is notable in the ancient civilizations along the Dead Sea in Israel, the pueblos of New Mexico, and Aboriginal rock art in Australia. Research on the management of the impact of tourism on these cultural sites is still in its infancy.

Forestry and Other Woody Production Systems The regions differ in the proportion of natural wooded ecosystems (natural regeneration) compared to human-created ecosystems (artificial regeneration); this is significant because human-created ecosystems (common in Israel) generally require a much larger investment of public resources to maintain than naturally regenerated ecosystems (more common in arid North America). There are also considerable differences in the utilization of the woody resources for primary wood products (fuelwood, poles and posts, building materials) and in the development and subsequent use of nonwoody products (extractives, wax, tannins, fiber, folk medicine) leading to the need for different types of management knowledge. The selective harvesting of a single species generally implies a need for knowledge about population responses and sustainable harvest levels, whereas broader uses, including small-scale agroforestry, tend to demand plant community knowledge for management locally and ecosystem knowledge to understand the impacts on the whole farming system.

The nature and intensity of agroforestry interventions differ in the regions. By definition, agroforestry combines woody production systems with livestock production, agricultural crop production, or combinations thereof (Nair 1993, Raintree 1987). In many dryland regions, including arid North America, silvipastoralism is a major form of agroforestry, as is windbreak planting. However, agroforestry other than for recreational and aesthetic purposes is minimal in the Negev because of space and water cost constraints and in arid Australia because of the low return on investment. It is more common on the semi-arid margins in Australia where it is regarded as an important management approach to salinization problems. There remain research questions about the detailed hydrological impacts of such treatments.

The assemblages of plants in home gardens, a major form of agroforestry in dryland ecosystems, also distinguishes the regions from one another. The pressures toward landscape enhancement in the form of urban forestry (amenity planting) are lowest in areal extent in Australia due to low population density while the highest investments in these activities are in Israel. Mexico is striving to regenerate degraded lands by planting trees over extensive areas. Amenity planting is rarely the subject of management research, but there are opportunities here; not only do many of the world's major weed problems escape from gardens (Humphries et al. 1991), but they also represent a major drain on water supplies in arid regions.

Water Resources Water management differs among regions owing to differences in natural supply, infiltration, topography, and competing uses. In the United States, irrigation and watershed drainage are both controlled. In most regions, drainage is a function of topography and perennial watercourses of the natural landscape, but in Israel, water is delivered through pressurized systems from the Jordan River. The cost of water differs hugely among regions; this helps to determine the type of irrigation systems used and the level of irrigation efficiency obtained (see "Water Policy" section below). Drip irrigation systems are most common in Israel, whereas flood and sprinkler methods are usually used in arid North America. Flooding is the irrigation method used most commonly in Australia. In the southwestern United States, the depth of affordable pumping is determined by electrical consumption, and this cost is not subsidized. In the face of escalating costs and shrinking water supplies, southwestern farmers are conserving water through technologically innovative approaches to irrigation scheduling and delivery. For example, farmers pumped 20% less water with little difference in yield when they converted from furrow to center-pivot irrigation. When center-pivot sprinkler heads were lowered and pointed downward to reduce water loss to the atmosphere, further gains in efficiency were obtained (Opie 1993). The cost of water and pumping are subsidized in Israel. Mixing waters of variable quality is common in Israel and is becoming an increasingly necessary alternative in the United States, leading to research questions about impacts.

Opportunities for the economic enhancement of water supplies (watershed management, water harvesting and spreading, salvage, and reclamation) differ among the regions. Water capture provides opportunities for seasonal irrigation of crops and trees, for livestock, and for replenishing subterranean supplies. Stratigraphy plays a key role in determining the efficiency of water capture, but this is also driven by the balance between the costs of local water and importing products from elsewhere. In Alice Springs, Australia, for example, attempts to grow significant quantities of local vegetables have consistently failed because the cost of local water is too expensive to undercut the costs of growing and transporting produce from the Murray-Darling Basin.

The quality and quantity of water supply is greatly affected by managers' decisions concerning the amount of wood harvesting, fire control, wildlife numbers, stocking rates, development of stock water tanks, and exclusion of stock from river banks. Subsequent chapters in this volume describe research on ecosystem responses to managers' decisions, such as to let a fire burn or reduce stocking rates, but our knowledge of these hydrological effects, which should be readily extrapolated, is still inadequate.

Mining Regulations on where, when, how, and to what extent mining is permitted differ among the regions. Surface mining is the most common type across the regions. Reclamation efforts vary across the regions in accordance with government regulations and standards. Relatively large expenditures are allocated to the reclamation of mined areas in the United States. In Israel, reclamation in the Negev is especially difficult, and mining activities underway before reclamation laws were created are often exempt from their regulation. Australia still has a strong bias toward mineral extraction and export as an industry, reflected, for example, in the fact that mining is permitted in some way under the laws that establish national parks in most states (e.g., in Kakadu National Park) (Hill and Press 1994). Reclamation efforts after mining are now common in Australia. With the help of a $200 million loan from the World Bank, 15 million acres of Mexico's state-owned land will be released for mining. Sonora and Chihuahua are receiving considerable attention from North American mining companies seeking land or exploration and development permits (Kamp 1993). Restoration ecology has certainly developed some generalities but remains dependent on region-specific research on appropriate plants and ecosystems.

Urbanization Zoning regulations and building codes related to urbanization differ among the regions, often reflecting differing policies of central and regional governments. Likewise, although air quality is not a public issue in arid regions of Australia or Israel, the United States is seeking to reduce the impact of dust from drylands on urban areas. For example, farm windbreaks are being established with drip irrigation in areas once considered too dry, and farm operations creating fine atmospheric dust are being regulated. Urban areas can, of course, create environmental problems for arid regions. However, problems associated with air pollution, such as acid rain, generally are more severe in wetter regions that have more polluting industries.

Migration rates into and out of urban centers differ. Massive relocations from Mexico's rural areas to urban centers continue to magnify unemployment and environmental problems. Within the southwestern United States, there are also demographic movements from rural to urban environments, enhanced by more general movements from the northern and eastern United States to the Southwest. In Israel, population dispersal is considered advantageous for reasons of national security, and rural living is supported by government. In Australia, rural communities are declining somewhat, but population movements are negligible in comparison to those in Mexico and the United States.

Overall, the urban interface with arid lands is most pronounced in the United States because it is the most urbanized. As Mexico and, to a lesser degree, Israel become more urbanized, this interface will undoubtedly become more important. Australia is sufficiently sparsely populated so that little change is anticipated in the near future. However, the "footprint" (Foran et al. 1995, Wackernagel et al. 1993) of urban areas extends far beyond town boundaries, through water use, waste disposal, greenbelt enhancement, recreational impacts, and so on. Thus, the effects of urban areas extend well beyond the arid regions themselves. Consequently, research on the sociological needs of people living in arid areas, and on alternative ways of meeting their needs without compromising landscape function, is bound to increase in importance throughout the world.

Military The level of military activity differs among regions. Issues relating to rehabilitation after military use are similar to those for mining (see above), although there is a greater opportunity for modifying the initial impact through control of use. To some extent, these opportunities parallel concerns relating to recreation, and there is a considerable opportunity to "cross-fertilize" these areas of research. Two approaches are to require special use permits and to require the presence of an environmental officer to direct activities away from sensitive areas. With other human activities being strongly excluded from them, these lands can effectively conserve biota—if military disturbance is minimal or very intermittent; in this case, opportunities arise for conservation research and management.

Mixed Land Uses A major hole in past research has been in handling complex systems with multiple uses and multiple management goals, whether pursued simultaneously or sequentially. The balance between single and multiple land uses varies among regions, and this has great implications for the form and level at which research on management might be done. Mixed uses may occur at the local level (e.g., mixtures of recreation, hunting, and grazing on U.S. ranches, mixed subsistence farming by Bedouins in Israel, and mixed traditional and commercial uses by Aboriginal people in Australia, mixed subsistence agriculture, livestock, and seasonal harvesting of wild plants for traditional and commercial uses in Mexico) or at the regional level (e.g., balancing conservation reserves with grazing, recreation, and hunting; this level is essential from the standpoint of meeting sustainability goals regionally in all areas). Mixed local uses mean balancing land-use objectives within a small community; regionally, the concern is with the landscape responses of ecological function, catchment management, and maintenance of biodiversity to different mixtures of land uses from the standpoint of policy (Morton et al. 1994). Both aspects require information at different scales about the impacts of one land use on another and on the effects of different levels of imports (water, nutrients, money) into the management system rather than the optimal level of one land use for its own sake. Past research has tended to focus on the latter.

Most management decisions are based on management criteria on multiple uses where resources include water, timber, forage, wildlife, and minerals and the products are irrigation, municipal and industrial water, recreation use, timber, livestock and wildlife populations, and minerals. Decisions about the production capacity of any one resource affect the other resources and products. For example, managing for optimal forage and livestock production may not optimize water quality. Ideally, resources should be managed regionally to maximize the total benefits to society, which may include reducing adverse effects such as flooding, water pollution, sedimentation, and loss of biodiversity. If economic values can be assigned to all production variables, and the production responses of all alternative decisions are known, then an economic analysis comparing existing conditions with other multiuse options can be undertaken. Such decisions can be based on professional opinion in the absence of quantitative data, but a major role of research should be to supply managers with new and better tools for evaluating these options. This is the ideal; in practice, decisions are affected not only by economics but also by the political process, which is a function of the form of government in the different regions. Adverse outcomes of this process can be minimized by improved objective information.

Human and Policy Components

The human component is fundamental to arid ecosystems. This section describes some of the historical and current dimensions of the social and political pressures that shape land-use priorities in arid lands that lead again to certain constraints and opportunities for research.

History of Settlement Histories of settlement in the arid regions of interest identify practices that largely directed plant successional trajectories prior to recent land-use changes. In this vein, the Middle East has experienced at least 4,000 years of grazing, dryland agriculture, and runoff harvesting, with a generally steadily increasing pressure on the land; one exception was a considerable intensification of wood harvesting during early industrial times. Disturbance associated with humans' uninterrupted use of the landscape supports the view that current grazing practices are probably not now worsening land conditions in most areas (although they may have shifted the state of the system to a considerably less productive condition in the past). The concern in these lands is, therefore, primarily focused on landscape enhancement rather than landscape maintenance.

By contrast, Australia was occupied by a hunter-gatherer civilization until two hundred years ago. Although Aboriginal people made extensive use of fire in hunting, clearing vegetation for new growth and for opening pathways, and signaling (Thomas 1993), they never actively cultivated crops or domesticated herbivores. Australian landscapes, therefore, experienced a considerable discontinuity in use in historical times, exacerbated by the fact that grazing by megaherbivores had been absent for at least 30,000 years. As a result, research and management have a strong focus on avoiding land degradation, that is, on landscape maintenance as well as restoration; indeed, there is still a target condition to which restoration should strive.

Rather abrupt discontinuities in land use in arid North America can be traced to the arrival of cattle and horses with Columbus in 1493. Horses brought to the New World soon became wild and migrated from Mexico to the Great Plains where they were domesticated by Native Americans (Ponting 1991). With the aid of horses, the plains Indians hunted buffalo (dominant large herbivore of the landscape) and abandoned agriculture. Domesticated sheep did not flourish in the Americas until they were taken to Mexico in the 1540s (Ponting 1991).

In the arid southwestern United States, Native Americans relied entirely on hunting and gathering until maize was introduced as a domesticated crop, sometime between 2000 and 1000 B.C. Thereafter, for at least another thousand years, the cultivation of maize and other plants was combined with traditional hunting and gathering (Thomas 1993). The earliest varieties of corn were not very productive but required little effort to cultivate and, therefore, did not disrupt traditional hunting and gathering activities. However, farming in this manner did provide a buffer against possible shortages of natural foodstuffs. This subsistence pattern is perhaps best illustrated by the western Apache of east-central Arizona where some of the earliest examples of plant cultivation in the Southwest can be found. Maize farming had spread to western North America from Mesoamerica. About 7000 B.C., the indigenous people of central Mexico began collecting plants more intensely, which eventually gave rise to squash, beans, and the wild ancestor of maize. In fact, permanent villages relying significantly on domesticated plants were thriving in some areas of Mexico by 2500 B.C. (Thomas 1993). Before the arrival of Europeans in the sixteenth century, the arid and semi-arid lands of Mexico and the American Southwest were occupied by a nomadic hunting-and-gathering civilization. Missionaries then founded small settlements to teach people farming techniques and introduced domestic animals.

In summary, regions with a continuous history of grazing by large herbivores will become adapted to such grazing, even though the dominant grazing animals and their grazing patterns have changed in recent times. Regions with a continuous history of relatively intensive land use may have suffered a historical loss of productivity but are probably more resilient in the face of modern pastoralism than "native" landscapes such as Australia. The patterns of crop culture used in these regions suggest that early farming did not exact a heavy cost from the people or the land. In view of the histories discussed, research should carefully address landscape impacts created by new and different uses. These are most likely to cause changes in ecosystem function. Research should focus on landscape enhancement where practices have been maintained for a long time.

Water Policy Water is the single most important limiting constraint for economic development, biological diversity, and sustainable agricultural production in arid lands. Thus, water policy has a tremendous impact on existing and future uses of such lands. Water policies are developed by public and private organizations that often have conflicting objectives and goals, and politics is dominant in water policy formulation. Political considerations vary in scope and scale at different levels of government. For example, local government, in a village or town, may be concerned about the impact of demands from recreational (swimming) or ornamental (lawns and gardens) irrigation on water supplies. In contrast, national concerns may include economic issues (balance of trade), national security (food supply), or social considerations (migration to the cities).

Government control and management of water resources tend to escalate in response to pressures imposed by citizens. The government of Israel has total control of water within its borders because water is such a critical natural resource for them. The United States is moving toward more centralized control of water resources on western public lands because of their mixed role in providing urban supplies, input to wildlife and conservation management, and irrigation production. Australia, with its sparse settlement, has few water-use conflicts in the arid interior but faces similar issues in the semi-arid water catchments where irrigation is common, such as the Murray-Darling Basin. In Mexico, water resources are controlled by the federal government, which gives concessions for use to private owners under Water Federal Law. Urban water supplies and water quality have high priority under government policy, which seeks to improve the efficiency of use and recycling of water, especially in arid lands (González-Villarreal and Garduño 1994).

Demographic issues are vital to water resources policy. The world population is projected to expand fastest in the semi-arid regions of the world. To keep pace with population growth, economies will have to grow in order to maintain standards of living, and adequate water supplies will be the key to economic development in arid regions. "Adequate" needs to be defined as an amalgam of the underlying resources, efficiency of use, degree of recycling, and demand. Efficiency of use and recycling of water resources is poor in Australia and the United States. The Middle East is a highly visible region of the world where water resources are scarce and competition among neighboring countries is keen. Peace will not be possible without a resolution of regional water conflicts. Population growth will also increasingly create conflicts between urban and rural water demands within regions. As Israel's population grows from 4.66 million in 1990 to 8.50 million by 2025 (Gleick 1994), existing water supplies will be sorely stretched. The Middle East countries most likely to suffer acute water shortages will be those unable to bear the cost of obtaining new water supplies from projects involving out-of-basin transfers or desalination. Unfortunately, Middle East countries outside the Persian Gulf region do not have access to inexpensive fossil fuels, making the desalination approach to unlimited freshwater a difficult, if not unreachable, goal (Gleick 1994).

The use of water resources is often restricted by quality as well as quantity. Both factors are dealt with in water treaties between the United States and Mexican governments in regard to the Colorado and Rio Grande Rivers, where significant investments in water-related infrastructure have been made by both countries. Serious conflicts can also arise among individual states within a given country. Longstanding litigation between states about water is common in the United States (e.g., Texas vs. New Mexico, California vs. Arizona). In Australia, conflicts over the use of water from the Murray-Darling Basin have arisen among New South Wales, Victoria, and South Australia, which is downstream. These disputes are now being managed by the creation of a single statutory Murray-Darling Basin Management Authority. Israel, as a singular state, has no such problem.

The conservation and efficient use of water resources is paramount in areas of limited freshwater resources. Mexico is facing increasing pressure to supply quality water to meet the needs of both population and industrial expansion. At present, some manufacturing companies are bearing the burden of financing water pretreatment plants to obtain water quality adequate for manufacturing. Mexico must expand its sanitation treatment plants to avoid further contamination of existing water supplies. Israel is self-sufficient in food production and maintains a high standard of living for its citizens despite using considerably less water per capita (200 m^3 per year) than most countries of the Middle East (Gleick 1994). However, Israel's efficient use of water is achieved by substantial investment in the infrastructure associated with the transport of water from its source to areas of use and by the extensive use of waste water for the irrigation of industrial- and livestock-related crops.

It is not surprising, then, that issues relating to water, and understanding and costing the hydrological cycle properly, are vital questions for research worldwide.

Desertification Desertification is a complex subject because it potentially reflects dynamic interactions among diverse factors such as climate, soils, socioeconomics, and government policy. Concerning human abuse, desertification can reflect poor management of irrigated cropland, overcultivation, deforestation, and even soil and vegetation disturbance by recreational vehicles and military campaigns (Thomas and Middleton 1994). However, the vast areas dedicated to rangeland make the sustainability of livestock raising the single most critical management factor related to desertification. In the United States, 11% (almost 20 million ha) of rangeland suffers at least moderate desertification. In Australia, 6% (6 million ha) of rangeland is affected (Grainger 1992).

Biodiversity Policy Arid lands are rich in biota. At the same time, they are fragile: minor environmental stresses can severely affect the biota and, hence, biodiversity. This problem has brought the issue of biodiversity to the forefront in all three regions. Public pressure is steadily increasing to conserve biological diversity for its intrinsic value and for a variety of uses, ranging from future genetic pools for producing pharmaceuticals to a basic understanding of the ecological complexity of biotic systems. Specialized genetic resources are rich in these arid lands; for example, arthropod venoms are currently being explored for curing human diseases, and there are many more research opportunities in this area.

The protection of the genetic diversity of endangered endemic species, especially vertebrates and vascular plants, has gained popularity in the United States and Israel and is currently gaining similar status in Mexico and Australia. Conservation associations are thriving in these countries and throughout most of the developed world. As the conservation of genetic resources gains importance, ancillary concerns about the importation of exotic species, associated quarantine issues, and the exploitation of biological specimens also acquire a higher profile. For example, Australia has a major problem of imported weeds and pests, often from the Eurasian region (Humphries et al. 1991, Lonsdale 1994), and there are major research efforts on control measures.

The exploitation of biodiversity can be categorized into commercial and noncommercial uses and into uses that harvest live material as opposed to those that kill the organisms. Commercial uses include animals captured for sale as pets (uncommon in Australia) or hunted for recreation (most common in the United States) or products such as skins (e.g., kangaroos in Australia, game in the United States); plants may be removed for gardens and landscaping (e.g., endemic cacti in the Sonoran and Chihuahuan deserts) or for processed resources (e.g., specialist timbers and artifacts in Australia, medicines, and genetic material in most countries). Noncommercial harvesting is most importantly related to subsistence uses in all countries. The potential for social conflict among groups in society means that understanding appropriate levels of harvesting should be a high priority research issue.

In Australia, the Negev desert, and North American deserts, management of biodiversity underpins the concepts of conservation management and the functional maintenance of recreational sites. To sustain these activities, a concerted well-funded effort is needed to develop an inventory of the biota of arid lands. In the United States, a new administrative entity, the National Biological Survey (NBS), has recently been established with the Department of the Interior. The role of the NBS in coordinating an assessment of the country's biological heritage has received considerable discussion (NRC 1993). Israel, Mexico, and Australia are also developing inventories. Australia remains the least well-known region of the four in this regard.

Subsidies Due to their low productivity, arid and semi-arid regions often receive various monetary subsidies to maintain agriculture, forestry, and other land uses. Direct subsidies from government depend on settlement priorities of the society concerned (e.g., Cohen 1993) and the lobbying power that people in these lands exert on their government. The rate of subsidy affects the value of regional products and can indirectly discourage or promote sustainable land use.

In Israel, limited land availability means that the government aims to settle the arid and semi-arid areas and spread the population that is otherwise concentrated in the central part of the country. Farmers are assisted in drought years by a safety net subsidy if they are located above the "poverty line" (the 250 mm rainfall isohyet). On the other hand, irrigation water is subsidized to the same price throughout the country, giving some advantage to the drier regions. National organizations provide infrastructure for agriculture and various landscape restoration measures; for example, some arid zone tree plantings cost up to $2,000 ha^{-1} to establish, and they are primarily for long-term amenity purposes with little or no overt financial return. Australia provides fewer agricultural subsidies than the United States, but in both countries irrigation water is cheap (e.g., $3–15 Ml^{-1} in Australia compared to $200 Ml^{-1} in Israel) and rangeland grazing rents are very low. Each country has support schemes for restoration work to prevent soil erosion. In Mexico, farmers are subsidized to produce maize and beans due to the concern of government to be self-sufficient in food production.

Australia has seen a major shift in policy away from subsidies during droughts in recent years, on the premise that most dry periods are a normal business risk for agricultural production. Drought subsidies were seen to support the pastoralists who were least prepared for dry years and most likely to cause land degradation. The current moves toward "self-reliance" in farming emphasize the need for preparation and the recognition of the need to live within the constraints of climatic variability. These moves have major implications for research, which must focus on assisting managers' abilities to cope with risk and on safe stocking rate strategies rather than on "engineering" higher productivity.

Subsidies, in general, tend to be capitalized into the land use that receives them and reduce the resilience of that land use to policy changes in the future. Where the benefits to society exceed the possible costs of this loss of flexibility, especially in the start-up phase of new uses, subsidies can be important; often, though, they increase dependence and are politically difficult to remove. Their existence can greatly affect research priorities, and research into their net social and environmental impact is especially important in arid regions that are often dependent on external inputs.

Land Tenure Tenure issues concern both public and private lands. The percentage of land under public sector stewardship differs substantially among the countries, being highest in Israel, decreasing in the order of United States, Australia, and Mexico. Management policies on private lands are mainly in the hands of the landowners who nevertheless must function within local zoning ordinances in most countries.

Public sector land is often administered at more than one level. For example, in Israel, arid lands are managed almost entirely under country-level policy making and implementation. Policy for Australian public lands is created and administered by the states, which leads to difficulties in obtaining nationwide consistency and to replicating efforts when implementing sustainable land use in the various states. In the United States, public arid lands fall under several jurisdictions: national monuments and parks, national forests, and public domain lands are federal, while a considerable amount of land, such as state parks, is state managed. There are also conservation and recreational areas under local stewardship. Broad-scale, openly debated national stewardship policy sometimes conflicts with local policy derived from a strong sense of ownership. Such contrasting views can provide complementarity and diversity in relation to the issue of land tenure, but too often the appropriate balance is poorly researched.

Mineral rights are rarely included in private land ownership, being purchased and exploited separately with compensation perhaps being paid to the users of the surface resources. Mining residues may mar the landscape, bringing into play restoration policies that are usually inspired by policies at a national level.

Land tenure is a vital part of land management policy (e.g., Cohen 1993). Private ownership provides greater interest but usually allows a less flexible role for societal intervention. In regions where the relationship between use and degradation is simple and predictable and damage is economically reversible, private ownership is likely to be preferable. In environments where damage is intermittent, hard to forecast precisely, and economically irreversible, the interests of society are probably better served by a flexible system of public ownership (Holmes 1994). Arid and semi-arid lands often fall into the second category, but research is needed to determine how degradation results from different land uses so as to identify when and which tenures are appropriate.

Indigenous People All three regions are occupied by indigenous people living more or less traditional lifestyles. A growing concern for equity for these people has had a different impact on land-use practices in the different regions. Notwithstanding major changes in population immigration, the Bedouins in Israel have had a continuous history of co-occupation of the land with other groups for millennia, so that impacts center on changes in the relative levels of occupation (e.g., Ben-David 1990). By contrast, European peoples arrived in the Americas and Australia about 500 and 200 years ago, respectively. In both regions, indigenous people have been increasingly granted rights over traditional lands. The philosophy underlying these moves in the past has often meant a greater concentration on the rights themselves than on sustainable land use, resulting in poor land management servicing in terms of agricultural extension. On both continents, this oversight is being rectified, and concerns are growing about the impacts of nontraditional land uses on harvesting of local food and medicinal resources.

There is a considerable difference in the degree and form of links between traditional land uses and the marketplace. The Bedouins sell livestock and can accumulate reasonable wealth, which allows a close feedback between effort and success; they often spend this wealth on improving the education and housing of the next generation. In Australia, Aboriginal people are gaining ownership of increasing amounts of land, but this is usually in the least productive areas and is mainly held under an inalienable form of tenure. Traditional management of ceremonially significant sites is still important, but off-road vehicles, guns, and nontraditional food supplies are a major part of Aboriginals' current lifestyle. The remoteness of most dryland Aboriginal communities makes it very difficult for most of the population to find employment, although income is received from mining royalties and through their own tourism-related activities, including craftwork. Some commercial or subsistence grazing of cattle is carried out, and the use of unmanaged food resources such as kangaroos and the more easily harvested plants remains common. The situation with Native Americans in the United States is similar, albeit with closer access to markets. In the arid parts of Mexico, the small population of Native Americans has access to limited parcels of public land assigned to their exclusive use; they subsist on grazing, traditional medicines, and craftwork aimed at the tourist market. In both Australia and America, indigenous people remain disadvantaged in terms of financial resources, health, and education.

Traditional communities tend to have a greater complexity of land-use goals within a single land management unit. While traditional authorities remain strong and methods for resolving these conflicts are well developed, there has still been loss of culture and the introduction of nontraditional land uses. There is an increasing risk that community conflict and land degradation will occur due to unresolved land-use interactions. Society at large is beginning to expect standards of sustainable land management on these lands that meet the requirements applied to mainstream land uses, so significant research is needed to provide these land users with a level of support comparable to other industries such as commercial pastoralism.

Conclusions and Implications

This chapter has established the land-use and management context within which subsequent chapters will develop the research needed to improve the sustainable use of the world's arid and semi-arid lands. Research priorities are all too often established on the basis of scientific curiosity without an explicit consideration of management needs. By examining case study regions, it is possible to examine the possible range of management problems, without becoming focused on just one region, and to look for common problems among regions that may indicate high priority for research. Many of the issues raised in this way are summarized in tables 9.1, 9.2, and 9.3. The regions dealt with here are biased toward the developed countries, but some of the issues are also applicable in less-developed countries where subsistence plays a much larger role in land use.

Some specific research topics have been raised in the preceding pages (see also tables 9.2 and 9.3). Here we identify some important principles for modern scientific research in arid regions that arise from these considerations. The principles fall into two categories: those related to the process and implementation of research and those related to the content of research itself.

The Approach to Research

Understanding Goals and Decisions Research must be based on a clear recognition of what goals different land users are seeking to meet. This, in turn, requires some dedication to participatory decision analysis techniques and a move away from the old research-extension-management linear transfer paradigm (Jiggins 1993). In particular, there is a growing recognition that different types of paradigms for knowledge development and transfer are appropriate for different circumstances. To meet the implications of this recognition, managers and other stakeholders must be involved in research-and-development groups at a local scale; while time-consuming, this exercise has great rewards in terms of uptake rates, eventual benefits, and even, consequently, subsequent research funding (not that this should be the principal motivation!).

Integrated Research A strong message arising from the management issues discussed above is the major trend toward mixed land uses, which requires a systemically interdisciplinary approach to research. In fact, such an approach is increasingly seen as necessary even within a single land use since managers, by their nature, integrate diverse sources of information to manage a whole system; research that focuses on one part of that system in isolation has repeatedly been found irrelevant when the interactions between parts of the system are considered (Stafford Smith and Foran 1988). The systems approach can provide trustworthy predictions of human impacts on the environment. Moreover, it can present an integrated view of the form and function of an entire landscape when sophisticated mathematical models are used to synthesize the best available information on the dynamics of the abiotic and biotic components of environmental systems being studied (Huggett 1993).

Scale of Research Another important issue in assessing most land uses is that plot-scale research is, at best, of marginal relevance to the scales at which managers of arid areas must operate; in fact, it can be very misleading if the emergent properties of the broader-scale landscape override findings at the small homogeneous plot level. For example, data on animal productivity relative to stocking rates collected according to the Jones-Sandlands model on small plots (Wilson and MacLeod 1991) omits the ability of animals to graze selectively in space, as well as among species, invalidating the relationships in large paddocks. Moving from time-series data at the small plot scale up to spatially cross-sectional studies is a great challenge but one that cannot be evaded. Remote sensing becomes a vital component of this (Pickup 1989).

Involving Socioeconomics The simplest decision analysis for any commercial (and even subsistence) manager highlights the fact that conservation ranks below numerous factors such as economics, family, debt, and other personal aspirations. Similarly, the availability and strength of markets for products is far more important to managers in the short term than is the health of the ecosystem. Of course, the state of the system is ultimately the constraint within which any management must operate but to claim this without giving due weight to the pressures of socioeconomic factors is to marginalize research in the eyes of managers by giving it an appearance of irrelevance. There are also circumstances where one must be able to recognize that social forces such as population growth simply overwhelm the value of an option that might otherwise be ecologically sensible.

Research Issues

Landscape Stratification The importance of patterns of nutrient distribution and water redistribution in arid and semi-arid lands at the scale at which landscapes are managed means that research, in relation to any use, must be sensitive to spatial heterogeneity. It becomes vital to have good landscape stratification in experimentation. A major focus of research needs to be on understanding how different landscape units can provide for and respond to different levels of use, whether in relation to levels of recreational impact, grazing production, or conservation protection. These uses are increasingly in conflict, and research is needed to resolve the levels of use at which the conflict is tolerable.

Water Use Although nutrients are probably the greatest limit to growth in most arid lands, after water is available, it is nonetheless the primary limiting factor in almost any use of arid lands. A better understanding of the functioning of aquifers, the surficial redistribution of water resources, and the net effects of different patterns of water supply on soil moisture availability through time are all crucial to land use decisions. Political issues such as the costing of water and the balance between urban and rural uses are also dependent on more accurate net valuations of the benefits to society.

Inventory Our knowledge of the arid biota in most parts of the world remains poor, and continued efforts toward inventory are vital. However, inventory can too easily become a self-serving goal, and it is important that such work should be targeted toward the aim of understanding system function and, in particular, understanding what elements of the system can be legitimately substituted without impairment (Irwin 1991). This question ultimately underlies predictions of the extent to which the maintenance of biodiversity is compatible with continued commercial production.

Handling Mixed Land Uses In the past, research has almost always focused on the processes involved in a single land use (and often only a small part of that). Today's challenge to research is to examine the interactions among land uses—for example, the effects of different grazing levels (not just its presence or absence) on conservation, of run-on agriculture on grazing in the catchment, or water use for tourism on traditional waterholes. Another component of this systems research is understanding the regional-scale consequences of multiple-location land uses such as transhumance and tourism. In short, we need information that allows local and regional optimization of mixed land uses (Morton et al. 1994) rather than the maximization of production from a single land use.

Resource Maintenance Compared to Enhancement The objectives of land use can be aimed at maintaining the current land condition and productivity or at enhancing it; the latter may be needed to restore land from past degradation or simply to improve the natural patterns of productivity. The objective is determined by whether the general goal of land management is the maintenance of ecological integrity or the maximization of resource outputs, and whether the intention is to have a functioning ecosystem of any type as opposed to one of a specific type and, if so, what type is desired and what level of function is acceptable. These different objectives color the way in which research should be carried out, where it should be located, and what variables should be monitored.††

Conclusions

This chapter has emphasized that each arid region has its own special character but that some issues are universal in importance. The previous section identifies the generalized approaches that are needed everywhere, while tables 9.1 and 9.2 illustrate some of the research topics that have more or less wide application. The result, of course, is a large number of issues interacting in a complex way. Unfortunately, this is usually reality in the world, both for management and for research. Because of this, it is useful to conclude by embracing the principles of a systems approach, made explicit by modeling and decision support systems if necessary. Because of the complex nature of the analysis that must be undertaken to support decisions relating to ecosystem sustainability, a systems approach is needed to assist managers to select the best of alternative management options. Decision support tools do not need to be computer-based, but computer simulations of ecosystem responses to alternative management, presented in user-friendly graphical formats, will be one of the decision tools of future managers. When researchers are faced with a plethora of possible research topics (e.g., tables 9.1 and 9.2), a systems approach is vital to setting objective research priorities and to developing a program that is both comprehensive and well balanced.

References

Australian Surveying and Land Information Group (AUSLIG). 1992. The Ausmap Atlas of Australia. Cambridge: Cambridge University Press.

Axelrod, D. I., and P. H. Raven. 1985. Origins of the Cordilleran flora. Journal of Biogeography 12:21–47.

Ben-David, J. 1990. The Negev Bedouin: From nomadism to agriculture. *In* R. Kark, ed. The land that became Israel: Studies in historical geography. New Haven, Conn.: Yale University Press. 181–95.

Boyko, H. 1949. On the climax vegetation of the Negev, with special reference to arid pasture problems. Palestine Journal of Botany [Rehovot, Israel] 7:17–35.

Brown, D. E., ed. 1982. Biotic communities of the American Southwest: United States and Mexico. Desert Plants 4:1–342.

Brown, J. H., and A. C. Gibson. 1983. Biogeography. St. Louis, Mo.: C. V. Mosby.

Buffington, L. C., and C. H. Herbel. 1965. Vegetational changes on a semidesert grassland range. Ecological Monographs 35:139–64.

Bye, R. 1993. The role of humans in the diversification of plants in Mexico. *In* T. P. Ramamoorthy, R. Bye, A. Lot, and J. Fa, eds. Biological diversity of Mexico: Origins and distribution. New York: Oxford University Press. 707–31.

Byington, K. E. 1990. Agroforestry in the temperate zone. *In* K. G.

MacDicken and N. T. Vergara, eds. Agroforestry: Classification and management. New York: John Wiley and Sons. 228–89.

Carr, M. E., C. T. Maso, and M. O. Bagby. 1986. Renewable resources from Arizona trees and shrubs. Forest Ecology and Management 16:155–67.

Caughley, G., N. Shepherd, and J. Short, eds. 1987. Kangaroos: Their ecology and management in the sheep rangelands of Australia. Cambridge: Cambridge University Press.

Chisholm, A. H., and R. G. Dumsday, eds. 1987. Land degradation: Problems and policies. Melbourne and Canberra: Cambridge University Press and Centre for Resource and Environmental Studies.

Cohen, S. E. 1993. The politics of planting. Chicago: University of Chicago Press.

Dick-Peddie, W. W. 1993. New Mexico vegetation: Past, present, and future. Albuquerque: University of New Mexico Press.

Downing, T. E., and P. F. Ffolliott. 1983. The social dimensions of rangeland management. *In* D. R. Patton, J. M. de la Puente, P. F. Ffolliott, S. Gallina, and E. T. Bartlett, eds. Wildlife and range research needs in northern Mexico and southwestern United States. Gen. Tech. Rep. WO-36. Washington, D.C.: USDA Forest Service. 19–23.

Dregne, H. E. 1983. Soil of semiarid regions. *In* E. Campos-Lopez and R. J. Anderson, eds. Natural resources and development of arid regions. Boulder, Colo.: Westview Press. 53–62.

Ehleringer, J. 1985. Annuals and perennials of warm deserts. *In* B. F. Chabot and H. A. Mooney, eds. Physiological ecology of North American plant communities. New York: Chapman and Hall. 162–80.

Foran, B. D., F. Crome, and L. Moore. 1995. Population growth and biodiversity loss in Australia. Proceedings of Demography Conference, Canberra.

Foran, B. D., M. H. Friedel, N. D. MacLeod, D. M. Stafford Smith, and A. D. Wilson. 1990. A policy for the future of Australia's rangelands. Canberra: CSIRO, Division of Wildlife and Ecology.

Foster, K. E., and R. G. Vardey. 1987. A regional center for new crops and agrisystems for dry lands in Mexico. *In* E. F. Aldon, V. C. Gonzalez, and W. H. Moir, eds. Strategies for classification and management of native vegetation for food production in arid zones. Gen. Tech. Rep. RM-150. Fort Collins, Colo.: USDA Forest Service, Rocky Mountain Station. 154–59.

Friedel, M. H., B. D. Foran, and D. M. Stafford Smith. 1990. Where the creeks run dry or ten feet high: Pastoral management in arid Australia. Proceedings of the Ecological Society of Australia 16:185–94.

Gethes, D. H. 1993. Water resources: A wider world. *In* L. J. MacDonnell and S. F. Bates, eds. Natural resources policy and law. Washington, D.C.: Island Press. 124–47.

Gleick, P. H. 1994. Water, war, and peace in the Middle East. Environment 36:6–15, 35–42.

González-Villarreal, F., and H. Garduño. 1994. Water resources planning and management in Mexico. Water Resources Development 10:239–55.

Goodin, J. R., E. Epstein, C. M. McKell, and J. W. O'Leary. 1990. Saline agriculture: Salt-tolerant plants for developing countries. Washington, D.C.: National Academy Press.

Grainger, A. 1992. Characterization and assessment of desertification processes. *In* G. P. Chapman, ed. Desertified grasslands: Their biology and management. London: Academic Press.

Hacker, R. B., D. Beurle, and G. Gardiner. 1990. Monitoring western Australia's rangelands. Western Australian Journal of Agriculture 31:33–38.

Herbel, C. H. 1979. Utilization of grass- and shrublands in the southwestern United States. *In* B. H. Walker, ed. Management of semiarid ecosystems. New York: Elsevier Scientific Publishing. 161–203.

Herrera-Toledano, S. 1992. The ecological factor in the NAFTA. Business Mexico 2:28–31.

Hill, M. A., and A. J. Press. 1994. Kakadu National Park: An Australian experience in comanagement. *In* D. Western, R. M. Wright, and S. C. Strum, eds. Natural connections: Perspectives in community-based conservation. Washington, D.C.: Island Press. 135–57.

Holechek, J. L., R. D. Pieper, and C. H. Herbel. 1989. Range management: Principles and practices. Englewood Cliffs, N.J.: Prentice-Hall.

Holland, L. 1989. Weapons and water in the Southwest. *In* Z. A. Smith, ed. Water and the future of the Southwest. Albuquerque: University of New Mexico Press. 133–49.

Holmes, J. H. 1994. Changing rangeland resource values: Implications for land tenure and rural settlement. *In* Proceedings, ABARE outlook conference, Canberra, Australia, February. 160–75.

Huggett, R. J. 1993. Modelling the human impact on nature: System analysis of environmental problems. New York: Oxford University Press.

Humphries, S. E., R. H. Groves, and D. S. Mitchell. 1991. Plant invasions of Australian ecosystems. ANPWS Project Report No. 58. Kowari Vol. 2. Canberra, Australia: Australian National Parks and Wildlife Services.

Irwin, M. E. 1991. The land use controversy: A resolvable dilemma. *In* R. L. Metcalf and S. B. Vinson, eds. Proceedings of the centennial symposium of the Entomological Society of America: Emerging technologies and challenges. 7–10.

Jiggins, J. 1993. From technology transfer to resource management. *In* Grasslands of our world. Wellington, New Zealand: SIR Publishing. 184–91.

Kamp, D. 1993. Mexico's bylines: Source of wealth or woe? Business Mexico 3:29–30.

Kelso, L. 1992. Mexico's environment. Business Mexico 2:32.

Kessler, W. B., H. Salwasser, C. W. Cartwright, and J. A. Caplan. 1992. New perspectives for sustainable natural resources management. Ecological Applications 2:221–25.

Kovda, V. A., E. M. Samoilova, J. L. Charley, and J. J. Skujins. 1979. Soil processes. *In* D. W. Goodall et al., eds. Arid land ecosystems. Vol. 1: Structure function and management. New York: Cambridge University Press. 439–70.

Lonsdale, W. M. 1994. Inviting trouble: Introduced pasture species in northern Australia. Australian Journal of Ecology 19:345–54.

MacMahon, J. A., and F. W. Wagner. 1985. The Mohave, Sonoran, and Chihuahuan deserts of North America. *In* N. Evenari, I. Noy-Meir, and D. W. Goodall, eds. Ecosystems of the world. Vol. 12A: Hot deserts and arid shrublands. Amsterdam: Elsevier. 105–201.

Manzanilla, H. 1992. Border environment and NAFTA: A commentary. *In* P. Ganster and E. O. Valenciano, eds. The Mexican-U.S. Border Region and the Free Trade Agreement. San Diego: San Diego State University and the Institute of Regional Studies of the Californias. 80–82.

Martin, C. S., and D. R. Patton. 1981. Characteristics, resources, and uses of hot deserts of the United States. *In* H. G. Lund, M. Caballero, R. H. Hamre, R. S. Driscoll, and W. Bonner, eds. Arid land resources inventories: Developing cost-efficient methods. Gen. Tech. Rep. WO-28. Washington, D.C.: USDA Forest Service. 21–27.

Medellin, L. F. 1983. Mexico's semi-arid zones. *In* E. Campos-Lopez and R. J. Anderson, eds. Natural resources and development in arid regions. Boulder, Colo.: Westview Press. 15–25.

Mentis, M. T., D. Grossman, M. B. Hardy, T. G. O'Connor, and P. J. O'Reagain. 1989. Paradigm shifts in South African range science, management, and administration. South African Journal of Science 85:684–87.

Minckley, W. L., and D. E. Brown. 1994. Wetlands. *In* D. E. Brown, eds. Biotic communities: Southwestern United States and northwestern Mexico. Salt Lake City: University of Utah Press. 223–68.

Morton, S. R. 1993. Determinants of diversity in animal communities of arid Australia. *In* R. E. Ricklefs and D. Schluter, eds. Species diversity in ecological communities. Chicago: University of Chicago Press. 159–69.

Morton, S. R., and C. D. James. 1988. The diversity and abundance of lizards in arid Australia: A new hypothesis. American Naturalist 132:237–56.

Morton, S. R., and G. Pickup. 1991. Sustainable land management in arid Australia: How can we achieve it? Search 23:66–68.

Morton, S. R., D. M. Stafford Smith, M. H. Friedel, G. F. Griffin, and G. Pickup. 1994. The stewardship of arid Australia: Ecology and landscape management. Journal of Environmental Management 43:195–218.

Mosino-Aleman, P. A., and E. Garcia. 1974. The climate of Mexico. *In* R. A. Bryson and F. K. Hare, eds. Climates of North America. Vol. 11: World survey of climatology. New York: Elsevier. 305–404.

Mott, J. J., and G. P. Edwards. 1992. Beyond beef: The search for a level grazing paddock. Search 23:223–25.

Mumme, S. P. 1992. System maintenance and environmental reform in Mexico. Latin American Perspectives 19:123–43.

Nabham, G. P., and R. S. Felger. 1985. Wild desert relatives of crops: Their direct uses as food. *In* G. E. Wickens, J. R. Goodin, and D. V. Fields, eds. Plants for arid lands. London: Allen and Unwin. 19–32.

Nair, P. K. R. 1993. An introduction to agroforestry. Dordrecht, The Netherlands: Kluwer Academic Publishers.

National Research Council (NRC). 1993. A biological survey for the nation. Washington, D.C.: National Academy Press.

Nicholls, N., and K. K. Wong. 1990. Dependence of rainfall variation on mean rainfall, latitude, and the Southern Oscillation. Journal of Climate 3:163–70.

Norton, B. G. 1988. The constancy of Leopold's land ethic. Conservation Biology 2:93–102.

Nuccio, R. A. 1991. The possibilities and limits of environmental protection in Mexico. *In* J. S. Tulchin, ed. Economic development and environmental protection in Latin America. Boulder, Colo.: Lynne Rienner Publishers. 109–22.

Nuccio, R. A., A. M. Ornelas, and I. Restrepo. 1990. Mexico's environment and the United States. *In* J. W. Brown, ed. In the U.S. interest. Boulder, Colo.: Westview Press. 19–58.

Odum, E. P. 1977. The emergence of ecology as a new integrative discipline. Science 195:1289–93.

Olivier, M. 1993. Shared responsibility. Business Mexico 3:43–45.

Opie, J. 1993. Ogallala: For a dry land. Lincoln: University of Nebraska Press.

Peet, R. K. 1988. Forests of the Rocky Mountains. *In* M. G. Barbour and W. D. Billings, eds. North American terrestrial vegetation. New York: Cambridge University Press. 64–101.

Pickup, G. 1989. New land degradation survey techniques for arid Australia: Problems and prospects. Australian Rangelands Journal 11:74–82.

Pickup, G., and D. M. Stafford Smith. 1993. Problems, prospects, and procedures for assessing the sustainability of pastoral land management in arid Australia. Journal of Biogeography 20:471–87.

Ponting, C. 1991. A green history of the world. New York: St. Martin's Press.

Purvis, J. R. 1986. Nurture the land: My philosophies of pastoral management in central Australia. Australian Rangelands Journal 8:110–17.

Raintree, J. B. 1987. The state of the art of agroforestry diagnosis and design. Agroforestry Systems 5:219–50.

Ross, H., E. Young, and L. Liddle. 1994. Mabo: An inspiration for Australian land management. Australian Journal of Environmental Management 1:1–19.

Rzedowski, J. 1993. Diversity and origins of the phanerogamic flora of Mexico. *In* T. P. Ramamoorthy, R. Bye, A. Lot, and J. Fa, eds. Biological diversity of Mexico: Origins and distribution. New York: Oxford University Press. 129–44.

Smith, J. G. 1899. Grazing problems in the Southwest and how to meet them. Bull. 16. USDA Division of Agrostology. Washington, D.C. 1–47.

Stafford Smith, D. M. 1995. Australian deserts. *In* W. A. Nierenberg, ed. Encyclopedia of environmental biology. Orlando, Fla.: Academic Press.

Stafford Smith, D. M., and B. D. Foran. 1988. Strategic decisions in pastoral management. Australian Rangelands Journal 10:82–95.

Stafford Smith, D. M., A. McNee, B. Rose, G. Snowdon, and C. Carter. 1994. Goals and strategies for Aboriginal cattle enterprises. Australian Rangelands Journal 16:77–93.

Stafford Smith, D. M., and S. R. Morton. 1990. A framework for the ecology of arid Australia. Journal of Arid Environments 18:255–78.

Thomas, D. E. 1993. Farmers of the New World. *In* G. Burenhult, ed. People of the Stone Age: Hunter-gatherers and early farmers. New York: HarperCollins. 163–85.

Thomas, D. S. G., and N. J. Middleton. 1994. Desertification: Exploding the myth. New York: John Wiley and Sons.

Toledo, V. M., and M. de Jesus Ordonez. 1993. The biodiversity scenario of Mexico: A review of terrestrial habitats. *In* T. P. Ramamoorthy, R. Bye, A. Lot, and J. Fa, eds. Biological diversity of Mexico: Origins and distribution. New York: Oxford University Press. 757–77.

Van Pollen, H. W., and J. R. Lacey. 1979. Herbage response to grazing systems and stocking intensities. Journal of Range Management 32:250–53.

Wackernagel, M., J. McIntosh, W. E. Rees, and R. Woolard. 1993. How big is our ecological footprint? A handbook for estimating a community's carrying capacity. Vancouver: Task Force on Planning Healthy and Sustainable Communities, University of British Columbia.

Ward, D., and L. Olsvig-Whittaker. 1993. Plant species diversity at the junction of two desert biogeographic zones. Biodiversity Letters 1:172–85.

Williams, O. B., and J. H. Calaby. 1985. The hot deserts of Australia. *In* M. Evenari, I. Noy-Meir, and D. W. Goodall, eds. Ecosystems of the world. Vol. 12A: Hot deserts and arid shrublands. Amsterdam: Elsevier. 269–312.

Wilson, A. D., and N. D. MacLeod. 1991. Overgrazing: Present or absent? Journal of Range Management 44:475–82.

Woods, L. E. 1984. Land degradation in Australia. Canberra: Australian Government Publishing Service.

World Conservation Monitoring Center (WCMC). 1992. Global biodiversity: Status of the earth's living resources. New York: Chapman and Hill.

Young, M. D. 1992. Sustainable investment and resource use: Equity, environmental integrity, and economic efficiency. Parkridge, N.J.: Parthenon Publishing Group.

Zohary, M. 1962. Plant life of Palestine. New York: Roland Press.

———. 1973. Geobotanical foundation of the Middle East. Stuttgart: Gustave Fischer.

10 A Planning Process

Peter F. Ffolliott, Itshack Moshe, and Theodore W. Sammis

Management's task is to make ecological systems productive in satisfying people's needs within the biophysical and socioeconomic limits imposed in sustaining the ecological system being managed. Planning is one of the most important functions of management (Drucker 1986, Hanna 1985). In small enterprises with simple goals and objectives to be achieved in short periods of time, the organization of human, capital, and natural resources to carry out the necessary tasks can require relatively little formal planning. However, some type of formal planning is necessary in larger enterprises with large amounts of diverse human, capital, and natural resources and multiple goals and objectives to be achieved over a long time period.

In this chapter, we present a process to help people in preparing a plan of action for the implementation of development projects in arid and semi-arid ecosystems. It is from such a process that project planning for specific forms of management activities evolves. *Development projects* are sets of practices that are undertaken on lands for which the management activities are suitable to achieve the specified goals and actions (Gregersen and Contreras 1992).†

An Overview

Planning involves an integration of three major sets of elements (Brooks et al. 1991, Bryson 1988, Jakes et al.1989, Pfeiffer et al. 1989). First, the *objectives* are established on the basis of a problem analysis for the proposed project area and directives from higher-level authorities. Second, consideration is given to the various physical-biological, social, cultural, financial, and political *constraints* associated with the specific situation confronted. Third, the *techniques* for carrying out the alternative management activities that are necessary to satisfy the objectives are delineated. *Management planning* requires organizing, analyzing, and integrating the objectives, constraints, and management techniques in such a way that decision making and implementation are done more efficiently and effectively than if an unplanned approach had been used.

Many of the constraints to management activities in arid and semi-arid ecological systems are physical (e.g., climate, land form, and soil conditions). But, just as many constraints are likely to be social, economic, and institutional in nature (Brooks et al. 1991); these constraints include cultural limitations, political acceptability, and available budget. A challenge for the planner is to bring all these factors into an efficient planning process using appropriate datasets where available and estimates in other cases.

In relatively few instances will a planner and manager, with whom the planner should closely work in the planning process, have enough information to make risk-free decisions concerning actions and reactions. Therefore, planning requires judgment and flexibility. If people have learned one thing from past experiences, it is that development projects seldom unfold as originally planned. Those who are ready to adjust to changes in conditions are further ahead than those who rigidly follow an original plan despite changes in conditions. The purpose of this chapter is to present a process so that the planner and manager can be better prepared to anticipate these needed adjustments, identify critical information and concepts, and work cooperatively on the inevitable changes in constraints and opportunities that will take place as planning and implementation proceed.

Each development-related problem that confronts a planner and manager has its own unique set of technical elements and characteristics. Therefore, each development project requires a different technical approach (such as identification and design of vegetative manipulations) in terms of physical planning; the same does not hold for the planning process itself, however. A similar planning process can be used regardless of the type of development project in question. Only the relative emphasis placed on each step in the process will differ, not the steps themselves. This chapter outlines the basic steps in planning. A fine-tuning of the planning process is discussed in terms of adjustments in emphasis and content needed to deal with particular problem types.

The Planning Context

Management activities are undertaken to produce benefits to people and sustain, to the extent possible, the functioning of the managed ecological systems (Drucker 1986, Stuth et al., 1991). For example, people do not necessarily plant trees and shrubs as an end objective in itself. Rather, people plant trees and shrubs to obtain benefits that have direct value to them, such as reducing soil loss from agricultural lands or increasing fuelwood supplies for a village. Therefore, planners, managers, and (importantly) the local people must understand the potential flows of effects and benefits, both *on site* and *off site*. Three major elements are considered in the framework of producing benefits for people through management activities:

1. management practices (inputs) and costs associated with them;
2. biophysical effects and environmental changes associated with the management practices; and
3. socioeconomic changes (both benefits and costs) associated with the biophysical effects and environmental changes (e.g, changes that can be associated with direct economic value to humans).

The relationships between management activities (inputs) and their physical effects (outputs) are discussed throughout this book.

Effects and benefits are defined as changes *with* and *without* implementing the management activities being considered (Gregersen et al. 1987). For example, by "increased productivity," we mean the differences in productivity between conditions with and without the development project. Productivity can still be declining after the project has been implemented but it may be at a slower rate than without the project. Nonetheless, there is a higher level of productivity (an increase) over what it would have been without the project, and this increase represents a benefit that can be valued.

Steps in the Planning Process

The planning process presented in this chapter (Brooks et al. 1991, Bryson 1988, Ffolliott and Thames 1983, Hanna 1985, Pfeiffer et al. 1989) includes the following steps:

- monitoring and evaluating past activities and identifying problems and opportunities;
- determining the main characteristics of the problems and opportunities identified, setting objectives necessary to overcome the problems or take advantage of the opportunities, developing strategies for action, and identifying constraints;
- identifying alternative actions to implement strategies within the defined constraints;
- appraising and evaluating the impacts of alternatives, including environmental, social, and economic effects, and assessing uncertainty associated with the results; and
- ranking the alternatives and recommending action when requested.

Therefore, a planner and manager are concerned with the questions:

- How do we decide that we want to do something?
- What is it that we want to do?
- How do we decide what alternatives we have available to do it?
- How do we decide which alternative(s) is (are) best given the circumstances?

Iterative Process of Planning

Planning is an iteration of successively refined approximations in which people learn from past experiences and incorporate that learning into the ongoing planning process (Brooks et al. 1991). Management planning, therefore, involves an incremental learning experience. It is often difficult to get immediately to the details of a project design, however. Rather, through a series of iterations, we move from a rapid, low-cost assessment of the situation through progressively more detailed design and appraisal stages until we arrive at the point where we have all the necessary, desired, and affordable information needed to make a decision.

This iterative process makes sense since one purpose of planning is to help decision makers and managers reject unsuitable options at an early stage. An experienced planner can focus almost immediately on an appropriate strategy and viable, suitable options for dealing with a given situation. In essence, the purpose here is to discard unsuitable options with a minimum expenditure of time and effort and then to spend more time studying the alternatives that experience has taught are the best ones in a given situation.

People often point out that, by only following rules of thumb and past experience, one risks missing more efficient or effective solutions to a particular development-related problem. It is frequently argued that better solutions might be identified if more time and resources were devoted to the effort. This possibility certainly can exist in many cases. However, funds, skills, and time are scarce in most situations, and, as a consequence, most planners and managers have time constraints and limited resources to devote to problem analyses, project designs and evaluations, and comparisons of alternatives. The optimum solution to the development-related problem is defined and constrained by all these factors.

The steps in the planning process are discussed in more detail in the following sections of this chapter. In keeping with the orientation of the chapter, we will attempt to emphasize the practical and avoid the impractical.

Monitoring and Problem Identification

The process of planning has no beginning and no end. Nevertheless, the manager-planner team must start somewhere. A logical starting point is at a point before a problem or opportunity has been identified (Brooks et al. 1991). Therefore, we will start the planning process with monitoring and problem identification (fig. 10.1). People are continuously collecting data and creating knowledge about ongoing changes in biological, physical, and socioeconomic environments throughout the world. We suggest that this information can be used in planning to determine whether a management problem exists and what might be done to resolve it.

The types of information needed to identify problems are numerous. For example, information needed to describe arid and semi-arid environments includes that related to climatology, aridity, soils, vegetation, and hydrology. In general, people collect data on both a long-term and continuing basis to define major problems and on a onetime basis to provide information to solve a particular crisis. Long-term monitoring activities ideally allow people to observe and analyze resource responses through time and thus help identify problems. To illustrate this point, precipitation and stream flow data must be collected on a continuous basis for many years before we can characterize drought, soil erosion conditions, and flooding probabilities. In contrast to long-term data acquisition, information is often collected on a onetime basis to analyze a known specific problem. Analysis of these data is undertaken to quantify the severity of a particular problem and to evaluate whether or not the implemented corrective actions are solving the problem.

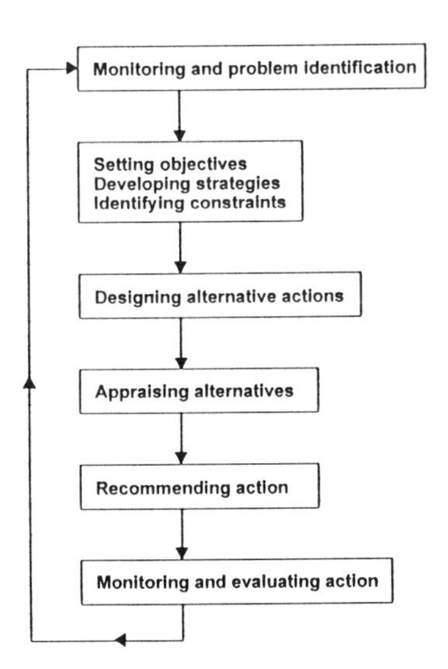

Figure 10.1. Steps in the planning process.

Different types of data and the levels of detail and accuracy for their collection are needed in different situations, depending largely upon the intensity of land use, the rates at which problems appear to be developing, the severity of an already identified problem, or combinations thereof. Often, no formal measuring or monitoring system is used to produce the information that leads to an identification of management concerns and eventual action. Rather, problems are observed directly after they have occurred. For example, when it becomes increasingly apparent that a reservoir is silting up rapidly, the scars of erosion begin to appear on the upstream landscape or floods become more frequent and serious.

Regardless of how problems and opportunities are noticed, their definition becomes one of the first steps in management planning. More than one solution to a problem is possible in many instances. For example, limited water supplies might be enhanced by either developing water harvesting schemes or reclaiming wastewater. In other cases, solutions can be mutually exclusive. Increased cultivation of agricultural crops or agroforestry practices to enhance the production of food can require that the livestock grazing in the area be discontinued entirely.

Not previously discussed in this chapter are the nonproduction-oriented objectives of management activities in arid and semi-arid ecosystems, such as enhancing scenic beauty or maintaining wildlife habitats. In many parts of the world, people set aside land as preserves that have unique wildlife habitats, landscape beauty, historical significance, or other amenity values. The establishment of these set-aside areas can be valid objectives and, as a consequence, must be considered in a planning effort together with adjacent or nearby lands that are managed for different purposes.

The types of problems that are important in management are many. In some cases, planning evolves into the implementation of actions to prevent a problem from occurring.

In other instances, a problem may already exist, and the planning task involves the development of a course of action to alleviate the problem or ameliorate the conditions that have caused it (Hanna 1985, Pflaum and Delmont 1987). While the managerial actions will likely differ in each case, the basic planning process should remain the same. The importance of monitoring and problem identification in planning cannot be overemphasized. The resultant problem statement will set the focus and orientation for the rest of the planning activity.

Setting Objectives, Developing Strategies, and Identifying Constraints

The next steps in the planning process involve setting objectives, developing strategies to solve the problems or respond to the opportunities presented, and identifying constraints to these strategies (Bryson 1988, Gregersen et al. 1989, Mager 1972, Pfeiffer et al. 1989). Objectives usually flow directly from the problem analysis. In their most general form, statements of objectives merely indicate that there is a need to develop a means of overcoming or preventing the identified problem or ameliorating the conditions causing it, eventually leading to the desired benefits. With greater levels of specification, objectives are translated into targets that are constrained (e.g., in terms of the riskiness of approaches taken, level of cost applied, and level of achievement of other objectives).

Setting Objectives

A *single objective* is not too difficult to deal with in the planning process, even when it is constrained in a number of ways. In most instances, where people are dealing with a single objective, clear decision criteria can be developed for determining the extent to which alternative project designs are acceptable and even how they rank relative to each other. As stated earlier, ranking is a common step in the decision-making process.

Difficulties arise when more than one objective exists and, as a result, *multiple objectives* have to be considered, a situation commonly encountered in the management of arid and semi-arid ecological systems. Decision-support systems have been developed to deal quantitatively (in theory at least) with multiple objectives (Cohon 1978, Dykstra 1984, Hof chap. 12, this volume, Loucks et al. 1981, Pickup and Stafford Smith chap. 11, this volume). However, the use of such systems has not always been successful. For decision-making purposes, it can be easier to focus on one main objective, with the other objectives then expressed as constraints on the main one. For example, a major objective of a development project might be to increase fuelwood production in a given area. Additional objectives, such as improved wildlife habitats and enhanced amenity values, might be expressed as constraints on the main objective. Within a specified budgetary level, therefore, the objective could be to maximize fuelwood production subject to an associated maximum improvement in wildlife habitats, enhancement of amenity values, and, in all likelihood, maintenance of environmental quality.

Planners can often avoid the problems of dealing quantitatively with multiple objectives by developing an array of statements on the effects of alternative project designs for various objectives. The decision maker then has to provide subjective weightings to compare different alternative combinations of outputs related to the various objectives.

Developing Strategies

Once objectives have been established and agreed upon by all concerned, possible strategies to achieve the objectives must be developed (Hanna 1985). Here, *strategy* is used to describe the general direction taken to satisfy the objectives. A *plan* further includes the magnitudes or targets to be achieved and timing of the actions to be taken to achieve them.

One might think of a number of alternative strategies to reduce soil erosion by looking at the problem statement, for example. Identifying acceptable and cost-effective land-use practices that decrease the rate of erosion can be one strategy. Regulations against land uses that, in all likelihood, intensify the erosive process with appropriate enforcement might be suggested in an autocratic environment. In other situations, the best strategy could be to leave the situation alone because the independent nature of the local people might preclude any chance for a successful intervention; this being the case, it would probably be better to spend scarce resources elsewhere. There are likely other scenarios that could be developed and transformed into strategy statements. The problem, therefore, is deciding upon feasible strategies to meet the set objectives within the framework of the constraints confronted.

Identifying Constraints

We can arrive at a logical strategy by looking systematically at information on constraints and the conditions that surround the problem (Brooks et al. 1991, Pfeiffer et al. 1989). For example, constraints that might have been considered in developing a strategy to reduce soil loss include:

1. Land form and soils may be such that only certain types of changes in land-use practices would have a good chance of success in the area.
2. There may not appear to be cultural constraints that stand in the way of changing land-use management practices, although tradition is important to the local people. Incentive mechanisms might have to be considered, at least initially.

3. Financing mechanisms available for long-term agricultural crop and livestock management may currently be inadequate. Since some tree or shrub cover will likely have to be established on steep slopes, improved credit mechanisms, alternate incomes, and other ways to support the establishment of trees or shrubs and other long-term management activities may be needed.

4. The literacy rate of the local people may be low, and mass communication facilities might not exist in the region. Furthermore, there may be no current extension organization in operation.

5. The political structure may be such that effective regulation and enforcement would be difficult. Therefore, an approach using incentives is likely needed.

Many other questions could be asked in this strategy formation stage. However, the idea is not necessarily to formulate precise courses of action, for example, how many hectares have to be treated with a certain treatment, how much of an incentive payment and what kind will be needed, and so forth. Instead, the planner should move quickly through the entire context of the development project in developing a strategy to:

- eliminate some approaches that would not work given the constraints identified;
- determine which constraints need to be addressed to have a reasonable chance for success; and
- identify the points that should be taken advantage of in project design and implementation.

Designing Alternative Actions

Once an acceptable strategy has been developed, the design of alternatives to implement the strategy becomes the primary focus of the planning process. The need here is identifying the various actions that could be used to implement the strategy and produce desired results (Brooks et al. 1991, Bryson 1988, Hanna 1985). This is where the technical expertise of agronomists, foresters, soil scientists, hydrologists, engineers, and other specialists comes into play. The expertise of social scientists, economists, politicians, and others dealing with social, economic, and cultural elements also enters the process at this time.

The types of actions that may be needed in different instances include:

- biological actions such as planting trees, shrubs, agricultural crops, or combinations thereof;
- physical and engineering actions (e.g., building terraces, dams, or gabions);
- regulatory measures to control specified actions including the harvesting of wood, controlling the grazing of livestock, and regulating water use;
- actions to create incentives, that is, providing free goods and services, subsidies, credit, outright cost sharing, and so forth; and
- educational activities including the preparing informational materials, presenting demonstrations, and offering training activities.

The task at this point in the planning process is to identify the options that are available given the constraints and conditions surrounding the development project. The manager should be familiar with a number of different managerial techniques and potential actions that are available to solve various types of problems. Remember, to do nothing, that is, not to implement a development project of any kind, is a valid alternative to be considered in the planning process.

Appraising Alternatives

While alternatives are being developed, they should also be undergoing appraisal. Appraisal is an ongoing activity in the planning process in most instances. In its broadest meaning, *appraisal* refers to the process of identifying, defining, and quantifying the likely or expected impacts of a practice, project, or program (Gregersen and Contreras 1992, Gregersen et al. 1987). Some of these impacts will be positive and some negative.

Nature of Impacts Appraised

Possible impacts of management activities, grouped into environmental, financial and economic, and social effects, are presented in table 10.1. Note that the groupings presented relate to different viewpoints of the impacts of a given change.

Table 10.1. Possible Impacts of Management Activities

Environmental effects on:
Soil protection
Ecological functioning, stability, and diversity
Water yield, timing of flow, quality
Wildlife protection
Financial and economical effects on:
Allocation of resources
Stability and distribution of income
Local, regional, and countrywide levels of production
Public- and private-sector budgets
Social effects on:
Local, regional, and countrywide employment
Migration flows
Cultural traditions
Political stability

Source: Adapted from OECD 1986.

Making Appraisals Useful

Appraisals of alternative development projects are useful only if they provide to decision makers relevant information in a way that they are comfortable with the information (Gregersen and Lundgren 1986, Gregersen et al. 1987). This means distinguishing between the technical analyst's considerations in choosing a good appraisal approach and the decision maker's view of what characterizes a good, acceptable, usable appraisal. The task of a planner is bringing these two sets of considerations together into the final appraisal product. Integration of perspectives of technical personnel and decision makers is a key to successful planning.

Appraisals should support the objective of spending the minimum amount of resources needed to reach an acceptable decision on how best to achieve an objective. This point is attained after a quick appraisal in some cases; that is, the supporting evidence and political agreement are so clear concerning the best alternative that one need go no further to reach a decision (Brooks et al. 1991). In other cases, a more detailed and thorough second-stage appraisal is needed because the evidence is not clear enough after the first-stage appraisal to make a judgment. A formal feasibility study is likely to be necessary to arrive at the point where a decision can be made in situations involving major commitments of resources. Sometimes a formal feasibility study is also required by the involved institutions (donors, management organizations, local community groups, etc.).

A question to ask in any appraisal is: What criteria of acceptance are we going to use? Suffice it to say that there are many criteria that can be used in most decisions. No single set of criteria will be enough to judge the applicability of a proposed project in most instances. Regardless of the criteria, appraisal results can be, and often should be, presented in different ways, depending upon the nature of the planning situation (Gregersen et al. 1987). In general, it is preferable to present a ranked set of alternatives or several rankings utilizing different evaluative criteria. However, the decision on which alternative is chosen ultimately can be made only by the responsible decision maker.

Risk and Uncertainty

One typically faces a situation of uncertainty rather than risk in most development projects to be undertaken in arid and semi-arid ecosystems. The distinction is simply that, in the case of *risk,* one can apply probabilities to various outcomes while quantitative measures of probabilities of occurrences can not be generated in the case of *uncertainty.* In a situation of uncertainty, one might develop subjective probability estimates for different aspects of a project that are of highest interest. However, such estimates often do more harm than good since subjectivity in the planning process should not be hidden. Therefore, the use of a straightforward analysis of how the measures of project worth (net present values, internal rates of return, benefit-cost analyses, etc.) or desirability would change under different assumptions concerning the values of key parameters is suggested.

Recommending Action

The planner's tasks often stop after the alternatives and implications of risk and uncertainty for the different options have been evaluated. However, in some cases, the planner is asked to make recommendations on which of the alternatives should be undertaken and the timing and approach of implementation (Brooks et al. 1991). To make these recommendations, it is necessary to establish *criteria of acceptance* against which the alternatives can be evaluated. It is likely that no single set of criteria will be sufficient for evaluating the applicability of the proposed alternatives. However, any criteria of acceptance can be based on:

- ecological guidelines, including providing sustained benefits while meeting the needs of the people, conserving and protecting the ecological systems, maintaining or improving soil productivity, or furnishing a range of multiple benefits;
- economic objectives, such as maximizing benefits, maximizing the returns on an investment, or achieving a specified production level at least cost; and
- social and cultural considerations (e.g., ensuring that livestock grazing patterns will not be disrupted in a way as to encourage hostility among the members of a community or making people available to plant agricultural crops or control the poaching of animals).

Criteria of acceptance that reflect the principles of *appropriate technology* should also be considered (Ffolliott and Thames 1983). These criteria may require a project to:

- make optimal use of locally available human resources, materials, and marketing opportunities;
- increase the potential for people's self-reliance in both the short and long terms;
- be compatible with available funding;
- make use of and, when appropriate, adapt traditional technologies and approaches to problem solving;
- have the potential for being maintained and sustained by the local community; and
- have a reasonable time frame for people to take responsibility for the project activities.

Continuous Process of Planning

What has been presented in this chapter is a simplified planning process, moving from problem identification through design to recommending action for implementation. In point of fact, the planning process is an ongoing, continuous process in most instances (Brooks et al. 1991, Hanna 1985, Stuth et al. 1991). As mentioned above, planning is also an iterative process, with information concerning the results of actions and emerging problems constantly being fed back into it (fig. 10.1). This information is frequently used to suggest incremental changes in the ongoing project. More formally, the process of collecting and disseminating information on ongoing projects is part of the monitoring and evaluation task referred to earlier. There should be continuing feedback of information on project activities that is synthesized, analyzed, and placed into the development of alternative strategies and suggestions for actions that are passed back to project management. This type of continuous process leads to useful interactions among planners, technical personnel, and managers of development management activities.

Role of Research in the Planning Process

There are frequently gaps in information that hinder planning and, as a result, should stimulate an investment in research (Gregersen and Lundgren 1986, Jakes et al. 1989, Rose et al. 1982). In deciding what information is needed and, therefore, what research should be undertaken, one might develop a *conceptual model* of the ecological and cultural setting in which the development project will be implemented to define the system in question and specify the information needed to fill the informational gaps. Very simply, a model is a *representation* of how that part of the world in which the development project will be implemented operates. A model can be in various forms, such as:

- simple statements;
- network diagrams; and
- sets of detailed mathematical expressions.

It is important to remember that the conceptual model should be as complete and accurate as possible. Information in the model, which can serve as a *baseline* when the project is evaluated, should include:

- ecological characterizations;
- cultural and social descriptions; and
- economic and, when appropriate, marketing conditions.

After developing a conceptual model, one can identify those processes that cannot be quantified with the information on hand, pointing to gaps in knowledge and the need for research (Brooks et al. 1991). The information needed to operate a *predictive model* that quantifies the impacts of a proposed project defines the data collection needs.

Concluding Comments

While the discussion presented in this chapter makes the planning process appear neat and orderly, planners and managers with even limited experience know that it is anything but that. Therefore, the steps presented here must be adapted to suit specific situations. Other procedures may be more appropriate or may be used to supplement the planning steps discussed in this chapter. Regardless of the quantitative techniques employed or the checklists used, the key to effective planning is to achieve flexibility within the established guidelines of acceptance.

Nevertheless, the principles behind the planning process presented can be important in most developmental situations. These principles are relevant to the development worker who is either present when the planning is just beginning or who arrives in the middle of implementing a project. While the specific steps can be changed, the principles of the planning process endure.

References

Brooks, K. N., P. F. Ffolliott, H. M. Gregersen, and J. L. Thames. 1991. Hydrology and the management of watersheds. Ames: Iowa State University Press.

Bryson, J. M. 1988. Strategic planning for public and nonprofit organizations: A guide to strengthening and sustaining organizational achievement. San Francisco: Jossey-Bass.

Cohon, J. L. 1978. Multiobjective programming and planning. New York: Academic Press.

Drucker, P. F. 1986. The practice of management. New York: Harper and Row.

Dykstra, D. P. 1984. Mathematical programming for natural resource management. New York: McGraw-Hill.

Ffolliott, P. F., and J. L. Thames. 1983. Environmentally sound small-scale forestry projects: Guidelines for planning. Arlington, Va.: VITA Publications.

Gregersen, H. M., K. N. Brooks, J. A. Dixon, and L. S. Hamilton. 1987. Guidelines for economic appraisal of watershed management projects. FAO Conservation Guide 16. Rome: Food and Agriculture Organization.

Gregersen, H., and A. Contreras. 1992. Economic assessment of forestry project impacts. FAO Forestry Paper 106. Rome: Food and Agriculture Organization.

Gregersen, H. M., and A. L. Lundgren. 1986. An evaluation framework. *In* Alternative approaches to forestry research evaluation and assessment. Gen. Tech. Rep. NC-110. St. Paul, Minn.: USDA Forest Service, North Central Forest Experiment Station. 2–6.

Gregersen, H. M., A. L. Lundgren, P. J. Jakes, and D. N. Bengston. 1989. Identifying emerging issues in forestry as a tool for research planning. Gen. Tech. Rep. NC-137. St. Paul, Minn.: USDA Forest Service.

Hanna, N. 1985. Strategic planning and management: A review of recent experience. World Bank Staff Working Paper No. 751. Washington, D.C.: World Bank.

Jakes, P. J., H. M. Gregersen, and A. L. Lundgren. 1989. Research needs, assessment, and evaluation: Identifying emerging issues as a key to forestry research planning. *In* A. L. Lundgren, ed. The management of large-scale forestry research programs and projects. Gen. Tech. Rep. NE-130. St. Paul, Minn.: USDA Forest Service. 107–13.

Loucks, D. P., J. R. Stedinger, and D. A. Haith. 1981. Water resource systems planning and analysis. Englewood Cliffs, N.J.: Prentice-Hall.
Mager, R. F. 1972. Goal analysis. Belmont, Calif.: Lear Siegler/Fearon Publishers.
Organization of Economic Cooperation and Development (OECD). 1986. The public management of forestry projects. Paris: OECD.
Pfeiffer, J. W., L. D. Goodstein, and T. M. Nolan. 1989. Shaping strategic planning. Glenview, Ill.: Scott, Foresman.
Pflaum, A., and T. Delmont. 1987. External scanning: A tool for planners. Journal of the American Planning Association 53:56–67.
Rose, D. W., H. M. Gregersen, A. R. Ek, and H. Hofanson. 1982. Planning with minimum data and technology. *In* M. C. Vodak, W. A. Leuschner, and D. I. Navon, eds. Symposium on forest management planning: Present practice and future decisions. FWS-1-81. Blacksburg: School of Forestry, Virginia Polytechnic Institute and State University. 188–97.
Stuth, J. W., J. R. Conner, and R. K. Heitschmidt. 1991. The decision-making environment and planning paradigm. *In* R. K. Heitschmidt and J. W. Stuth, eds. Grazing management: An ecological perspective. Portland, Oreg.: Timber Press. 201–23.

11 Management of Arid Lands: A Simulation Approach

Geoffrey Pickup and Mark Stafford Smith

Arid zone management, whether for grazing, conservation, or both, is fraught with difficulty. The basic driving force, rainfall, is variable and unpredictable, and the response of the landscape can be both nonlinear and spatially heterogeneous. At the same time, grazing management has to operate in an economic environment in which it is a price taker rather than a price setter, adding another element of unpredictability. Given these complexities, simulation modeling can be a useful technique for exploring management alternatives, identifying risk levels in both ecology and economics and for planning future activities. This chapter examines the potential of modeling and describes a number of applications that have been used for land management in arid Australia, at the broad scale that is appropriate for extensive rangelands. In doing so, it describes not only model structures, types, and applications but also some of the criteria that have to be met to encourage people to apply models to real problems.†

Concepts in Simulation

Purposes of Modeling

Simulation is a process in which selected aspects of system behavior can be represented by mathematical relationships and then reproduced by applying those relationships. It is a particularly useful technique for dealing with situations where there are many uncertainties or where behavior is so complex that simple analytical solutions to posed questions or problems are unavailable. Simulation techniques are, therefore, potentially useful in the management of arid lands where ecosystem behavior varies in time and space, where management has to cope with both complexity and uncertainty, and where externally imposed market conditions add yet another layer of variability for production systems.

Simulation is most commonly used to answer "what if" questions in which to evaluate the consequences of change in the system's characteristics, behavior, or utilization. Such questions might include: What will happen to livestock production if we experience a period of below-average rainfalls? What are the effects of increased land degradation on forage production? How is a change in stocking level likely to affect profitability over the next ten years? Answering such questions by simulation requires the identification of key issues, the development of an appropriate mathematical model, the creation of appropriate scenarios to run with the model, and a capability to present often complex results in an understandable format. Most simulation studies emphasize model development at the expense of scenario creation and presentation of results. As a consequence, they get very little use.

Broadly speaking, there are two reasons for building a model. First, there may be a need to understand the functioning of a complex system by mimicking its behavior. In this case, it is appropriate to build a model that represents system processes as fully as possible and to investigate changes in behavior by changing equations, parameter values and paths, or structures in the model. Here we are interested in model behavior per se rather than in reproducing the behavior of a specific system, and the outcome is usually improved scientific understanding of the system rather than extensive use by managers. The second reason for modeling is to simulate the behavior of a particular system in response to changes in input or to evaluate that behavior over a longer period than can be done with available records. The aim is then to model behavior as accurately as possible, often from a minimal dataset. Detailed internal model structure becomes less important and is replaced by a need for parsimony in parameters, inputs and structures, a capability for calibration from limited data, and appropriate outputs and interpretability for end users.

Types of Models and Their Implementation

Simulation models fall into two basic types: univariate models that describe the behavior of a single variable through time or space and transfer function models that transform values of one or more input variables into a set of output variable values.

Univariate models tend to be based on the statistical characteristics of the variable in question and are frequently expressed as a stochastic process. They are commonly used to extend existing records by generating a long synthetic series whose characteristics may be studied or used as input to a transfer function model (e.g., Yevjevich 1972). Univariate models are widely used in hydrology to investigate extreme events. For example, in the case of drought, a thousand-year daily rainfall record may be generated from the probability distribution and autocorrelation statistics of a relatively short existent record using pseudo-random number techniques. The resultant synthetic series is then analyzed to determine the duration, intensity, and statistical characteristics of periods of low rainfall.

Transfer function models have a wide variety of uses but are most commonly employed to determine the impact of changes in the values of an input variable on a chosen output characteristic. They are also used to investigate how changes in system characteristics (represented by changes in model structure or in model equation parameter values) affect the behavior of an output variable. For example, what does an increase in rainfall mean for the grass/tree ratio in a pasture or, given a particular rainfall sequence, what does a change in the proportion of unpalatable plants in a pasture mean for animal production?

Transfer function models vary in sophistication from simple mathematical "black boxes," such as regression equations, to sophisticated process models in which most of the physical and biological processes operating in a system are explicitly represented by mathematical relationships (Stafford Smith 1988, Walters 1986). Depending on their purpose, both types of models have advantages and disadvantages. Simple black-box models are ideal for situations in which data are available for fitting, input data used in subsequent simulations are similar to those used for fitting, and system behavior is unlikely to change. They are often unsuitable for transferring between systems that behave differently to one another, for situations in which there are not enough data to give a proper fit, and for situations in which threshold effects occur beyond which system behavior markedly changes.

Process-based models are more versatile. Individual elements of system behavior may be modified by changing particular equations because such models provide a fuller representation of internal system behavior. This may ease transfer between systems and may permit them to be applied to input variable values that are not normally observed; this is important in examining the effects of climate change, for example, where some combinations of condition may not yet be observable. While simple process-based models are highly versatile and easy to transfer, more complex models may be less useful than expected for a number of reasons. First, they contain many equations whose parameter values may be unknown—the parameters may be as diverse as the blood temperature that triggers shade-seeking behavior in sheep, the growth rate of a grass species, or the influence of surface roughness on the sediment loading of runoff. Sometimes these parameter values are available from published data or can be estimated from easily measured system characteristics, but often they must be laboriously obtained by experiment, and then it becomes uneconomic to use the model in situations other than those for which it was developed. Second, complex process models of landscapes tend to need a great deal of initialization data, whether in the form of digital elevation models, detailed vegetation maps, or spatially distributed rainfall inputs, for example. Such data requirements may be met in research conditions for a single locality but be impracticable over whole regions. Thus, while there are some excellent whole-system process models for the research environment (e.g., Freeman and Benyon 1983, Noble 1975), there are few examples of them being used on an operational basis.

The most frequently used transfer function models take a middle course and represent only those processes that are necessary to achieve a realistic transform from input to output. We term these intermediate models "gray box." Ideally, the parameter values for these models should be readily available, rapidly measured, or heuristically estimable, but they may be dimensionless or have no direct physical meaning (Pickup 1977). In a few cases (e.g., CREAMS [Knisel 1980], WEPP [Lane and Nearing 1989]), such parameter values are available from published maps or tables. Often such information is not available and models have to be calibrated using nonlinear programming algorithms that "tune" parameter values to achieve a close match between observed and modeled outputs. If the parameter values have no physical meaning, different values of the same parameter set may produce identical results. Individual parameter values in a given parameter set are, therefore, usually not independent of each other and cannot be interpreted outside the context of the whole interacting parameter set. This is not always the case; for example, Pickup (1995) shows that it is possible to produce independent and, to some extent, physically meaningful parameter estimates with a carefully constructed model. If so, there is the possibility of correlation with more easily measured system characteristics, making it possible to estimate parameter values without the need for model calibration.

The principal limitation on the use of these gray-box models for management in grazing lands has been the lack of suitable data for calibration and validation in all but a few cases, especially over the large areas relevant to landscape models. This situation is now changing; in particular, considerable progress is being made in model calibration using remotely sensed data. It is, therefore, becoming possible to determine sets of parameter values over large areas quickly and cheaply. Several models that are calibrated using remotely sensed data are described in the sections that follow.

Management Issues in Arid and Semi-Arid Lands

Arid Australia as a Case Study: Managing for Sustainability

We use our experiences in arid Australia as a basis for exploring how different forms of models can be useful in different circumstances. One principle requirement of modeling is a clear objective for the model, since all models are simplifications—and simplifying for the wrong purpose will usually give an irrelevant answer. As a consequence, it is important to understand the context in which our modeling is undertaken; this is reviewed in Fisher and colleagues (chap. 9, this volume), but some key points (Stafford Smith 1994b) are:

1. Australia's arid and semi-arid lands extend over 5.6 million km^2, with about two-thirds used for commercial pastoralism, but with a population of only 300,000 people and only some 4,000 pastoral properties; thus, investment and labor is limited.

2. The ancient, flat, weathered nature of the continent gives rise to generally poor soils, low productivity, and no competent drainage systems; management units ("paddocks" within properties) are, therefore, very large (e.g., 400 km^2) and encompass much landscape heterogeneity rather than break it up.

3. Australia is generally subject to a particularly uncertain climate with the effects of the El Niño–Southern Oscillation added to the normal increased variability experienced by arid areas at low latitudes; dealing with temporal variability and, especially, the effects of extreme events is, therefore, of major concern to managers and scientists alike.

The primary aim of management in arid and semi-arid areas should be to ensure that human land use is carried out on a sustainable basis (Foran et al. 1990). However, sustainability can be defined in a variety of ways, depending on society's objective for that land use and the spatial and temporal scale considered. Thus, in some societies, the principal goal of land use may be animal production, in which case goals such as the maintenance of biodiversity may be secondary or even completely discounted. In others, conservation, through the maintenance of natural habitat, may be the primary goal. In this chapter, we are dealing with management for sustainable commercial livestock production, although it has been suggested that conservative pastoral management is also good conservation practice (Curry and Hacker 1990, Morton et al. 1995).

Commercial livestock production in arid lands must fulfill two separate criteria to be classed as locally sustainable (Pickup and Stafford Smith 1993). First, it must maintain the biological system's long-term capacity to produce forage from rainfall, although both species' composition of the forage and capacity to produce it in the short term may vary. Second, it must produce acceptable financial and nonfinancial returns for the manager, dependents, and future generations at a reasonable standard of living. Achieving these ends requires an understanding of both the ecological and economic functioning of the system and a set of management strategies and tactics for dealing with it.

Ecological Functioning

Forage production in arid zone systems is both episodic and rainfall driven. It occurs as a series of growth pulses followed by periods of decay during which plant biomass is consumed by grazers or converted to litter and consumed by detritivores. The timing and magnitude of these growth pulses may be both erratic and unpredictable, except on a probabilistic basis. The supply of forage is, therefore, highly variable in time, which can lead to significant management problems given the continuous need to feed grazing animals.

Forage production varies across the landscape, as well as through time, because of spatial variability in soil water-holding capacity, rainfall intensity, and water runoff and run-on. These processes also affect risk of degradation and, in some areas, generate patterns of landscape structure that may be intensified by wind erosion during dry times. With the flat landscapes and noncompetent drainage systems of inland Australia, such patterns are common and have been termed *erosion cell mosaics* (Pickup 1985). The mosaics contain both active areas and zones of apparent stability. The active parts of the landscape may be divided into sediment *production* zones, sediment *transfer* zones, and *sinks*. The production zone consists of the eroded area itself; lateral flow of both water and sediment moves out of this zone whenever there is rainfall of sufficient magnitude. Most of the sediment from the production zone passes through the transfer zone, downslope of the production zone, but there is some temporary and intermittent storage of material. The sink occurs downslope of the transfer zone and is an area of sediment accumulation. Taken together, the production, transfer, and sink zones make up an erosion cell. Erosion cells may be fully developed, partly developed, or latent depending on the climatic and geomorphic history of the landscape. They may also exist at a variety of spatial scales, with the smaller cells embedded in the larger ones to form a complete erosion cell mosaic (for more details, see Pickup 1985).

Grazing animals do not uniformly use the large paddocks common in Australia's arid lands, which adds another element of spatial variability. Instead, a radial pattern of use develops around the artificial water points to which animals must return at regular intervals to drink, with the heaviest paddock use close to the water points. The basic radial pattern can be modified by several other factors. If the paddock contains different landscape types, the ones with more palatable forage will be favored for grazing. This will distort the radial pattern, as corridors of greater activity develop in areas of more palatable forage and around the water point (Pickup and Chewings 1988a). Water and pasture salinity may have an effect since both limit the distance animals can travel to grazing areas before needing to return to drink (Cridland and Stafford Smith 1993, Stafford Smith 1988). Animals also prefer fresh water, so it is quite possible to find different distribution patterns around waters with different salinity levels in the same paddock. Animals of different species, breeds, and even in different physiological states (e.g., pregnant) have different speeds of movement, grazing selection strategies, heat tolerances, and water requirements, among other factors, that interact with the environment to alter grazing patterns. Finally, the weather has a direct effect, particularly with sheep, which tend to graze into the wind, with the result that the radial pattern may be extended or foreshortened, depending on whether it is toward or away from the prevailing wind direction.

The interaction between grazing distributions and erosion cell mosaics can produce very complex patterns of land degradation risk (Pickup and Chewings 1994). In some sites, a radial grazing pattern centered on water points translates into a similar pattern of land degradation. In others, it produces more redistribution of water and eroded sediment and intensifies the rate of activity within existing erosion cell mosaics. Existing trends in landscape development then become reinforced so that sediment production zones extend and generate more sediment, resulting in increased deposition in associated sinks. Transfer zones experience more rapid and more intense variation as an increased volume of sediment passes through. As erosional activity increases, production zones will become larger and may extend upslope, laterally, and downslope into the transfer zone. The smaller sinks may lose their ability to trap sediment and begin themselves to erode. With further intensification, production zones will expand even more, and the smaller sinks can no longer be sustained; deposition will then be concentrated only in the largest sinks. At that point, the landscape is dominated by large areas of bare ground, separated by smaller areas with very dense vegetation cover that receive most of the runoff, nutrients, and transported seed generated elsewhere in the landscape.

Thus, the underlying system characteristics mean that useful models in the arid lands must cope with temporal variability in forage production and grazing, spatial variability in soils, rainfall, and grazing, and the interaction of all of these. For some purposes, these can be dealt with independently, but both short and long time frames must also be considered. We will seek to illustrate most of these combinations, but first it is important to discuss who might use the results.

Management Problems

Ultimately, land managers must cope with spatial and temporal heterogeneity in the ecological functioning of the system. For example, grazing is a spatially variable process that operates on a nonuniform landscape where forage quantity and quality, vegetation response to rainfall, and risk of degradation all vary. It is, therefore, possible to overgraze some areas while simultaneously undergrazing others and to cause degradation in some parts of the landscape while having ecologically sustainable operations elsewhere. While variability tends to reduce homogeneous options and decrease mean outputs, it also provides opportunities for those who understand it. Thus, for example, spatial variability that leads to runoff and consequent run-on enhances the productivity of small depositional patches sufficiently to allow millet cropping in the Indian Thar desert; comparable resource concentrations are vital, as we have noted, in the less-intensified erosion cell patterns of the Australian rangelands. Modifications of the uneven spatial use of the landscape, through paddock design and water point location, allow managers considerable scope for buffering the ecosystem against spatial variability. A major question for science is to define what level of spatial heterogeneity, natural or engineered, is most efficient given different objectives for the use of a landscape and given its particular characteristics of slope, soils, vegetation responses, and weather patterns.

Managers are often more immediately concerned with temporal variability, however. Part of this variability is based on markets since rangeland areas, being relatively unproductive, are price takers rather than price setters. The remainder is driven by variation in climate. Droughts, based on climatic downturns but often exacerbated by short- and long-term grazing history, are an inevitable feature of rangelands management, and risk-handling strategies to deal with this variability are essential. Various studies have examined stocking rate strategies, including the dichotomy between a climate-tracking "trader" and a drought-avoiding "low stock" approach (Foran and Stafford Smith 1991), comparable to that of Sandford (1983) in Africa. The implications of these caricatured extreme strategies for Australian conditions are summarized in table 11.1.

These clearly represent a gross simplification of the options open to managers (Stafford Smith 1994a). Management goals vary widely, even within one land use such as commercial pastoralism (e.g., table 11.2); these can be expanded still further with other land uses, even within Australia (e.g., table 11.3), and yet further when other regions and industries are considered (table 11.4). Some of these

Table 11.1. A Summary of the Features of Trader and Low Stock Management Strategies

	Low Stock	Trader
Background strategy		
Stocking rates	Low	Track climate
Production rates	High	Good
Drought production	Persistent	Low
Destocking	Rare	Every dry year
Profits and coping with drought		
Profits	Moderate, consistent	High, variable
Drought reserve	Fodder bank/land condition	Financial investments
Riskiness and management style		
Decision times	Long	Short
Effect of error	Minor	Severe
Ecological impact	Light	Risky
Risk attitude (marketing/ ecological)	Avoid	Embrace
Impact of uncontrollable factors ("externalities")		
Climate	Low	High (but managed)
Markets	Moderate	High
Interest rates, etc.	Low	High
Effect of everyone else doing the same	Low	Hopeless

Source: Stafford Smith et al. 1994a.
Note: There are other factors to consider, such as capital values, loan repayments, and return on capital. In reality, most managers adopt a combination of the low stock and trade strategies. The analysis of economic outcomes from which these conclusions are partly drawn is shown in figure 11.6.

goals are compatible and could be met simultaneously, but most involve subtle or major differences in the resulting management priorities and strategies. It should be noted that most goals involve economic values, so it is important to understand how economic risk is tied to ecological risk (Pickup and Stafford Smith 1993).

The key messages from this section are twofold. First, managers (and, hence, their advisers) have a vital interest in understanding spatial and temporal heterogeneity in ecological processes in order to improve their ability to manage risk. This leads to challenges in understanding the best levels and uses of spatial heterogeneity and the best ways of coping with temporal variability. Second, there is a wide range of alternative goals that managers might be aiming to meet in dealing with risk and that must be under-

Table 11.2. Alternative Theoretical Goals for Pastoral Management in Australian Rangelands

1. A constant forage utilization level.
2. A constant "safe" stocking rate.
3. A constant target stocking rate with regular adjustments.
4. A guaranteed minimum level of plant cover.
5. The maximum grass or palatable perennial plant productivity.
6. The best possible responsiveness to prices.
7. A constant year-to-year income.
8. A constant year-to-year grazing exclusion.
9. The least cost of errors in management.
10. Protection of special areas of the property.

Source: Based on Stafford Smith and Hope 1992.

Table 11.3. Some Key Goals and Subgoals That Aboriginal People Have Expressed for the Management of Country

Goals related to traditional management:
1. Living on the country
    - Access to water and living areas
    - Road access
    - Maintenance of gates and grids
2. Looking after the country traditionally
    - Keeping rockholes/waterholes clean
    - Burning for traditional purposes
    - Burning for land management purposes
    - Access
    - Sacred site protection
    - Reintroduction of traditional species
    - Time for ceremonial activity
3. Hunting animals for food
4. Gathering plants for food

Goals related to a pastoral enterprise:

5. Minimizing capital outlay (cattle costs)
6. Optimal financial return
7. Stable financial return
8. Employment for community members
9. "Recreation"—something to do
10. Security for the future (if outside funding ceases)
11. Prestige as a cattleman
12. Stable supply of predators
13. Flexibility of management action timing

Other land-use goals:

14. Tourism
15. Feral animal harvesting
16. Horticulture
17. Mining operations

Table 11.4. Options for Diversification Encountered during a Survey in Rangelands Australia

Diversify on Farm	
Other animals	
Harvest	Kangaroos
	Goats
	Pigs[a]
Cultivate	Emus
	Quarter horses, etc.
	Fish
	Yabbies[a]
Plants	
Harvest natives	Native seed
	Cut wildflowers
	Bush tucker
	Tree oils
	Sandalwood[a]
	Gutta percha, etc.[a]
Cropping (for sale)	Grains
	Fodder
Horticulture[a]	Grapes
	Mangoes
	Dates
	Quandongs
	Melons
	Cut flowers
People[a]	
Recreation	Hunting rights
	Fishing
Tourism	Tours
	Accommodation
Other	
Conservation	Private parks
	Reimbursal for management
Diversify off Farm	
Land-based	
Management	Rehabilitation
	Consultant manager
Equipment	Earthmoving/roads
	Mining
Employment	
With neighbors[b]	Mustering/shearing
	Water drilling
In nearby town	Wide variety
	School bus
In distant town	Wide variety
Own enterprise	
Skill-dependent	Computers
	Accounting, etc.

Note: Excludes financial investments and changes within the domestic stock enterprise—i.e., sheep/cattle balances or herd and flock structures. See Stafford Smith 1994b.
a. These "on-farm items" may be independent of climate, depending on water sources and recreational forms.
b. These "off-farm" items may be very climate dependent.

stood to set the goals for modeling. Given the complexity and uncertainty in the process of landscape management, there are many solutions—and no one gets all of it right all of the time. Incorporating such complexity into a single model is impossible. It can, however, be useful to test the extent to which different subsets of goals are met by different actions, and here simulation can play a role in developing both the strategy and tactics of management. In the sections that follow, we describe various tools that may assist in this process.

Simulation and Modeling Approaches

Handling Spatial Complexity

The spatial complexity of rangelands has been a significant impediment to modeling in the past. Only a few models deal with it explicitly; others ignore it altogether and express their results in spatially lumped forms such as paddock averages or "representative" values. This is adequate for some purposes but not for others. Where spatial variability must be dealt with in a model, there are several ways of coping. One possibility is to derive individual parameter sets for different areas and then combine the results to produce a larger picture, assuming spatial independence between landscape units. A slightly more complex approach is to express the variability in the form of a spatial trend with respect to factors such as distance to water; this captures some aspects of spatial dependence across the landscape, as with the animal distribution models described below. Finally, the model can be designed to operate in a fully distributed manner whereby the behavior of each location or landscape unit is simulated, taking interactions with surrounding locations into account to lesser or greater degrees. This approach is exemplified by the erosion forecasting example.

Animal Distribution Models Animal distributions have considerable significance for both production and land degradation. Poorly designed paddocks, where substantial areas are too far from water, may have zones that are substantially undergrazed while the areas closer to water are overgrazed. Paddocks redesigned to promote more even grazing can result in substantial productivity increases. For example, Cridland and Stafford Smith (1993) quote claims of increases in wool production of 0.5 kg/animal/year from the rangelands of western Australia when grazing distributions were improved. Other situations may occur where the juxtaposition of water points and areas of palatable forage intensifies trampling and defoliation on vulnerable areas and greatly intensifies erosion.

While there are many rules of thumb for designing paddock layouts, there was little scientific study until Senft and colleagues (1983) and Stafford Smith (1984, 1988) pioneered animal distribution modeling. This technique makes it possible to test alternative paddock layouts and determine how uneven grazing pressure is likely to be. The models have been released in commercial form as the *Paddock* module of the *RANGEPACK* decision-support system (Stafford Smith and Hope 1992). Figure 11.1 shows an example of the use of *Paddock* to design a sheep paddock.

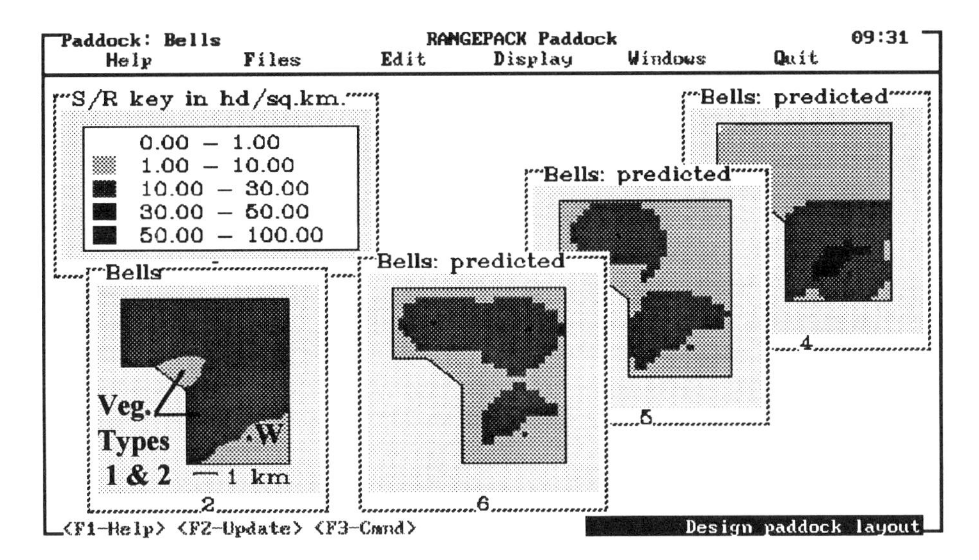

Figure 11.1. A screen from the program *RANGEPACK Paddock* showing a comparison of different water point locations in a real paddock, "Bells," using models described in the text. Inside the menu area of the program are five windows; the bottom left window (2) shows the original paddock layout with two vegetation types and a water point (W) toward the southeast corner (the vegetation type with the larger area is more preferred than the other in this case). The top right window (4) shows the predicted long-term pattern of grazing impact, scaled into stocking rate equivalents (key in top left window) for the original design with one water point. The other two windows (5 and 6) show the predicted pattern if, respectively, one or two additional waterpoints were located in the northern half of the paddock. As expected, the grazing pattern evens out, removing extremely high and low impacts; the important point is that the extent of this improvement can be quantified (other windows, not shown here, can display statistics on these changes).

The sheep distribution models were originally based on a complex simulation model (Noble 1975, Stafford Smith 1984); the understanding derived from this has allowed a robust simplification to a regression equation with the form:

$$y = a + b.e^{(c.D)}$$

in which *a, b,* and *c* are constants, *D* is distance from water, and *y* is an index of animal activity that is then standardized across a management unit relative to its total animal numbers. This simple exponential decay model can be modified to allow for factors such as water and feed salinity (which affects *c*) and prevailing wind direction (since sheep graze into the wind).

The principal limitation in using this model had been the need to calibrate it (i.e., fit appropriate values to parameters such as *a, b, c*) for different vegetation types, water salinities, climatic regions, and animal breeds by ground or aerial survey, radio tracking of animals, or the use of dung counts. The problem of calibration has been greatly reduced by the use of remote sensing, which can potentially detect both short-term (Pickup and Chewings 1988a) and long-term (Pickup and Chewings 1994) grazing effects. In this approach, the landscape is stratified into different landscape types and distances from water. For short-term grazing effects, a vegetation cover index (for details, see Pickup and Chewings 1988a) is calculated for Landsat multispectral scanner images for a period shortly after rain and after a subsequent period of grazing. If the second set of vegetation index values is subtracted from the first, the result indicates loss of cover that is partly due to grazing and partly due to natural decay. When the loss of cover is averaged over the area at each distance from water and plotted against distance from water, a gradually decreasing loss of cover at greater distances from water occurs until a point is reached beyond which the loss remains effectively constant. The gradually varying component of cover loss is interpreted as a grazing effect, which should decrease with distance from water, while the constant component is treated as loss by natural decay.

This approach has been used successfully to model observed cattle distributions in the Northern Territory (Pickup and Chewings 1988a) and sheep distributions in the rangelands of western Australia (Cridland and Stafford Smith 1993) and shows a rich variety of patterns. It also indicates that the simple sheep distribution model described above will not work for cattle in the highly variable central and northern rangelands of arid Australia. Indeed, Loza and colleagues (1992) in the United States have shown that the interaction between climate and the thermal characteristics of cattle also requires modifications to the original sheep model. Current developments in Australia, therefore, center on using remotely sensed data to produce more appropriate distribution models for cattle.

In this context, Pickup (1994) has developed a model based on a convection-diffusion analogue that can be calibrated from remotely sensed data and can handle the effects of varying land conditions. The model formulation assumes that vegetation cover loss over time is a function of initial cover, exposure to grazing, and pasture palatability. The basic form of the model is shown in figure 11.2 and contains optional terms describing loss of cover due to natural decay and the effects of changing pasture palatability at different distances from water. So far, it has proved capable of handling a very wide range of spatial distribution patterns, often of great complexity (more details in Pickup 1994).

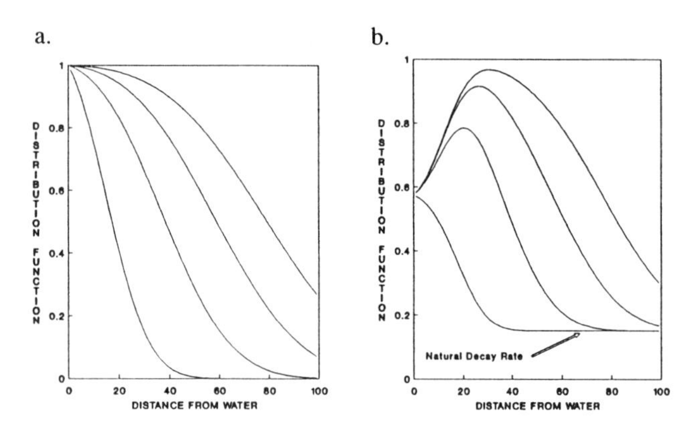

Figure 11.2. Animal distribution model forms using the convection-diffusion model approach of Pickup (1994). Graph (a) shows the relative distribution model using the basic equation. Graph (b) shows forms modified to allow for an increase in the proportion of unpalatable herbage or trees and shrubs close to water and for loss of vegetation cover over time due to natural decay.

Erosion Forecasting Erosion forecasting is a technique for determining the likely distribution of erosion and deposition in a large area, such as a paddock, if the erosion process intensifies. Areas of high erosion risk can be identified in the landscape with these models; this can be useful when laying out fence lines and water points to avoid having areas of high animal usage in zones that are prone to erosion. The models make use of algorithms developed by Kashyap and Chellappa (1983) for describing and generating synthetic textures using stochastic spatial process models. The procedure was designed for use with information from remote sensing satellites; other data requirements are minimal, unlike the needs of more physically based erosion models (e.g., Biot 1990, Lane and Nearing 1989). It also implicitly incorporates the biological influences and feedbacks that dominate erosion processes in Australia's arid lands yet are poorly handled by other models. Detailed descriptions of the procedure have been published elsewhere (Pickup and Chewings 1988b). This description is summarized from Pickup (1988).

The spatial process models used in erosion forecasting are based on the erosion cell concept (see definition earlier). A prerequisite for the modeling of erosion cell behavior is the ability to map erosion and deposition of a landscape and express it as a single variable. An approximate measure of the intensity of erosion and deposition can be derived from Landsat multispectral scanner data using an index developed by Pickup and Nelson (1984). The state of an erosion cell mosaic may be described by the location, mean, frequency distribution, and spatial autocorrelation function (SACF) of the land stability index values in an area (Pickup and Chewings 1988b). The changes that occur in the mean, frequency distribution, and SACF as a landscape becomes more eroded are illustrated in figure 11.3. In the relatively stable landscape, the mean is high, the frequency distribution has a relatively narrow range, and the SACF declines rapidly with increasing spatial lag. This indicates that only a small part of the landscape is occupied by areas of extreme erosion and deposition and that there is great spatial diversity. Large erosion cell structures have, therefore, not developed, and much of the landscape is both stable and resilient. In the partially degraded landscape, the mean is slightly smaller, the frequency distribution has a wider range, and the SCAF decreases less rapidly with lag. This suggests a shift toward more erosion and deposition and greater spatial uniformity, implying expansion of the zones within the various erosion cells. In the degraded landscape, these trends continue. The frequency distribution spreads further,

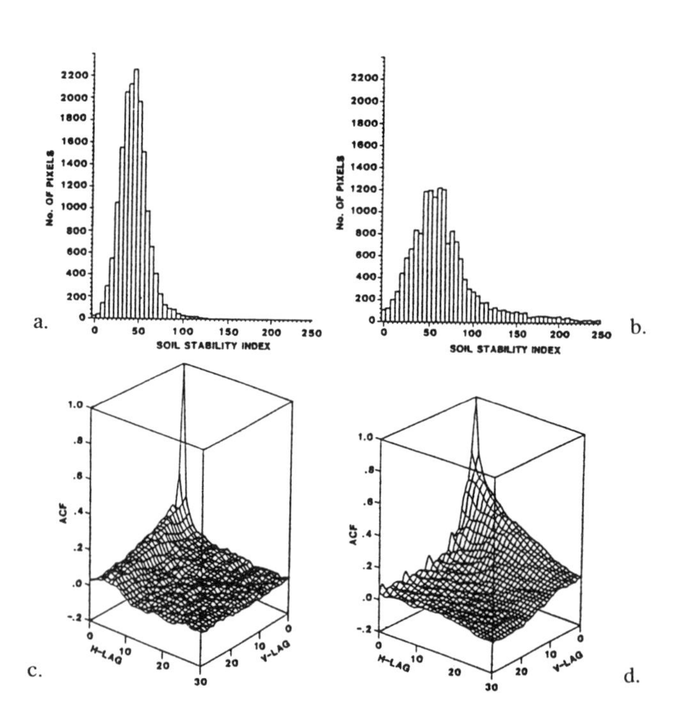

Figure 11.3. Changes in frequency distribution and spatial autocorrelation function (*sacf*) of soil stability index values as an area becomes more eroded as a result of grazing. Graphs (a) and (c) show conditions on a central Australian pasture in 1972, while graphs (b) and (d) are for the same area in 1984 (from Pickup 1988).

indicating that most of the landscape is made up of areas of erosion or deposition. The SCAF also increases at each spatial lag greater than zero. The erosion cell structures are, therefore, becoming increasingly large.

Shifts in the mean, variance, and autocorrelation structure of a landscape in which the pattern of existing erosion cells remains largely in place can be modeled using a two-dimensional autoregressive random field model defined over a finite lattice (Kashyap and Chellappa 1983). The random field model expresses each observation in an area as a weighted sum of the surrounding observations plus a noise component. A model structure that appears to adequately describe erosion cell mosaic behavior is the first-order, eight-neighborhood version (Pickup and Chewings 1988b):

$$Y_{ij} = \theta_1 y_{i-1,j} + \theta_2 y_{i+1,j} + \theta_3 y_{i,j-1} + \theta_4 y_{i,j+1} + \theta_5 y_{i-1,j-1} + \theta_6 y_{i-1,j+1} + \theta_7 y_{i+1,j+1} + \theta_8 y_{i+1,j-1} + \sqrt{\rho w_{ij}} + Y_\mu$$

where Y is the erosion index value, $\theta_1, \ldots, \theta_8$ are a set of weighting parameters, w is a noise series consisting of random variates scaled between 0 and 1, ρ is the noise series variance, i and j are grid coordinates, and y values are deviations about the mean Y_μ. Changes in the overall system state are represented by shifts in the mean. Changes in the frequency distribution of system states and the spread or breakup of patches within the mosaic are handled by the θ and ρ parameters. The w_{ij} series contains information on the geographical location of erosion and deposition but is not a true white noise process as used by Kashyap and Chellappa (1983). Instead, it is assumed to remain unchanged over time and to represent the underlying pattern in the landscape.

The steps involved in modeling the way in which a landscape changes as erosion proceeds are as follows. First, a model is fitted to a particular area. The model is then used as an inverse filter to obtain the underlying pattern series w_{ij}. A prototype area is then selected that represents the same type of landscape in a more eroded condition, and a second model is fitted. Filtering the underlying pattern series with the prototype area modeling then provides a forecast of potential change (fig. 11.4).

This approach may appear simple, but tests show that it is capable of producing accurate forecasts of change for large areas with very complex patterns of erosion and deposition (Pickup and Chewings 1988b). It seems to work because the

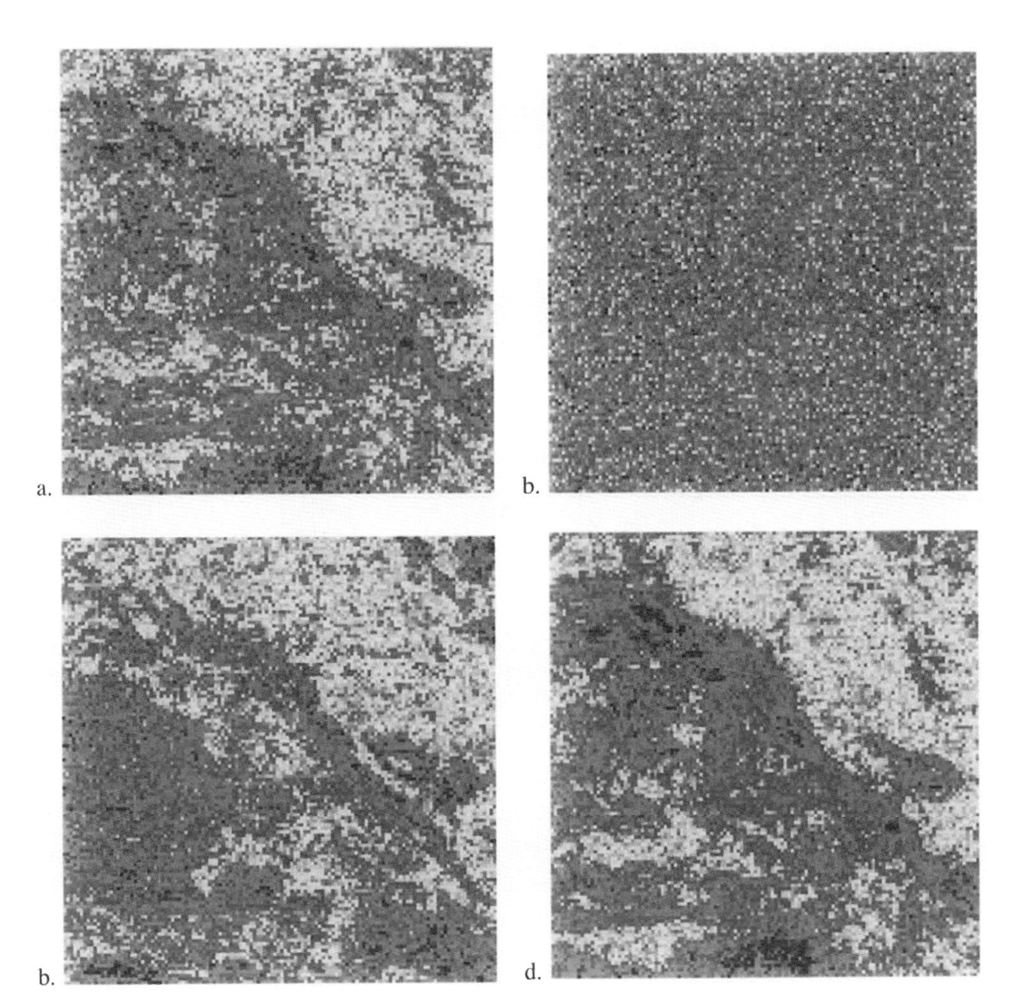

Figure 11.4. An example of erosion forecasting: (a) the state of a central Australian landscape in the early 1970s, (b) the underlying pattern series obtained from this landscape by inverse filtering, and (c) a prototype from another area that has been more heavily eroded. Applying the model from (c) to the underlying pattern series results in (d), which shows how (a) would change as it becomes more eroded. (Key to landscape conditions: light gray = heavily eroded; white = mild erosion; dark gray = stable; black = intense deposition and shrub encroachment.)

underlying pattern series w_{ij} contains most of the information on the likely pattern of soil movement. Also, because the model parameters used to generate a more eroded condition are derived from a similar or neighboring area, they contain the averaged response of the landscape, bypassing the need for detailed modeling.

The use of prototypes is analogous to model calibration but takes the process a step further. Instead of using a whole data set in which the system may pass through a number of states to calibrate a single set of model parameters, individual system states are explicitly recognized and a model is derived for each. This process is made possible by the existence of a large body of commercially available Landsat data going back to 1972 for many areas. The skill, of course, lies in recognizing and selecting data representative of each system state. The modeller should also be sure the prototypes chosen for a particular system are truly representative.

The erosion forecasting procedure relies on the assumption that latent or incipient erosion-deposition sequences are present in the landscape and may be detected in Landsat data. This assumption is reasonable in many landscapes because of their geomorphic and climatic history. Virtually all arid zone landscapes contain features resulting from geomorphic processes that operated in the past but are less or no longer active today. Many of these features were associated with more active erosion and deposition, although the reason for that activity is still a matter of speculation. When grazing animals produce erosion, they tend to reactivate the old patterns of activity that intensify and expand to occupy a larger proportion of the landscape. This makes it possible to use the latent pattern of erosion to forecast the new one. The method breaks down when a completely new pattern develops that is unrelated to past conditions.

Handling Temporal Variability

Although the issue of handling variability through time cannot be fully separated from dealing with spatial heterogeneity, many models do treat the environment as a single point (effectively assuming homogeneity of any spatial processes) or as a set of aggregate responses (accepting the possibility of spatial heterogeneity but parameterizing at a level at which this can be assumed to be trivial). We present examples of each.

Pasture Production Models: Point Models of Temporal Change The traditional agronomic approach to modeling plant and animal production is to develop a detailed process model of soil moisture, plant growth, and, in some cases, animal intake and consequent animal production. This has worked well in cropping systems with high levels of information, even though the applicability of these process models to specific on-farm conditions is still fraught with data problems. These problems become greater with application to semi-arid and arid pastures, but there are several successful examples. An outcome of the International Biological Program in the United States was the SPUR model (Hanson et al. 1988) that can simulate a number of different pasture types (recent developments handle spatial aspects as well—see next section); in Australia, a decade of data collection in Queensland has allowed the GRASP model to be parameterized for most common pasture types there (McKeon et al. 1990). Similar models have been developed for semi-arid areas of South Africa (e.g., Richardson et al. 1991, Snyman and Fouché 1993), although mainly for higher production pastures, as is GrazPlan in temperate Australia (Moore et al. 1991). Hobbs and colleagues (1994), Ludwig and colleagues (1994—SEESAW), and Hacker and colleagues (1991—IMAGES) have also derived plant growth models for other systems in Australia, with the latter concentrating on palatable chenopod shrublands where the recruitment and mortality of shrubs is a particularly important feature.

In their original forms, all of these models that were aimed at grass-dominated systems concentrated on simulating the processes of pasture growth on a point basis. They tend to incorporate at least a three-layer soil moisture store model, use the rainfall or water-use efficiency of photosynthesis to predict dry matter production based on transpiration, allow in some way for nitrogen use and dilution, and include leaf senescence and detachment rates. Other factors depend on the environment in which the model was conceived. GRASP, for example, also considers the influence of trees on production, the effects of fire, and maintains a grass basal area that is affected by grazing as some measure of pasture condition. Both GRASP and SPUR then model animal liveweight gain based on preferential access to the higher quality leaf material. Slightly different approaches are taken in the different models (see also Moore et al. 1991).

The great value of process-based approaches to modeling is that the detailed incorporation of process means that the model can be used to test conditions that are not part of the input validation dataset, such as new management treatments or climate change; that is, they can be used to *predict* the effects of significant changes in the system. The corresponding problem, of course, is that they are computationally expensive and often require very large numbers of parameters to be applied to a new pasture. For example, the ideal minimum field dataset for fitting GRASP to a new pasture involves 1–2 years of collection of soil and vegetation parameters on a site subjected to at least cutting and no-cutting treatments.

Despite this, GRASP has been converted to a decision-support package, GRASSMAN (Clewett et al. 1991), by running GRASP to obtain simplified production curves for different areas according to different management regimes. Similarly, SPUR was run to produce a probability distribution of pasture production which was then incorporated in the simpler decision aids, STEERRISK and STEERRISKIER (Hart and Hanson 1993).

An alternative approach has recently been described by Pickup (1995) in which a much simpler model (with understanding based on the ground-derived model of Hobbs et al. 1994) is fitted to observed plant production data derived from remotely sensed measures of vegetation cover. The model determines herbage growth from daily rainfall using estimated evapotranspiration and a single soil moisture store. Growth rates are determined via a moisture-use efficiency term that allows for hidden processes, such as moisture losses due to runoff, deep percolation, and evaporation, as opposed to transpiration in a single step. Loss of herbage due to grazing and natural decline is represented as an exponential decay process, and both herbage growth and loss are limited by the percentage of tree and shrub cover present.

Once this simple model is calibrated, it needs only local rainfall records or potential evapotranspiration data as input. It gives good predictions of variations in observed cover even though only three parameters are needed to describe a given landscape type in given a condition. It thus has the advantage of being much simpler and more cheaply executed and, given the initial remotely sensed dataset, more quickly fitted to a new landscape type. The limitation of this type of model, of course, is that no process other than a relationship with rainfall is explicitly incorporated in the model, so there is no reason to suppose that it would apply in conditions where any of the climatic, edaphic, or management parameters are decoupled from their arrangements in the source dataset. However, this does not matter greatly because ease of calibration and the wide availability of remotely sensed data make it possible empirically to obtain parameter sets for a wide range of individual environmental conditions or combinations of conditions.

An example of the use of this simple model in a real situation is shown in figure 11.5. The model was initially calibrated for the period 1982–89 on a set of subareas at 1 km increments from water in a 570 km² paddock in the Muller Land System of central Australia. The paddock is heavily degraded, and this is reflected by progressively lower shrub cover and higher water-use efficiency away from the water point where there is less grazing-induced impact. When the forage production and consumption are modeled over a fifty-year period using observed model parameters and compared with similar results modeled using parameters for the same land system in a less-degraded condition, the losses in both forage production and consumption are clearly visible. This illustrates the opportunity cost of land degradation and could be used when costing preventative measures compared to doing nothing.

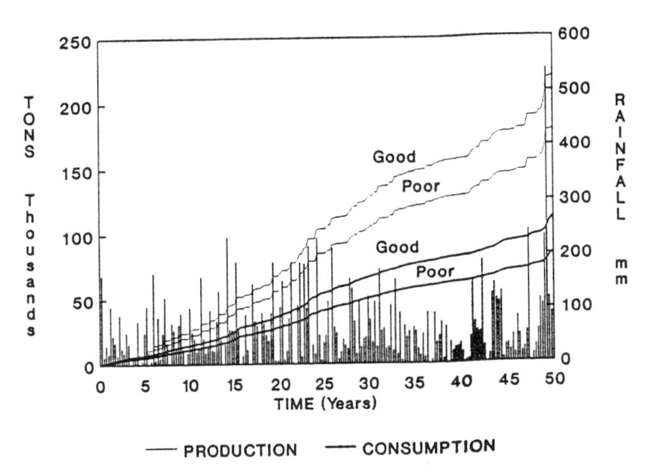

Figure 11.5. Biomass production and consumption for landscapes in good and poor (degraded) condition on the Muller Land System of central Australia for the period 1925–74. Results were modeled from observed rainfalls using Pickup's (1995) model.

Economic Strategies: Aggregated Models of Temporal Change

The previous section described models that arose primarily from the need for scientific understanding and then sought application. In this section, we examine a model of economics and herd dynamics that has almost precisely the reverse genesis. The *RANGEPACK* project was conceived in 1986 as a result of concerns that research on rangelands management was not reaching users. Initially, there were moves to develop a production model base comparable to SPUR or GRASP, but early consultations with producers highlighted the fact that their decision making was driven by the economic end of the system. However, their problems were dominated by the effects of temporal variability (expressed more vernacularly as "drought"!), and yet all economic packages at the time dealt in static, equilibrium optimizations. Hence *RANGEPACK Herd-Econ* was developed (Stafford Smith and Foran 1990).

Herd-Econ is a simulation package, but not a complex one; it could be implemented clumsily in a spreadsheet but was instead written as a stand-alone package to make it easier to set up and use. It takes the animal class and age group numbers of a grazing enterprise and couples them with herd dynamics (birth, death, and growth rates), management strategies (primarily buying and selling activities), and property economics (cash flows, financial returns, and equity). Its significant innovation was to encourage users to implement the herd dynamics and management strategies for a number of different "year types": recognizing the general lack of information available on farm, it restricted these originally to only four year types rather than trying to implement continuous relationships between stocking rates and herd dynamics.

Herd-Econ was implemented as a commercial package but has seen use in many scientific studies of the impacts of climatic variability on rangeland production systems (e.g., Foran et al. 1990, Foran and Stafford Smith 1991, Stafford Smith and Foran 1992, Stockwell et al. 1991). One example is given in figure 11.6, where the data obtained from two properties in central Australia were used to compare and contrast two extremes on the continuum of management strategies from low and constant stocking rates to high but very variable stocking rates (thus necessitating trading to maintain viability). This continuum has been noted before (e.g., Sandford 1983), but *Herd-Econ* provides a means to analyze the effects of the strategies objectively on real properties, leading to conclusions outlined in table 11.1.

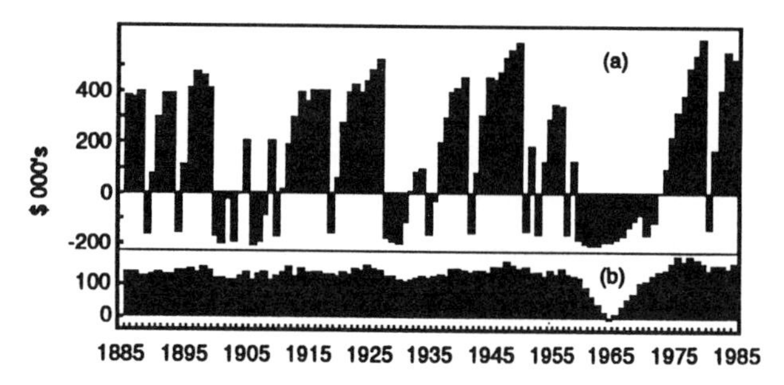

Figure 11.6. Predicted annual cash surpluses (before tax and debt payments) for (a) trader and (b) low stock strategies for the past century of climatic conditions, with constant 1988 costs and prices. The mean annual surplus for trader is a little higher than for low stock, but it is far more variable and risky (table 11.1); the results are affected by interest rates and taxes (more details in Foran and Stafford Smith 1991). Note that the 1960s were the driest decade on record and the 1970s the wettest; the low stock approach cannot capitalize on the good years (like trader can) but does survive the very worst droughts in a way that trader, with real interest rates, could not.

The major limitation of *Herd-Econ* is the lack of feedbacks to the underlying system—there is no means of checking whether the presumed production levels can indeed be met under a particular management strategy. Links are now being made between *Herd-Econ* and GRASP (see previous section) to rectify this (Stafford Smith et al. 1995); however, the resulting simulation tool will slip back into the mold of scientific study rather than being a readily accessible decision-support tool for producers.

Handling Space and Time

The final challenge is to link approaches that handle space with those that handle time. The immediate problem is that traditional approaches to such links are likely to create parameter-hungry monstrosities that are only appropriate for scientific studies. The natural process-oriented way to approach the problem would be to create a spatially explicit model that simulated the flow of spatial information through time.

An example of one that did just this, GRZMOD (Noble 1975; Stafford Smith 1984, 1988) is a model of a single sheep paddock (about 1,500 ha) that is broken into 500 x 500 m² cells. Runoff and run-on occurs among these cells, with consequent germination and growth of a number of plant groups (six by default) calculated on a daily basis. Over the top of this landscape, a sheep flock runs in subgroups that have their own physiological status in terms of hunger, thirst, and temperature load. On an hourly basis, they satisfy these physiological demands through certain activities (grazing, walking, drinking, shading, resting in the open) that involve explicitly modeled movement among cells around the paddock; grazing off-take and nitrogen redistribution feeds back to the plant populations in a spatially localized fashion. Management activities are superimposed on top of this, again, in the removal of the animals for shearing once a year and possible manipulation of water supplies.

The model works surprisingly well, both in terms of long-term patterns and short-term animal activities. However, it contains so many parameters and so many hidden interactions that it could never be said to be truly validated. A full sensitivity analysis of the model would be quite impracticable. Despite this, the development of this model (always seen as a knowledge-enhancing endeavor) was extremely valuable in identifying processes that could be simplified for more practical purposes. It gave rise to regression-based animal distribution models (Stafford Smith 1988), for example, as well as a quantified understanding of the impact of water salinity on grazing distribution and of shading behavior (cf. Stafford Smith et al. 1985).

The U.S. model SPUR has been used in a spatially explicit way, akin to GRZMOD. SPUR has also been used for a major study of possible climate change impacts on a local government-area basis across the western United States (Baker et al. 1993) that involved simulating many points through time in a way that builds up a spatial pattern. GRASP is now being used in Queensland, Australia, in a similar way (Brook et al. 1992), having been parameterized for every 5 x 5 km grid cell of the state; this capability is currently being extended to the whole of the continent, using the power of today's parallel computing systems. This will obviously not be a tool to be used on farm, but is intended to give farmers forecasts of drought and degradation risks, comparable to weather forecasts, and produced on the same timely basis. Similar tools may become available on farm eventually: PHYGROW, a plant production model developed at Texas A&M University, is now being run remotely by logging into a central parallel computer (J. Stuth personal communication).

If space-time modeling of complex grazing systems is to become a practical tool, the models must be quickly and easily applicable to real situations. For example, simple models of forage production and consumption calibrated from remotely sensed data can be used to investigate the effects of land degradation on productivity. To illustrate this process, the model of Pickup (1995, see above) has been calibrated for areas of two central Australian cattle paddocks in different stages of degradation. Model parameter values were derived for each landscape type within the paddock and at 1-km distance increments from water. Different levels of land degradation were indicated by lower moisture-use efficiency values and increased tree and woody shrub cover close to water. After calibration, the model was run for a fifty-year period on the degraded paddock using rainfalls for 1925–74 and with parameter values derived for the existing land condition. A subsequent model run for the same period and the same paddock but using parameter values for the less degraded paddock shows how productivity is affected by degradation (fig. 11.5). Such efforts are starting to move toward the handling of the long-term impacts of grazing on sustainability, an area of essential research that still depends on linking models of short- to medium-term dynamics, as described here, with state-and-transition models (Westoby et al. 1989) of long-term system changes. A different approach to similar goals is being developed by Bellamy and colleagues (1993) in northern Australian landscapes; this model, LandAssess, is based explicitly on discontinuous ecosystem states and land unit mapping but without the same resolution of calibration obtainable from remote sensing. There is a great opportunity for future synergy between the two approaches.

Discussion

This chapter has reviewed some approaches to simulation that have special relevance to arid lands. There are, of course, many other models of processes—populations, genetics, energy flows, and so on—that can play a part, but we have focused on the need to handle the ecological characteristics of temporal and spatial heterogeneity that are particularly notable in arid lands at the scale of management.

We return in this final section to reemphasize the need for models to be appropriately targeted. As the extensive literature on adaptive management of natural resources (e.g., Ludwig et al. 1993, Walters 1986) repeatedly illustrates, there are many more models of systems than there are *useful* models, let alone models *in use* (Seligman 1993). There is a legitimate role for models in the extension of understanding, but all too often the creators of such models seek to apply their models in inappropriate ways. This is not surprising—noone likes to create a model that appears to have no practical purpose, and funding agencies look ever less favorably at such efforts.

However, application models must involve a very different development process from research models with very different goals. Where models containing a level of detail more appropriate to the research task do contribute to practical application in hands other than those of the developers, often it is through a process of extreme simplification—classic examples given earlier were the models of the spatial distribution of grazing impact and the development of STEERRISK from SPUR. The importance of the complex models is to develop the appropriate understanding to ensure that the simplifications are justified. Sometimes it is not efficient to travel this path if the original principal purpose was to develop the application.

Model application involves one other vital step. The developer must identify and work with a user who actually wants the model and is able to use it. This sounds trivially obvious, but the failure to consider this step adequately is the single greatest impediment to the success of decision-support systems (Stuth et al. 1993). "Simulation" conjures up an image of a FORTRAN programmer, tapping away into the wee hours to fit relationships to data that he has, at least one hopes, collected himself in the field and, therefore, can understand its ecological limitations. Such people are not necessarily the most able to communicate with down-to-earth managers or even policy makers. From this image has come the unfortunate terminology of "hard systems" contrasted with the "soft systems" approach of social scientists (Ison 1993). In fact, the two approaches are synergistic when practiced in the correct way (Stuth and Stafford Smith 1993). The soft systems approaches seek to empower managers to recognize their problems and define their decision-making needs clearly; the hard systems approaches, with the appropriately well-designed models, can provide some of the tools that may answer some of these questions. To an increasing extent, the concept of a decision-support system is leaving behind the image of computers and incorporating a much wider range of activities and materials (and, sometimes, computer-based products), all aimed at improving decision-making abilities (Clark and Filet 1994, Stafford Smith et al. 1994b).

A final part of defining the appropriate approach to simulation is more scientific. It is absolutely vital to have the right conceptual model before spending effort to design a detailed process model. All models are simplifications; good models incorporate the necessary processes while discarding irrelevant ones. Given the complexity of ecological relationships, this requires sharp insights into the system. We emphasize again that a system includes not only the detailed biology but also the way in which the human element interacts with it.

References

Baker, B. B., J. D. Hanson, R. M. Bourdon, and J. B. Eckert. 1993. The potential effects of climate change on ecosystem processes and cattle production on U.S. rangelands. Climatic Change 25:97–117.

Bellamy, J. A., D. Lowes, and N. D. MacLeod. 1993. A decision-support system for the sustainable use of Australian monsoonal tallgrass woodlands. *In* M. J. Baker, ed. Proceedings of the seventeenth International Grassland Congress, New Zealand, February. Wellington, New Zealand: SIR Publishing. 1929–30.

Biot, Y. 1990. THEPROM: An erosion productivity model. *In* J. Boardman, I. D. L. Foster, and J. A. Dearing, eds. Soil erosion on agricultural land. Chichester: John Wiley and Sons. 465–79.

Brook, K. D., J. O. Carter, T. J. Danaher, G. M. McKeon, N. R. Flood, and A. Peacock. 1992. SWARD: Statewide Analysis of Risks of Land Degradation in Queensland. Agricultural Systems and Information Technology Newsletter 4(2):9–11.

Clark, R., and P. Filet. 1994. Local best practice, participatory problem solving, and benchmarking to improve rangeland management. Proceedings of eighth biennial conference, Australian Rangelands Society, Katherine, June. 70–75.

Clewett, J. F., J. M. Cavaye, G. M. McKeon, I. J. Partridge, and J. C. Scanlan. 1991. Decision-support software as an aid to managing pasture systems. Tropical Grasslands 25:159–64.

Cridland, S., and D. M. Stafford Smith. 1993. Development and dissemination of design methods for rangeland paddocks which maximise animal production and minimise land degradation. Misc. Publ. 42/93. Western Australian Department of Agriculture. South Perth.

Curry, P. J., and R. B. Hacker. 1990. Can pastoral grazing management satisfy endorsed conservation objectives in arid western Australia? Journal of Environmental Management 30:295–320.

Foran, B. D., M. H. Friedel, N. D. MacLeod, D. M. Stafford Smith, and A. D. Wilson. 1990. The future of Australia's rangelands. Canberra, Australia: Commonwealth Scientific Industrial and Research Organisation (CSIRO), Division of Wildlife and Ecology.

Foran, B. D., and D. M. Stafford Smith. 1991. Risk, biology, and drought management strategies for cattle stations in central Australia. Journal of Environmental Management 33:17–33.

Freeman, T. G., and P. R. Benyon, eds. 1983. Pastoral and social problems in a semi-arid environment: A simulation model. Canberra, Australia: CSIRO and UNESCO's Man and the Biosphere Program.

Hacker, R. B., R.-M. Wang, G. S. Richmond, and R. K. Lindner. 1991. IMAGES: An integrated model of an arid grazing ecological system. Agricultural Systems 37:119–63.

Hanson, J. D., J. W. Skiles, and W. J. Parton. 1988. A multi-species model for rangeland plant communities. Ecological Modelling 44:89–123.

Hart, R. H., and J. D. Hanson. 1993. Managing for economic and ecological stability of range and range-improved grassland systems with the SPUR2 model and the STEERRISKIER spreadsheet. *In* M. J. Baker, ed. Proceedings of the seventeenth International Grassland Congress, New Zealand, February. Wellington, New Zealand: SIR Publishing. 580–85.

Hobbs, T. J., A. D. Sparrow, and J. J. Landsberg. 1994. A model of soil moisture balance and herbage growth in the arid rangelands of central Australia. Journal of Arid Environments 28:281–98.

Ison, R. L. 1993. Soft systems: A non-computer view of decision support. *In* J. W. Stuth and B. Lyons, eds. Emerging issues for decision-support systems for management of grazing lands. Paris: MAB-UNESCO. 83–121.

Kashyap, R. L., and R. Chellappa. 1983. Estimation and choice of neighbours in spatial interaction models of images. I.E.E.E. Transactions on Information Theory 29:60–72.

Knisel, W. G. 1980. CREAMS: A field scale model for chemicals, runoff and erosion from agricultural management systems. Cons. Res. Rep. 26. USDA. 529.

Lane, L. J., and M. A. Nearing, eds. 1989. USDA water erosion prediction project: Hillslope profile model documentation. NSERL Rep. 2. USDA Agricultural Research Service. 629.

Loza, H. J., W. E. Grant, J. W. Stuth, and T. D. A. Forbes. 1992. Physiologically based landscape use model for large herbivores. Ecological Modelling 61:227–52.

Ludwig, D., R. Hilborn, and C. Walters. 1993. Uncertainty, resource exploitation, and conservation: Lessons from history. Ecological Applications 3:547–49.

Ludwig, J. A., D. J. Tongway, and S. G. Marsden. 1994. A flow filter model for simulating the conservation of limited resources in spatially heterogeneous semi-arid landscapes. Pacific Conservation Biology 1:209–13.

McKeon, G. M., K. A. Day, S. M. Howden, J. J. Mott, D. M. Orr, W. J. Scattini, and E. J. Weston. 1990. Northern Australian savannas: Management for pastoral production. Journal of Biogeography 17:355–72.

Moore, A. D., J. R. Donnelly, and M. Freer. 1991. GrazPlan: An Australian DSS for enterprises based on grazed pastures. *In* J. W. Stuth and B. G. Lyons, eds. Proceedings of international conference on decision-support systems for resource management. College Station: Texas A&M University. 23–26.

Morton, S. R., and G. Pickup. 1991. Sustainable land management in arid Australia: How can we achieve it? Search 23:66–68.

Morton, S. R., D. M. Stafford Smith, M. H. Friedel, G. F. Griffin, and G. Pickup. 1995. The stewardship of arid Australia: Ecology and landscape management. Journal of Environmental Management 43:195–217.

Noble, I. R. 1975. Computer simulations of sheep grazing in the arid zone. Ph.D. diss. Adelaide: University of Adelaide, Australia.

Pickup, G. 1977. Testing the efficiency of algorithms and strategies for automatic calibration of rainfall-runoff models. Hydrological Sciences Bulletin 22:257–74.

———. 1985. The erosion cell: A geomorphic approach to landscape classification in range assessment. Australian Rangelands Journal 7:114–21.

———. 1988. Hydrology and sediment models. *In* M. G. Anderson, ed. Modelling geomorphological systems. Chichester: John Wiley and Sons. 153–215.

———. 1994. Modelling patterns of defoliation by grazing animals in rangelands. Journal of Applied Ecology 31:231–46.

———. 1995. A simple model for predicting herbage production from rainfall in rangelands and its calibration using remotely-sensed data. Journal of Arid Environments 30:227–45.

Pickup, G., and V. H. Chewings. 1988a. Estimating the distribution of grazing and patterns of cattle movement in a large arid zone paddock: An approach using animal distribution models and Landsat imagery. International Journal of Remote Sensing 9:1469–90.

———. 1988b. Forecasting patterns of erosion in arid lands from Landsat MSS data. International Journal of Remote Sensing 9:69–84.

———. 1994. A grazing gradient approach to land degradation assessment in arid areas from remotely-sensed data. International Journal of Remote Sensing 15:597–617.

Pickup, G., and D. J. Nelson. 1984. Use of Landsat radiance parameters to distinguish soil erosion, stability, and deposition in arid central Australia. Remote Sensing of Environment 16:195–209.

Pickup, G., and D. M. Stafford Smith. 1993. Problems, prospects, and procedures for assessing sustainability of pastoral land management in arid Australia. Journal of Biogeography 20:471–87.

Richardson, F. D., B. D. Haln, and P. I. Wilke. 1991. A model for the evaluation of different production strategies for animal production from rangeland in developing areas: An overview. Journal of the Grassland Society of South Africa 8:153–59.

Sandford, S. 1983. Management of pastoral development in the Third World. Chichester: John Wiley and Sons.

Seligman, N. G. 1993. Modelling as a tool for grassland science progress. *In* M. J. Baker, ed. Proceedings of the seventeenth International Grassland Congress, New Zealand, February. Wellington, New Zealand: SIR Publishing. 228–33.

Senft, R. L., L. R. Rittenhouse, and R. G. Woodmansee. 1983. The use of regression models to predict spatial patterns of cattle behavior. Journal of Range Management 36:553–57.

Snyman, H. A., and H. J. Fouché. 1993. Estimating seasonal herbage production of a semi-arid grassland based on veld condition, rainfall, and evapotranspiration. African Journal of Range Forage Science 10:21–24.

Stafford Smith, D. M. 1984. Behavioural ecology of sheep in the Australian arid zone. Ph.D. diss. Canberra: Australian National University.

———. 1988. Modelling: Three approaches to predicting how herbivore impact is distributed in rangelands. Reg. Res. Rep. 628. Las Cruces: New Mexico State University, Agriculture Experiment Station.

———. 1994a. A regional framework for managing the variability of production in the rangelands of Australia. Rep. No. 1. Alice Springs, Australia: CSIRO/RIRDC Project.

———. 1994b. Sustainable production systems and natural resource management in the rangelands. Proceedings of the ABARE outlook conference, Canberra, February. 148–59.

Stafford Smith, D. M., A. McNee, B. Rose, G. Snowdon, and C. Carter. 1994. Goals and strategies for Aboriginal cattle enterprises. Australian Rangelands Journal 16:77–93.

Stafford Smith, D. M., and B. D. Foran. 1990. RANGEPACK: The philosophy underlying the development of a microcomputer-based decision-support system for pastoral land management. Journal of Biogeography 17:541–46.

———. 1992. An approach to assessing the economic risk of different drought management tactics on a South Australian pastoral sheep station. Agricultural Systems 39:83–105.

Stafford Smith, D. M., and M. L. Hope. 1992. RANGEPACK Paddock version 1: User's guide. Alice Springs, Australia: CSIRO.

Stafford Smith, D. M., I. R. Noble, and G. K. Jones. 1985. A heat balance model for sheep and its use to predict shade-seeking behavior in hot conditions. Journal of Applied Ecology 22:753–74.

Stafford Smith, D. M., J. F. Clewett, A. M. Moore, G. M. McKeon, and R. Clark. 1994b. DroughtPlan: Grazier-based profitable and sustainable strategies for managing climatic variability. DroughtPlan Working Paper No. 1. Alice Springs, Australia: CSIRO.

Stafford Smith, D. M., N. Milham, R. Douglas, N. Tapp, J. Breen, R. Buxton, and G. McKeon. 1995. Whole farm modelling and ecological sustainability: A practical application in the NSW rangelands. Proceedings of the inaugural conference of Australian and New Zealand Ecological Economics Society, Coffs Harbour, Australia, November. 243–49.

Stockwell, T. G. H., P. C. Smith, D. M. Stafford Smith, and D. J. Hirst. 1991. Sustaining productive pastures in the tropics. Pt. 9: Managing cattle. Tropical Grasslands 25:137–44.

Stuth, J. W., W. T. Hamilton, J. C. Conner, and D. P. Sheehy. 1993. Decision-support systems in the transfer of grassland technology. *In* M. J. Baker, ed. Proceedings of the seventeenth International Grassland Congress, New Zealand, February. Wellington, New Zealand: SIR Publishing. 234–42.

Stuth, J. W., and D. M. Stafford Smith. 1993. Decision support for grazing lands: An overview. *In* J. W. Stuth and B. Lyons, eds. Emerging issues for decision-support systems for management of grazing lands. Paris: MAB-UNESCO. 1–35.

Walters, C. 1986. Adaptive management of renewable resources. New York: Macmillan.

Westoby, M., B. H. Walker, and I. Noy-Meir. 1989. Opportunistic management for rangelands not at equilibrium. Journal of Range Management 42:266–74.

Yevjevich, V. 1972. Stochastic process in hydrology. Fort Collins, Colo.: Water Resources Publications.

12 Arid Lands Management: An Operations Research Approach

John Hof

Why Optimization?

The traditional tool of "management science" or "operations research" is optimization. In the simplest of terms, any time that more of a good thing is good or less of a bad thing is good, an optimization problem is at hand. Because we live in a finite world, practical problems usually involve constrained optimization where we wish to maximize or minimize an objective function, subject to constraints that reflect physical or other limitations to our options. A solution to these constrained optimization problems usually involves linear, nonlinear, integer, dynamic, or some other form of mathematical programming.

In land management and planning, the problem typically revolves around the timing and physical placement of various management actions. The problem is generally constrained by the biological production limits of the managed ecosystem, policy limitations that give primary weight to certain criteria such as sustainability or predetermined land designations (e.g., wilderness), economic limitations such as a budget constraint, and any other constraining factors. There is typically no single objective, economic or biological, that everyone agrees upon; but, from any given perspective, there are good things we want to maximize and bad things we want to minimize. Often, in planning settings, a variety of objectives are used to define the "decision space" and to identify trade-offs between alternatives.

Even if we view resource planning as an optimization problem, it can still be addressed with simulation (or other) analysis procedures. A variety of management schemes could be used as driving variables in a simulation analysis, for example, and the simulated impacts could then be judged in terms of conformance to constraints and attainment of whatever objectives are of concern. The problem is that many, many management schemes may be possible. A simple example might be useful. Imagine a rectangular watershed as depicted in figure 12.1. For simplicity, assume that because a low level of resolution is required, we segment the watershed into twenty-five equal sections (fig. 12.1). Also assume that we have only one management action (e.g., clear-cutting) to consider versus doing nothing. Even in this very simple example, if each of the twenty-five sections could be clear-cut, 2^{25} or over 33 million spatial configura-

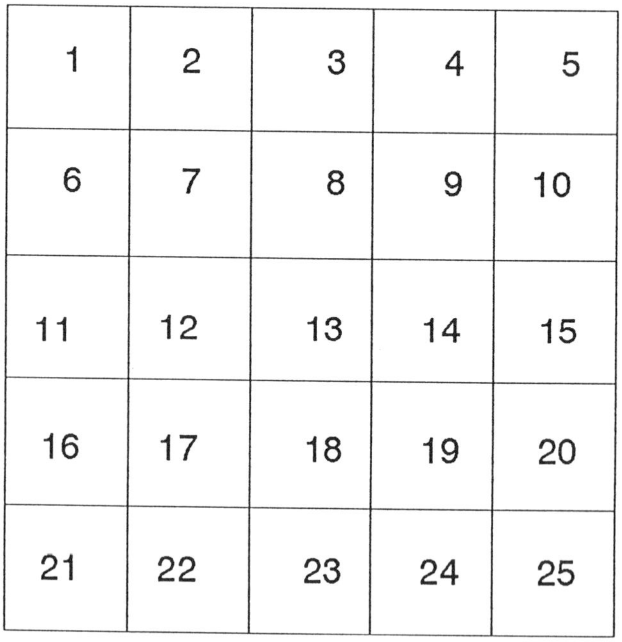

1	2	3	4	5
6	7	8	9	10
11	12	13	14	15
16	17	18	19	20
21	22	23	24	25

Figure 12.1. A 25-cell planning area.

tions are possible. With more potential management actions or with finer resolution, the number of spatial possibilities approaches google numbers very rapidly. Add the scheduling component (dynamic considerations) and the problem becomes larger still. The power of the mathematical programming approaches is that all of these many schemes are efficiently enumerated (whether explicitly or implicitly). The simple example above could be formulated with twenty-five choice variables (0–1), and the problem solution would reflect an implicit evaluation of all 33 million management schemes. As will be seen below, however, formulating these programs to capture the desired ecosystem limitations can be difficult.†

Arid Lands Management

When considering the special management setting for arid lands, the most immediate consideration might be that they are not likely to be terribly good timber-growing lands, such that the management objective will typically not emphasize the commercial production of fiber. In fact, the fragile nature of arid lands suggests an emphasis on ecosystem condition rather than any output production. This implies two particularly important modifications to the traditional timber-oriented optimization approaches: (1) different objective functions are implied that capture the ecological concerns instead of the commodity-production emphasis (this may also imply tracking different variables) and (2) spatial considerations will be much more important. The remainder of this chapter discusses these two modifications using wildlife concerns as an example; however, the methods discussed are applicable to a variety of nontimber concerns. This chapter is based on previously published papers with coauthors as cited.

A Species-Richness Objective

There are quite a number of potential biological objectives in arid lands management. In order to serve as an objective function in an optimization model, these objectives must be captured in a closed-form function whose arguments are the choice variables available to management. This section explores a few promising examples, but it is obvious that strong simplifying assumptions are necessary to capture a biological objective with a closed-form objective function.

One of the biological concerns that is of most interest is *biodiversity*. This concept, however, has not been rigorously defined in a way that allows its formulation in forest management optimization models. I will focus here on wildlife diversity measured by a probabilistic statement of species richness. This section is based on Hof and Raphael (1993). Let us first assume that the total number of animals in the ith species (S_i) is determined elsewhere in the model. Assume that T_i is the maximum number of animals in the ith species that can be obtained. Let us then assume that the probability of a species being viable in the study area over a given period of time (V_i) is a function of the number of animals in that species (Mace and Lande 1991, Marcot et al. 1986) and that, if T_i animals are present, the probability of viability is near 100%. A reasonable functional form for the V_i would be logistic:

$$V_i = \frac{100}{1 + \exp\{[(T_i / 2) - S_i] / (T_i / 8)\}} \qquad [1]$$

These logistic V_i functions have the following properties: $V_i = 98.2$ for $S_i = T_i$; $V_i = 50$ for $S_i = T_i i/2$; and $V_i = 1.8$ for $S_i = 0$.

There are three obvious ways to construct a species-richness objective function from the V_i: (1) maximize the expected value of the number of viable species, (2) maximize the minimum probability of viability across species, and (3) maximize the joint probability of all species being viable. To maximize the expected value of the number of viable species, we maximize $\sum^n_{i=1} V_i$ where n is the number of animal species considered. To maximize the minimum probability of viability across species, we maximize λ subject to $\lambda \leq V_i$, $\forall i$ where λ is the minimum probability of viability across species. To maximize the joint probability of all species being viable, we maximize $\Pi^n_{i=1} V_i$.

The first approach is the most "efficiency oriented" in terms of maintaining the highest possible number of species. The second approach maximizes evenness of abundance because the poorest probability of viability is maximized. The third approach is somewhat of a middle ground because some species are allowed to have low viability probabilities, but this is penalized because the entire objective function value approaches zero if the viability probability of any one species approaches zero. Bevers and colleagues (1995) discuss piecewise approximation of these objective functions for use with linear solution methods.

A dynamic version of this basic objective function might be to define $V_{it} = f(S_{it})$, $\forall_i$ where V_{it} is the probability of viability for the ith species based on each time period's population. Then we could maximize $\sum^n_{i=1} \rho_i$ subject to $\rho_i \leq V_{it}$, $\forall_t$, maximize $\Pi^n_{i=1} \rho_i$ subject to $\rho_i \leq V_{it}$, $\forall_t$, or maximize λ subject to $\lambda \leq \rho_i$, $\forall_i$; $\rho_i \leq V_{it}$, $\forall_t$. This would be based on an assumption that a given species' effective probability of viability is based on the minimum population over the time periods included. We could also maximize $\Pi_i \Pi_t V_{it}$ as the joint probability of viability across all species and time periods. Or we could maximize λ subject to $\lambda \leq \Pi^n_{i=1} V_{it}$, $\forall_t$, which would maximize the minimum time period's joint probability of viability. Any number of other possibilities might also have appeal. These dynamic species richness objective functions capture at least one type of sustainability in that they involve the viability of species over time. In order for this objective to capture the sustainability element, however, lengthy time horizons may have to be included in the modeling analysis. Bevers and colleagues (1995) suggest that it may require time horizons in excess of five hundred years to confirm a sustainable cycle of management actions in this context. There are certainly many other types of biological objectives that might be the basis for an objective function—those suggested here are only examples meant to demonstrate the type of assumptions and formulations that might be required.

Spatial Considerations

For simplicity, the previous discussion ignored how the species populations are determined in the model. To do this, we must recognize the importance of the spatial layout of management actions. If management prescriptions are defined on a per-acre basis, the linear program determines the number of acres to which each management prescription applies. The problem is that the (nonlinear) response to different sizes and shapes of the management action is lost in a fixed per-acre production coefficient.

Many authors (e.g., Clements et al. 1990, Nelson and Brodie 1990, Nelson and Finn 1991, O'Hara et al. 1989, Roise 1990) have defined the management variables in terms of timber stands that are treated discretely and are preserved as discrete units in solution. The spatial considerations are then typically viewed in terms of either nonadjacency constraints over time or constraints that limit the size of contiguous cutover areas at any given time. Considerable progress has been made in solving the problem viewed this way, but the approach is limited by accepting and preserving the initial stand definitions. Also, this approach avoids "spatial anomalies" but does not account for the nonlinear response of many forest outputs (such as wildlife and fish, water, aesthetics, etc.) to different sizes, shapes, and arrangements of management actions. It would thus be difficult to argue that it finds "spatially optimal" solutions for all outputs of concern (as opposed to just timber).

In terms of these nontimber outputs, it is often more important how a management action (for example, a timber harvest) is spatially laid out than how many acres are involved (Diamond 1975, Franklin and Forman 1987, Harris 1984, Saunders et al. 1991). Some applications of Version 2 FORPLAN have attempted to deal with this problem by making the LP behave more like an integer program and by defining management prescriptions not on a per-acre or stand basis but on an entire watershed (or similar land unit) with a defined spatial configuration. This allows the prediction of the resulting outputs to take a given spatial layout into account, but a serious problem remains: within the practical limits of the LP, only a tiny number of the possible spatial layouts can be considered. The simple example in figure 12.1 demonstrates how many layouts are often possible.

Static Spatial Optimization

This section is based on a different approach that is described in Hof and Joyce (1993). Let us begin with the static case, focusing on integer programming approaches that optimize spatial layout, per se, for a single time period and that have the property that the number of integer choice variables increases linearly with the level of spatial resolution. The static formulations developed are not terribly powerful unless the spatial layout created by the choice variables is relatively permanent and the static characterization of spatial layout is important to nontimber resources. Recent issues, such as those involving old-growth harvesting, have a more semipermanent character than traditional forest-regulation problems. It is assumed that wildlife populates the habitat to the degree permitted by the spatial layout (edge, fragmentation, and size thresholds). This would be something of an "equilibrium" population pattern based on that spatial layout. Because this is a static model, dynamic *processes,* such as population movement and birth and death rates, are not modeled directly.

First, we divide the land area into M "cells" as depicted in figure 12.1. Choice variables, C_i, are defined such that $C_i = 0$ if the ith cell is harvested and $C_i = 1$ if it is left in old growth. As a simple example, assume that the objective function, as discussed above, would involve two species:

G = population of an edge-dependent species, perhaps goshawk; and

P = population of an area-dependent species, perhaps pine marten.

The problem is thus to determine the amount of G and P from the choice variables, C_i. The next section discusses the calculation of edge (E), and it will be assumed that G is a linear function of edge:

$$G = gE \qquad [2]$$

where:

g = the number of goshawk from each km of edge.

The following section will then discuss the calculation of P so as to account for fragmentation effects. A size threshold on habitat area will also be formulated.

Edge Effects

The goshawk population is assumed to be determined by the amount of edge between a mature stand of timber and a cutover area. Such an animal is termed an *edge-dependent* here. An edge-dependent might be a small mammal that has a narrow range and has habitat requirements involving both the old growth and cutover areas or an animal such as goshawk that preys on those small mammals. The amount of edge can be accounted for with the following:

$$N = \sum_{i=1}^{M} C_i \quad [3]$$

$$E = N(4 \cdot L) - L\left(\sum_{i=1}^{M} D_i\right) \quad [4]$$

$$D_i \leq \propto_i C_i \quad \forall_i \quad [5]$$

$$D_i \geq \left(\sum_{j=1}^{M} \delta_{ij} C_j\right) - \propto_i (1 - C_i) \quad \forall i \quad [6]$$

where:

C_i = an integer variable that is 0 if cell *i* is harvested and 1 if cell *i* is left in old growth;
D_i = a continuous, nonnegative dummy variable for each *i*, which indicates the number of deductions (of *L*) taken for each *i* in calculating *E;*
$\propto_i$ = an arbitrarily large constant (it must be set so: $\propto_i \geq \sum_{j=1}^{M} \delta_{ij} C_j \quad \forall i$);
N = the number of cells left in old growth;
E = the amount of edge between old growth and harvested areas;
L = the length of one cell's side; and
δ_{ij} = 1 if cell *i* shares a side with *j*, 0 otherwise.

It is assumed that the area around the planning area is already cutover, so that when the C_i for the border cells are equal to 1, the outer perimeter is edge. The approach could easily be modified for different conditions. Constraint [4] calculates the amount of edge as follows. The amount of edge from each C_i that is equal to 1 is $(4 \cdot L)$ if all adjacent C_i are 0. Thus, constraint [4] starts with $N(4 \cdot L)$. For each adjacent C_i that is not 0, the edge will be overestimated by $(2 \cdot L)$.
Each pair of equations such as:

$$D_1 \leq \propto_1 C_1 \quad [7]$$

$$D_1 \geq \sum_{j=1}^{M} \delta_{1j} C_j - \propto_1 (1 - C_1) \quad [8]$$

taken with [3] and [4], serves the purpose of deducting the edge overestimates. If, for example, C_1 is 0, then [7] forces D_1 to be 0 (no deductions) and [8] is "switched off" because $(1 - C_1)$ is 1. When C_1 is 1, then [7] is "switched off" and [8] becomes:

$$D_1 \geq \sum_{j=1}^{M} \delta_{1j} C_j$$

Because D_1 is deducted from the edge calculation and E will be positively valued (through G) in the maximization objective function, D_1 will solve equal to $\sum \delta_{ij} C_j$, and the correct amount of edge overestimate will be deducted.

It should also be noted that in the "geometric" approach in Hof and Joyce (1992) we included "rings" of forage around old-growth areas that serve as the effective feeding area around the (old-growth) cover area for animals such as deer. In the formulation in [3]–[6], this feeding area could be approximated as a linear function of edge. For example, for every km of edge, there could be ⅓ km² of feeding area within a ⅓ km distance of the old growth.

Habitat Fragmentation Effects

Again, the problem addressed in this section is to determine a semipermanent optimal layout of habitat versus nonhabitat such that wildlife can then use (populate) the habitat. The static concept of habitat connectivity should thus reflect the long-term movement behavior of wildlife, even though we are not modeling wildlife population movement (or dynamics) per se.

If dynamic wildlife movements are random, then the static condition of habitat connectivity is also random. That is, if a group of areas are not perfectly conjoined with corridors, the probability of all areas being populated in any amount of time is diminished but not necessarily zero. Likewise, even a habitat system perfectly conjoined with corridors may not have a 100% probability of being totally populated within any given amount of time. We will thus define the static condition of *connectivity* of habitat areas as a probabilistic condition that reflects equilibrium dispersal patterns, given the permanence of management actions (which defines the single time period for the analysis in the first place). The ecological literature (e.g., Diamond 1975, Soulé 1991) typically indicates that habitat areas grouped, for example, in a triangle or square are preferable to those grouped in a row. Likewise, a circle is a better shape than an oblong, and one big area is better than several smaller connected ones. These preferences are based on a notion that wildlife populations radiate in a more or less directionless (360°) fashion and, thus, "clumped" configurations are easier to populate. This distinction would be lost in a movement-corridor analysis.

Put another way, there would be some probability that a given habitat area around any other habitat area would be connected, and this probability would diminish as the distance increases between the two habitat areas. With many habitat areas, the probability of a given area being "connected" would be a function of the number of other habitat areas nearby and the distances to them. We assume that the probability of each area being connected to a group of areas is the joint probability that the area is connected to *any* (not all) of the areas in the group. We also assume independence between the individual connectivity probabilities. Thus, the joint probability (PR_i) of each cell i being connected would be:

$$PR_i = 1 - \left[\prod_{j=1}^{M}(1 - pr_{ij}C_i)\right] \quad \forall i \qquad [9]$$

(pr_{ii} assumed to be 0) where pr_{ij} is the probability that cell i is connected to cell j. Presumably, pr_{ij} would be smaller, the farther cell j is from cell i. Equation [9] simply calculates the joint probability that cell i will not be connected to any of the $j = 1, \ldots, M$ ($j \neq 1$) cells, and then calculates PR_i as the converse of that joint probability. In other words, wildlife habitat connectivity is similar to a gravity model—an area would have a very high probability of being connected if it is surrounded closely by other areas of habitat and a lower probability with fewer surrounding habitat areas or with greater distances between areas. At some distance, the probability of two habitat areas being connected would be effectively zero. Thus, when an area is left as habitat, it has a certain probability of being connected that is determined by the number and location of other habitat areas, and it also contributes to the probability of other areas being connected in an equivalent manner. Cells of old growth only contribute to effective habitat to the degree that they are probabilistically connected.

We can approximate the formulation in [9] with a mixed-integer linear set of constraints. Let us assume that the probability of a cell being connected to an adjacent cell is 0.5, that the probability of it being connected to the next one over is 0.15, and that there is zero probability of connectivity if the cell is more than one cell removed. For convenience, define the set Ω_i as the indexes of the cells that immediately surround the ith cell and the set θ_i as the indexes of the cells that surround Ω_i. Because the planning area is surrounded by cutover area and we will use Ω_i and θ_i for habitat connectivity, we will limit Ω_i and θ_i to the planning area. Equation [9] could then be approximated by:

$$PR_i \leq \sum_{p \in \Omega_i} .3C_p + \sum_{q \in \theta_i} .1C_q \quad \forall i \qquad [10]$$

$$PR_i \leq C_i \quad \forall i \qquad [11]$$

The coefficients (0.3 and 0.1) were chosen to approximate equation [9] with the probabilities of 0.5 and 0.15 for the two sets of surrounding cells. Constraint [11] prevents PR_i from exceeding 1 (because $C_i \leq 1$). Also, if C_i is 0 in the solution, then PR_i is forced to 0 by [11].

The expected value of the total pine marten population, $E(P)$, would thus be:

$$E(P) = \sum_{i=1}^{M} PR_i \cdot p_i \qquad [12]$$

where the p_i is now interpreted as the pine marten yield from C_i if $C_i = 1$ and $PR_i = 1$.

Habitat Size Thresholds

The area-dependent animal might very well have a minimum habitat size requirement below which no population is maintained. This can be accounted for by adding:

$$E(P) \leq \gamma \cdot S \qquad [13]$$

$$S \leq (1/TH) \cdot N \qquad [14]$$

where:

S = a 0–1 integer variable;
TH = the size threshold in cell units (number of cells needed to meet the minimum size requirement);
γ = an arbitrarily large number (set so γ exceeds the maximum possible value of $E(P)$); and
P and N are as defined above.

If N is less than TH, then S is forced by [14] to be 0 and $E(P)$ is forced by [13] to be 0. If, on the other hand, N is greater than or equal to TH, then S can equal 1, which makes [13] nonbinding on $E(P)$. Note that [13] and [14] together require one additional integer variable (S).

Dynamic Spatial Optimization

This section, based on Hof and colleagues (1994), focuses on wildlife habitat connectivity/fragmentation (Saunders et al. 1991) by modeling the processes of wildlife growth and dispersal. Thus, it addresses the dynamic problem where management activities must be scheduled over time, wildlife habitat (timber age classes) must be tracked as forest stands age and grow, and individual wildlife species respond differently to those habitats. We again divide the land into cells as in figure 12.1. We then define a set of 0–1 choice variables for each cell, each one of which represents a complete, scheduled management prescription. Initial timber age classes are assigned to each cell, as well as initial population numbers for each wildlife species included. The model then chooses one management prescription for each cell.

In order to optimize the spatial layout of management actions that indirectly accounts for habitat connectivity over time, some rather specific assumptions regarding wildlife population growth and dispersal are necessary. First, we assumed that wildlife populations in any land cell are limited (and therefore determined) by either the carrying capacity of that cell or the combination of growth and dispersal that connects habitat cells from one time period to the next (Kareiva 1990). We assumed that each species has an r-value that indicates a growth potential per time period in the absence of other limiting factors (in particular, carrying capacity). In addition, we assumed that wildlife disperses between one time period and the next in a random fashion. That is, it radiates in all directions (360°) according to some probability density function that relates probability of dispersal to distance. Thus, if carrying capacity is not limiting, the population of each cell in any time period after the first is determined as follows: (1) The *r*-value is applied to each cell's population in the previous time period to determine the unconstrained net growth in population between time periods. (2) It is assumed that each of these animals has a fixed probability of remaining in the given cell and of dispersing to each other cell between time periods and that those probabilities decrease with distance according to the probability density function. This density function is defined such that the sum of all these probabilities over some distance in all directions is equal to one, so all animals are accounted for. The basic constraint set for the model is thus:

$$\sum_{h=1}^{q_h} X_{kh} = 1 \quad \forall h \qquad [15]$$

$$S_{ih1} = N_{ih} \quad \forall i \;\; \forall h \qquad [16]$$

$$S_{iht} \le \sum_k a_{ihtk} X_{kh} \quad \forall i \;\; \forall h \;\; t = 2, \ldots, T \qquad [17]$$

$$S_{iht} \le \sum_n g_{inh}\left[(1 + r_i) S_{in(t-1)}\right] \quad \forall i \;\; \forall h \;\; t = 2, \ldots, T \qquad [18]$$

$$F_{it} = \sum_h S_{iht} \quad \forall i \;\; \forall t \qquad [19]$$

where:

- i = indexes species;
- k = indexes the management prescription ($k = 1, \ldots, q_h$) where q_h is the number of potential management prescriptions for the hth cell;
- h = indexes the cells, as does n;
- t = indexes the time period;
- T = the number of time periods;
- $X_{kh} \in \{0, 1\}$ = the kth potential management prescription to be considered for the hth cell;
- S_{iht} = the expected population of the animals in species i in cell h at time period t;
- a_{ihtk} = a coefficient set that gives the expected carrying capacity of animal species i in cell h at time period t if management prescription k is implemented. These numbers are based on the timber age class of cell h in time period t if management prescription k is implemented;
- N_{ih} = initial population numbers for species i in cell h;
- g_{inh} = the probability that an animal of species i will disperse from cell n in any time period to cell h in the subsequent time period. This includes a probability for $n = h$ thus, $\sum_n g_{inh} = 1$ for each combination of n and i;
- r_i = r-value population growth rate for species i;
- F_{it} = the total species population for species i in time period t.

Equation [15] forces the selection of one and only one management prescription for each cell. The management prescriptions are defined with no action in the first time period, which is used simply to set initial conditions. Equation [16] sets the initial ($t = 1$) population numbers for each species by cell. The S_{iht} (expected population by species, by cell, for $t = 2, \ldots, T$) will be determined by whichever of [17] or [18] is binding. One or both of them will always be binding because of the optimization framework. Constraint set [17] limits each cell's population to the carrying capacity of the habitat in that cell, determined by timber age class. Constraint set [18] limits each cell's population according to the growth and dispersal from other cells and *itself* in the previous time period. The growth (r_i) and dispersal (g_{inh}) characteristics of each species are reflected by the parameters in constraint set [18]. Constraint set [18] adds up the expected value of the population dispersing from all cells in the previous time period to the given cell in the given time period. It is important to note that whenever [17] is binding for a cell, some of the animals assumed to disperse into that cell are lost because of limited carrying capacity. Thus, actual population growth is determined by a combination of potential growth and spatially located limiting carrying capacities. Constraint set [19] defines the total population F_{it} of each species in each time period.

A wide variety of objective functions, biologically oriented or not, could be utilized with either the static or dynamic spatial formulation just described. The examples of species richness objective functions discussed earlier would be directly applicable. These approaches constitute a fundamentally different definition of the problem and model structure than the traditional timber-oriented optimization models. The formulations presented here require many simplifying assumptions and only capture certain aspects of an enormously complex problem. They do, however, provide an initial example of how this fundamentally different problem might be addressed with optimization methodology.

Conclusion

We are currently expanding this type of work into the areas of water flow management, management of forest pathogens, and management of spatially defined risk and uncertainty. We are also working on several "real-life" case studies with much larger model dimensions. The software that we have written to build these types of models is being included in a system called "Spectrum" that is being jointly developed with the Ecosystem Management (formerly Land Management Planning) Staff of the USDA Forest Service for use in future forest planning analyses. Clearly, though, much work remains to be done in the challenging area of spatial optimization of managed forest ecosystems. We have really only begun.

References

Bevers, M., J. Hof, M. Raphael, and B. Kent. 1995. Sustainable forest management for optimal multi-species wildlife habitat age class distribution: A coastal Douglas-fir example. Natural Resource Modeling 9:1–23.

Clements, S. E., P. L. Dallain, and M. S. Jamnick. 1990. An operational, spatially constrained harvest scheduling model. Canadian Journal of Forest Research 20:1438–47.

Diamond, J. M. 1975. The island dilemma: Lessons of modern biogeographic studies for the design of natural reserves. Biological Conservation 7:129–46.

Franklin, J. F., and R. R. Forman. 1987. Creating landscape patterns by forest cutting: Ecological consequences and principles. Landscape Ecology 1:5–18.

Harris, L. D. 1984. The fragmented forest. Chicago: University of Chicago Press.

Hof, J., M. Bever, L. Joyce, and B. Kent. 1994. An integer programming approach for spatially and temporally optimizing wildlife populations. Forest Science 40(1):177–91.

Hof, J. G., and L. A. Joyce. 1992. Spatial optimization for wildlife and timber in managed forest ecosystems. Forest Science 38(3):489–508.

———. 1993. A mixed integer linear programming approach for spatially optimizing wildlife and timber in managed forest ecosystems. Forest Science 39(4):816–34.

Hof, J. G., and M. G. Raphael. 1993. Some mathematical programming approaches for optimizing timber age class distributions to meet multispecies wildlife population objectives. Canadian Journal of Forest Research 23:828–34.

Kareiva, P. 1990. Population dynamics in spatially complex environments: Theory and data. Philosophical Transactions of the Royal Society of London B 330:175–90.

Mace, G. M., and R. Lande. 1991. Assessing extinction threats: Toward a reevaluation of IUCN threatened species categories. Conservation Biology 5:148–57.

Marcot, B. G., R. Holthausen, and H. Salwasser. 1986. Viable population planning. *In* B. A. Wilcox, P. F. Brussard, and B. G. Marcot, eds. The management of viable populations: Theory, applications, and case studies. Stanford, Calif.: Center for Conservation Biology, Stanford University Press. 49–61.

Nelson, J., and J. D. Brodie. 1990. Comparison of a random search algorithm and mixed integer programming for solving area-based forest plans. Canadian Journal of Forest Research 20:934–42.

Nelson, J. D., and S. T. Finn. 1991. The influence of cut block size and adjacency rules on harvest levels and road network. Canadian Journal of Forest Research 21:595–600.

O'Hara, A. J., B. H. Faaland, and B. B. Bare. 1989. Spatially constrained timber harvest scheduling. Canadian Journal of Forest Research 19:715–24.

Roise, J. P. 1990. Multicriteria nonlinear programming for optimal spatial allocation of stands. Forest Science 36(3):487–501.

Saunders, D. A., R. J. Hobbs, and C. R. Margules. 1991. Biological consequences of ecosystem fragmentation: A review. Conservation Biology 5:18–32.

Soulé, M. E. 1991. Land use planning and wildlife maintenance. Journal of the American Planning Association 57(3):313–23.

PART 4 *Progress toward Ecological Sustainability*

Introduction

This section's chapters consider the various authors' concepts of sustainability that have developed through their experience and research. This is with a general understanding that human intervention in biophysical ecological systems occurs in at least three ways: (1) removing elements that influence the system structure and function (water or timber), introducing elements that influence system structure and function (livestock or exotic plants), and replacing parts of the biophysical ecological system with elements that are preferred by humans (houses, roads, etc.). Management and sustainablity of arid lands biophysical systems is dependent upon our ability to manage: (1) biophysical systems that are controlled by humans and (2) a landscape mosaic that includes both biophysical ecological systems and human-dominated ecological systems. One basic component for sustainable management of arid lands is an understanding of water flow and its distribution in time and space. Chapter 13 by Joseph Morin, Daniel Rosenfeld, and Eyal Amitai and chapter 14 by A. Yair provide examples of the status of understanding and management of water flow for sustainability. Morin and colleagues address arid lands water systems from a regional perspective with implications for sustainability in terms of water harvesting and heterogeneity of water input to the system. Yair presents water input-output relationships for a watershed and their implications for management of sustainable biophysical systems in terms of runoff harvesting, management of a landscape mosaic, and the modification of landscape structure and function at the microsite around the individual tree.

A second basic component is the understanding and management of human-dominated biophysical systems. The chapter by Avi Perevolotsky presents the ecological implications of natural biophysical ecological systems that are influenced by grazing for community and ecological systems.

A third component for sustainability and management of arid lands systems is understanding the relationships between natural biophysical ecological systems and human-dominated ecological systems within the framework of the same landscape mosaic. Chapter 16 by Roger Farrow and chapter 17 by Menachem Sachs and Itshack Moshe are examples of a conceptual construct for management of these mixed ecological systems. Farrow discusses the implications of transforming natural biophysical systems into agroecological systems through grazing from a landscape perspective. Sachs and Moshe present the ecological implications of transforming desertified biophysical ecological systems into regional savannah ecological systems from a landscape perspective with consideration of productivity and biological diversity.

All the processes of landscape ecological systems are understood within the framework of a heterogeneous environment. Natural forces and human activities are changing the landscape mosaic and, therefore, changing the interactions among landscape units. Moshe Shachak, Steward T. A. Pickett, Bertrand Boeken, and Eli Zaady present the role of patchiness in the functioning of arid ecosystems and the relationship between patchiness and management. Their chapter discusses the ecological implications of natural and human-induced landscape mosaics for population, community, and ecosystem processes within the landscape system.

13 Radar Rainfield Evaluation and Use

Joseph Morin, Daniel Rosenfeld, and Eyal Amitai

The objective of this study is to show how accurate radar-estimated rainfall with good temporal and spatial resolution might be used for hydrological purposes. This is demonstrated with several case studies where the impact of rain intensity variability on runoff is calculated (e.g., calculating the daily "excess rainfall" while assuming constant infiltration rates and surface storage results with 300% differences between radar rain cells that are 1 km apart). Maximum storm rain intensities for different time durations depend strongly on the area size but also on the differences in the segment's intensity sequences. The accurate radar "rainfield" in timing will allow dynamic calculation type along the storm path.

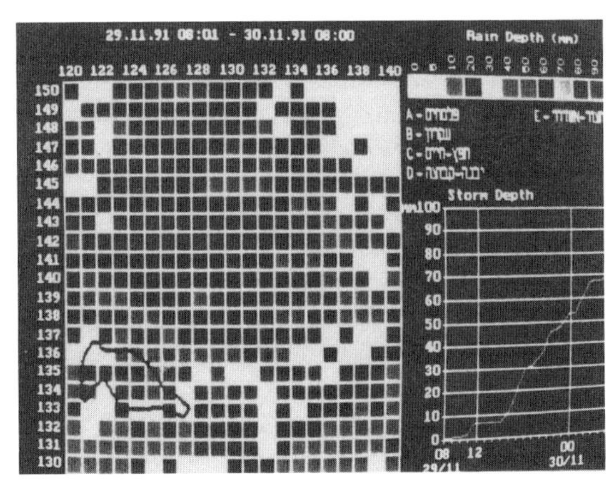

Figure 13.1. Distribution of 24-hr radar rain depth (mm) for area of 20 x 20 km² (29 November 1991).

The Study Area

Areal Rainfall Distribution Data and Evaluation

In this section, the distribution of rainfall intensity and amount will be presented for a target area of 20 x 20 km². The area is located 30 km south of Tel Aviv in the central plane of the coastal region. Figure 13.1 demonstrates the 29 November 1991 rainfall amount as analyzed for 400 cells, each 1 km².

Daily rainfall amount distribution for the Yavne watershed is illustrated in figure 13.2, which demonstrates the large areal variability even for this small 15-km² watershed. The tremendous spatial variability in space and time of the rainfall intensities is illustrated by figures 13.3 and 13.4 for the Yavne region. Intensity rates in mm/hr are written in each of the 1-km² cells. In figure 13.3, the only operating rain recorder in the watershed is marked by the letter A. Assuming the accuracy of the radar-estimated rainfall, we can

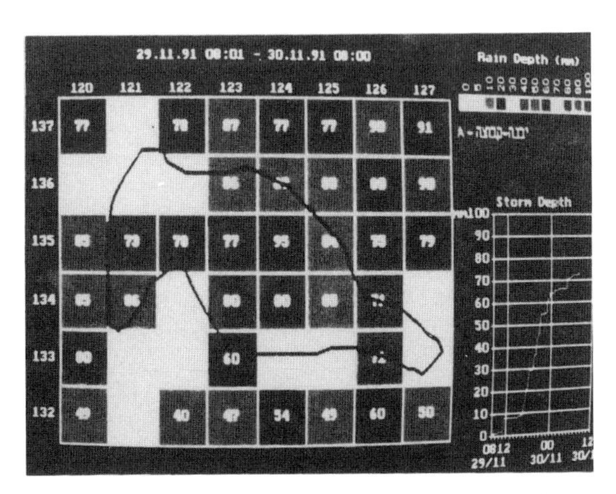

Figure 13.2. Distribution of 24-hr radar rain depth (mm) for Yavne watershed (29 November 1991).

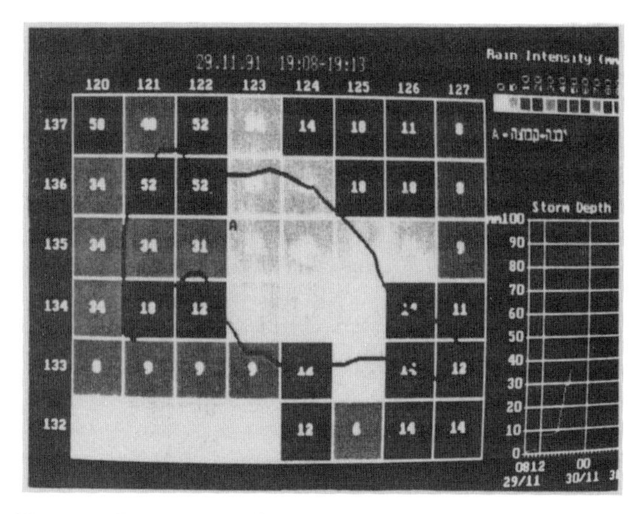

Figure 13.3. Five-minute radar rain intensity (mm/hr) for 29 November 1991, 19:08–19:13.

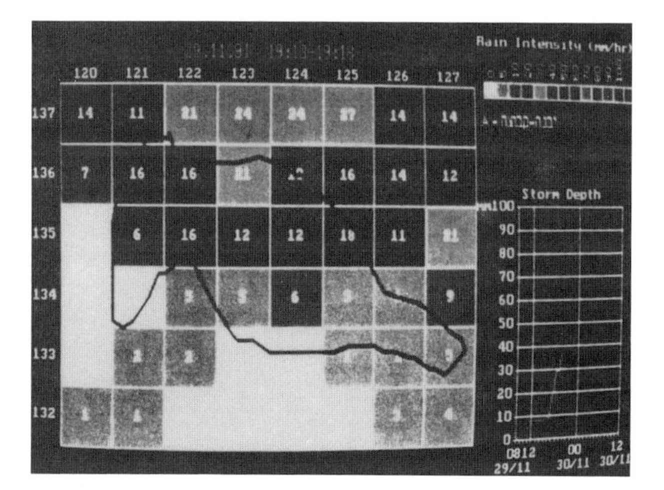

Figure 13.4. Five-minute radar rain intensity (mm/hr) for 29 November 1991, 19:13–19:18.

see that this recorder did not sense the high rainfall intensity of 52 mm/hr, which was twice the recorder's intensity and was falling only a short distance away (fig. 13.3). Storm intensity sequences and rainfall hydrology parameters, such as the maximum intensity for different duration, can now be evaluated for any target area using a combination of the radar cells. For comparison, the rainfall of 29 November 1991 was analyzed from a rain recorder and from radar mount estimated over the gauges. Figure 13.5 presents the storm rain segments as well as the maximum intensities for chosen time durations. Figure 13.5a shows measurements from the rain recorder. Figure 13.5b shows radar measurements for the 1 km² of the recorder location, and figure 13.5c indicates radar measurements for the entire watershed area. Figure 13.5 shows that the maximum intensities for thirty minutes (I30), for example, are 38, 28, and 21 mm/hr, respectively. I30 serves not only the hydrology calculations but also is an important factor in soil erosion prediction. It is obvious from the differences in the charts that watershed rain intensity disposition is presented poorly by the single recorder gauge.

Radar calibration by the WPMM in the Yavne region was based on five recording rain gauges. A comparison between rain amounts for twenty-four hours of the radar cells and rain gauges is presented in figures 13.6 and 13.7 and table 13.1. A total storm rainfall amount of 330 mm (average) was collected from seventeen accumulating rain gauges from the 29 November to 4 December 1991 storm (table 13.1). Figure 13.7 presents the location of the gauges as well as the differences in percentage in each cell between the radar measurements and the ground gauge measurements of the storm's total rainfall. Figure 13.6 presents regression and correlation analysis between daily rain amount collected from rain gauges and that estimated by the radar over the same locations.

From the figures and table, it can be seen that the regional average amount and standard deviation of the seventeen stations are similar for the gauges and the corresponding radar cells. A regression coefficient of $r^2 = 0.9$ and a standard deviation of 9.6 mm were obtained by the daily rain amount regression analysis. The above comparison analysis was made between a set of rain gauges with an orifice of 200 cm² and cells of 1 km² each. Calculated differences between radar and ground measurements may result from:

1. errors in matching the radar points with the rain gauge in time and space;
2. inaccuracies in the Ze-R relationship;
3. variations due to large differences in the measured rain area; and
4. errors in gauge readings between the radar and the rain gauge (mostly the result of different reading times).

At this stage in the research, we cannot quantify the relative contribution of the different sources of errors. However, timing and placement problems are canceled out through integration over a catchment area and over time.

The Importance of an Accurate Rainfield

The relation between rainfall and surface runoff are complicated. The main factors controlling it are submitted to strong variability in time and space. Besides the "rainfall field" variability as introduced before, soil infiltration, for example, varied dramatically by the difference in types of vegetation and density cover. The same variability is true for the soil surface and different types of subsoil (Sharma et al. 1980). The *hydrology rain* (the rain that actually hit the unit area) de-

MAXIMUM INTENSITIES FOR CONTINOUS CHOSEN TIME DURATION:

a.

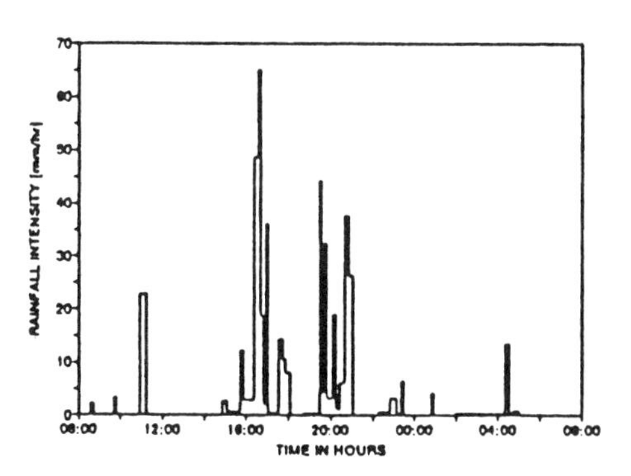

DURATION(min)	INTENSITY(mm/hr)
10	62
15	57
30	38
60	23
120	14
150	13
180	11

b.

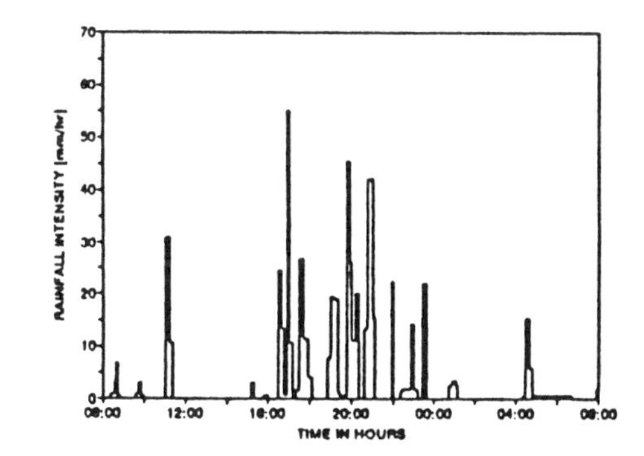

DURATION(min)	INTENSITY(mm/hr)
10	52
15	39
30	28
60	19
120	17
150	14
180	12

c.

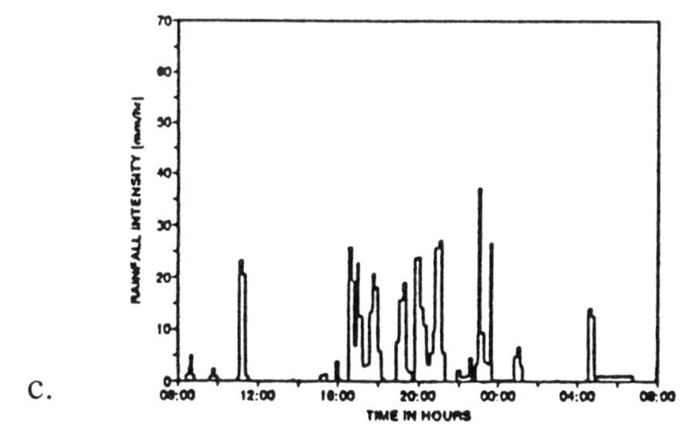

DURATION(min)	INTENSITY(mm/hr)
10	33
15	27
30	21
60	14
120	13
150	12
180	10

Figure 13.5. Recording rain gauge and radar-estimated rain intensity for 24 hours starting at 8 A.M. LST, 29 November 1991: (a) Yavne rain recorder, (b) Yavne 1 km^2 radar grid point, and (c) Yavne watershed averaged radar grid points.

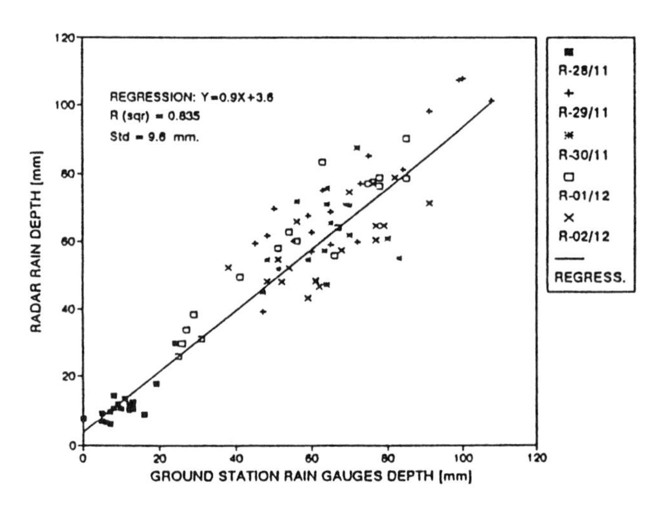

Figure 13.6. Daily rain depth regression between rain gauges and radar estimation over the gauges for 28 November–2 December 1991.

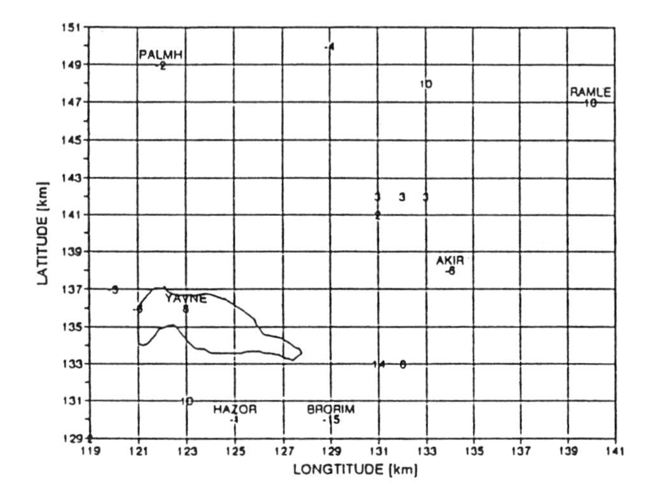

Figure 13.7. Rain depth differences in percentage for 20 x 20 km² area for 28 November–2 December 1991 rainstorm. Average rain depth: recorder—328 mm; radar—321 mm.

pends certainly on the rainfield but also on the wind velocity and direction as well as the specific topography of the unit area (Sharon 1980, Sharon and Arazi 1993). Runoff in the outlet of any watershed depends not only on the contribution of its small unit area but also on the concentration and routing system as well. Therefore, rainfall cannot be "translated" to the runoff "product" in a simple way. Nevertheless, accurate knowledge of the rain intensity distribution in time and space, the true rainfield, is probably the most important factor in this complex evolution. Possible and potential use of the radar rainfield will be discussed here.

Table 13.1. Differences between Rain Gauges and Estimated Radar Rain Depth for 28 November–4 December 1991 Rainstorm

Station Name	Longitude (km)	Latitude (km)	Rain Gauge (mm)	Radar (mm)	Gauge-Radar
Havazelet	132	142	410	398	12
Netaim	129	150	247	258	-11
Eqron	134	138	311	329	-18
Brorim	129	130	293	338	-45
Palmahim	122	149	288	293	-5
Gan Shlomo	131	142	411	398	13
Bnei Darom	121	136	302	320	-18
Hafez Haim	131	133	335	287	48
Kfar Bilu	133	142	386	374	12
Nir Galim	120	137	304	314	-10
Ramie Nesher	140	147	319	287	32
Givat Brener	131	141	408	400	8
Nezer Sireni	133	148	313	283	30
Bet Hilkia	132	133	295	273	22
Hazor	123	131	306	275	32
Yavne-Kvutza	123	136	342	314	29
Hazor-Ashdod	125	130	308	313	-5
Average (mm)			328	321	7.6
Std. (mm)			47	45	22.9
CV%			22.9.628 = 7		

Note: Longitude and latitude based on Israel grid coordinates.

Excess Rain Potential Working on each radar cell rain information separately can be done easily by assuming constant infiltration rates (f) and maximum surface storage and detention (SD_{max}) according to the following equation (Morin 1993):

$$\sum_{i=1}^{n} RP_i = (\Delta t_i R_i + SD_{i-1}) - (\Delta t_i f + SD_{max})$$

Figure 13.8 demonstrates the potential runoff variability for the Yavne watershed for the 30 November–1 December 1993 storm. Rain intensity (P_i) for each 1-km² unit area was taken by the WPMM method at time intervals (t_i) of five minutes. Infiltration rates (f) of 8 mm/hr were taken as constant in time for all the unit area cells. SD_{max} of 1 mm was taken as constant also. Figure 13.8 presents the potential runoff that accumulates during the entire storm. We can see the variability throughout the watershed and sometimes within very short distances. The variability in runoff intensity for the unit area is even much greater. Figures 13.9, 13.10, and 13.11, which show runoff intensities for continuous five-minute intervals, demonstrate this. Potential runoff as demonstrated in figures 13.9–13.11 changed according to the variability of the rain intensities only. No variations of infiltration in time and space or differences in soil surface storage were considered here. We have to stress here that, even for this simplified case, we cannot use the sum potential runoff of all its unit area cells as the watershed runoff. Runoff coming from the upper parts might be absorbed in

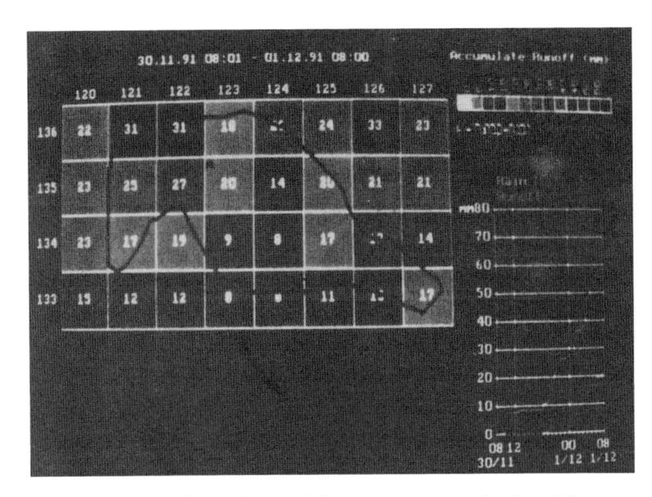

Figure 13.8. Runoff depth (mm) for 24 hours as calculated from radar rainfield (30 November 1991).

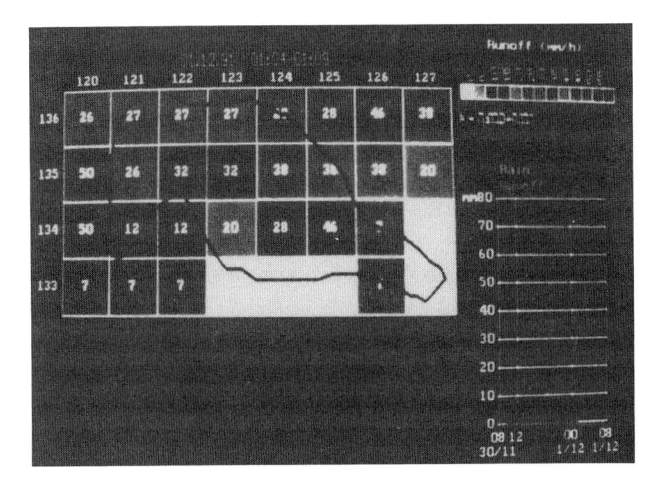

Figure 13.9. Runoff intensity (mm/hr) for five-minute interval for 01:04–01:09, 1 December 1991.

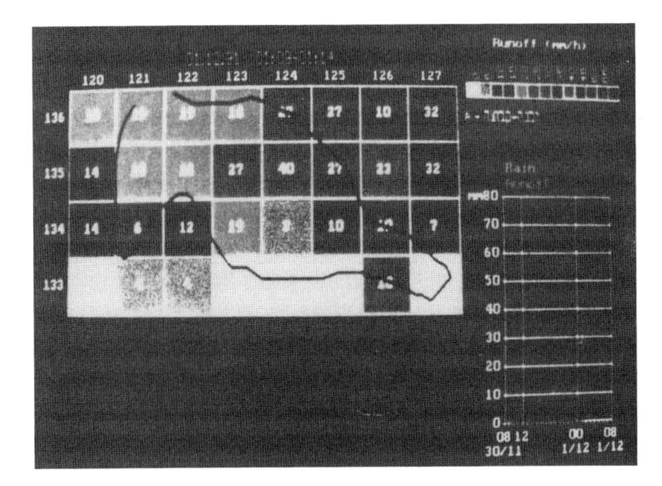

Figure 13.10. Runoff intensity (mm/hr) for five-minute interval for 01:09–01:14, 1 December 1991.

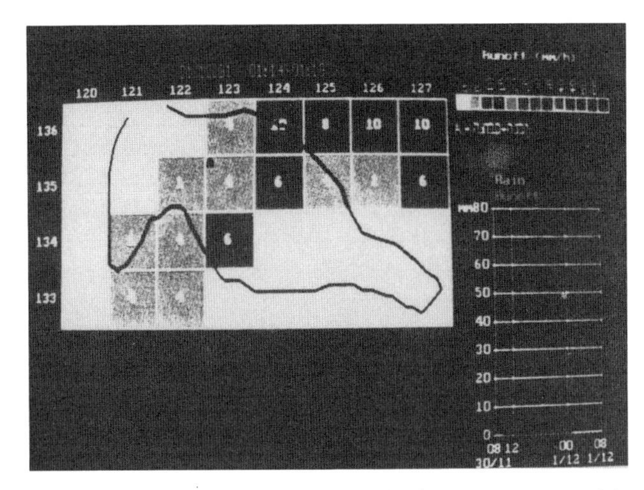

Figure 13.11. Runoff intensity (mm/hr) for five-minute interval for 01:14–01:19, 1 December 1991.

the down cells since, as demonstrated in figure 13.11, only some cells "produce" runoff in many storm time segments.

Rain Intensity Sequences The importance of getting a dense network of rain intensities for hydrology study is illustrated for the Sa'ad region (located near Gaza [city]). Two regions 2.5 km apart, each of an area of 2 km^2, were selected. In both areas, the total rain amount for the 1 December 1991 rain was 77 mm. Rain similarity for the two areas in intensities and energy is illustrated in figure 13.12, which shows only small differences. On the other hand, as will be demonstrated later, runoff can differ considerably. Overland runoff is generated whenever the rainfall intensities exceed the infiltration rates of the soil surface, providing that surface storage and detention were filled; the decaying nature of the soil infiltration was described by Morin (1993) as a function of accumulated rain amount. Figures 13.13 and 13.14 demonstrate the importance of rain intensity sequences in generating runoff. By applying the Morin method (Morin 1993, Morin and Benyamini 1988) to calculate rainfall storm runoff, a total runoff of 28 mm (36%) was generated in the lower Sa'ad area (fig. 13.13). Using exactly the same infiltration parameters for the upper Sa'ad area generates only 18 mm (26%) runoff. The higher runoff in lower Sa'ad was generated by the higher rain intensity segments at the later part of the storm when the infiltration rates were already very low. In the upper Sa'ad, high intensity segments were at the beginning of the storm when the infiltration rates were still high. This kind of detailed area-intensity relations could not be done without reliable radar rainfall estimates at small scales. Rain intensity sequences from a single-point rain recorder were usually projected to areas typically larger than 50 km^2. In the best cases, rates were corrected for different locations by applying linear relation to known total rain depth (Diskin et al. 1984).

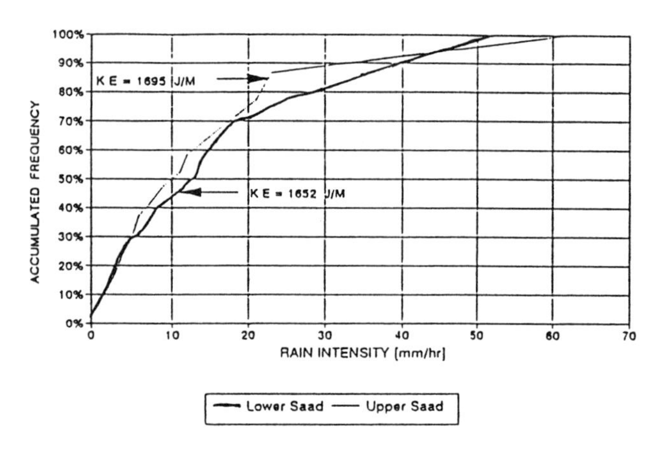

Figure 13.12. Rain intensity distribution for 24 hours (1 December 1991) in upper and lower Sa'ad regions.

Inf. Equ. It= 5+(40-5) *exp -(0.15*pt) S.D(max)-1.0 m

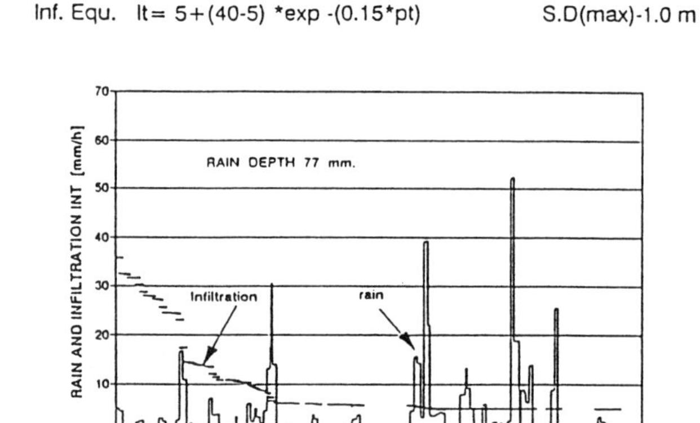

a.

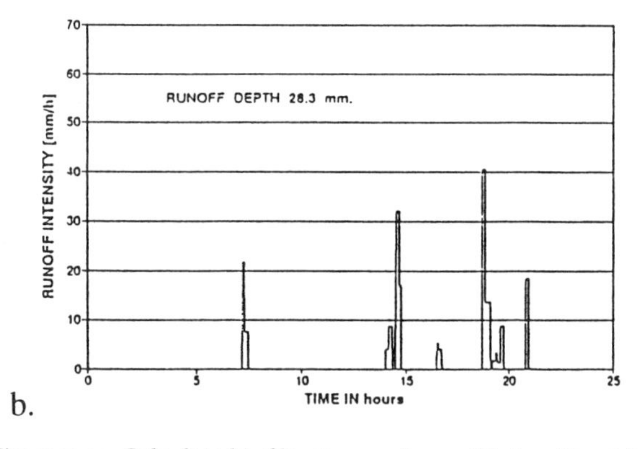

b.

Figure 13.13. Calculated infiltration and runoff intensities (a) and radar-estimated rain intensity (b) for lower Sa'ad region for 24 hours (1 December 1991).

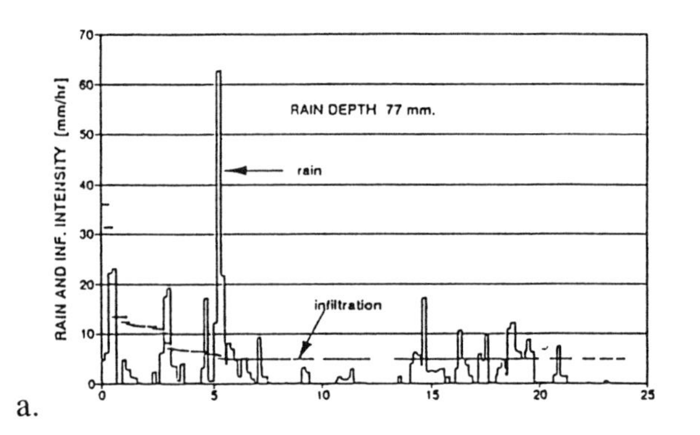

a.

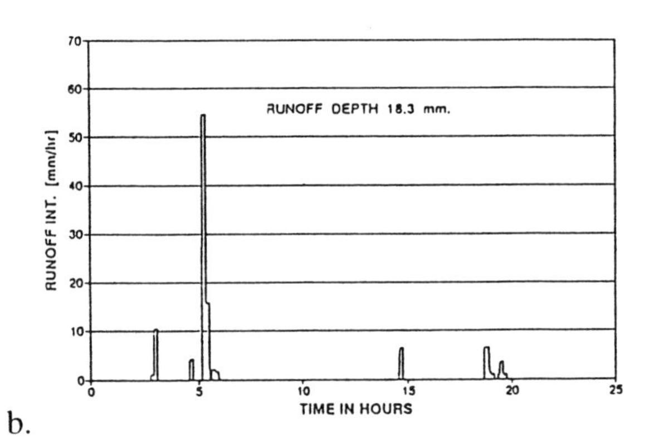

b.

Figure 13.14. Calculated infiltration and runoff intensities (a) and radar-estimated rain intensity (b) for upper Sa'ad region for 24 hours (1 December 1991).

Future Use of the Radar Rainfield in Hydrology

As stated before, the complexity of rainfall runoff relations comes from the tremendous variability in time and space of its major controlling factors. Cell models that divide the watershed area into a combined complex of cells according to the topography, such as the Diskin models (Diskin and Simpson 1978, Diskin et al. 1984) or the KINEROS Model (Woolhizer et al. 1990), can benefit directly from this intensive rainfield information.

The intention of this group is to use the tremendous power of modern geographic information system (GIS) methods for combining the radar rainfield with the other major controlling factors (e.g., infiltration, vegetation, topography, wind, and routing).

The recent improvement in radar rainfall estimation at all scales opens a new possible strategy for monitoring outlier events on ungauged watersheds. During a given time period over a large region, a small sample of low-probability events (outlier) will often occur with a higher frequency on isolated watersheds. Identifying the outlier events and analyzing the rainfield data for these isolated watersheds can greatly reduce the time and expenditure required for obtaining long-term data from a permanently instrumented watershed. Runoff generated for this extreme rainfall event can be evaluated later by indirect discharge methods in the ephemeral regions. A combined research effort of this kind has already started at the Hydrology Department of the University of Arizona.

References

Diskin, M. H., and E. S. Simpson. 1978. A quasi-linear spatially distributed model for the surface system. Water Resources Bulletin 14(4):903–18.

Diskin, M. H., G. Wyseure, and J. Feyen. 1984. Application of a cell model to the Bellebeek watershed. Nordic Hydrology 15:241–56.

Morin, J. 1993. Rainfall intensity analyses for tillage decisions. Soil and Tillage Research 27:241–52.

Morin, J., and Y. Benyamini. 1988. Tillage method selections based on runoff modeling. *In* Challenges in dry land agriculture: A global perspective. Proceedings of the international conference on dry land farming, Amarillo, Tex. 251–54.

Sharma, M. L., G. A. Gander, and C. G. Hunt. 1980. Spatial variability of infiltration in a watershed. Journal of Hydrology 45:101–22.

Sharon, D. 1980. The distribution of hydrologically effective rainfall incident on sloping ground. Journal of Hydrology 46:165–88.

Sharon, D., and A. Arazi. 1993. The distribution of wind-driven rainfall and the reconstructed local flow affecting it in the experimental watershed near Lehavim. Final Sci. Rep. Hebrew University of Jerusalem, Institute of Earth Science.

Woolhizer, D. A., R. E. Smith, and D. C. Goodrich. 1990. KINEROS: A kinematic runoff and erosion model. Documentation and user manual. ARS-77. Washington, D.C.: U.S. Department of Agriculture.

14 Spatial Variability in the Runoff Generated in Small Arid Watersheds: Implications for Water Harvesting

A. Yair

Extensive, ancient, sophisticated agricultural systems are located in the Negev Highlands, where average annual rainfall is 75–100 mm. Some of the fields are more than 2,500 years old. Hydrological works conducted in the area have led different authors to the conclusion that the very existence of sedentary agriculture in this very dry area was made possible by ingenious water harvesting techniques (Kedar 1957, 1967; Tadmor et al. 1958; Shanan et al. 1971). Detailed mapping of whole farm units revealed two different agricultural systems: (1) fields in large wadis irrigated by diverting channel runoff into cultivated fields and (2) fields in small wadis with flat, terraced, valley bottoms, irrigated by collecting overland flow from adjoining hill slopes. The ratio of the catchment area to the cultivated area in the small wadis varies from 17:1 to 30:1, the average being about 20:1. The runoff coefficient was estimated as 15% of the annual precipitation. Each cultivated unit in the valley could be expected to receive, in an average rainfall year, an amount of water equivalent to 300 mm in addition to some 100 mm of direct rainfall.

The considerations presented above were mainly derived from hydrological data collected at the reconstructed Avdat farm. Hydrological units monitored represent the area drained by the different water conduits as built by the Nabateans and not the natural, undisturbed, hydrological response units. The hydrological setup and a summary of data obtained are presented in Shanan and Schick (1980). However, the interesting pioneer work conducted at the Avdat farm does not address the following issue: Cultivated fields were limited to flat and terraced valley bottoms that occur in small- to medium-sized watersheds. No attempt was made to plant trees on the rocky hillslopes or in very small narrow watersheds where flat valley bottoms are lacking. Is it possible, within the Negev Highlands, to extend the area where trees can be planted to such narrow watersheds and hillslopes?

Aim of Present Work

The present chapter has two objectives: (1) to review the approach adopted for the study of hydrological processes in a small rocky, undisturbed, arid watershed and (2) to discuss the implication of data obtained for water purposes at the local and regional scale.

The Study Area

The study was conducted at the Sede Boqer experimental site. The site is located in the Negev Highlands (fig. 14.1). Average annual rainfall is 95 mm, with recorded extremes ranging from 30 mm to 183 mm. The number of rain days varies from 15 to 42. Rain is limited to the winter season from October to April. Mean monthly temperatures vary from 9 °C in January to 25 °C in August. Potential evaporation is 2,200 mm (Zangvill and Druian 1983).

Geologically, the area is composed of limestones and chalks of Turonian age. The area is dissected into small watersheds. Hillslopes are relatively steep (up to 29°) and subdivided into two distinct sections: an upper part that is mainly barren, having a stepped rocky limestone surface and a shallow patched soil cover and a lower part consisting of a stony unconsolidated colluvial soil cover that thickens downslope. A similar subdivision is observed along the channels. The upper channel is rocky, and the lower channel is covered with an alluvial fill.

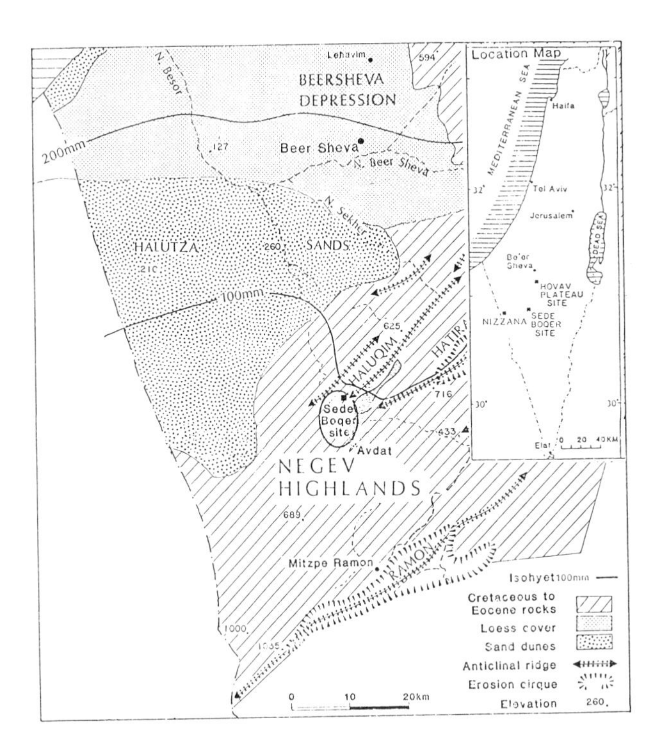

Figure 14.1. Physiography and location map.

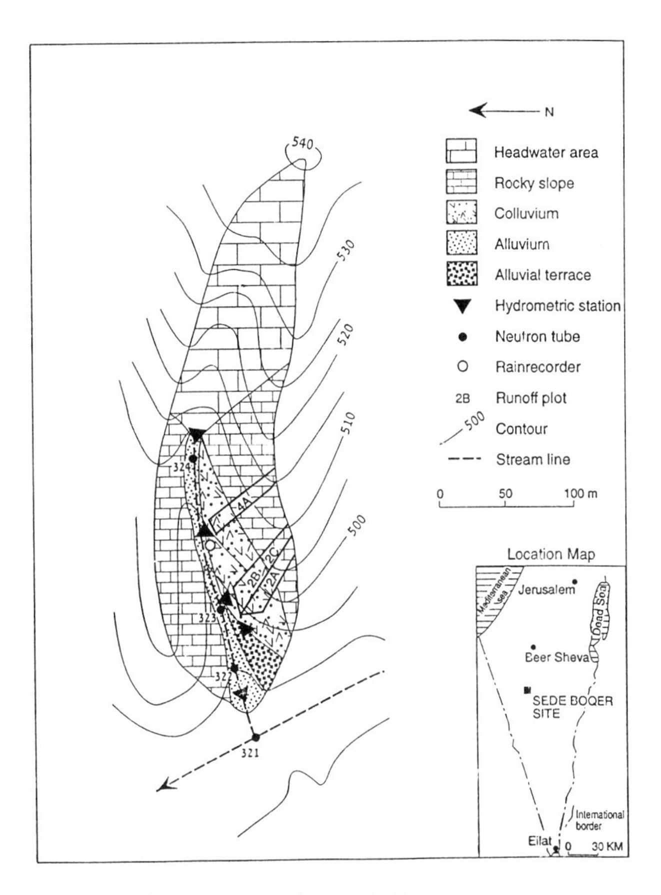

Figure 14.2. The instrumented watershed.

The Experimental Watershed

Geomorphic-Hydrologic Response Units

The experimental watershed represents a first-order drainage basin that extends over an area of 2.15 ha. The watershed can be subdivided into five distinct geomorphic-hydrologic response units (fig. 14.2).

Unit 1: Headwater Area This unit covers the upper part of the watershed and extends over 0.81 ha. It is composed of massive, flinty limestones. Very extensive rocky outcrops, almost devoid of any soil cover, form most of the surface.

Unit 2: Rocky-Colluvial Valley Side Slopes This unit extends over 0.64 ha with slopes ranging from 55 m to 72 m in length. The upper part is made of massive limestone and displays a stepped topography with soil strips at the base of the steps. The massive limestone is underlain by a thinly bedded and densely jointed limestone. Most of the latter formation is covered by a gravelly colluvial mantle that thickens quickly downslope.

Unit 3: Rocky Valley Side Slopes These slopes are much shorter than the opposite ones. They are carved in the thinly bedded and densely jointed limestone. The rock weathers into gravels and cobbles but lacks a colluvial cover. This unit extends over 0.37 ha.

Unit 4: Alluvial Terrace This low terrace separates part of the colluvial area from the present-day active channel. It covers an area of 0.13 ha.

Unit 5: Alluvial Reach The alluvial fill extends from the upper rocky headwater area down to the mouth of the valley. It is up to 50 cm thick and mainly composed of fine-grained material of loessic origin with few gravels. The alluvial fill covers 0.20 ha.

The Experimental Setup

The experimental watershed is equipped for automatic and simultaneous measurement of rainfall, hillslope, and channel runoff (fig. 14.2). In addition, soil moisture down to a depth of 3 m was measured with a neutron probe. Rainfall is measured with a rain recorder located at the lower part of the drainage basin. The location of the hillslope and channel hydrometric stations is based on the geomorphic-hydrologic units described above. Hillslope runoff is measured at two plots. Plot 4A drains a whole slope whose upper part is rocky and lower part is colluvial. Plot 2 is subdivided into three subplots. Subplot 2C drains the rocky slope section, subplot 2B the colluvial slope section, and plot 2A the adjoining whole slope from top to bottom. This design allows us to study the specific response of colluvial and rocky surfaces to rainfall as well as the response of a combined rocky-colluvial slope. The hydrometric stations for plots 4A and subplots 2A and 2B are located at the interface between the colluvium and the adjoining alluvial terrace.

Channel runoff is measured at two stations. The upper one is located at the transition from the headwater area into the alluvial reach. The second station is located downstream, close to the mouth of the drainage basin. The distance between the two stations is 200 m. All hydrometric stations are equipped with pressure transducers connected to data loggers as well as to the rain recorder.

In addition to rainfall and runoff, systematic measurements of soil moisture down to a depth of 3 m were conducted with a neutron probe along two lines: (1) along a colluvial slope section and (2) along the alluvial channel of the instrumented watershed (fig. 14.3).

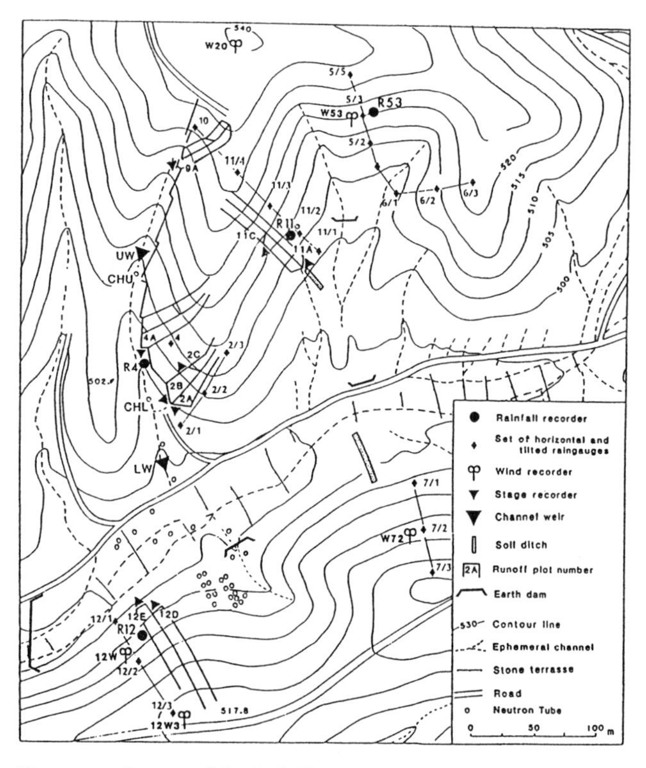

Figure 14.3. Layout of the Sede Boqer experimental site.

Results

Watershed Hydrology

Semi-arid and arid areas belong to the morphogenetic zone of the globe where physical weathering predominates over chemical weathering. The resulting landscape is often rugged and subdivided into two main process response units: (1) rocky areas with a limited patchy and thin soil cover. The extensive bedrock outcrops are characterized by a compact, dense structure with a low to very low permeability. They form the upper part of hillslopes and of small watersheds. (2) Unconsolidated surficial sediments such as scree slopes and colluvial or alluvial units composed of a mixture of various-sized particles. Their porosity and, therefore, water absorption capacity is much higher than that of the bare rocky surfaces. The extent of rocky surfaces increases with increasing aridity while the extent of soil-covered surfaces increases toward the more humid areas, where loess material of aeolian origin may predominate. The subdivision of rocky deserts into two distinct surface units, rocky and debris-mantled areas, creates two environments whose response to rainfall is basically different (Yair 1992). Following is a detailed analysis of runoff generation and runoff rates at various scales, ranging from 1.5 m^2 to 21.5 m^2. The latter area represents a small first-order watershed.

Small Plot Scale (1.5 m^2) Figure 14.4 presents the hydrological response of a rocky and a colluvial surface at the Sede Boqer site to a simulated rainfall. Intensity applied was 26 mm/hr with a duration of sixty minutes. This represents rather rare conditions in terms of rain intensity and especially duration. Runoff started almost immediately over the rocky plot, and infiltration dropped to zero after ten minutes. Runoff at the soil-covered plot started after six minutes, but the infiltration rate remained at a level of 18 mm/hr after one hour. Rain amount applied represents about 27% of the average annual rainfall. Furthermore, rain intensities below 15 mm/hr represent 90% of the rain recorded in the area for the last twenty years. Under such conditions, runoff frequency and magnitude can be expected to be

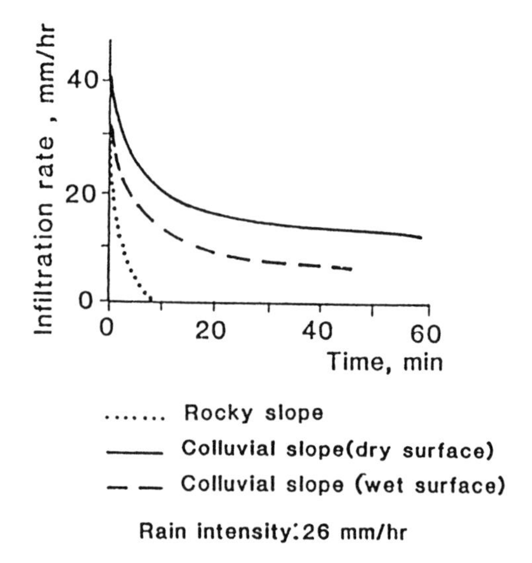

Figure 14.4. Infiltration curves for rocky and soil-covered areas (1.5 m^2).

higher on rocky areas than over areas covered with loose sediments, such as colluvial or alluvial materials.

Hillslope Scale (440 m²) The distribution of rocky and sediment-covered surfaces is not random in arid watersheds. Rocky outcrops always extend over the headwater area of the watershed and over the upper part of hillslopes, whereas colluvial mantles cover the lower part of the hillslopes and alluvial fills cover valley bottoms. Considering the difference in the hydrological response of the two basic types of surfaces as well as their spatial distribution, the sediment-covered surfaces can be expected to act as a filter or sink that separates the rocky source area of runoff from the channel, limiting or even inhibiting the possibility of runoff generated upslope from reaching the channel and contributing to channel flow.

Figure 14.5 demonstrates clearly the expected phenomenon of flow discontinuity described above. A sprinkling experiment was conducted over a whole slope 60 m long, whose upper part is rocky and lower part colluvial. Three runs were performed. The first, under dry conditions, had an intensity of 15 mm/hr and a duration of thirty minutes. Rain intensity was doubled on each of the following days for the same duration. Soluble dyes were used to map the flow lines. During the first run, runoff started over the upper rocky area four minutes after the onset of rain. No runoff could be observed over the colluvial slope section, and all of the runoff generated upslope was absorbed immediately on reaching the top of the colluvium. The same phenomenon was observed on the following day despite the higher initial soil moisture conditions. It was only at the last run and under very wet soil conditions that a continuous flow was detected ten minutes after the beginning of the rain. Data obtained indicate that a continuous flow along a whole slope can be expected only under rainstorms with a rain amount in excess of 30 mm and an intensity of 60 mm/hr for at least ten minutes. Rain intensities of the order of 30–40 mm/hr were recorded several times during the last twenty-three years; however, their duration never exceeded three minutes. The phenomenon of flow discontinuity described above was attributed to the pronounced difference in the infiltration rate between rocky and sediment-covered areas. This phenomenon can be greatly enhanced by the rain characteristics. All rainstorms in the study area bear an intermittent character. Each storm actually consists of several separate showers, often of short duration, each one totaling a rainfall depth of a few millimeters (fig. 14.6). Consecutive showers are separated by time intervals of from a few minutes to more than one hour, during which time the unconsolidated sediment can drain and dry. The brief duration of the effective rain showers causes runoff also to be short-term and intermittent (fig. 14.6). This duration is most often shorter than the concentration time required for a continuous flow along an entire slope from top to bottom or along the entire channel length.

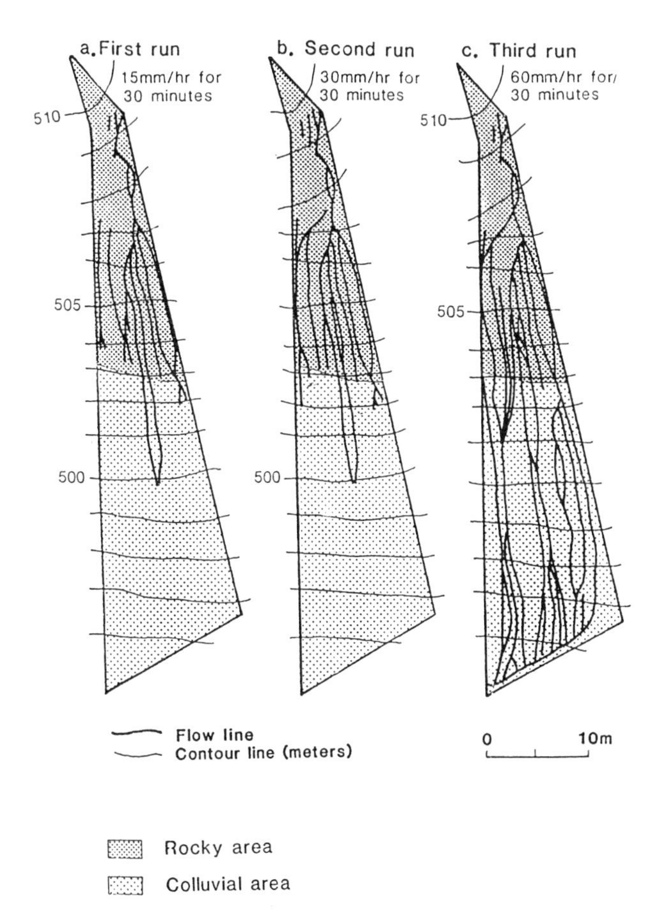

Figure 14.5. Flow discontinuity at the interface rocky-colluvial slope.

Watershed Scale (21,500 m²) The systematic and simultaneous study of hillslope and channel hydrology started in 1988. It took place within the instrumented watershed described earlier. Hillslope data exist for a longer period. The research period 1988–94 is typical of the high variability in rainfall and runoff to be expected in an arid environment. The two years of 1991 and 1992 were among the wettest recorded since 1951 and experienced the highest runoff rates recorded over the last twenty years. The two following years were relatively dry with the lowest runoff for the same period. Numerous storms had been recorded during the period considered. Data presented will cover the response of the watershed to three storms of different magnitudes:

1. low intensity storms (peak rain intensity below 10 mm/hr);
2. medium intensity storms (peak rain intensity of 10–20 mm/hr); and
3. extreme storms recorded in the area in the last years.

The analysis of the storms will be followed by a summary of the annual records available for hillslope and channel flow.

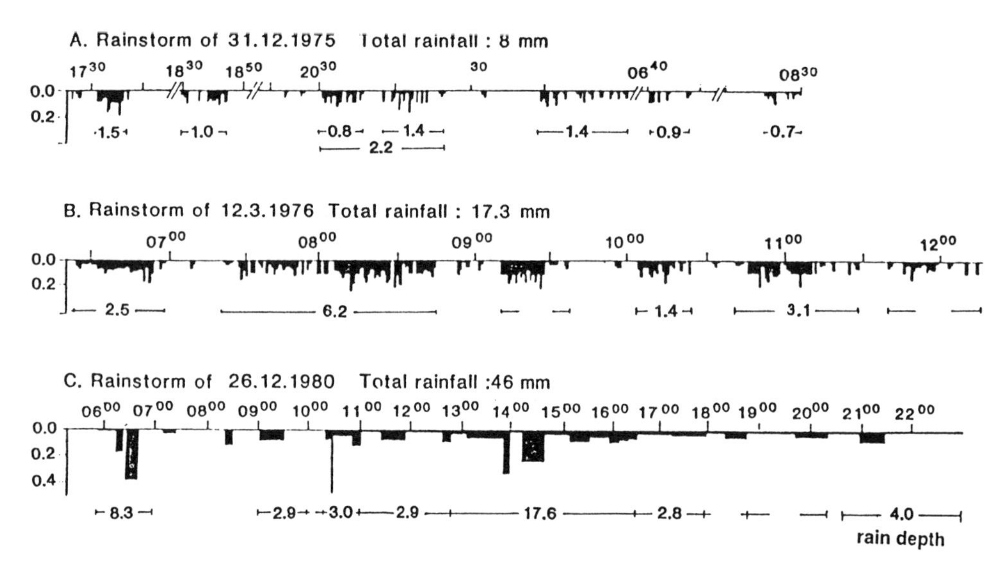

Figure 14.6. The intermittent character of rainstorms.

Low Intensity Storms Such a storm was recorded in December 1988. It lasted three consecutive days. Total rainfall amounted to 35.1 mm with intermittent rainfall resulting in several separate flows (fig. 14.7). Rain intensities were initially very low and increased toward the end of the storm but never exceeded 10 mm/hr (table 14.1). The area most responsive to rainfall was the headwater area represented here by the upper channel hydrometric station. Some runoff developed here on the first day when 16.1 mm fell on the area. The limited runoff occurred during a short burst when rain intensity reached the value of 4.5 mm/hr. Despite the high daily rain amount, no runoff was generated at any of the other plots. The following day the area received 10.3 mm. Because of a slight increase in maximum rain intensity and an already wet surface, runoff developed on the headwater area and over the rocky plot, being higher on the former than on the latter. However, no runoff was recorded at plots 2A and 4A, whose upper part is rocky and lower part colluvial. Most of the runoff was generated on the third day when the highest rain intensity (9.5 mm/hr) occurred on a very wet surface. At this stage, the response of the area to rainfall was quite general. Very significant differences in specific runoff yields, however, can be observed. The highest runoff per unit was obtained for the headwater area, followed by the rocky slope. The lowest runoff values were obtained for the colluvial plot 2B and especially for the alluvial channel. It is worth noting that, even at peak flow, discharge was higher at the upper than at the lower channel (fig. 14.7), clearly indicating that the highest transmission losses take place along the alluvial reach (table 14.1). A volume of 27.7 m^3 was recorded at the upper channel but only 3 m^3 at the lower channel. A dense vegetation covers the channel over a distance of about 30 m downstream of the upper channel station. Vegetation density decreases in the downstream direction, indicating that most of the runoff generated over the headwater area is often absorbed within a short distance of 30–70 m, as observed several times in the field during flow events.

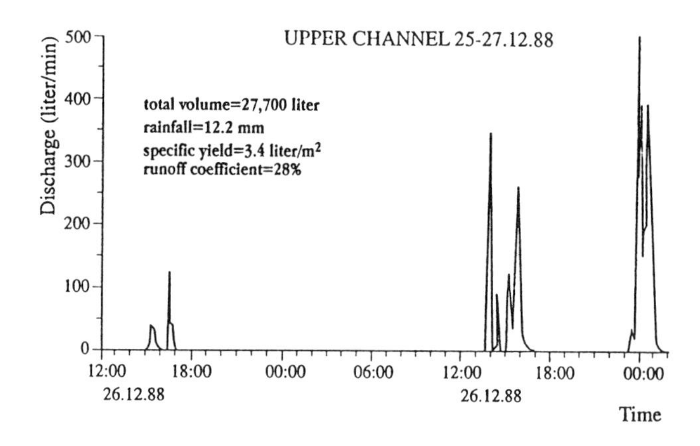

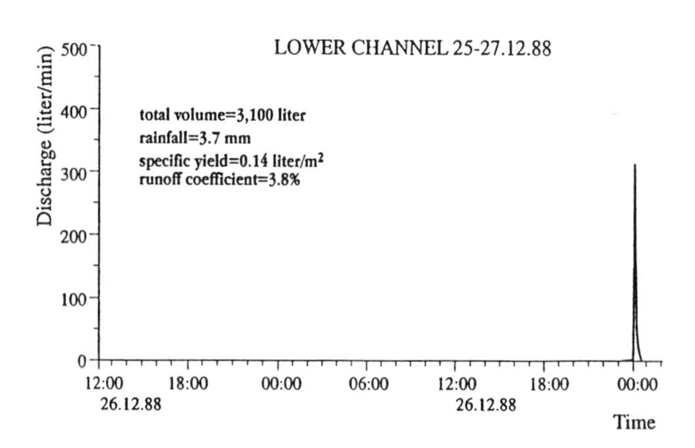

Figure 14.7. Storm of December 1988.

Table 14.1. Rainfall-Runoff Relationships for Selected Rainstorms, Sede Boqer Experimental Site

			Rocky Slope		Colluvial Slope		Upper Channel		Lower Channel	
Date	Rain (mm)	I MAX for 3 min (in mm/hr)	V (l)	V/A (l/m²)	V (l)	V/A (l/m²)	V (l)	V/A (l/m²)	V (l)	V/A (l/m²)
12/24/88	16.1	4.5	0		0		269	0.1	0	
12/25/88	10.3	7.7	0		0		2,122	0.3	0	
12/26/88	8.7	9.5	313	1.7	293	0.9	25,945	3.2	3,001	0.1
1/25-26/90	20.4	10.2	117	0.6	74	0.2	10,982	1.4	0	
3/22-23/92	14.4	17.2	646	3.6	379	1.1	25,086	3.1	11,325	0.5
1/24-25/91	51.5	18	3,258	18	1,516	4.5	182,470	22.4	255,000	11.9
2/1-2/90	8.4	31.8	118	0.7	251	0.7	8,558	1.1	1,992	0.1
3/21-23/91	45.7	34.2	4,235	12.6	240,000	29.5	376,000	17.5		
1/1-3/92	53.9	22.2	2,563	14.2	1,667	5.0	249,856	30.7	253,024	11.8

Key: V = volume; A = area

The importance of antecedent moisture conditions for flow generation over sediment-covered areas and at such low rain intensities is clearly demonstrated by the storm recorded on 25–26 January 1990. Total rain amount for this storm was 20.4 mm, with a maximum rain intensity of 10.2 mm/hr for a short duration of only one minute. No runoff was recorded over the colluvial plot or at the lower channel. Runoff at the headwater area was 1.43 l/m² and for the rocky plot 0.73 l/m².

Medium Intensity Storm A medium intensity storm was recorded on 22–23 March 1992 at the very end of the most rainy season since 1972. The storm was relatively short (fig. 14.8). The highest rain intensity (17.2 mm/hr) was recorded at the very beginning of the major rain shower and lasted three minutes. Runoff was generated all over the area. Three individual flows were recorded over the rocky areas but only one over the sediment-covered areas. Runoff was 6–7 times higher on former than on latter areas (table 14.1) with very high transmission losses along the alluvial reach.

High Intensity Rainstorm A high intensity storm was recorded on 21–23 March 1991. This is an extreme storm in both rain amount and intensity. Rain amount was 45.6 mm and maximum rain intensity reached 32.5 mm/hr for the short duration of one minute. The storm consisted of three main rain spells (fig. 14.9). Most of the runoff was recorded at the last rain spell whose intensity was the highest. Runoff per unit area was in the range of 23–30 l/m² for rocky areas but only 10 l/m² for rocky-colluvial hillslopes, indicating a transmission loss in the order of 60% on passing from the upper rocky slope section into the downslope colluvial section. Losses along the channel were relatively lower. For the first time in seven years, runoff volume recorded at the lower channel was higher than that at the upper channel (table 14.1), pointing at an important contribution from the hillslopes. However, the difference in runoff per unit area

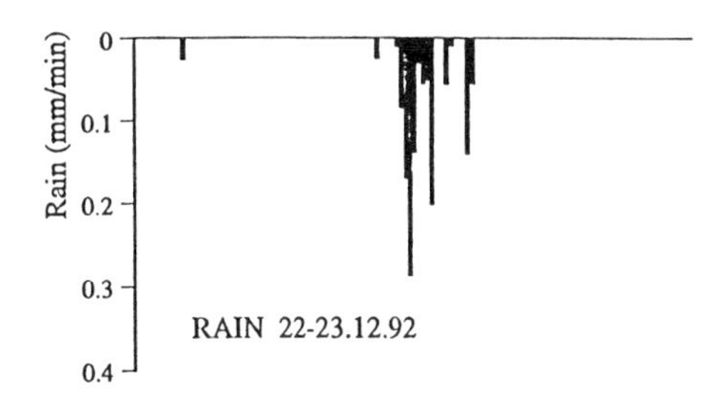

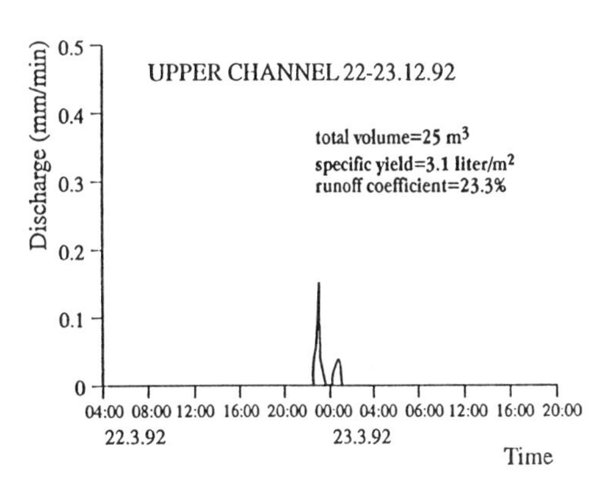

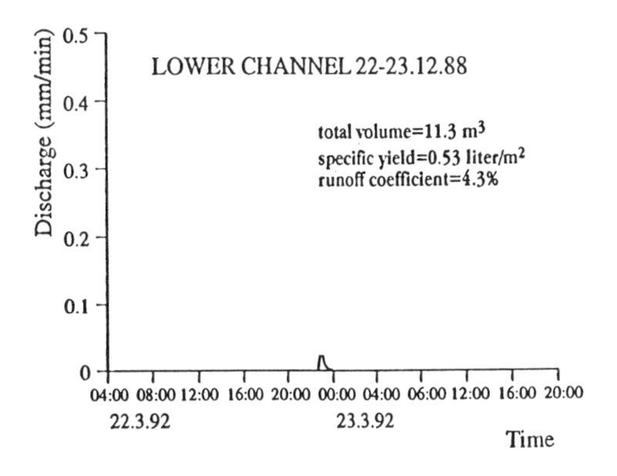

Figure 14.8. Storm of March 1992.

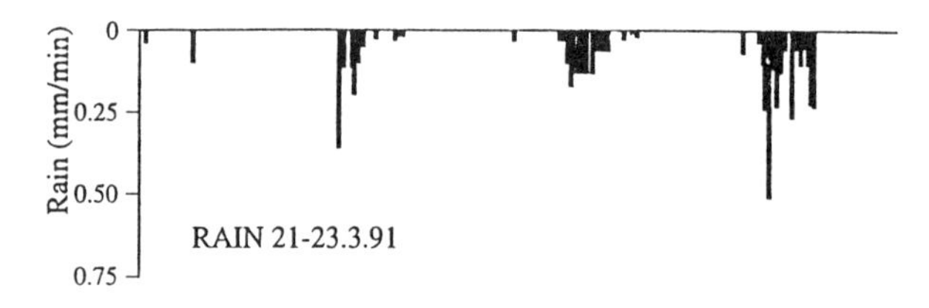

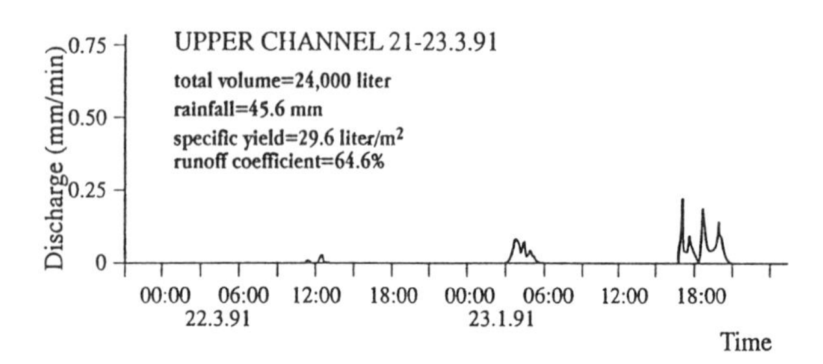

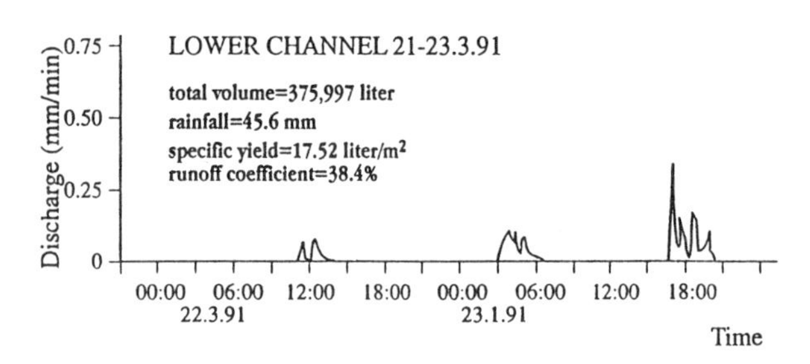

Figure 14.9. Storm of March 1991.

was still noticeable, being 17.6 l/m^2 for the lower channel and 29.6 l/m^2 for the upper one.

Summary of Hydrological Results

Data presented can be summarized as follows: rocky areas composed of a compact, dense limestone are characterized by a low infiltration rate. The threshold rain intensity for runoff generation is as low as 5 mm/hr for a short duration of three minutes only. Therefore, rocky areas respond quickly to rainfall and frequently generate a high runoff yield per unit area. At the same time, sediment-covered areas with their high porosity and high infiltration rate absorb all of the water of most storms, producing little or no runoff per unit area. As these latter areas extend always downslope or downstream of rocky areas, a frequent discontinuity in runoff is observed on passing from the rocky to the sediment-covered areas. Runoff generated over the rocky areas is usually absorbed within a narrow belt of the soil-covered unit. The process of surface flow discontinuity is greatly enhanced by the low intensity and intermittent character of the rain prevailing in the area.

The phenomenon of surface flow discontinuity leads to the concentration of water from a large contributing area into a small collecting area, resulting locally in wetting depth far beyond those expected from the limited direct rainfall. The deeply infiltrated waters are protected from evaporation and are, thus, available for salt leaching and building up a water reservoir available for plants. Three distinct locations of flow discontinuity and, hence, water concentration were identified at the Sede Boqer experimental site (fig. 14.10):

1. *The transition zone between the upper rocky and the lower colluvial slope:* Hydrological, pedological, and botanical studies indicate that the amount of water that infiltrates into the upper colluvium is usually 2–3 times higher than the amount derived from direct rainfall (Wieder et al. 1985, Yair and Danin 1980, Yair and Lavee 1985).

2. *The transition zone from the rocky headwater area into the alluvial reach:* Runoff data derived from the two hydrometric stations, one located at the transition from the rocky to the alluvial channel and the second 200 m downstream within the alluvial reach, indicate high infiltration losses between the two stations at most rainstorms. On the average, four flow events are recorded at the upper station as compared to only one event downstream. Flows did not occur at the lower station in the years 1992–94.

3. *The confluence of a small tributary with the main channel:* Quite often runoff from a small tributary infiltrates over a

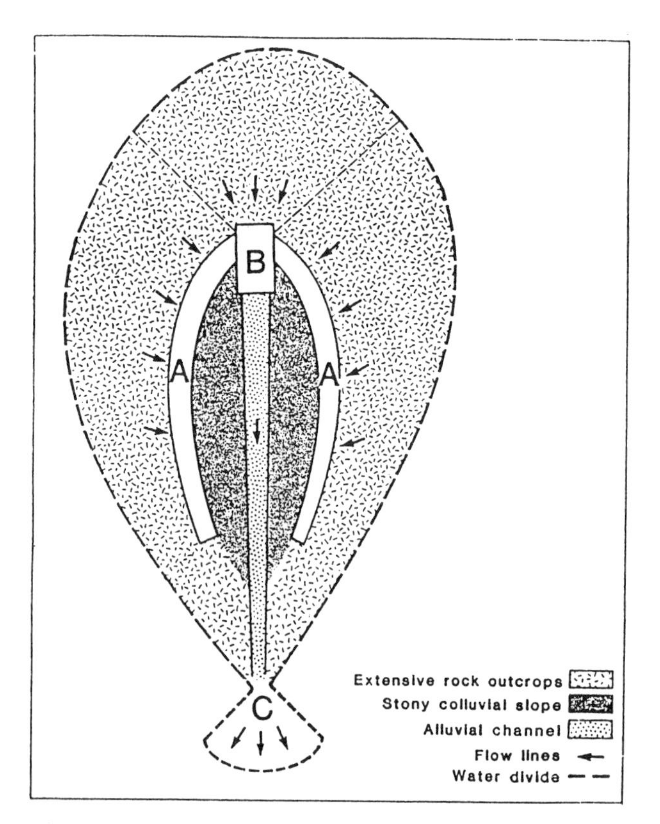

Figure 14.10. Areas of water concentration in a rocky arid watershed.

short distance at the confluence with the main channel, forming a sort of a water fan. It is worth noting that only three runoff events had been recorded in the main channel during the last seven years. Two of them occurred during the rainy year 1991–92, but no flow was observed on the following two years.

Watershed Hydrology: Implications for Water Harvesting

Location of Water Concentration Sites

The identification of the sites where runoff water concentration occurs under natural rainfall conditions led to the idea that the positive effect of water concentration could be further increased artificially. This would allow tree plantings over rocky hillslopes and rocky, narrow, small watersheds where trees never grew before. For this purpose, small earthen dams for runoff water collection were constructed at the following sites of frequent flow discontinuity:

1. *Interface rocky-colluvial slope sections:* Small furrows dug at this interface can collect a water volume of 1,500–2,000 liters (fig. 14.11). The ratio of contributing to collecting area is in the range of 80–100 to 1, much higher than the 20 to 1 ratio practiced by the ancient farmers. This is supposed to raise the amount of water harvested under natural conditions (250–300 mm) to a level exceeding 500 mm, which is assumed to be suitable for supporting trees adapted to arid conditions. The furrows were named *minicatchments* (Yair and Enzel 1987). One tree is planted in each of them.

2. *Interface rocky-alluvial channel:* A small earthen dam was built across the narrow channel at a small distance downstream of the rocky area (fig. 14.3). A group of 5–8 trees can be planted behind the earthen dam. Such collectors are named *limans* by the Forestry Service. The ratio of contributing to collecting area is of the order of 150:1.

3. *Limans at confluence with main channel:* In a similar way, earthen dams were constructed at the outlet of small watersheds. The area suitable for planting is slightly higher than at the upstream dam, allowing for the plantation of a small grove of 8–15 trees. The ratio of contributing to collecting areas is of the order of 200–230 to 1.

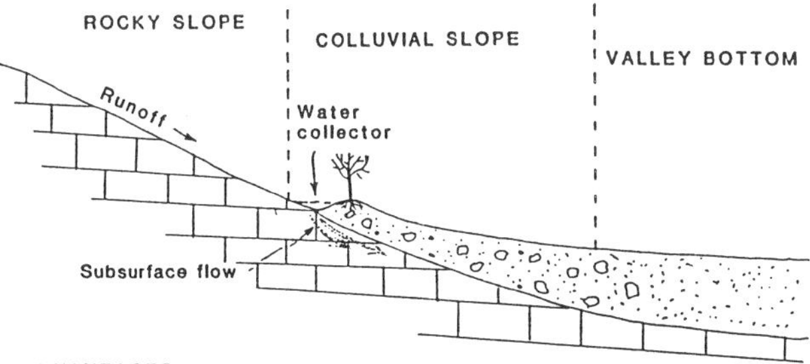

Figure 14.11. Location of a minicatchment.

Water Harvesting Efficiency at Water Concentration Sites

The period considered (1982–94) is typical of the high annual variability in rainfall and in runoff to be expected in an arid environment. Most years had rain amounts close to the long-term average. However, the last four years (1990–94) were extreme in opposite directions. The two consecutive years of 1991–92 were rainy years with large rainstorms and the highest runoff rates since 1972. But the two following years, 1993–94, were among the driest, with the lowest runoff rate for the same period. A summary of data collected is presented in table 14.2.

Table 14.2. Runoff Water Harvesting at Sites of Flow Discontinuity

Year	Annual Rainfall (mm)	Runoff in Cubic Meters		
		Minicatchment	Upper Channel	Lower Channel
1982	136.00	6.75		
1983	72.00	1.45		
1984	79.00	1.40		
1985	138.00	9.82		
1986	64.30	1.83		
1987	96.60	3.03		
1988	95.60	1.38	10.25	2.27
1989	73.30	0.97	6.25	1.22
1990	139.40	12.90	74.07	63.27
1991	163.90	10.20	70.03	38.74
1992	66.80	0.42	0.46	0.00
1993	63.20	0.56	0.46	0.00
Ave.	99.00	4.26	26.92	10.55

Note: A single tree is planted in a minicatchment, 5–6 trees at an upper liman, and 7–10 trees at a lower liman.

Minicatchment Site The average annual runoff volume collected 1982–92 by a minicatchment that drains an area of 250 m^2 was of the order of 5 m^3 per tree. This is 4–5 times more than the annual needs of a tree (Herwitz et al. 1988) and 10 times more than the water amount claimed by Evenari and colleagues (1971) for the ancient Nabatean installations. Nevertheless, a very high annual variability in runoff water input was recorded. During this decade, runoff volume collected was at least equal to and usually much higher than the trees' needs. On the last two dry years (1993–94), runoff collected by the minicatchments was the lowest on record, approximately one-half of the estimated water needs. Yet, so far, no damage could be observed. None of the trees planted before 1988 died. This is probably due to the water reserve accumulated during the better runoff years and/or an overestimation of tree water needs.

Upper Liman Site The record for the upper and lower limans is much shorter (table 14.2). It covers two very rainy years, two very dry years, and two intermediate years. Runoff water harvesting per tree is most efficient on normal and rainy years, far beyond the needs of the trees. This extremely high efficiency is due to the fact that runoff frequency and magnitude are the highest over the headwater area as well as to the fact that, at the interface rocky-alluvial channel, the area suitable for planting is relatively limited. Runoff collection per unit area is, therefore, high. On the last two dry years, runoff collection per tree was of the same order of magnitude as for the minicatchment.

Lower Liman Site Water harvesting efficiency, on average or more than average rainfall and runoff years, is slightly higher here than at the minicatchments. Flow frequency is always lower. However, a single medium to large magnitude flow per year compensates for the low runoff frequency. Water harvesting efficiency is always lower than at the upper liman. Runoff was not recorded at the lower channel station for the last two consecutive years. By the end of the first dry year, no signs of stress were observed. This first dry year followed two consecutive very wet years when a very large amount of water had accumulated at depth. However, signs of stress appeared in the summer of 1994, expressed as yellow leaves that were not seen before. The yellow color disappeared on the following rainy season.

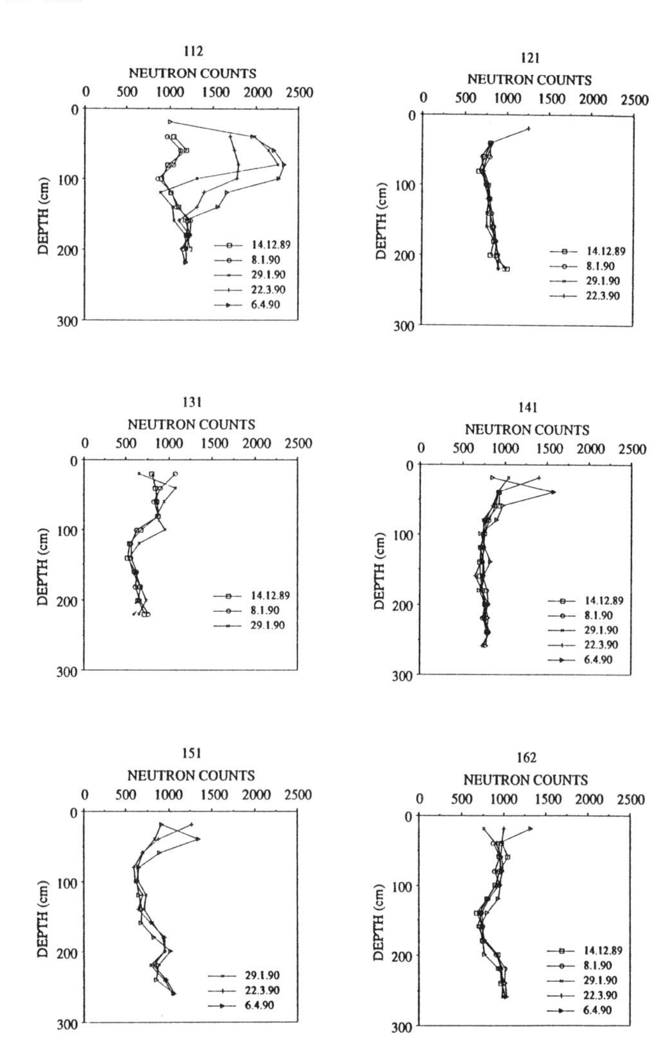

Figure 14.12. Moisture profiles along the colluvial slope (1989-90).

MOISTURE CONDITIONS

As indicated in the section on the experimental design, moisture measurements were taken with a neutron probe along a colluvial slope and along the channel of the instrumented watershed (fig. 14.3). Moisture data are presented in neutron count units as obtained in the field. Data presented here will cover a relatively dry year (1989-90), with 73 mm of annual rainfall and limited runoff, and a very wet year (1991-92), with 164 mm and high runoff.

Rainfall Year, 1989-90

Colluvial Slope Figure 14.12 displays changes in water content along the colluvium. An increase in water content is observed in all boreholes following the rainstorms at the end of January, beginning of February, and in April 1990. The highest vertical depth of water penetration, 160 cm, was obtained at borehole 112, where a minicatchment collects runoff from the rocky hillslope. In all other boreholes, the depth of water penetration was lower, in the range of 50-80 cm.

Alluvial Section Two boreholes were drilled before the beginning of the rainfall season. One borehole (323) is located at mid-distance between boreholes 324 and 322, and the second (borehole 321) just below the lower dam (fig. 14.2). A vertical infiltration of 120 cm was observed at borehole 324, located below the upper channel weir. Vertical infiltration in the three other boreholes is shallower, in the range of 50-70 cm (fig. 14.13).

Rainfall Year, 1991-92

Colluvial Slope Data obtained are presented in figure 14.14. This very rainy year, together with flow events that lasted for several consecutive hours (fig. 14.9), allowed a deep water penetration over most of the area. Depth of water penetration exceeded 200 cm along the whole of the colluvial section, with signs of subsurface flow at boreholes 122, 131, 132, and 133. Subsurface flow was noted at a depth of 150-180 cm.

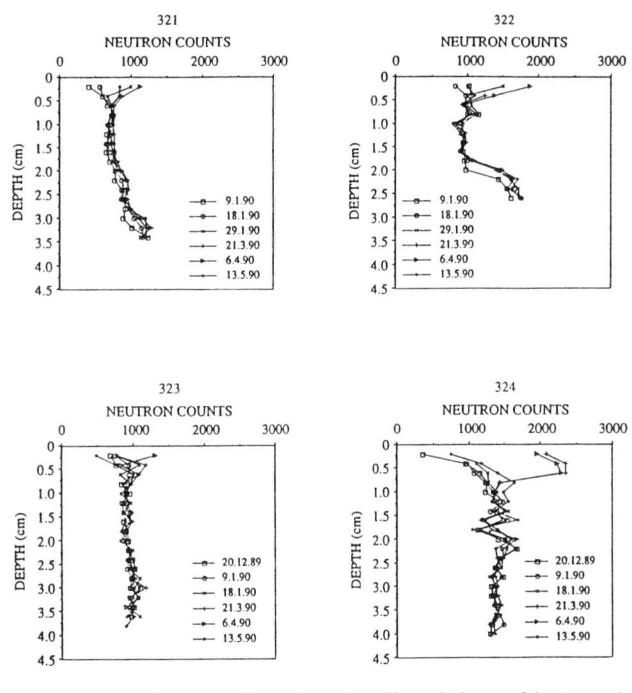

Figure 14.13. Moisture profiles along the alluvial channel (1989–90).

Alluvial Section Data obtained are presented in figure 14.15. Penetration depths in excess of 400 cm were recorded at the upper boreholes 324 and 323. Wetting depth and moisture content decreased slightly in the downstream direction. The lowest penetration was recorded at borehole 321, located below the lower dam. Signs of subsurface flow can be observed at a depth of 2 m at boreholes 324 and 321.

Salt Leaching

The positive effects of water harvesting at water concentration sites are also well expressed by the soil salinity profiles. The depth of water infiltration into the minicatchments and limans is on the order of 1.5 to more than 4 m. On very wet years (e.g., 1992), infiltration is deeper along the channel than at the top of the colluvium. This deep infiltration allows the creation of a water reservoir at depth available for tree growth during the long, hot summers. The deep infiltration also allows deep leaching of salts from the root zone (fig. 14.16). Exchange capacity (EC) down to 250 cm is on the order of 1 ms/cm, which is a very low figure in this arid environment. The great importance of the two consecutive rainy years (1991–92) is that, in a very short period of time, salts are leached down to a great depth, allowing the development of nonsaline conditions within the root zone in an arid environment.

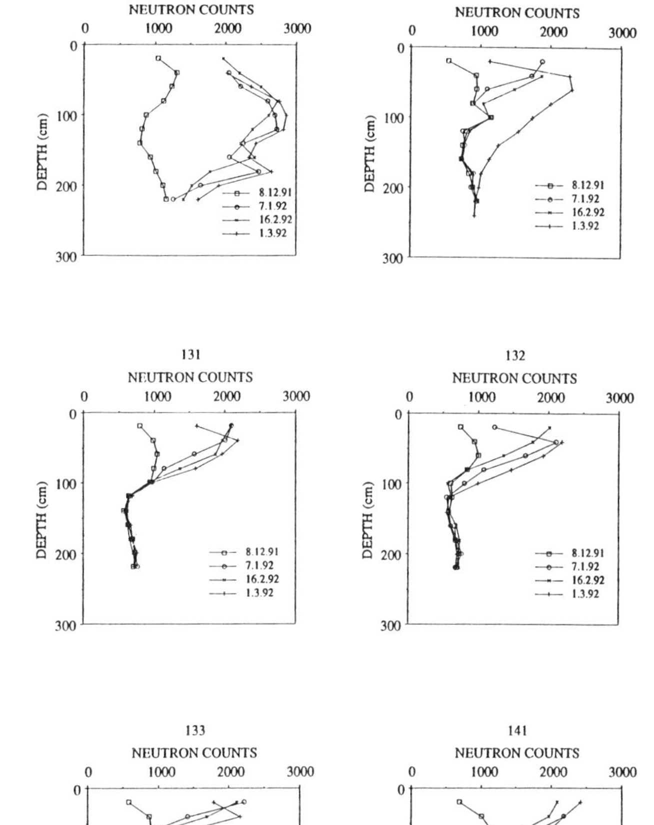

Figure 14.14. Moisture profiles along the colluvial slope (1991–92).

Conclusions

Data collected during the research period (1982–94) lead to the following main conclusions:

1. On ten out of the twelve years, runoff water harvesting at the water concentration sites identified was high enough to support the needs of the trees. Water harvesting efficiency is up to one order of magnitude higher than tree needs on years with high runoff. The deeply infiltrated waters allow a deep leaching of salts far beyond the root zone.

2. On dry rainfall and runoff years (1993–94), water harvesting efficiency is very low. Water harvesting is limited to the rocky-colluvial interface where the volume of water collected is approximately half of the estimated water needs. Trees at this site did not show any sign of stress. No runoff was collected at the lower channel station for two consecutive years, and initial signs of stress appeared on some of the trees. These signs disappeared one year later.

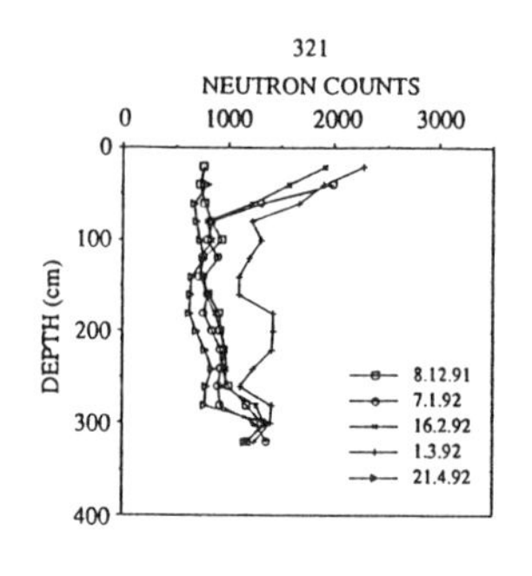

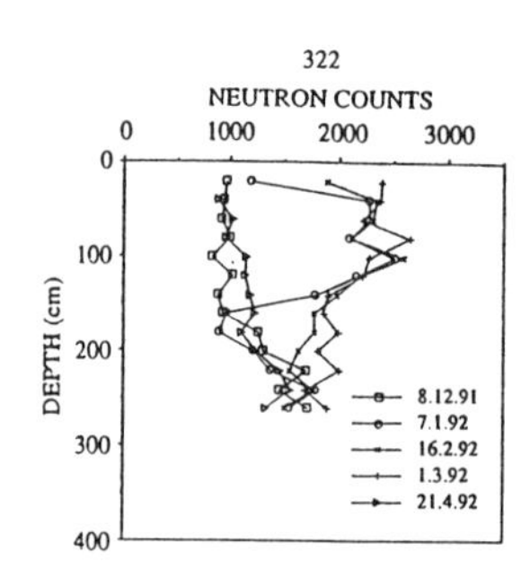

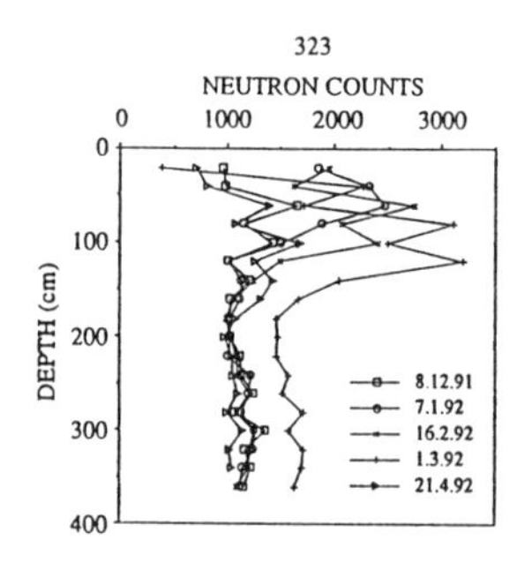

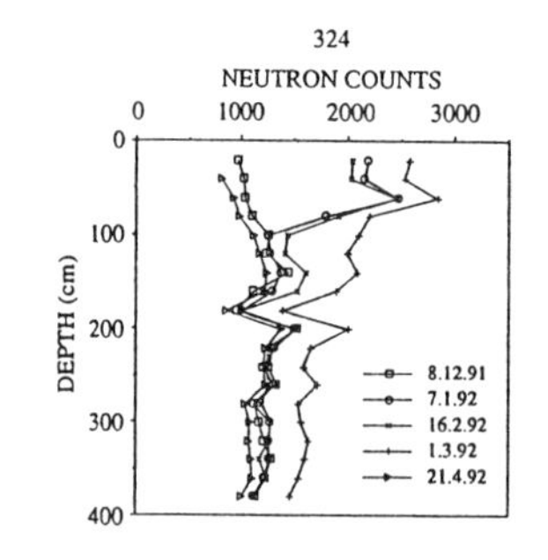

Figure 14.15. Moisture profiles along the alluvial channel (1991–92).

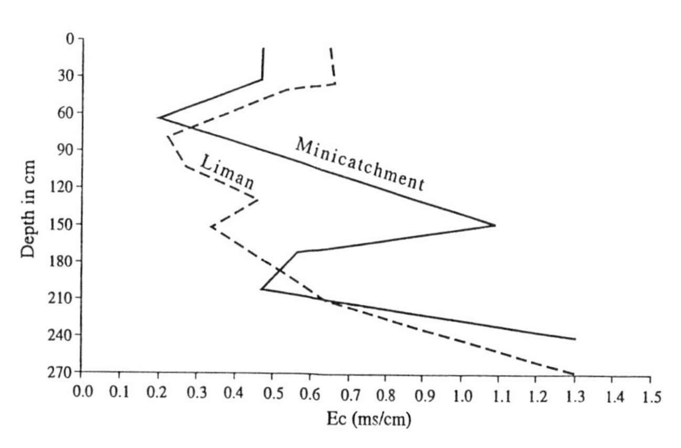

Figure 14.16. EC profiles (ms/cm) in a minicatchment and a liman (August 1992).

3. The two consecutive dry years highlight two points: (1) more than two dry years in a row might limit the sustainability of water harvesting practices at the water concentration sites identified and (2) sustainability might be different for the different sites. Data obtained show that the most favorable site is the headwater-alluvial interface. Next is the rocky-colluvial interface, and the least favorable is the lower alluvial reach where a first-order stream watershed enters into the main channel.

4. Finally, results obtained show that, in view of the high interannual variability in rainfall and runoff, the assessment of sustainable afforestation practices in such a dry environment must be based on a long-term approach.

Note

This study was conducted with the help of the Arid Ecosystems Research of the Hebrew University of Jerusalem. The center provided the technical support for the fieldwork. The study is also part of both a project on groundwater recharge in arid areas supported by the German-Israel Foundation and a project on afforestation in arid areas supported by the Alton Jons Foundation (U.S.). I am grateful to Mr. E. Sachs for his help with data collection and analysis at the Sede Boqer site and to Mss. Bloch-Altman and T. Soffer for drawing the illustrations.

References

Evenari, M., L. Shanan, and N. H. Tadmor. 1971. The Negev: The challenge of a desert. Cambridge, Mass.: Harvard University Press.

Herwitz, S. R., A. Yair, and M. Shachak. 1988. Water use patterns of introduced carob trees (*Ceratonia siliqua* L.) on rocky hillslopes in the Negev desert. Journal of Arid Environments 14:83–92.

Morin, J., and H. S. Jarosch. 1977. Rainfall-runoff analysis for bare soils. Pamphlet No. 164. Beit Dagan, Israel: Volcani Center, Division of Scientific Publications.

Shanan, L., and A. P. Schick. 1980. A hydrological model for the Negev desert highlands: Effects of infiltration, runoff, and ancient agriculture. Hydrological Science Bulletin 25(3):269–82.

Wieder, M., A. Yair, and A. Arzi. 1985. Catenary soil relationship on arid hillslopes. Catena Supplement 6:41–57.

Yair, A. 1983. Hillslope hydrology, water harvesting, and areal distribution of ancient agricultural fields in the northern Negev desert. Journal of Arid Environments 6:283–301.

———. 1992. The control of headwater area on channel runoff in a small arid watershed. *In* T. Parsons and A. Abrahams, eds. Overland flow. London: University College Press. 53–66.

Yair, A., and A. Danin. 1980. Spatial variations in vegetation as related to the soil moisture regime over an arid limestone hillside, northern Negev, Israel. Oecologia 47:83–88.

Yair, A., and Y. Enzel. 1987. The relationship between annual rainfall and sediment yield in arid and semi-arid areas: The case of the northern Negev. Catena Supplement 10:121–35.

Yair, A., and H. Lavee. 1985. Runoff generation in arid and semi-arid areas. *In* M. G. Anderson and T. P. Burt, eds. Hydrological forecasting. New York: John Wiley and Sons. 183–220.

Zangvill, A., and P. Druian. 1983. Meteorological data for Sede Boker. Desert Met. Papers, Series A. No. 8. Blaustein Institute for Desert Research.

15 Natural Conservation, Reclamation, and Livestock Grazing in the Northern Negev: Contradictory or Complementary Concepts?

Avi Perevolotsky

There are two apparent contradictions in the title of this chapter. Grazing by domestic livestock is often considered an activity hostile to nature conservation, and the ongoing conflict between ranchers and environmentalists in the western United States (Gillis 1991) is just one contemporary example of many cases all over the world, especially in semi-arid environments (Fleischner 1994). Conservation and development (in the sense of reclamation) also seem to be contradictory, since the latter concerns changing or manipulating the natural environments while the former tends to exclude or prevent it. This chapter will explore the relationships among these concepts under the conditions of some Old World semi-arid ecosystems. Under semi-arid conditions, grazing by domesticated animals is a major ecosystem component and, as such, should be perceived as an integral part of either conservation or development schemes. In much of the Old World, the environmental values that are the object of conservation or reclamation are the result of a long-term history of grazing and settlement with the associated human impact.

Nature conservation deals, by definition, with minimizing human manipulation of the ecosystem in order to conserve a desired situation. In this context, livestock grazing is viewed as a gross intervention that threatens the structure and function of natural ecosystems. Some argue that such activity may cause soil compaction and deformation, result in vegetation removal and modification, increase runoff and soil erosion, export minerals from the system, and contribute to climatic changes on both micro- and macrolevels (for a review of the relevant literature on herbivory impact on semi-arid ecosystems, see Joyce et al. chap. 7, this volume). It is not surprising, therefore, to find in a recent text on environmental issues that "plowing and pastoralism are responsible for many of our most serious environmental problems" (Goudie 1990).

One of the ecological similarities between the contradictory concepts of conservation and development, in particular at their primary establishment stage, is that both consider exclusion of livestock as an essential condition for their success. Consequently, such exclusion is one of the first actions to be taken when either a nature reserve is established or a reclamation project is initiated. In terms of conservation in arid zones, livestock exclusion is often the only active management action taken (bringing with it a reduction in firewood gathering and other activities that accompany a pastoralist lifestyle).

This chapter will discuss the impact and consequences of grazing exclusion from both conservation and reclamation practices. The notion of pastoral ecosystems (PES) will be presented, and its implications for any planned change—minimal, as in the case of conservation, or massive, as in some reclamation projects—will be discussed. PES are natural ecosystems exposed for thousands of years to grazing by domesticated livestock in numbers large enough to influence their structure and function. Grazing and associated forage removal may have been intermittent or continuous and generally varied between light and heavy, sometimes very heavy. Consequently, grazing became one of the driving forces that have determined the character of the ecosystem (e.g., replacing palatable perennial species with unpalatable ones). The fact that these systems are composed of natural components allows us to treat them as ecosystems, but the significant impact of pastoral activity on their nature entitles them to a somewhat contradictory name—pastoral ecosystems. PES are usually related to the Old World, referring mainly to the old civilizations of the Near Middle

East. The domestication of most common ruminants took place in this region 9,000–7,000 years ago, mostly during the Neolite (Stone Age) period. Consequently, some of the most ancient nomadic-pastoral societies evolved in this region.

Moreover, applying one of the criteria of sustainability—long-term independent persistence—will reveal that both the biological and human components of PES have survived for a long period of time and through severe environmental fluctuations, indicating high sustainability; however, this ambiguous term will finally be defined (Brklacich et al. 1991). Whether modern conservation or development programs under similar environmental conditions will be as sustainable is yet to be shown.

The following discussion focuses on prevailing misconceptions of the ecological relationships between livestock grazing and natural vegetative communities that have led to management decisions that have failed to meet their goals. Often the role of grazing in the sustainable management of natural reserves or developed sites in semi-arid landscapes has been ignored or denied. The present discussion draws primarily on observations and studies conducted in the northern semi-arid Negev desert of Israel—a typical Old World region. However, it is suggested that most of the conclusions are relevant to other traditional pastoral ecosystems.

Conservation and Development

While nature conservation as a sociocultural value has gained much public support in the last decades, its professional foundation is much weaker. There is, perhaps, a consensus that *conservation* means an attempt to preserve the *structure* (the native biological diversity) and *function* (the network of ecological processes) of a specific natural ecosystem, but the term *natural* remains vaguely defined (Warren 1993). A *natural state* is usually considered as representing an ancient ecosystem free of any human intervention, but such ecosystems are quite rare after the continuous, massive human activity of the past centuries. Peterken (1981, 1991) suggested that rehabilitation of damaged natural aspects (past naturalness) will be included among the conservation goals. However, the feasibility of achieving this ambitious goal has raised much doubt and concern in light of the substantial dynamic nature of environmental conditions, such as climatic changes, biological evolution, or drastic modifications imposed by previous human activities (Warren 1993). Livestock domestication and its continuous impact on rangelands over millennia is a good example of such a modification. In addition, the emerging view that considers ecosystems as "open," dynamic, and in a state of nonequilibrium (Pickett et al. 1992) makes operational conservation goals and criteria even more elusive.

Nature conservation is aimed at defending natural ecosystems against unnatural intervention and disturbance. *Development* is a more general term but, in any form, refers to changes induced in an attempt to achieve a designated goal. Some development efforts are more "natural," meaning that they concern human-made manipulation of the components of a natural unit while exploiting some of its natural resources, not necessarily according to ecological (ecosystem) guidelines. In this chapter, development is closely related to the concept of *reclamation,* defined as a "deliberate attempt to return a damaged ecosystem to some kind of productive use or socially accepted condition" (Jordan et al. 1988).

One example of a reclamation project is the ambitious and original Savannization Project recently initiated in the northern Negev desert in Israel. This project has attempted, and largely succeeded in extending the afforested area in southern Israel well into the semi-arid area (250–300 mm rainfall). Contour embankments have been built with heavy equipment to collect runoff from slopes covered with crusted loessic soil (aeolian sediments of silt). On the moist microhabitats behind the embankments, trees are planted to create an open forest. The area was previously grazed by nomadic or semi-nomadic Bedouin pastoralists.

The present study was motivated by observations on the Savannization Project. A promising start produced an efficient water collection and water storage system based on limited landscape intervention. Livestock, mostly Bedouin sheep and goats, were excluded from the project area prior to the initial development. Various tree species were planted after the physical infrastructure was completed. Establishment and primary development of the trees were successful. However, within 4–5 years, many water collection areas ceased to contribute runoff to the trees due to the accumulation of dense stands of herbaceous vegetation. Sheep were reintroduced, while scientists and managers discussed possible adequate, sustainable management schemes.

Pastoral Ecosystems

In the 1970s, it was common in ecology to define ecosystems on the basis of their principal driving force: a water-controlled ecosystem in arid climates (Noy-Meir 1973) or a fire-determined ecosystem in Mediterranean climates (Naveh 1974). Later, researchers realized that the complexity of most ecosystems requires a multivariate approach (Naveh 1984). However, many semi-arid ecosystems of the Old World are pastoral ecosystems. Livestock grazing over thousands of years has modified these systems so profoundly that such grazing must be regarded as one of their principal ecological driving forces. Livestock grazing is obviously not the sole influence on semi-arid ecosystems: variable and unpredictable precipitation regimes and a limited amount of nutrients in the soil also interact with grazing to determine the features of arid ecosystems (Noy-Meir 1985). Similarly, the structure and function of arid and semi-arid savanna ecosystems are determined by a combined effect of soil moisture, soil nutrients, fire, and herbivory (Walker 1987).

Most ecosystems in the Levant have evolved over a long period (about 7,000 years) under grazing by domesticated livestock and its associated human-made disturbances (Noy-Meir and Seligman 1979, Zohary 1983). In the Sinai and Negev regions, socioeconomic symbiotic-dimorphic relationships between desert pastoralism and semi-arid dry agriculture have created a stable subsistence model over thousands of years and over an extensive area (Finkelstein and Perevolotsky 1990). Consequently, many of the present ecological features (e.g., community structure, soil cover, nutrient status, and cycling) have evolved under the influence of grazing livestock.

Nothing in this argument is intended to deny the fact that the systems may well have been different prior to grazing by domestic livestock or that, if grazing had been excluded, another significant change would have occurred in the structure or function of the vegetative community. The literature is full of examples of substantial changes in the vegetation of rangelands following the exclusion or control of grazing (Ayyad et al. 1990, Floret 1981, Huntly 1991, McNaughton 1979, Noy-Meir 1990, Skarpe 1991, Vickery 1981). Pastoral ecosystems must have undergone substantial changes after the time of livestock domestication 7,000 years ago (Dayan et al. 1986) and especially during the development of the pastoral mode of subsistence. Zohary (1983) described the effect of the newly domesticated livestock on natural ecosystems in a rather vivid, though weakly supported, statement: "Thousands of species have disappeared from the scene before they could become known to science. . . . Enormous areas of grazed lands lost their palatable plants and became altogether deprived of their grazing potentials. These are now vegetated by plant communities dominated by anti-pastoral, stubborn and highly aggressive plants."

Statements such as: "Degradation of vegetation due to overgrazing is manifested in almost all deserts in the world. No areas are found where the present plant growth might be considered as the climax composition" (Batanouny 1983), are common in the professional literature and express the prevailing perception of the impact of the pastoral activity on the ecosystem.

Ecologically, the impact of a pastoral economy on dry pastoral ecosystems is an example covered by the state-and-transition model (Westoby et al. 1989). Under the premise of this model, rangeland vegetation can appear in a variety of discrete "states" while various "transitions" may drive it from one state to another. Transitions may be triggered by either natural or anthropogenic causes. In this sense, domestication and the initiation of pastoral activity may have caused a drastic transition from the previous "undisturbed" or less disturbed natural ecological state to a disturbed state with "an alternative [less 'natural'] stable vegetation state" (Westoby et al. 1989). It is most likely that the ecosystem-level changes imposed by the continuous disturbance (pastoral activity) drove the system over a threshold with little probability for reversibility, as proposed by the state-and-threshold model (Holling 1973) and adopted by range scientists (Friedel 1991, Laycock 1991) to describe grazing impacts on rangeland vegetation. In other words, a drastic (and most probably irreversible) change in the structure and function of the ecosystem accompanied the appearance of pastoral societies in much of the semi-arid and arid Old World. Such changes in vegetation formation, composition, and productivity as a result of human impact (grazing, fire prevention) have been documented also in the New World, where relatively heavy grazing by domestic livestock was introduced only 400 years ago to ecosystems exposed previously only to light wildlife herbivory.

In the northern Negev, small-ruminant pastoralism has been a dominant and organized mode of subsistence since the Pre-Pottery Neolithic B period (almost 8,000 years ago) (Avner et al. 1994, Finkelstein and Perevolotsky 1990) and continued, with minor modifications, until a few decades ago. A quantitative archaeological survey carried out in the southern-most arid part of the Negev revealed 1,400 ancient sites within 1,200 km², reflecting a continuous sequence of human activity from the eighth millennium B.C. to the modern era (Avner et al. 1994). The continuity of the impact within the context of a harsh environment probably affected the vegetation community. Moreover, until the domestication of the camel at the end of the second millennium B.C., pastoralism in the Levant was only semi-nomadic or enclosed nomadic and took place on a rather limited space because movement of humans and livestock was limited (Khazanov 1984, Rowten 1974). Consequently, the impact of grazing should have been stronger than during later periods in which seasonal migration became the common practice.

The drastic changes imposed on Bedouin pastoralism in the Negev during the second half of the twentieth century through their incorporation into a modern, capitalistic society made little difference to the livestock impact on the environment. Bedouin long-term migration was replaced by heavy grain supplementation during the winter. Exclusion from some traditional grazing territories that became military training areas or intensively cultivated necessitated the use of alternative pasture that, with water, is available now almost everywhere including stubble and other sources of readily available fodder. Access to veterinary care is easy and the market is usually good (Ginguld 1994), so animal numbers are maintained and even increased and grazing pressure on remaining rangeland is quite high.

It is impossible to reconstruct the ecological changes imposed by human activities in Old World PES, but in regions where such intervention is recent, the impact can still be documented. Vetaas (1993), for instance, reported on changes in the vegetation community in northeastern Sudan, resulting from the transition from transhumance and seasonal browsing to permanent settlements and year-long grazing during the last forty years. Many similar examples are known from New World rangelands in the western United States and in Australia.

The ecological scenario following the domestication of livestock and the development of pastoralism can be summarized as follows:

- Natural prepastoral vegetation formations and primary successional processes have long disappeared over extensive regions and beyond any possibility of reconstruction.
- A landscape well adapted to heavy grazing became dominant in the pastoral ecosystems.
- Heavy use over extended periods induced an adapted vegetation, stable and resilient, that had to be sustainable over a long time.

In traditional African pastoral groups, the livestock populations appeared to be in a nonequilibrium but persistent state (Ellis and Swift 1988), while the pastures they use seemed to be stable at a low equilibrium level, according to Noy-Meir (1975), with consequently low livestock productivity.

Semi-arid pastoral ecosystems can be abused. Heavy grazing pressure may cause undesirable changes and even irreversible damage to the ecosystem. However, many of the accusations concerning resource degradation in traditional pastoral systems have not been substantiated. Old World PES are resilient and well adapted to livestock grazing due to a long mutualistic evolution. This does not mean that they can sustain any degree of *disturbance* (here defined according to Grime [1979] as mechanisms that limit plant biomass by causing its partial or total destruction). The "redline" separating use and abuse in PES needs to be determined objectively. Since such guidelines are not yet available, it remains a challenge for the new generation of experts.

What Determines Structure and Function in Dry Pastoral Communities and Ecosystems?

The ecological interrelationships in semi-arid PES are complex and include feedback mechanisms that may operate in opposite directions (fig. 15.1 provides an overview of the most relevant factors). In many cases, ecological and managerial examination of rangeland conditions uses the immediate-proximate perspective. Moreover, it emphasizes (and sometimes overemphasizes) the impact of grazing on range productivity while ignoring other factors. Figure 15.1 presents, on the left side, the well-documented effects of grazing (livestock type and characteristic behavior, stocking rate, management) on plant survival and production. However, it emphasizes the importance of *future productivity* of the rangeland in its portrayal of overgrazing and degradation. If productivity returns to the pregrazing (or ungrazed) level within a short period (a few years) after grazing removal, then overgrazing or range deterioration cannot be substantiated. Such analysis should simultaneously account for the impact of climatic fluctuations, community relationships (mainly shrub-grasses), and ecosystem processes (nutrient availability, soil erosion, and soil moisture dynamics) on rangeland productivity in addition to that of the grazing regime.

However, such integration is not simple or straightforward. Walker and colleagues (1981), for example, could not determine whether continuous heavy grazing or climatic conditions are more significant in the evolution of savanna-like vegetation.

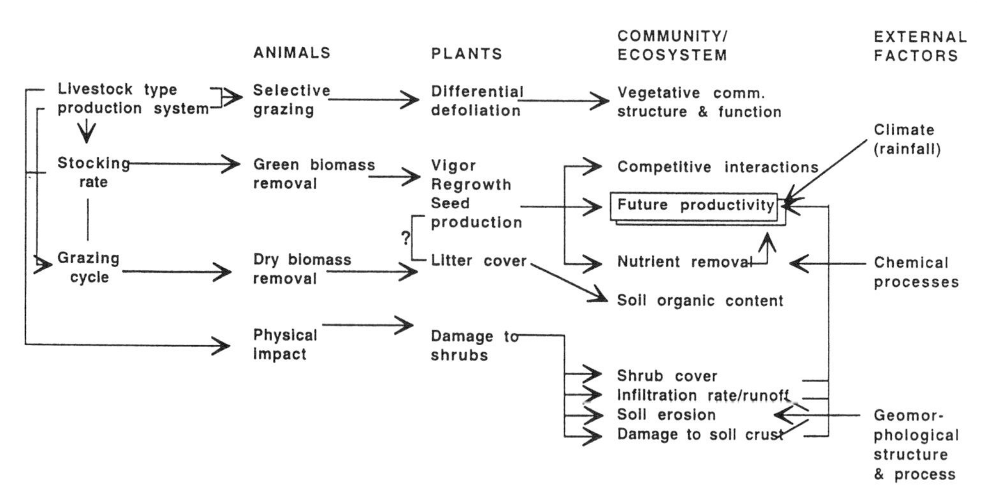

Figure 15.1. Relevant processes and interactions in the analysis of semi-arid GSD ecological status.

Climatic Effects

Climate, especially the specific rainfall regime, has an obvious effect on biomass and species composition of vegetative communities in dry ecosystems (Noy-Meir 1973). Walker (1988) suggested that, under such unpredictable conditions, different annual grasses have developed different adaptive strategies, each of which best fits a particular set of conditions. As a result, every year there will be a different group of species best adapted to the existing conditions that will become dominant. Evenari and Gutterman (1976) demonstrated how density, species composition, and dominance of annuals fluctuated greatly from year to year in the Negev. A similar phenomenon was reported for the arid steppe of Tunisia (Floret 1981).

A review of seventy-two long-term (more than five years) grazing experiments conducted in South Africa revealed that the effect of climate on productivity was by far more significant than that of the grazing treatments (O'Conner, cited in Walker 1988). Frost and Smith (1991) suggested that productivity on semi-arid rangeland is determined primarily by soil moisture that, in turn, is determined by site characteristics (e.g., local precipitation, topography, soil formation).

Data collected during the last six years at Lehavim, located in the northern Negev, support this contention (figs. 15.2a and 15.2b). Habitat is the most significant factor determining herbaceous productivity (the wadi produced 2–3 times more biomass than the south-facing slope), accounting for 64% of the variance in primary production. Grazing pressure accounts for only 6% (Ungar et al. 1995). Adding to the habitat and grazing pressure, the year effect (mostly in terms of rainfall amount and distribution) also helps to explain 82% of the variance.

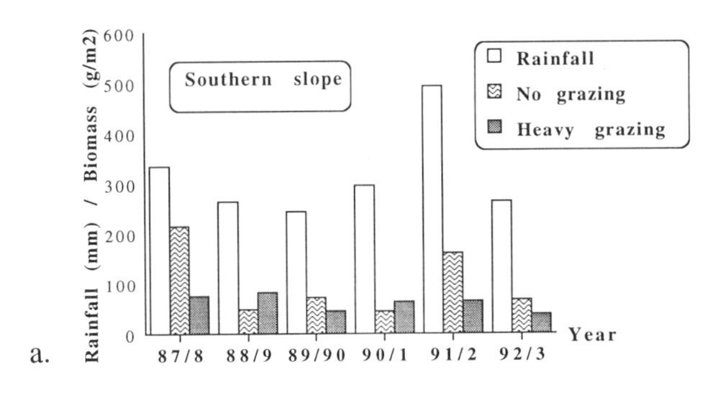

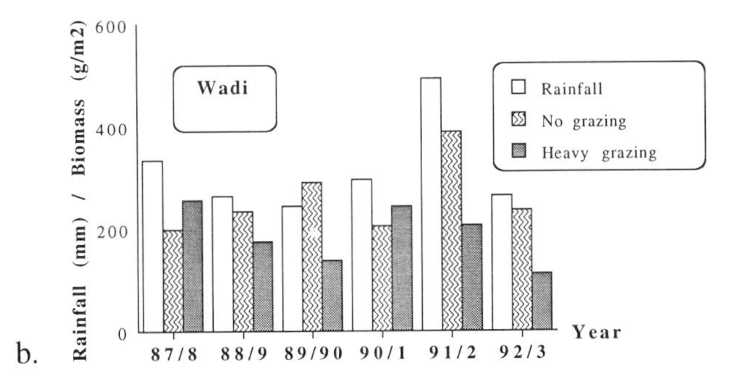

Figure 15.2. The effect of climatic conditions (by year), grazing, and habitat on rangeland productivity in the northern Negev.

Total amount of precipitation affected range production but not in a linear fashion (e.g., compare production on control plots in 1987–88 versus southern slope in 1991–92 in fig. 15.2). But the most significant observation is that heavy grazing did not have a negative effect on the production over the years. Production in each year was the outcome of the specific conditions (rainfall amount and distribution, solar radiation, and nutrient availability) and not of biological damage imposed by previous heavy grazing.

Nutrient Availability

The central role of moisture in determining the vegetative characteristics of dry ecosystems is obvious, but the realization that nutrients may be as important as water in determining productivity in these systems is not yet widely recognized (Ludwig 1987, West 1991). Breman and de Wit (1983) concluded that low soil fertility in the semi-arid Sahel contributes no less than limited precipitation to the low primary production. The fact that herbaceous vegetation ceases to grow while soil moisture is still around 20% of the absorbed amount indicates that low availability of minerals may be even more limiting than low water availability. Additional water may double biomass production in the Sahel, but fertilization increased it up to five times and improved pasture quality as well (Breman 1992).

In nutrient-poor environments, grazing is expected to have a smaller quantitative impact than nutrient deficiency (Oksanen and Oksanen 1989). In the northern Negev, nutrient amelioration with NPK fertilizer (50 kg/ha) increased pasture yield by 90–340% depending on the habitat (increase on southern slope > northern slope > wadi). In comparison, changes of 50–100 mm in the annual rainfall (out of an average of 270 mm) caused an increase of up to 100% in herbage biomass but rarely more than that. Moreover, the fertilizer application hastened early season development of the pasture, which attained 100 kg/dry matter/ha 6–7 weeks earlier than in unfertilized plots (Perevolotsky et al. 1990). Aside from the potential for range improvement, these results indicate that the growth and development of vegetation in the semi-arid range system is strongly controlled by nutrient availability. High nitrogen availability may also affect the successional dynamics and community structure of semi-arid shrublands (McLendon and Redente 1991).

Grazing Effects

There is ample literature on the effect of livestock grazing on the cover and composition of the vegetative community, especially under high stocking rates. In general, reports indicate that communities subjected to natural and anthropogenic disturbances are dominated by annual species and characterized by rapid and often irregular changes in species composition (Symonides 1988). Both the Mediterranean region in Israel and nearby desert environments—two classic PES—were found to contain very rich annual communities (Shmida 1981), a feature attributed to the continuous disturbance regime. Naveh and Whittaker (1980) proposed a model that combines grazing stress and evolutionary time (relating to the length of grazing history) to account for the very high values of species diversity recorded in northern Israel.

Noy-Meir (1990) reported on limited changes in production and species composition of annuals compared with a rapid change in production and biomass of most perennials following 6–7 years of cessation of grazing in the northern Negev. He related the former observation to climatic and edaphic conditions and the latter to grazing release. In the arid steppe of Tunisia, no change in species composition or successional stage was recorded after seven years of protection against grazing (Floret 1981).

Human-caused physical disturbance in dry habitats generated an increase in microscale species richness as well as in biomass production (Holzaphel et al. 1992). Olsvig-Whittaker and colleagues (1993) observed changes in the plant community composition in response to grazing and to soil texture. The effect of soil texture, in terms of species composition, was "at least as significant as the effects of a fairly extreme gradient of grazing intensity" (Olsvig-Whittaker et al. 1993). From the management and conservation points of view, their conclusion suggests that these communities have become very well adapted to heavy grazing intensity. The authors' expression of "surprise" reflects, most probably, the commonly held assumption that heavy livestock grazing is a proximate cause of community degradation.

The continuous effect of grazing on the following year's productivity has been studied extensively in productive systems where the perennial herbaceous species play an important role (Hodgkinson 1992). Naturally, in these systems, productivity is a function of previous grazing pressure that has affected plant vigor and physiological status as well as the seedbank. However, the picture is far from being so clear in dry systems where most of the herbaceous biomass consists of annual species or graminoids and short-cycle perennials that complete their life cycle very rapidly, an adaptation to the dry climate and short growing season (Shmida and Burgess 1988).

In California, heavy grazing caused a decline in productivity of annual vegetation in the following year, but this effect seemed to be very temporary (Pitt and Heady 1979). Harrington and Johns (1990) found large but temporary changes in herbaceous biomass during three years of livestock exclusion following six years of extremely heavy grazing in a semi-arid savanna woodland in Australia. Their conclusion was that, following the environmental impact of heavy grazing, the vegetation shifted from one stable state to another less productive but more stable state. PES, which have evolved over a much longer time, have most probably experienced the same ecological trend albeit thousands of years ago.

In summary, it is the interactions between autecological features of the dominant species that have evolved under long-term continuous grazing, fluctuating climate, and soil characteristics (including nutrients) that have determined the structure and function of the vegetative community in a semi-arid environment. Almost by definition, excluding grazing means changing the evolved natural state. Whether this change is desirable is obviously a value-loaded question.

A Scale-Dependent Perspective

Any account of the impact of livestock grazing (or its exclusion) on natural environments should be studied on a quantitative, long-term basis, adopting a flexible hierarchical-scale perspective (Hoekstra et al. 1991, O'Neill et al. 1986, Wiens 1989). The ecological role of shrubs in the northern Negev is a good example of such an approach. As in many of the dry PES, the vegetation here is composed principally of unpalatable shrubs and ephemeral annuals. There are mutual feedback relations between the two major vegetative components: shrubs provide a grazing refuge and thus favored conditions for the development of the herbaceous layer underneath its canopy (Noy-Meir 1985). In Lehavim, herbaceous biomass underneath the moderate-sized shrubs (*Thymelaea hirsuta*) was 50% higher compared to the biomass growing 50–100 cm outside of the shrub crown. Seeds produced by the protected annuals contribute to the grazed open patches, while the availability of abundant palatable green forage during the growing season releases the less palatable shrubs from high grazing pressure, allowing them to regrow. Belsky and colleagues (1989) and Jackson and colleagues (1990) have reported examples of similar interactions of savanna trees and shrubs with herbaceous vegetation.

One may consider the individual shrubs as having a positive feedback on the herbaceous component, especially under grazing conditions. However, when the observational scale is altered, different effects are detected. At the long-term research site near Lehavim, an increase of 30–50% in shrub cover was observed eight years after grazing pressure was reduced from very heavy to low-moderate and the area was fenced. This change was partly a result of the regeneration of relatively palatable perennial species, principally *Noaea mucronata*—one of the very few palatable shrubs in this community (Noy-Meir 1990, Perevolotsky and Landau 1988). The partial release of other unpalatable shrubs from continuous physical damage by the grazing livestock also helped to increase the shrub cover. The overall outcome has been a reduction of space available for the establishment and development of herbaceous vegetation and, hence, a significant decrease in area productivity. The decrease in shrub cover as a result of increasing grazing pressure is in contrast to many observations from various dry ecosystems of the world where heavy grazing has resulted in shrub encroachment (Skapere 1990, and many references cited therein). The reasons for these differences are beyond the scope of this chapter.

Ecosystem Management

This chapter deals with issues in ecosystem management. This complex issue has recently become a focus of much public and political attention, possibly as a prelude to the scientific solidification of the concept.

Perhaps the most popular current concept in ecosystem management is that of *desired future conditions* (DFC). It stems from the realization that simplistic models (e.g., succession based) have not brought us to the promised land of "stable ecosystem" and, most certainly, are not capable of doing so. It seems that, in the wake of massive environmental changes that have occurred in most ecosystems, any management benchmark based on a hypothetical state is irrelevant. The DFC idea is radically different from any previous management model that pretended to be based on scientific objectivity by the fact that it is, by definition, a value-laden concept. Society, or the public, determines (through professional agencies? by democratic means?) the nature and character of a future desired ecosystem (Bonnicksen and Stone 1985). Based on their understanding of the existing ecological conditions, ecologists can propose a method for attaining the desired future state. Then professional managers can pursue this goal with ecologically sound procedures (Kessler et al. 1992).

Along these lines, the concept of *desired plant community* (DPC) has been developed within the context of range management (TGU 1991), and it is within this context that it makes sense. Once the goal is defined as an objectively quantified index, such as maximizing livestock production or performance, the DPC can be customized and, in certain cases, can also be attained through management (Borman and Pyke 1994). This concept also plays a role in succession-based management models. Here, obviously, the DPC will follow the composition of the desired successional stage, usually the climax, that traditionally has served as a management goal (Smith 1989). In this case, DPC becomes more of a theoretical and, in many instances, illusory criterion rather than a feasible management target, either due to partial understanding of the natural succession process or to a lack of knowledge on how exactly to achieve it through management.

It is suggested that in PES the guideline for management, in terms of species composition, is the one that has evolved over a long period of time under heavy grazing. Any attempt to use a succession-based model or to assume a self-return to predisturbance community conditions has no substantiated grounds and is, therefore, misleading. Hence, the reconstruction of the "pristine" ecosystem (prior to domestic grazing) is definitely not a feasible conservation goal. One should note that, in the dry Mediterranean region, botanists still argue about whether the climax formation was a forest dominated by pine and oak trees—as can still be observed at favorable sites—or a woodland/scrubland dominated by tall evergreen shrubs.

The prevailing conservation approach considers grazing by domesticated animals as destructively negative and says it should be excluded from natural reserves (and beyond; on the campaign against the black goat, see Perevolotsky 1991b) in PES and elsewhere. Hidden in this approach is the belief that, once grazing is excluded, the "natural," "real," "good," "adequate," "healthy" vegetation will return and take over. There is very limited information on long-term changes in the community/ecosystem level following grazing exclusion in PES and, as a result, no scientific discussion on whether such changes are really desired. Moreover, no one can propose, based on existing scientific knowledge, what the desired vegetative community structure on PES is or should be. At best, the vague concept of managing for high species diversity is evoked. West (1993) provided an extensive review of the concept, including its weaknesses, within the context of range management, while Noss (1992) raised significant questions concerning its applicability to nature conservation.

It remains to be seen whether livestock grazing will be reintroduced to nature reserves in southern Israel after researchers have established that grazing increases species diversity (Olsvig-Whittaker et al. 1993). Until a clear set of ecologically based management guidelines is available for nature conservation in PES, development is bound to commit drastic errors even if the intention is to avoid them.

In Israel (as well as in many neighboring countries), grazing was excluded from natural reserves in the Mediterranean region by a law that was effectively enforced. The recovery of the woody vegetation was rapid and extensive, but it also created new headaches (Perevolotsky 1991b):

- development of dense scrubland with declining species diversity, especially of the herbaceous and low growth forms;
- decrease in forage potential for livestock or wildlife; and
- a dramatic increase in fire hazard due to accumulation of dry woody biomass.

It took a few decades to realize that the ecological changes in the Mediterranean ecosystem are not necessarily all in the desired direction. Recently, a new management policy has been implemented in which grazing, and sometimes heavy grazing, plays a central role in creating biological firebreaks. It has become evident that undergrazing rather than overgrazing is a more serious ecological threat to Mediterranean ecosystems (Seligman and Perevolotsky 1994). Naveh (1984) claimed that the conservation of the Mediterranean landscape and its variety can be ensured only by a continuation and/or simulation of the traditional agropastoral functions under which this landscape evolved. Naveh and Whittaker (1980) claimed that, in systems with a long history of pastoralism (PES), grazing is required to maintain species diversity. We recommend extending these conclusions to dry PES as well.

The continuation of traditional management (grazing, coppicing, hunting) as an active conservation practice was proposed in the 1960s (Ovington 1964) and is still advocated (Peterken 1991), not necessarily with less controversy or more support (Perevolotsky 1994). In light of the above discussion, the active role of livestock grazing in the conservation management of Old World PES should be given earnest consideration.

Long-term and large-scale studies of PES are required in order to uncover the mechanisms and rules governing the relationship between grazing animals and dry environments. For the unexpected results of livestock exclusion in the semi-arid northern Negev, see Noy-Meir (1990). It is impossible to reconstruct the past (predomestication) in these systems, but we can take advantage of the conservation and reclamation projects and establish ongoing research programs. The concept of learning from ongoing management experience (that is, using adaptive management [Walters 1986]) may be very helpful in both conservation and reclamation in PES. In this sense, Rosenzweig's (1987) suggestion to use restoration projects in a large-scale perturbation context to study the organization and operation of natural ecosystems should be extended to include, in PES, the perturbation itself—livestock grazing.

Conclusions

Some interesting conclusions that seem to contradict in part the conventional wisdom can be drawn from the above discussion:

1. Conservation in Old World PES should relate, at least on some spatial scale, to maintaining the grazing factor as an essential objective since it is the "most natural" situation historically. This implies keeping the prevailing grazing regime or an ecological equivalent.

2. Grazing and even heavy temporal grazing that has been practiced traditionally has not induced a significant degradation in PES. These systems appear to be well adapted to such land use.

3. Claims of overgrazing and resource degradation under the above conditions are a priori suspect and should be substantiated scientifically rather than manipulated politically (Perevolotsky 1991a).

4. Environmental factors of PES, such as soil moisture and nutrient availability, seem to be more significant in determining vegetative community features (species composition, primary production) than grazing impact.

5. Detailed long-term studies of ecological processes following grazing exclusion should be conducted to determine the outcome of a no-grazing conservation policy. Decisions should be based on these findings and not on the belief that, once livestock is removed, "nature will take care of itself."

6. Development and conservation in PES should take into consideration the role of grazing in molding the existing natural system and provide management solutions for the outcome of livestock exclusion.†

Note

This chapter was written during a sabbatical leave spent at Utah State University. The hospitality of the Range Science Department, the interactions with various faculty members of that department, and a set of enlightening seminars in ecology organized by the College of Natural Resources at USU all helped me considerably in transforming vague ideas into tangible arguments. Financial support by the Jewish National Fund and the Range Management Advisory Board of Israel and a Quinney Fellowship is acknowledged.

The chapter has benefited greatly from comments made by Greg Perrier and Neil West. I give special thanks to No'am Seligman for comments and style improvements. The assistance of R. Yonatan, S. Talker, and E. Zaddy in collecting field data and of Hagit Baram and Eugene Ungar in the data analysis is gratefully acknowledged.

References

Avner, U., I. Carmi, and D. Segal. 1994. Neolithic to Bronze Age settlement of the Negev and Sinai in light of radiocarbon dating: A view from the southern Negev. Radiocarbon 36:265–300.

Ayyad, M. A., R. El-Ghareeb, and M. S. Gaballah. 1990. Effect of protection on the phenology and primary production of some common annuals in the western coastal desert of Egypt. Journal of Arid Environments 18:295–300.

Batanouny, K. H. 1983. Human impact on desert vegetation. *In* W. Holzner, M. J. A. Werger, and I. Ikusima, eds. Man's impact on vegetation. The Hague: Dr. W. Junk Publishers. 139–50.

Belsky, A. J., R. G. Amundson, J. M. Duxbury, S. J. Riha, A. R. Ali, and S. M. Nwonga. 1989. The effects of trees on their physical, chemical, and biological environments in semi-arid savanna in Kenya. Journal of Applied Ecology 26:1005–24.

Bonnicksen, T. M., and E. C. Stone. 1985. Restoring naturalness to national parks. Environmental Management 6:109–22.

Borman, M. M., and D. A. Pyke. 1994. Successional theory and the desired plant community approach. Rangeland 16:82–84.

Breman, H. 1992. Desertification control, the West African case: Prevention is better than cure. Biotropica 24:328–34.

Breman, H., and C. T. de Wit. 1983. Rangeland productivity and exploitation in the Sahel. Science 221:1341–47.
Brklacich, M., C. R. Bryant, and B. Smit. 1991. Review and appraisal of the concept of sustainable food production systems. Environmental Management 15:1–14.
Dayan, T., E. Tchernov, O. Bar-Yosef, and Y. Yom-Tov. 1986. Animal exploitation in Ujrat El-Mehed, a neolithic site in southern Sinai. Paléorient 12:105–16.
Ellis, J. E., and D. M. Swift. 1988. Stability of African pastoral ecosystems: Alternate paradigms and implications for development. Journal of Range Management 41:450–59.
Evenari, M., and Y. Gutterman. 1976. Observations on the secondary succession of three plant communities in the Negev desert, Israel. *In* R. Jacques, ed. Etudes de biologie vegetale. Gif-sur-Yvette: CHRS.
Finkelstein, I., and A. Perevolotsky. 1990. Process of sedentarization and nomadization in the history of Sinai and the Negev. Bulletin of the American Schools of Oriental Research 279:67–88.
Fleischner, T. L. 1994. Ecological costs of livestock in western North America. Conservation Biology 8:629–44.
Floret, C. 1981. The effects of protection on steppic vegetation in the Mediterranean arid zone of southern Tunisia. Vegetatio 46:117–29.
Friedel, M. H. 1991. Range condition assessment and the concept of thresholds: A viewpoint. Journal of Range Management 44:422–26.
Frost, W. E., and E. L. Smith. 1991. Biomass productivity and range condition on range sites in southern Arizona. Journal of Range Management 44:64–68.
Gillis, A. M. 1991. Should cows chew cheatgrass on commonlands? BioScience 41:668–75.
Ginguld, M. 1994. Managing herds and households: Management practices and livelihood strategies of sheep-owning Bedouin households in the Negev region of Israel. Master's thesis. The Hague: Institute of Social Studies.
Goudie, A. 1990. The human impact on the natural environment. Cambridge, Mass.: MIT Press.
Grime, J. P. 1979. Plant strategies and vegetation processes. Chichester: John Wiley and Sons.
Harrington, G. N., and G. G. Johns. 1990. Herbaceous biomass in a eucalyptus savanna woodland after removing trees and/or shrubs. Journal of Applied Ecology 27:775–87.
Hodgkinson, K. C. 1992. Elements of grazing strategies for perennial grass management in rangelands. *In* G. P. Chapman, ed. Desertified grasslands: Their biology and management. London: Academic Press. 77–94.
Hoekstra, T. W., T. F. H. Allen, and C. H. Flather. 1991. Implicit scaling in ecological research. BioScience 41:148–54.
Holling, C. S. 1973. Resilience and stability of ecological systems. Annual Review of Ecology and Systematics 4:1–23.
Holzaphel, C., W. Schmidt, and A. Shmida. 1992. Effects of human-caused disturbance on the flora along a Mediterranean-desert gradient. Flora 186:261–70.
Huntly, N. 1991. Herbivores and the dynamics of communities and ecosystems. Annual Review of Ecology and Systematics 22:477–503.
Jackson, L. E., R. B. Strauss, M. K. Firestone, and J. W. Bartolome. 1990. Influence of tree canopies on grassland productivity and nitrogen dynamics in deciduous oak savanna. Agriculture, Ecosystems, and Environment 32:89–105.
Jordan, W. R., R. L. Peters, and E. B. Allen. 1988. Ecological restoration as a strategy for conserving biological diversity. Environmental Management 12:55–72.
Kessler, W. B., H. Salwasser, C. W. Cartwright, and J. A. Caplan. 1992. New perspectives for sustainable natural resources management. Ecological Applications 2:221–25.
Khazanov, A. M. 1984. Nomads and the outside world. Cambridge: Cambridge University Press.
Laycock, W. A. 1991. Stable states and thresholds of range condition on North American rangelands: A viewpoint. Journal of Range Management 44:427–33.
Ludwig, J. A. 1987. Primary productivity in arid lands: Myth and realities. Journal of Arid Environments 13:1–7.
McLendon, T., and E. F. Redente. 1991. Nitrogen and phosphorus effects on secondary succession dynamics on a semi-arid sagebrush site. Ecology 72:2016–24.
McNaughton, S. J. 1979. Grassland-herbivore dynamics. *In* A. R. E. Sinclair and M. Norton-Griffiths, eds. Serengeti: Dynamics of an ecosystem. Chicago: University of Chicago Press.
Naveh, Z. 1974. Effects of fire in the Mediterranean region. *In* T. T. Kozlowski and C. E. Ahlgren, eds. Fire and ecosystems. New York: Academic Press.
———. 1984. Mediterranean landscape evolution and degradation as multivariate biofunctions: Theoretical and practical implications. Landscape Ecology 9:125–46.
Naveh, Z., and R. H. Whittaker. 1980. Structural and floristic diversity of shrublands and woodlands in northern Israel and other Mediterranean areas. Vegetatio 41:171–90.
Noss, F. R. 1992. Issues of scale in conservation biology. *In* P. L. Fiedler and K. S. Jain, eds. Conservation biology: The theory and practice of nature conservation, preservation, and management. New York: Chapman and Hall. 250–57.
Noy-Meir, I. 1973. Desert ecosystems: Environment and producers. Annual Review of Ecology and Systematics 4:25–51.
———. 1975. Stability of grazing systems: An application of predator-prey graphs. Journal of Ecology 65:459–81.
———. 1985. Desert ecosystem structure and function. *In* M. Evenari, I. Noy-Meir, and D. W. Goodall, eds. Ecosystems of the world. Vol. 12A: Hot deserts and arid shrublands. Amsterdam: Elsevier.
———. 1990. Responses of two semiarid rangeland communities to protection from grazing. Israel Journal of Botany 39:431–42.
Noy-Meir, I., and N. G. Seligman. 1979. Management of semi-arid ecosystems in Israel. *In* B. H. Walker, ed. Management of semi-arid ecosystems. Amsterdam: Elsevier. 113–59.
Oksanen, L., and T. Oksanen. 1989. Natural grazing as a factor shaping our barren landscapes. Journal of Arid Environments 17:219–33.
Olsvig-Whittaker, L. S., P. E. Holsten, I. Marcus, and E. Shochat. 1993. Influence of grazing on sand field vegetation in the Negev desert. Journal of Arid Environments 24:81–93.
O'Neill, R. V., D. DeAngelis, J. B. Waide, and T. F. H. Allen. 1986. A hierarchical concept of ecosystems. Princeton, N.J.: Princeton University Press.
Ovington, J. D. 1964. The ecological basis of the management of woodland nature reserves in Great Britain. Journal of Ecology 52(suppl.):29–37.
Perevolotsky, A. 1991a. Goats and scapegoats: The overgrazing controversy in Piura, Peru. Small Ruminant Research 6:199–215.
———. 1991b. Rehabilitation of the black goat. Hassadeh 71:619–22. [Hebrew, with English summary]
———. 1994. Nature preservation or landscape conservation: Theory and practice in the management of nature reserves in the 2000s. Ecology and Environment 3. [Hebrew, with English summary]
Perevolotsky, A., and S. Landau. 1988. Improving and developing sheep production among the northern Negev Bedouin: A scientific report on the activities in the Bedouin demonstration farm (1982–1988). Bet Dagan: ARO.
Perevolotsky, A., N. Seligman, R. Yonathan, and S. Talker. 1990. An ecological analysis of fertilization effects on the yield of natural

pasture in the northern Negev. Hassadeh 70:957–59. [Hebrew, with English summary]
Peterken, G. F. 1981. Woodland conservation and management. London: Chapman and Hall.
———. 1991. Ecological issues in the management of woodland nature reserves. *In* I. F. Spellerberg, F. B. Goldsmith, and M. G. Morris, eds. The scientific management of temperate communities for conservation. Oxford: Basil Blackwell. 245–72.
Pickett, S. T. A., T. V. Parker, and L. P. Fiedler. 1992. The new paradigm in ecology: Implications for conservation biology above the species level. *In* P. L. Fiedler and K. S. Jain, eds. Conservation biology: The theory and practice of nature conservation, preservation, and management. New York: Chapman and Hall. 85–110.
Pitt, M. D., and H. F. Heady. 1979. The effects of grazing intensity on annual vegetation. Journal of Range Management 32:109–14.
Rosenzweig, M. L. 1987. Restoration ecology: A tool to study population interactions. *In* W. R. Jordan, M. E. Gilpin, and J. D. Aber, eds. Cambridge restoration ecology: A synthetic approach to ecological research. Cambridge: Cambridge University Press.
Rowten, M. B. 1974. Enclosed nomadism. Journal of the Economic and Social History of the Orient 17:1–30.
Seligman, N. G., and A. Perevolotsky. 1994. Has intensive grazing by domestic livestock degraded the Old World Mediterranean rangelands? *In* M. Arianoutsou and R. H. Groves, eds. Plant-animal interactions in Mediterranean-type ecosystems. Dordrecht, The Netherlands: Kluwer Academic Publishers. 93–103.
Shmida, A. 1981. Mediterranean vegetation in California and Israel: Similarities and differences. Israel Journal of Botany 30:105–23.
Shmida, A., and T. L. Burgess. 1988. Plant growth-form strategies and vegetation in arid environments. *In* M. J. A. Werger, P. J. M. van der Art, H. J. During, and J. T. A. Verhoeven, eds. Plant form and vegetation structure. The Hague: Dr. W. Junk Publishers.
Skapere, C. 1990. Shrub layer dynamics under different herbivore densities in an arid savanna, Botswana. Journal of Applied Ecology 27:873–85.
———. 1991. Impact of grazing in savanna ecosystems. Ambio 20:351–56.
Smith, L. E. 1989. Range condition and secondary succession: A critique. *In* W. K. Lauenroth and W. A. Laycock, eds. Secondary succession and the evaluation of rangeland condition. Boulder, Colo.: Westview Press.
Symonides, E. 1988. On the ecology and evolution of annual plants in disturbed environments. Vegetatio 77:21–31.
Task Group on Unity in Concepts and Terminology (TGU). 1991. New direction in range condition assessment. Report to the board of directors, Society of Range Management. North Platte, Nebr.
Ungar, E. D., A. Perevolotsky, N. G. Seligman, R. D. Yonatan, D. Barkai, and Y. Hafetz. 1995. Principal factors determining primary production on hilly semi-arid rangelands in the Negev desert, Israel. Proceedings of the fifth International Rangeland Congress, Salt Lake City, Utah, August 23–28.
Vetaas, O. R. 1993. Spatial and temporal vegetation changes along a moisture gradient in northeastern Sudan. Biotropica 25:164–75.
Vickery, P. J. 1981. Pasture growth under grazing. *In* F. H. W. Morely, ed. Grazing animals. Amsterdam: Elsvier.
Walker, B. H. 1987. A general model of savanna structure and function. *In* B. H. Walker, ed. Determinants of tropical savannas. Oxford: IRL Press.
———. 1988. Autecology, synecology, climate, and livestock as agents of rangeland dynamics. Australian Rangelands Journal 10:69–75.
Walker, B. H., D. Ludwig, C. S. Holling, and R. M. Peterman. 1981. Stability of semi-arid savanna grazing systems. Journal of Ecology 69:473–98.
Walters, C. J. 1986. Adaptive management of natural resources. New York: McGraw-Hill.
Warren, A. 1993. Naturalness: A geomorphological approach. *In* F. B. Goldsmith and A. Warren, eds. Conservation in progress. Chichester: John Wiley and Sons.
West, N. 1991. Nutrient cycling in soils of semi-arid and arid regions. *In* J. Skujins, ed. Semiarid lands and deserts: Soil resources and reclamation. New York: Marcel Dekker.
———. 1993. Biodiversity of rangelands. Journal of Range Management 46:2–13.
Westoby, M., B. Walker, and I. Noy-Meir. 1989. Opportunistic management for rangelands not at equilibrium. Journal of Range Management 42:266–74.
Wiens, J. A. 1989. Spatial scaling in ecology. Functional Ecology 3:385–97.
Zohary, M. 1983. Man and vegetation in the Middle East. *In* W. Holzner, M. J. A. Werger, and I. Ikusima, eds. Man's impact on vegetation. The Hague: Dr. W. Junk Publishers. 287–96.

16 Managing Rural Dieback of Eucalypts to Achieve Sustainable Dryland Agroecosystems

Roger Farrow

The earth's drylands have been affected by human activities for a longer period than any other ecoclimatic zone. Pastoralism and cropping have been recorded for over 11,000 years in the Middle East (Naveh and Dan 1973) and were preceded by a long period under the influence of hunter-gatherers using fire. Consequently, dryland ecosystems essentially function as agricultural ecosystems (de Vos 1975). The cumulative effects of overgrazing the pasture and shrub resources of drylands have caused soils to degrade, often irreversibly, through erosion and other processes (de Vos 1975, Roberts 1981, Williams et al. 1993). This, in turn, has changed the composition of the vegetation and fauna, also often irreversibly (Bourlière and Hadley 1970, de Vos 1975, Hobbs and Hopkins 1990, Naveh 1967). The destruction and loss of the native megafauna and its replacement by domesticated megaherbivores to exploit the plant biomass has also occurred as humans have manipulated the system to maximize productivity. This has adversely affected the ecosystem processes and biological diversities of many drylands (de Vos 1975, Hobbs and Hopkins 1990). The conversion of rangeland to cropping in areas often marginal for sustained yields adds to the degradation, largely as a result of erosion, loss of soil nutrients, and introduction of weeds, as shown in Australia by Williams and colleagues (1993) and Hobbs and Hopkins (1990).

Australian dryland ecosystems have been subject to European farming methods for less than two centuries (Barr and Cary 1992), compared with the 11,000 years in the Middle East (Naveh and Dan 1973). During this time, Australian ecosystems have experienced accelerated rates of change comparable with those occurring over a much longer time scale in Eurasia. Given the long period of intensive land use in Eurasia, it has been suggested that Eurasian ecosystems are now very resilient to disturbance because nonresilient components have been eliminated over time (Le Houérou 1981, Naveh 1967). Such ecosystems have reached a new equilibrium or metastable state (Hobbs and Hopkins 1990), defined as a quasi-stable, subclimax state maintained by anthropogenic pressures, which, although differing markedly from the original condition, are inherently sustainable in terms of productivity (Perevolotsky chap. 15, this volume). In Australia, this long period of coevolution has not occurred and the drylands contain nonresilient components that are highly susceptible to disturbance and subsequent degradation (Hobbs and Hopkins 1990). This is supported by early explorers' accounts of the rapid decline in the cover and diversity of native pasture plants once intensive grazing by sheep started (Barr and Cary 1992). Furthermore, the trees and shrubs associated with farming systems in Eurasia in hedgerows and parkland (defined as a farmed landscape of scattered mature trees growing among pasture or crops) are often very different in species composition and community structure from natural woodlands, whereas in Australia both are dominated by the same species. Current agricultural exploitation of Australian drylands is not sustainable in terms of agricultural production because of continuing loss of topsoil, declining fertility, and progressive acidification (Campbell 1993, Williams et al. 1993). Ecological processes are also degraded due to the loss of biodiversity (Leigh et al. 1984), particularly of soil organisms involved in nutrient cycling (Greenslade 1992). Despite high inputs, agricultural production of crops and livestock continues to fall in Australia's drylands (Williams et al. 1993), suggesting that a metastable equilibrium has not yet been reached.

The dominant trees of the Australian drylands belong to the Myrtaceous genus, *Eucalyptus*. Eucalypt woodlands have been extensively cleared for agricultural production. Eucalypt decline or dieback is now widespread in Australian dryland farmland (fig. 16.1) and has become a matter of great concern over the last fifteen years (Final report 1979, Farrow and Floyd 1995, Heatwole and Lowman 1986, Landsberg and Wylie 1988, Old et al. 1981). There is little natural regeneration of eucalypts in farmland, and the loss of trees from old age and dieback has been instrumental in influencing the government to fund a national tree planting program. This chapter discusses the causes of dieback with special reference to farming practice, the establishment of replacement trees, and the agricultural and ecological sustainability of such an approach. The argument is developed within an assumption that tree health is an important indicator of the productive state of dryland agroecosystems in Australia.

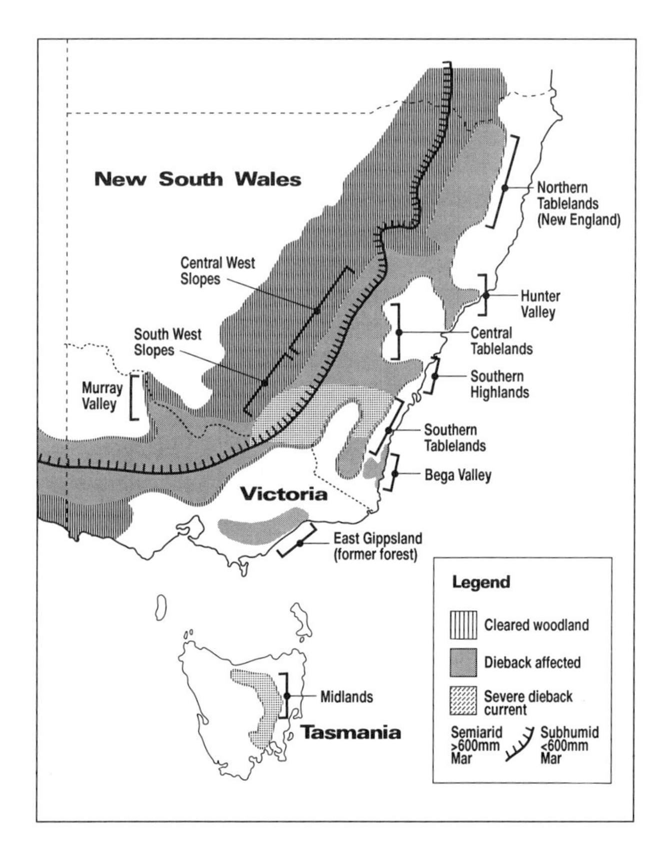

Figure 16.1. Map of cleared woodlands (generally intensively managed farmlands), dieback-affected farmlands, and dryland zone border in southeastern Australia. Regions with a known history of eucalypt dieback are indicated.

Dryland Degradation in Relation to Humans and Other Animals

It could be argued that the environmental changes observed in the drylands as a result of human activity are part of an evolutionary process exerted by the most successful species in the earth's history. It has been suggested, for example, that the Mediterranean steppe has existed in its present form for millennia, has reached a sustainable, metastable (albeit degraded) state by grazing (Perevolotsky chap. 15, this volume), and is not in a transitional stage of increasing degradation. Most animal or plant species have the potential reproductive capacity to dominate numerically their ecosystem to the possible detriment of ecosystem stability. In nature, this rarely happens due to the natural balances and buffering of most natural ecosystems, including those with simple food chains. For example, dryland ecosystems are characterized by outbreaks of species, such as locusts, armyworms, and rodents, that have the potential to consume large amounts of plant material. However, these outbreaks are usually initiated by unusually favorable weather conditions when food resources are not limiting and collapse after weather conditions return to normal. Such outbreaks rarely cause any adverse long-term effects to the condition of such dryland ecosystems and to the other organisms living in it. On the other hand, degradation of dryland ecosystems by humans through pastoralism and shifting cultivation has caused the development of chronic infestations of species, such as grasshoppers (Roffey 1972) and some rodents (de Vos 1975), that are favored by degraded environments. This results in a positive feedback that can exacerbate degradation processes and cause a shift to a lower metastable state that may be irreversible (Williams et al. 1993). The deliberate and accidental introduction of alien species to new environments in which the arrivals outcompete indigenous species and are not susceptible to control by indigenous natural enemies have also led to population explosions. These have resulted in substantial environmental degradation and displacement of native components of the ecosystem in drylands (e.g., the introduction of *Opuntia* and rabbits to Australia and *Acacia* and *Hakea* species to South Africa). This situation has been exacerbated by the return of some domestic animals to the wild as feral pests such as goats, camels, donkeys, and pigs in Australia.

One characteristic of many species inhabiting dryland ecosystems is the propensity for populations to fluctuate over wide ranges and to exist in outbreak and recession modes (Farrow and Longstaff 1986). This appears to be a response to fluctuations in weather and, in particular, to rainfall. The amplitude of this variation is particularly high in Australia's drylands. The flushing of plant growth following drought-breaking rains is frequently followed by outbreaks of insect herbivores that temporarily escape from natural enemy control (Farrow 1977). There is less convincing evidence that the onset of droughts also induces outbreaks through changes to host-plant quality, as proposed by White (1974). For example, outbreaks of a defoliating caterpillar, the gumleaf skeletonizer (*Uraba lugens*, Lepidoptera: Noctuidae), have occurred in river red gum (*E. camaldulensis*) forests in the Murray Valley in the semi-arid zone of western New South Wales at irregular intervals of about ten years since the start of records (Campbell 1962). The outbreaks that have been correlated with long-term fluctuations in river flooding cause complete defoliation but are short lived (about one season), and the trees soon refoliate with only minimal branch dieback apparent. Outbreaks of psyllids on several species of eucalypt have also been reported since the start of records in the nineteenth century but always appear to have been short lived (Froggatt 1923).

Benefits of Trees

The contribution that trees make to the sustainability of dryland agroecosystems can be measured in terms of their effect on agricultural productivity (Williams et al. 1993). Far from competing with pasture and crops, it is now known that strategically placed shelterbelts increase animal and plant productivity through changes to the microclimate (Bird et al. 1984, Lynch and Donnelly 1980, Reid and Bird 1990). Trees reduce water and wind erosion of the soil (Marshall 1990), and some species improve soil fertility (de Vos 1975) and reduce soil acidification possibly by capturing nitrate leachate (Borough 1990). Eucalypts can reduce deep infiltration of water in recharge areas and lower water tables, preventing the rise of salt to the surface (Hughes 1984). Fast-growing, introduced species of *Eucalyptus* with high water use appear to be better at this than the indigenous, slow-growing, drought-adapted, woodland species of the semi-arid zone. It is ironic that agricultural policy and taxation incentives have encouraged tree clearing for all but the last ten years, despite increasing evidence of the adverse effects of such action (Campbell 1993).

The dieback-affected area discussed in this review occurs across former eucalypt woodland, now largely pasture and cropping parkland, in the temperate semi-arid zone of southeastern Australia. The area is characterized by winter dominant rainfall, varying from 400 mm to 600 mm per annum, pan evaporation of 1,500–2,000 mm per annum, and extreme climate variability. Mean rainfall is skewed by periodic heavy falls such that most years receive below average rainfall (Atlas 1990a). Droughts can be severe and protracted, lasting for several years, and lead to severe pasture deterioration if grazing pressure is not reduced.

A Brief History of Exploitation of Dryland Woodlands

The historical perspective of land use, which has changed so radically over the 200–400-year life span of the surviving eucalypts, is integral to our understanding of the causes of dieback and of the apparent lack of resilience of this ecosystem to environmental change. When Europeans arrived in Australia some two hundred years ago, they faced an unfamiliar, often inhospitable, landscape. The environment was not totally pristine as it had been regularly burned for many thousands of years by Aboriginal inhabitants. This practice appears to have sustained the eucalypt woodland/perennial grassland ecosystem (Barr and Cary 1992). The vegetation encountered by the new settlers was quite alien and lacked the familiar cycles of growth and dormancy of the deciduous landscapes of Europe. Only the grasslands under the eucalypt canopy represented something familiar to the colonists. Their perceptions of the weather tended to be colored by the periods of abundant rains and luxurious pasture growth rather than by the long drought periods later shown to be typical of much of the drylands to which the native plants were highly adapted (Hobbs and Hopkins 1990). In the ecologically sustainable system that they first encountered, the pastoralists failed to realize that soil fertility was very low compared with European soils and that native vegetation was highly adapted to this condition. Nevertheless, the difficulties faced by the first colonists were not seen as an impediment to introducing European farming systems but as a challenge to tame the bush and to "green a brown land" (Barr and Cary 1992, Hobbs and Hopkins 1990). They pursued this course doggedly through massive clearing of forests and woodlands during the next century and a half (fig. 16.2)—introducing European farming methods with little thought to developing a more appropriate and sustainable form of agriculture (Campbell 1993).

The interpretation of eyewitness accounts in Barr and Cary (1992) indicates that the early explorers quickly realized the potential of the grasslands of the eucalypt woodlands for sheep raising, notwithstanding their inexperience in animal husbandry and agronomy. Sheep were introduced and herds were quickly multiplied to exploit this resource. However, the native grasslands were ill adapted to these close-grazing, cloven-hoofed herbivores, and the perennial

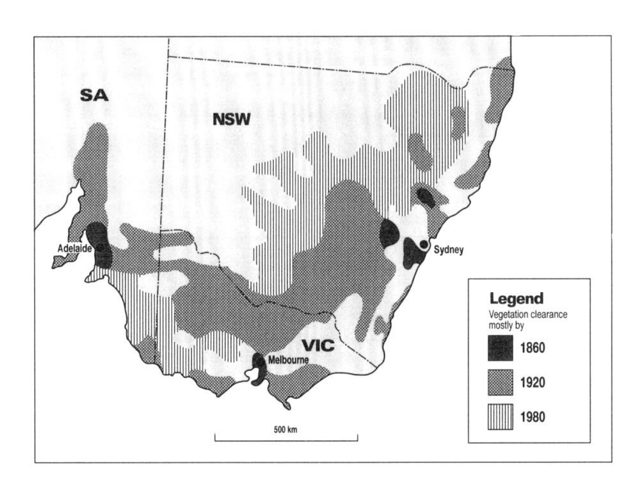

Figure 16.2. Map of history of woodland clearing.

grasses rapidly declined. With the removal of the regular burning regimes of the indigenous inhabitants, that controlled the regrowth of woody plants and maintained the vigor of the native pasture, thickets of eucalypts, acacias, and other shrubs sprang up while the older trees were reported to be attacked by plagues of insects as early as 1861. The settlers' next step was wholesale clearing of trees, usually leaving a few scattered mature eucalypts for shade, and cultivation to remove eucalypt lignotubers and to establish European grasses and legumes that had incidentally coevolved over the millennia with grazing ungulates in Eurasia. The first introductions of grasses and herbs were from England, but these species did not establish well in the semi-arid zone. The most successful introductions came from the Mediterranean region and included the nitrogen-fixing medics (Fabacae, *Medicago* spp.) that were winter active. However, the rate at which animal products were removed from this system did not enable this exotic pasture community to be sustained; because of the low nutrient status of the soils, weedy communities developed. This is not surprising as it has been estimated that ten thousand tons of phosphorus were removed annually in animal products from Victoria, a quite unsustainable practice from such nutrient-poor soils. In order to support this production system, phosphate was applied in large quantities as superphosphate. This supported the legumes that fixed nitrogen, and when they decayed, soil organic matter increased and nitrates and other nutrients were released to the soil that could then support cereal rotations without the addition of fertilizer. However, leaching of nitrate has become a serious problem over the last decade, causing soil acidification and manganese and aluminum toxicity (Williams et al. 1993). Soil physical properties have been adversely affected by compaction caused by hoofed livestock (Graetz and Tongway 1986, Hobbs and Hopkins 1990), although cultivation appears to have had the greatest impact on declining soil structure (Williams et al. 1993). Throughout this 200-year period, stocking rates were generally maintained at overoptimistic levels corresponding to the best seasons rather than the worst, causing severe episodes of degradation during drought years (Harrington et al. 1984, Roberts 1981).

Livestock grazing, competition by introduced pasture plants, and plowing for pasture and cereal establishment prevent eucalypt regeneration on farmland and have created the characteristic rural Australian landscape of scattered, mature eucalypts, many of which were present before European settlement. This landscape is termed parkland (defined earlier). The eucalypt woodlands in the semi-arid zone west of the Divide, dominated by the box-type eucalypts, yellow, white, and gray box (*E. melliodora, E. albens,* and *E. microcarpa),* have been the most extensively cleared of all woodland eucalypt communities (Atlas 1990b). The few remaining remnants are largely in undisturbed cemeteries, railway easements, and the like. However, even these remnant communities have been invaded by exotic weeds, feral herbivores, such as goats, and pests, such as rabbits, that have reduced recruitment of native trees and shrubs (Hobbs and Hopkins 1990). This is not to suggest that eucalypt regeneration is a frequent event in environments where disturbance and degradation has been minimal or where at least the soils and native pasture remain undisturbed. In eucalypt communities, regeneration events are often separated by decades so that most stands tend to be evenly aged. Regeneration occurs only after a favorable combination of environmental factors, such as fire and rain, form a suitable seedbed that coincides with a fall of seed so that a mass germination occurs (Cremer et al. 1990). In native pastures, successful germinants of eucalypts also accumulate over time as lignotubers in which annual growth is suppressed by fire, vertebrate grazing, and competition from mature trees. When these pressures are relaxed, substantial regrowth can occur.

The loss of trees and perennial grasses has had one additional and possibly unforeseen impact on the environment in many parts of the semi-arid zone, which is only now coming to the forefront. This is the problem of rising water tables, waterlogging, and the discharge of saline water into many catchments due to reduced evapotranspiration and increased recharge of the aquifers and to irrigation (fig. 16.3) (Morris and Jenkin 1990, Walker 1986, Williams et al. 1993). Far from having to cope with drought conditions, native plant life is now having to deal with frequent waterlogging and anaerobic conditions close to the soil surface. This is exacerbated in irrigation schemes in the semi-arid areas of the Murray Basin where water tables are as close as a meter from the surface and may be saline. The indigenous eucalypts, such as gray box (*E. microcarpa*), are declining rapidly in irrigated areas, and farmland is progressively going out of agricultural production in this area (Barr and Cary 1992). Substantial replants of fast-growing introduced eucalypts with high water use and tolerance of waterlogging may help lower water tables (Heupermann et al. 1984, Schofield and Bari 1990) and return abandoned farmland to production although the density of trees needed for rehabilitation may preclude intensive agriculture.

The effects of the change from native to exotic pastures on the biological diversity of this agroecosystem have been substantial. The indigenous arthropod fauna of the native grasslands has proved ill adapted to the introduced pastures and cereal rotations. These have been invaded by exotic species (Greenslade 1992) that evolved in the more resilient European systems. Most native phytophagous species have become restricted to patches of remnant vegetation (Key 1959), except those feeding on eucalypts. The numbers and species diversity of birds are much reduced on farmland compared with adjacent bushland, and this is exacerbated by dieback (Ford and Bell 1981).

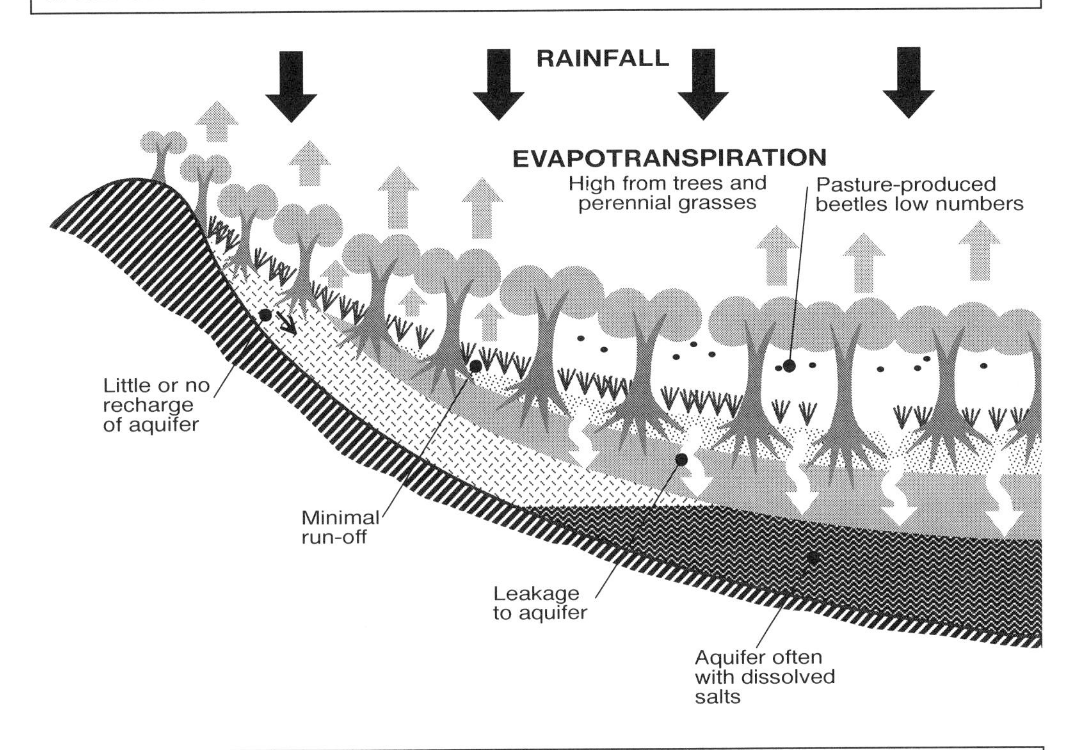

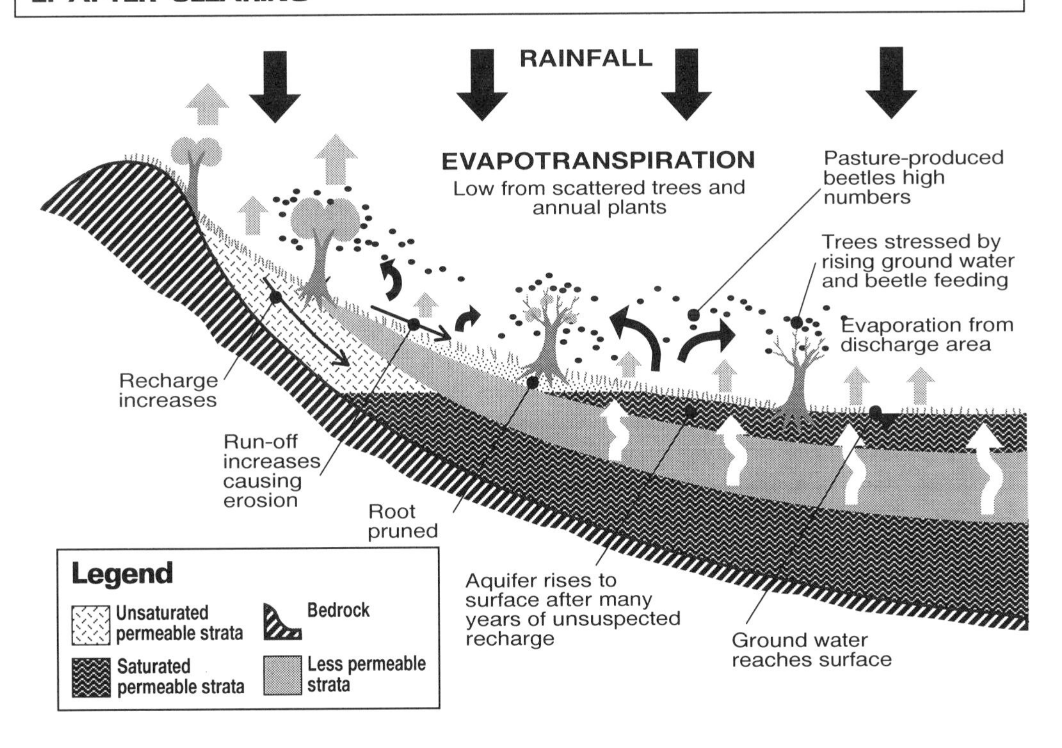

Figure 16.3. Diagram of effect of clearing on water tables.

Clearing has also affected the relative abundance of eucalypts remaining because of the preferential felling of undesirable species, such as *E. bridgesiana,* causing a rise in the relative abundance of species such as *E. blakelyi,* which could affect insect community structure in the parkland ecosystem (Clark 1964).

EUCALYPT DIEBACK

The state of health of rural trees in Australia's drylands is an obvious but often neglected sign that all is not well with the eucalypt parklands and the agricultural ecosystems of which they are a part. In many areas, eucalypts on farmland managed for intensive livestock production are visibly affected by progressive dieback of branches over and above natural senescence (Farrow and Floyd 1995, Heatwole and Lowman 1986) (fig. 16.4). Dieback tends to occur in regional episodes followed by intermissions when little change is observed; thus its intensity varies temporally, depending on the stage of the dieback cycle in a particular area, and spatially, as some localities are more prone to dieback than others and enter dieback phases at different times (Clark et al. 1981). It also varies according to the susceptibility of the eucalypt species present (Lowman and Heatwole 1992). On some occasions, the decline is so severe that large numbers of trees may die over short periods (White 1986), suggesting a climatic cause. Dieback is not necessarily confined to mature trees: young trees may be as affected as their parents (Clark et al. 1981, Mackay 1978). Although dieback is often regarded as a contemporary phenomenon, it was observed as far back as the 1860s at the time of early land clearing and intensive grazing.

Three approaches have been used to study dieback: (1) the spatial approach of correlating landscape and land-use characteristics with dieback intensity, usually over a short time frame, (2) the temporal approach of correlating climatic variables with specific dieback episodes, and (3) the intensive approach of examining the factors affecting the health of individual trees over 2–3-year periods. An approach that could do more to resolve the causes of dieback, namely, the long-term monitoring of the health of individual trees in relation to environmental factors during and between dieback episodes, has never been undertaken.

Correlations of dieback intensity with landscape and land-use values have sometimes proved equivocal (Jones et al. 1990), but more recent studies using different methods show that the more intensive the management, the worse the dieback (Lowman and Heatwole 1992). That is, dieback is worst in those areas where exotic pasture species have been sown for intensive livestock production and where soils have high levels of phosphorus (by supplementation), nitrogen (by nitrogen-fixing legumes), and organic matter (decay of annual pasture plants) and are affected by increasing acidity (by nitrification processes), rising water tables (through reduced evapotranspiration from loss of trees and perennial grasses), and declining structural properties (through cultivation and compaction by hoofed livestock). It is not surprising that eucalypts are stressed by such conditions (Landsberg and Wylie 1988) and may be root-pruned in wet anaerobic soil conditions (White 1986), leading to branch dieback and subsequent epicormic regrowth. Low-

Figure 16.4. Dieback-affected trees.

man and colleagues (1987) showed that the belowground biomass (by dry weight) of a dieback-affected tree was 20% of that of a comparable healthy tree. In the healthy trees, belowground and aboveground biomass were approximately equal, whereas in dieback-affected trees the roots were only a sixth of aboveground biomass (mostly of dead wood). However, even in these situations, it is difficult to determine whether branch dieback takes place first and causes root dieback as underground reserves are exhausted in attempting to replace lost foliage or whether the reverse occurs with root death initiating aboveground dieback.

Cycles of dieback have been correlated with climatic fluctuations of both wet periods (White 1986), that may promote root pruning by waterlogging, and severe drought (Pook 1981), although in this case dieback was confined to natural woodlands and did not affect the trees growing on farmland (Landsberg 1985). The principle cause of death of drought-stressed trees in the woodlands was destruction of the cambium by the eucalypt borer (*Phoracantha* sp.) (Coleoptera: Cerambycidae: Phoracanthini) (Pook 1981).

Intensive studies have demonstrated that the foliage of dieback-affected trees invariably shows extensive feeding damage with large amounts of leaf area lost to chewing insects or necrotized by sap-feeders (Clark 1962, Kile 1981, Landsberg 1988). Functional leaf area may be as little as 10% of normal in dieback-affected trees in parklands (Floyd et al. 1994, Lowman and Heatwole 1992) compared with more than 90% in undisturbed woodlands (Ohmart 1984). Nevertheless, if the shoots remain alive after an initial defoliation, this foliage can be replaced with no obvious residual effects. After repeated defoliations, however, the tree's starch reserves are exhausted and the shoots die back (Landsberg and Wylie 1988). The tree's response to branch dieback is to activate epicormic buds in the bark to create bunches of epicormic shoots carrying juvenile leaves. This is also the normal response to fire damage in eucalypts, enabling trees to refoliate after fire. However, in farmland trees, the appearance of substantial epicormic regrowth is generally a precursor to further defoliation, decline, and eventual death through cycles of regrowth and herbivory in a positive feedback interaction (Landsberg and Wylie 1988).

Insect diversity on eucalypts is very large (CSIRO 1991). Species representing most orders are present and they fall into a wide range of feeding strategies such as leaf tiers, miners, leaf chewers, sap-feeders on the leaves and stems, cambial and heartwood borers, and stem and leaf gallers. It is not always clear whether entire insect herbivore guilds are responsible for damage or whether a few key species cause the high rates of defoliation and necrosis found in farmland trees. There are considerable practical difficulties in studying the numbers and species composition of insect herbivores in the canopy of eucalypts, and many observations are carried out on saplings where defoliation may be caused by the combined effects of guilds of species in different functional groups. How far this reflects activities in the canopy of mature trees is not always clear. Nevertheless, there are a restricted number of species and species groups that are consistently associated with dieback events. They belong to two distinct life-history types: those with pasture-dwelling larvae and eucalypt-feeding adults and those whose entire life cycle is spent on eucalypt foliage except for periods of dispersal.

The patches of natural woodlands still remaining are not immune from dieback, although they may be minimally affected by farming practices. Dieback occurs in these remnants because defoliating and necrosis-causing insects can migrate from surrounding farmland trees and infest woodlands, particularly during outbreaks when numbers are very high. Most remnant woodlands are grazed by stock, resulting in changes in the nutrient status of the soils and tree foliage that could affect the numbers of defoliating insects (Landsberg et al. 1990). The health of the woodlands may also be adversely affected by exotic weeds and animals that prevent natural regeneration of the native plant community.

Clearly, dryland agricultural systems are not sustainable while dieback proceeds at such an intensity. The causes of dieback need to be understood if we are to improve our management of the agricultural ecosystems of Australian drylands, given the importance of trees in maintaining agricultural productivity.

Causes of Elevated Insect Herbivore Numbers

I have described how eucalypt dieback in farmland is associated with the intensity of both management practice and insect defoliation. What is not clear is whether the trees are primarily stressed by the farmland environment (leading to dieback and defoliation) that makes them more susceptible to attack by insects (possibly because of improvements to foliage quality), or whether high numbers of phytophagous insects are produced by the farming environment itself and are primary initiators of defoliation that then stresses the trees and leads to dieback. The observation that dieback-affected eucalypts recover when insect defoliators are kept off by stem injection of insecticide (Clark et al. 1981) supports the latter hypothesis for at least some sites. Eucalypt defoliators can be divided into two groups: those with a larval stage spent in the soil under pasture and those whose larvae feed on eucalypt foliage. The former could be expected to be directly influenced by the effects of farming practice while the latter would be affected only indirectly by ground conditions.

Pasture-Produced Defoliators

Increases in soil organic matter and grass root biomass, through the introduction of productive European pasture species and regular fertilizer applications (termed pasture improvement), have benefited development and survival of larvae of ruteline scarabs (Coleoptera: Scarabaeidae: Rutelini), such as *Anoplognathus* spp., *Sericesthes* spp., and several others (Davidson and Roberts 1968, Hutchinson and King 1980). Of these, *A. montanus* is the most abundant at the wetter edge of the semi-arid zone (Carne et al. 1981). The adults are commonly called Christmas beetles because they emerge in mid-December to feed on the spring flush of eucalypt foliage. Long-term fluctuations in scarab numbers appear to be controlled by fluctuations in rainfall (Carne et al. 1981) and would interact with the stress induced by waterlogging. One scenario could be that drought-breaking rains cause simultaneous waterlogging and outbreaks of scarab beetles (White 1986).

In outbreak years, parkland eucalypts are completely defoliated by scarabs over large areas of the countryside, often for several years in succession, although there is considerable variation in susceptibility to attack between and within species (Edwards et al. 1993). A basic interpretation of scarab productivity in a pasture and its impact on eucalypt foliage is given in table 16.1, which suggests that the defoliation caused by scarab plagues cannot be absorbed by any increase in density of susceptible trees compatible with current farming practice. A level of 10% tree cover of susceptible trees will still be significantly defoliated by recession levels of scarab beetles (i.e., < 1 larvae m^{-2}), whereas at moderate scarab densities of 10 larvae m^{-2} all trees will be completely defoliated. The beneficial effects on animal production of pasture improvement have, therefore, caused a scarab-induced defoliation of eucalypts that is independent of the degree of tree stress, although repeated defoliations will induce stress in the long term through exhaustion of starch reserves (Clark et al. 1981). Beetle production from native pastures and woodlands is relatively low, and adults rarely cause significant defoliation in those areas unless there has been invasion from neighboring pastures. The importance of scarab beetles to dieback has been confirmed by stem injection of insecticide of individual affected trees, which allowed trees to recover after insect feeding was prevented (Clark et al. 1981).

Foliage-Produced Insects

It is more difficult to explain the causes of prolonged outbreaks and chronic herbivory caused by this diverse group of insect defoliators and sap-feeders. Dieback is associated with insects as diverse as psyllids (Hemiptera: Psyllidae), leaf beetles (Coleoptera: Chrysomelidae: Paropsina), defoliating weevils (Coleoptera: Curculionidae: Gonipterini), armored scales causing witch's broom (Hemiptera: Diaspididae), and gall-forming wasps (various families of Hymenoptera). Other notable defoliators include the pergid sawflies (Hymenoptera: Pergidae), but these affect mostly young trees and are not implicated in dieback although they have a substantial impact on farm plantings and natural regrowth as do suites of other species that prefer more juvenile foliage. The current wave of tree decline in the parklands of western New South Wales is dominated by dieback to Blakely's red gum (*E. blakelyi*), which is associated with a prolonged outbreak of a necrosis-causing psyllid, *Cardiaspina albitextura* (Hemiptera: Psyllidae) (Floyd and Farrow 1994).

Outbreaks of phytophagous insects on eucalypts have been related to the stress-induced host reactions, escape from natural enemies, and fluctuations in weather. White (1974) proposed that the reason trees under stress suffer more insect damage is because such trees mobilize nitrogen (N), which improves the survival of early instar larvae

Table 16.1. Potential Defoliation of Eucalypts by Scarabs at Different Densities in Farmlands

	Density			Numbers of Scarabs per Farm of 1,000 ha with 100 ha (10%) under Eucalypts	
	Low	Medium	High		
1. Density of larvae m^{-2} [a]	1	10	100		10
2. Density of larvae ha^{-1}	10,000	100,000	1,000,000	Larvae on 900 ha	90,000,000
3. Density of adults $ha^{-1}yr^{1}$ (biennial life cycle)	5,000	50,000	500,000	Adults over 1000 $ha^{-1}yr^{-1}$	45,000,000
4. Food consumption @ 10g per adult[d]	50 kg	500 kg	5,000 kg	Potential weight of foliage consumed	450 tons
5. Trees/hectare	0.1	1	5	100 ha @ 10 ha^{-1}	1,000 trees
6. Eucalypt foliage @ 100 kg per tree[b,d]	10 kg	100 kg	500 kg	Total weight of foliage at 100 kg/tree	100 tons
7. Equivalent trees defoliated by adults in line 4[c]	0.5	5	50		4,500 trees

a. Roberts et al. 1982a.

b. Lowman et al. 1987.

c. This assumes that scarab larval density is unaffected by tree density for the purposes of this exercise. But see Roberts et al. 1982a.

d. Foliage and food consumption as fresh weight.

(White 1978). It has been difficult to confirm this hypothesis either in pot studies by Landsberg (1990) or at a more general level (Larsson 1989); nevertheless, the stressed, defoliated state of many eucalypts in farmlands, especially in recharge sites susceptible to waterlogging, make White's hypothesis attractive. Outbreaks could also occur on stressed trees if females prefer to oviposit on them rather than on healthy trees. It has also been shown that the leaves of eucalypts growing in improved pastures contain higher levels of nitrogen than those in woodlands, due to the higher levels of soil nutrients (Landsberg et al. 1990), although not all such trees show higher levels of herbivory. This may be because the nitrogen available through stress, in the form of mobilized amino acids, affects insects in a different way than elevated levels of nitrogen existing as bound proteins. Recent studies show that the foliage of psyllid-infested *E. blakelyi* had higher nitrogen levels than those trees with low psyllid numbers (Farrow and Floyd unpublished data). In both situations, it is difficult to ascertain whether this is a cause or effect since replacement foliage of defoliated trees tends to be more juvenile and is higher in nitrogen. This situation could potentially establish a positive feedback cycle (Landsberg and Wylie 1988) that could improve the survival of newly hatched larvae of leaf chewers requiring young soft leaves with relatively high nitrogen levels (Larsson and Ohmart 1988). However, in order for these responses to lead to high numbers of defoliators, insect survival and fecundity must be substantially higher on such trees. Apart from faster development and weight gain on nitrogen-rich foliage (Lansberg et al. 1990, Larsson and Ohmart 1988), there is little evidence of a numerical response large enough to give rise to outbreaks, particularly if there is a strong response from natural enemies such as insect parasitoids and predators. Furthermore, eucalypts are widely fertilized in plantations or are irrigated with sewage effluent, and there is no evidence that trees in these environments are any more susceptible to insect attack, although they are able to replace lost foliage rapidly because of their high growth rates.

Outbreaks can also result from a drop in the control exerted by predators and parasites (Clark 1962, Readshaw 1976). Davidson (1980) suggested that the numbers of natural enemies of insect herbivores are lower on farmland in parkland ecosystems than in remnant natural woodlands due to the lack of energy sources derived from nectar produced by flowering shrubs that are absent from parklands. There is as yet little evidence for this dependency in many parasitoids, and in any case, nectar is available in eucalypt flowers and from the honeydew secreted by sap-feeding insects. Furthermore, many semi-arid woodland communities originally lacked an understory, as shown by early explorers' accounts, which did not appear until the perennial grasses were destroyed by grazing (Barr and Cary 1992). It is more likely that the number and diversity of natural enemies are adversely affected by the isolation and exposure of parkland trees. In plantations, the herbivore populations studied appear well controlled by natural enemies that colonized the plantings together with their hosts (Farrow unpublished data, Rosario et al. 1992). Some population studies (e.g., Farrow 1982) have shown that the natural enemies of insects often act in an inversely density-dependent way with respect to prey numbers and fail to prevent upsurges and outbreaks of herbivorous insects that can remain at outbreak levels for extended periods. With respect to the psyllid, *C. albitextura,* involved in dieback of red gums, Clark (1962, 1964) proposed that cool, wet winters reduced natural enemy effectiveness, which led to outbreaks and persistent high-density populations.

The numbers of insectivorous birds are also known to be lower in farmlands than in bushlands (Ford and Bell 1981), and it has been suggested that this reduces the control of insect populations (Davidson 1980, Heatwole and Lowman 1986). The evidence for this is largely circumstantial because, although food consumption rates by birds are known, their effects have never been integrated into the population dynamics of the insect herbivores and partitioned among all the mortality sources. Birds need to show strong functional or numerical responses to variations in prey numbers that could prevent prey upsurges resulting in chronic defoliation. A poor response is probably why insect outbreaks by psyllids and other defoliators persist in woodlands as much as in farmlands, even though bird populations may be much higher in the former. Not all interactions with birds may be positive; chronic infestations of psyllids have been associated with the presence of miners (*Manorina* sp.) that drive other insectivorous birds out of their territories in remnant woodland patches (Loyn 1985, Loyn et al. 1983).

In a few long-term studies of insect herbivore abundance, annual fluctuations in numbers have been correlated with seasonal variations in rainfall (Carne 1969, Carne et al. 1981, Roberts et al. 1982b). This appears to operate through the effects of drought on larval survival, cold on overwinter survival, and weather-induced changes in foliage quality on survival and natality. This is not surprising for species in the semi-arid zone where availability of water limits so many processes. Once initiated by changes in weather, outbreaks of insect herbivores are usually widespread and may persist in the outbreak mode through many generations, as in the case of the psyllids, causing severe defoliation and intense dieback episodes.

The leaves of eucalypts contain large amounts of plant metabolites such as terpenoid oils and phenolics. The evolutionary significance of these compounds has not been resolved (Boland et al. 1991), although they reduce the digestibility of the leaves for arthropod and vertebrate herbivores, and variations in the composition of the terpenoids can affect insect feeding (Edwards et al. 1993, Floyd and Farrow 1994). It is not clear whether the terpenoids have evolved as *constitutive defenses* against insect feeding (Edwards 1989) or vertebrate browsing, are involved in the biochemical pathways of basic metabolic processes (Erman 1985, Siegler 1977), are carbon sinks (Penfold and Willis 1955, Tuomi et al. 1988), have allelopathic effects (Del Moral and Muller 1969, Whittaker and Feeny 1971), or enhance the flammability of the foliage that contributes to the dominance of eucalypts in forests and woodlands (Mount 1964). Feeding is also reduced by leaf hardness, waxiness, and low nitrogen, which are related to other environmental constraints: these are termed *neutral defenses* (Edwards 1989). Several eucalypt herbivores have been observed snipping the petioles of partially eaten leaves (Edwards and Wanjura 1989), suggesting they are responding to changes in leaf chemistry, involving the possible production of *induced defenses. Delayed-induced defenses* have been shown to lead to cycles of defoliation in deciduous trees in the northern hemisphere (Haukioja et al. 1994), but there is no evidence that changes in constitutive and induced defenses play a role in insect outbreaks on eucalypts in Australia.

We still do not have conclusive explanations for the causes of chronic herbivory in this group of eucalypt herbivores. The most likely hypothesis is that climate variability plays a major role in initiating outbreaks of herbivorous insects on eucalypts in farmlands. However, the severity and duration of outbreaks appears to be intensified by qualitative changes in eucalypt foliage, possibly due to the stress-related effects of farming practices that favor herbivore survival and by the poor numerical response of natural enemies. The collapse of outbreaks of insects such as the psyllids, *Cardiaspina* spp., result from the loss of resources caused by widespread, synchronized defoliation and the delay in the appearance of replacement leaves suitable for oviposition (Clark 1962), although reinfestation may be rapid through reinvasion. Healthy plantation trees can survive repeated defoliations without developing branch dieback (Carne et al. 1974). In some farm plantings, trees become bush-like through persistent insect feeding without dying back (Farrow unpublished data), although nearby naturally occurring trees may be seriously affected by dieback. These observations suggest that the imposition of environmental change on existing trees as a result of human activity is of underlying importance in rural dieback.

Developing a Sustainable Agroforestry Ecosystem in the Semi-Arid Zone in Southeastern Australia

It is increasingly clear that indigenous woodland eucalypts appear poorly adapted to the environmental changes brought about by European farming systems in the semi-arid zone (Farrow and Floyd 1995), notwithstanding the success of eucalypts when introduced to degraded lands elsewhere in the world. However, species and provenances (of local populations) vary phenotypically in their susceptibility to many of the factors involved in degradation, including salinization, waterlogging, and insect feeding (Cremer 1990, Floyd and Farrow 1994). Salt tolerant eucalypts, such as *E. camaldulensis* (indigenous to dryland water courses) and *E. occidentalis* (introduced from western Australia), among others, have been long used in salt-affected sites (Morris and Thompson 1983). *E. camaldulensis* and *E. grandis* (native to east-coast Australia) have been planted in high water table sites in the semi-arid zone to lower the water levels. However, there has been some reluctance to extend this approach to insect resistance although many insect-resistant species, provenances, and individual biotypes are now known (Floyd and Farrow 1994). The problem was first recognized thirty years ago by Pryor (1952) who stated that incorporating resistance to scarab beetles was of extreme importance if commercial planting of eucalypts on farmlands was to be successful.

Resistance to Insects

Resistance mechanisms fall into three functional categories (Painter 1951): (1) *antixenosis*—insects are deterred from feeding or ovipositing on host tree, (2) *antibiosis*—development and survival are adversely affected by host plant chemicals, and (3) *tolerance*—trees withstand and recover from insect damage.

Antixenosis Host preferences are an example of *antixenosis.* Studies of both natural and planted communities of eucalypt species have shown that more than 50% of the eucalypt fauna is restricted to two or less host species (Edwards and Wanjura 1990, Morrow 1977). The dieback-causing psyllid, *C. albitextura,* is restricted to two closely related red gums (*E. blakelyi* and *E. camaldulensis*), and adults do not colonize yellow box (*E. melliodora*) or other eucalypt species in the same community. Yellow box is host to a different species of *Cardiaspina, C. tenuitela,* which is rarely a pest in dryland communities. Studies on the scarab, *Anoplognathus montanus,* have demonstrated that significant variations in susceptibility to its attack occur between and within eucalypt species (Edwards et al. 1990, 1993) due to differences in the composition of leaf terpenoids. Flying beetles are deterred from landing on resistant trees due presumably to the deterrent effect of the terpenoids evaporating from the foliage. Species may be dimorphic or polymorphic for these traits and may occur sympatrically as genetically distinct chemotypes or allopatrically as chemical races (Boland et al. 1991). A range of species of *Anoplognathus* (Scarabaeidae) are deterred from feeding on foliage with high cineole content (Edwards et al. 1990, 1993). Some leaf feeders have evolved specific mechanisms for detoxifying or sequestering the terpenoids (Morrow et al. 1976, Ohmart and Larsson 1989), their survival is less likely to be affected by variations in the terpenoids, and, in this context, it is important to note that the variations in total terpenoids have little impact on development and survival of leaf-feeding larvae (Morrow and Fox 1980).

Antibiosis There are many examples of the effects of host-plant defenses on the development and survival of phytophagous insects. The psyllid, *C. retator,* shows considerable differences in development and survival between and within provenances of *E. camaldulensis* due probably to differences in leaf wax thickness (Floyd et al. 1994). Larvae of the autumn gum moth (*Mnesampela privata*) showed differential survival on the juvenile foliage of different provenances of blue gum (*E. globulus*) planted to control water tables in an irrigated area of the semi-arid zone.

Tolerance Although some provenances and species of eucalypt can be heavily attacked by insects, this may not have a major impact on growth because of tree vigor. For example, the Lake Albacutya provenance of *E. camaldulensis* is generally susceptible to psyllids, and foliage can be heavily necrotized at times, but this provenance can still outgrow less susceptible provenances in trial (Floyd and Farrow unpublished data); species introduced from montane forests in higher rainfall areas often perform better on farmlands than the local woodland species (Farrow unpublished data). Although heavily defoliated at times, such species generally grow faster (fig. 16.5), have a longer growing season, and recover more quickly from defoliation, especially in winter when most indigenous woodland species are dormant probably because the forest species do not possess conservative drought-adapted strategies. This could make them less tolerant of prolonged droughts when planted in dryland areas.

The value of insect-resistant eucalypts in dieback-prone

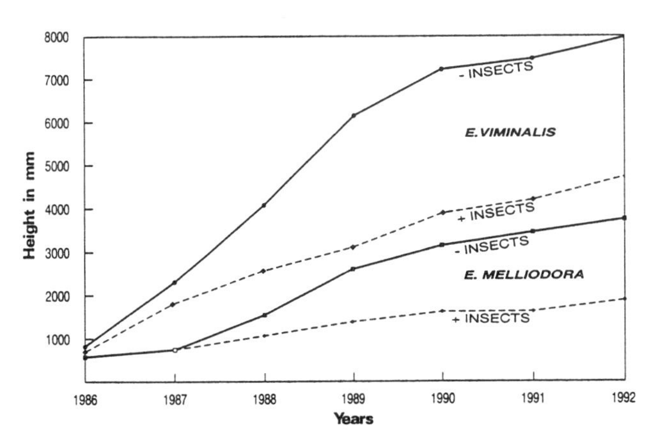

Figure 16.5. Graph of growth rates of two eucalypt species in trial: an indigenous woodland species (*E. melliodora*) versus an introduced forest species (*E. viminalis*), protected and exposed to insect attack. Protection was achieved by regular applications of a systemic/contact insecticide blend until year four (1991).

areas would be enhanced if cross-resistance against a range of pest species occurred. It appears unlikely that cross-resistance between sap-feeders, such a psyllids, and leaf chewers, such as scarab beetles, will occur because of the different resistance mechanisms involved; it is possible, however, that this could be achieved by controlled crosses between insect-resistant genotypes. In *E. globulus,* cross-resistance to both a leaf miner and a foliage chewer has been observed at the provenance level. On the other hand, a provenance of *E. camaldulensis,* which was resistant to a psyllid and a leaf miner, was not resistant to sawfly (Farrow unpublished data). Differences in the geographical distribution of the principal insect species contributing to dieback can also simplify the choice of genotype for replanting. An important consideration is the likelihood of counter-adaptation by insects to overcome host plant resistance. The genetic heterogeneity of eucalypts controlling a varying complex of plant chemicals is considered to present a sufficiently unpredictable resource that reduces the chances of counter-adaptation (Schultz 1983). The selection of outbred rather than clonal material is more likely to prolong the resistance of introductions of selected lines with high resistance.

The planting of exotic natives or of selected genotypes is not always accepted by conservationists who consider that the local species is the best adapted to the local environment and should be promoted for revegetating farmlands, notwithstanding the vast changes that have occurred to that environment over the past hundred years but within the lifetime of many mature trees, as discussed earlier. Is it any more valid to plant a local species in a novel habitat than to introduce one from afar? How is one to decide the geographical boundary of what is local and what is introduced? If natural regeneration was allowed to operate in farming systems by excluding stock and competitive pasture plants, the results of natural selection under insect pressure in existing eucalypt communities during regeneration cycles would be essentially similar to the current strategy of planting selected genotypes from the pool of naturally occurring genotypes. In at least some of our trials in areas subject to heavy insect attack, the local provenance is more susceptible to such attack than provenances introduced from other parts of the species range (Floyd and Farrow 1994), presumably due to local coevolution between host plant and insect such that current interactions favor the insect.

Discussion and Conclusions

The development of sustainable grazing and cropping agroecosystems is of key importance in maintaining the agricultural productivity of dryland areas of southeastern Australia (Williams et al. 1993). The role of trees in sustaining productive farming systems is being increasingly recognized (Campbell 1993); however, these agroecosystems need to be resilient and cannot rely on species that will be stressed by the environmental changes that farming has caused. Obviously, farmlands cannot revert to the original woodland climax community, and small-scale attempts to recreate this community are probably doomed to failure given the irreversible changes to the pasture/soil environment. Maintenance of the existing parkland structure appears equally doomed as indigenous trees not only decline from dieback but die from natural senescence at 200–400 years of age without producing any progeny (Farrow and Floyd 1995). Currently, the red gum, *E. blakelyi,* is suffering the greatest decline, especially in the drylands of the southwest slopes of New South Wales (Schmedding 1994) (fig. 16.1). This species is affected by a major leaf necrosis caused by a prolonged psyllid outbreak, and surviving leaves are often consumed by pasture scarabs and other insects. The poor health of the tree in discharge sites and during wet winters (fig. 16.3) suggests that this species is severely stressed by rising water tables and other environmental changes. Variations in climate are likely to initiate further dieback episodes in the drylands of southeastern Australia through their interactions with the cumulative effects of farming practices on the drylands environment and with the guilds of herbivorous insects. New species may become important, such as the psyllid, *Lasiophylla rotundipennis,* that has recently infested yellow box (*E. melliodora*) in some dryland areas in the central western slopes of New South Wales where it is inducing dieback.

Restoration of degraded farmlands generally focuses on the establishment of networks of shelterbelts and woodlots designed to alleviate problems of water recharge and discharge and of soil erosion and salinization, to provide shelter from prevailing winds, and to provide wood products (Barr and Cary 1992). The level of tree cover established must be compatible with the area set aside for an economic return from animals and crops. However, such programs are not currently exploiting the tree species and genotypes best adapted to the current farming environment (Parker 1994, Pryor 1952). In general, insect-resistant, nonindigenous eucalypt species from the subhumid zone, tolerant of waterlogging and exposure, are more successful than the drought-adapted local species of the semi-arid zone, although selected provenances of a riverine species of the semi-arid zone, *E. camaldulensis* (river red gum), are tolerant of a wide range of conditions and stresses and are relatively insect resistant. As already indicated, some local species and provenances may be more susceptible to insect attack than introductions. Current research in the eucalypt program at the Commonwealth of Scientific and Industrial Research Organization's Division of Entomology focuses on the introduction to farmlands of naturally occurring, insect-resistant genotypes of local and exotic species through propagation in seed orchards on farms.

As stated earlier, remnants of the original semi-arid woodlands still occur and represent the most important source of biodiversity for this zone. These communities are in urgent need of protection from domestic stock, feral animals, and weeds. Using shelterbelts to connect native remnants via corridors may assist the movement of vertebrates but may be less important for arthropods that disperse across open country by flight and drift.

Most trees are being established on existing improved pastures of exotic species of grasses and legumes that have been progressively acidifying soils. Agronomic practice has now turned full circle to the point that the potential of acid-tolerant, deep-rooted, perennial native grasses is being examined to determine if they can reduce nitrate leaching and produce a more sustainable agroecosystem, particularly when combined with plantings of native trees and shrubs. Such changes may also alleviate the scarab beetle problem by making the larval environment less favorable.

What is the value of plantations for biodiversity? Do they constitute an ecologically sustainable system as distinct from a *productive* agroecosystem (as defined by Williams et al. 1993)? As already described, about half the herbivorous species on eucalypts are stenophagous or monophagous. The polyphagous species comprise most of the common species including those involved in dieback, except for the psyllids. Plantings of exotic eucalypts and acacias are readily colonized by the local arthropod fauna, and the guilds established appear to be superficially similar to those in neighboring indigenous vegetation (Farrow unpublished data) and comprise the common polyphagous herbivores, a range of stenophagous species, according to the species mix of eucalypts (Edwards and Wanjura 1990), and the insect predators, parasites, and pathogens that in at least some plantings probably prevent outbreaks of herbivores. Insect-resistant genotypes still support low numbers of primary herbivores and their natural enemies, and their insect communities probably differ little in terms of diversity from those comprising susceptible species. Little is known of the epigeal fauna or of the biological activity involved in recycling nutrients in plantations. Apart from the establishment procedures, including the addition of nutrients and water and herbage control, it is not known whether these agroforestry systems take out more from the ecosystem than is returned by natural processes, particularly if trees are harvested at some stage. This area clearly warrants further research to determine the extent to which these systems are ecologically sustainable.†

Note

I thank Penelope Greenslade, Robert Floyd, and Barry Longstaff for their helpful comments in developing this essay and David Wainhouse for critical comments on the manuscript.

References

Atlas of Australian resources: Meteorology. 1990a. Canberra: Australian Government Publication Service.

Atlas of Australian resources: Vegetation. 1990b. Canberra: Australian Government Publication Service.

Barr, N., and J. Cary. 1992. Greening a brown land. Melbourne: Macmillan.

Bird, P. R., J. J. Lynch, and J. M. Obst. 1984. Effect of shelter on plant and animal production. Proceedings of the Australia Society of Animal Production 15:270–73.

Boland, D. J., J. J. Brophy, and A. P. N. House. 1991. Eucalyptus leaf oils. Melbourne: Inkata Press.

Borough, C. J. 1990. Soil acidity. *In* Cremer, ed. Trees for rural Australia. 367–68.

Bourlière, F., and M. Hadley. 1970. The ecology of tropical savannas. Annual Review of Ecological Systems 1:125–52.

Campbell, C. A. 1993. Can Australian farming systems be more sylvan? Agroforestry Systems 22:17–24.

Campbell, K. G. 1962. The biology of *Roeselia lugens* Walk the gum leaf skeletonizer moth with particular reference to the *Eucalyptus camaldulensis* Dehn River red gum forests of the Murray Valley Region. Proceedings of the Linnean Society, New South Wales. 87:316–38.

Carne, P. B. 1969. On the population dynamics of the eucalypt feeding sawfly *Perga affinis affinis* Kirby Hymenoptera. Australian Journal of Zoology 17:113–41.

Carne, P. B., R. T. G. Greaves, and R. S. McInnes. 1974. Insect damage to plantation-grown eucalypts in north coastal New South Wales with special reference to Christmas beetles. Journal of Australian Entomology Society 13:189–206.

Carne, P. B., R. S. McInnes, and J. P. Green. 1981. Seasonal fluctuations in the abundance of two leaf-eating insects. *In* Old et al., eds. Eucalypt dieback in forests and woodlands. 121–26.

Clark, L. R. 1962. The general biology of *Cardiaspina albitextura* Psyllidae and its abundance in relation to weather and parasitism. Australian Journal of Zoology 10:537–86.

———. 1964. The population dynamics of *Cardiaspina albitextura* Psyllidae. Australian Journal of Zoology 12:362–80.

Clark, R. V., D. W. Nicholson, S. M. Mackay, P. R. Lind, and F. R. Humphreys. 1981. A broadscale survey of land use and site factors associated with native tree dieback in the New England Tablelands. *In* Old et al., eds. Eucalypt dieback in forests and woodlands. 71–73.

Commonwealth of Scientific and Industrial Research Organization (CSIRO). 1991. Insects of Australia. Melbourne: Melbourne University Press.

Cremer, K. W., ed. 1990. Trees for rural Australia. Melbourne: Inkata Press.

Cremer, K. W., G. K. Unwin, and J. G. Tracey. 1990. Natural regeneration. *In* Cremer, ed. Trees for rural Australia. 107–29.

Davidson, R. L. 1980. Local management. *In* N. M. Oates, P. J. Greig, D. G. Hill, P. A. Langley, and A. J. Reid, eds. Focus on farm trees. Victoria: Capitol Press. 89–98.

Davidson, R. L., and R. J. Roberts. 1968. Influence of plants, manure, and soil moisture on survival and liveweight gain of two scarabaeid larvae. Entomology Experiments and Applications 11:305–14.

Del Moral, and C. H. Muller. 1969. Fog drip: A mechanism of toxin transport from *Eucalyptus globulus*. Bulletin of Torrey Botany Club 96:467–75.

de Vos, A. 1975. Africa the devastated continent? The Hague: Dr. W. Junk Publishers.

Edwards, P. B., and W. J. Wanjura. 1989. Eucalypt-feeding insects bite off more than they can chew: Sabotage of induced defences? Oikos 54:246–48.

———. 1990. Physical attributes of eucalypt leaves and the host range of chrysomelid beetles. Symposium on Biological Hung 39:227–36.

Edwards, P. B., W. J. Wanjura, and W. V. Brown. 1993. Selective herbivory by Christmas beetles in response to intraspecific variation in *Eucalyptus* terpenoids. Oecologia 95:551–57.

Edwards, P. B., W. J. Wanjura, W. V. Brown, and J. M. Dearn. 1990. Mosaic resistance in plants. Nature 347:434–35.

Edwards, P. J. 1989. Insect herbivory and plant defence theory. *In* P. J. Grubb and J. B. Whittaker, eds. Towards a more exact ecology. Oxford: Basil Blackwell. 275–88.

Erman, W. F. 1985. Chemistry of the monoterpenes. *In* P. G. Gassman, ed. Dekker studies in organic chemistry. Vol. 2 A and B. New York.

Farrow, R. A. 1977. Origin and decline of the 1973 plague locust outbreak in central western New South Wales. Australian Journal of Zoology 25:455–89.

———. 1982. Population dynamics of the Australian plague locust *Chortoicetes terminifera* Walker in central western New South Wales. Pt. 3: Analysis of population processes. Australian Journal of Zoology 30:69–79.

Farrow, R. A., and R. B. Floyd. 1995. Effects of changing land use on eucalypt dieback in Australia in relation to insect phytophagy and tree re-establishment. *In* N. Stork and R. Harrington, eds. Insects and a changing environment. London: Academic Press. 456–60.

Farrow, R. A., and B. C. Longstaff. 1986. Comparison of the intrinsic rates of natural increase of locusts in relation to the incidence of plagues. Oikos 46:207–22.

Final report to the Minister for Conservation and Natural Resources on the New England Dieback. 1979. Report to the Parliament of New South Wales. Sydney: Australian Government Publication Service.

Floyd, R. B., and R. A. Farrow. 1994. The potential role of natural insect resistance in the integrated pest management of eucalypt plantations in Australia. *In* S. C. Halos, F. F. Natividad, L. J. Escote-Carlson, G. L. Enriquez, and I. Umboh, eds. Proceedings of the symposium on biotechnological and environmental approaches to forest pest and disease management, Quezon City, Philippines, April. Bogor, Indonesia: Seameo Biotrop. 55–76.

Floyd, R. B., R. A. Farrow, and F. G. Neumann. 1994. Inter- and intra-provenance variation in resistance of red gum foliage to insect feeding. Australian Forestry 57:45–48.

Ford, H. A., and H. Bell. 1981. Density of birds in eucalypt woodland affected to varying degrees by dieback. Emu 81:202–8.

Froggatt, W. W. 1923. Forest insects of Australia. Sydney: Australian Government Publication Service.

Graetz, R. D., and D. J. Tongway. 1986. Influence of grazing management on vegetation soil structure and nutrient distribution and the infiltration of applied rainfall in a semi-arid chenopod shrubland. Australian Journal of Ecology 11:347–60.

Greenslade, P. J. M. 1992. Conserving invertebrate biodiversity in agricultural, forestry, and natural ecosystems in Australia. Agricultural Ecosystem Environment 40:297–312.

Harrington, G. N., A. D. Wilson, and M. D. Young. 1984. Management of rangelands ecosystems. *In* G. N. Harrington, A. D. Wilson, and M. D. Young, eds. Management of Australia's rangelands. Melbourne: CSIRO. 3–13.

Haukioja, E., S. Hanhimäki, and G. H. Walter. 1994. Can we learn about herbivory on eucalypts from research on birches or how general are general plant-herbivore theories? Australian Journal of Ecology 19:1–9.

Heatwole, H., and M. Lowman. 1986. Dieback: Death of an Australian landscape. Sydney: Reed Book.
Heupermann, A., H. Stewart, and R. Wildes. 1984. The effect of eucalypts on water tables in an irrigation area of northern Victoria. Water Talk 52:4–8.
Hobbs, R. J., and A. J. M. Hopkins. 1990. From frontier to fragments: European impact on Australia's vegetation. *In* D. A. Saunders, A. J. M. Hopkins, and R. A. How, eds. Australian ecosystems: 200 years of utilization, degradation, and reconstruction. Proceedings of the Ecological Society of Australia 16:93–114.
Hughes, K. K. 1984. Trees and salinity. Queensland Agricultural Journal 110:13–14.
Hutchinson, K. J., and K. L. King. 1980. The effects of sheep stocking level on invertebrate abundance, biomass, and energy utilisation in a temperate, sown grassland. Journal of Applied Ecology 17:369–87.
Jones, A. D., H. I. Davies, and J. A. Sinden. 1990. Relationship between eucalypt dieback land and land use in southern New England, New South Wales. Australian Forestry 53:13–23.
Key, K. H. L. 1959. The ecology and biogeography of Australian grasshoppers and locusts. *In* A. Keast, R. L. Crocker, and C. S. Christian, eds. Biogeography and ecology in Australia. Monthly Biology 8:192–210.
Kile, G. A. 1981. An overview of eucalypt dieback in rural Australia. *In* Old et al., eds. Eucalypt dieback in forests and woodlands. 13–26.
Landsberg, J. 1985. Drought and the dieback of rural eucalypts. Australian Journal of Ecology 10:87–90.
———. 1988. Dieback of rural eucalypts: Response of foliar dietary quality and herbivory to defoliation. Australian Journal of Ecology 15:89–96.
———. 1990. Dieback of rural eucalypts: Tree phenology and damage caused by leaf-eating insects. Australian Journal of Ecology 13:251–67.
Landsberg, J., J. Morse, and P. Khanna. 1990. Tree dieback and insect dynamics in remnants of native woodlands on farms. Proceedings of the Ecological Society of Australia 16:149–65.
Landsberg, J., and F. R. Wylie. 1988. Dieback of rural trees in Australia. Geojournal 17:231–37.
Larsson, S. 1989. Stressful times for the plant stress–insect performance hypothesis. Oikos 56:277–83.
Larsson, S., and C. P. Ohmart. 1988. Leaf age and larval performance of the leaf beetle *Paropsis atomaria.* Ecological Entomology 13:19–24.
Le Houérou, H. N. 1981. Impact of man and his animals on Mediterranean vegetation. *In* F. di Castri, D. W. Goodall, and R. L. Specht, eds. Mediterranean-type shrublands. Amsterdam: Elsevier. 479–521.
Leigh, J., R. Boden, and J. Briggs. 1984. Extinct and endangered plants in Australia. Sydney: Macmillan.
Lowman, M. D., A. D. Burgess, and W. D. Higgins. 1987. The biomass of New England peppermint *Eucalyptus nova-anglica* in relation to insect damage. Australian Journal of Ecology 12:361–71.
Lowman, M. D., and H. Heatwole. 1992. Spatial and temporal variability in defoliation of Australian eucalypts. Ecology 73:129–42.
Loyn, R. H. 1985. Birds in fragmented forests in Gippsland, Victoria. *In* A. Keast, H. F. Recher, H. A. Ford, and D. Saunders, eds. Birds of eucalypt forests and woodlands: Ecology, conservation, management. Sydney: Surrey Beatty. 323–31.
Loyn, R. H., R. G. Runnalls, G. Y. Forward, and J. Tyers. 1983. Territorial bell miners and other birds affecting populations of insect prey. Science 221:1411–13.
Lynch, J. J., and J. B. Donnelly. 1980. Changes in pasture and animal production resulting from the use of windbreaks. Australian Journal of Agricultural Research 31:967–79.
Mackay, S. M. 1978. Dying eucalypts of the New England Tablelands. Forests and Timber 14:18–20.
Marshall, C. J. 1990. Control of erosion. *In* Cremer, ed. Trees for rural Australia. 369–76.
Morris, J. D., and J. J. Jenkin. 1990. Trees in salinity control. *In* Cremer, ed. Trees for rural Australia. 357–66.
Morris, J. D., and L. A. J. Thompson. 1983. The role of trees in dryland salinity control. Proceedings of Royal Society, Victoria 95:123–31.
Morrow, P. A. 1977. Host specificity of insects in a community of three co-dominant *Eucalyptus* species. Australian Journal of Ecology 2:89–106.
Morrow, P. A., T. E. Bellas, and T. Eisner. 1976. Eucalyptus oils in the defensive oral discharge of Australian sawfly larvae Hymenoptera: Pergidae. Oecologia 24:193–206.
Morrow, P. A., and L. R. Fox. 1980. Effects of variation in *Eucalyptus* essential oil yield on insect growth and grazing damage. Oecologia 45:209–19.
Mount, A. B. 1964. The interdependence of the eucalypts and forest fires in southern Australia. Australian Forestry 28:166–72.
Naveh, Z. 1967. Mediterranean ecosystems and vegetation types in Israel and California. Ecology 48:345–59.
Naveh, Z., and J. Dan. 1973. The human degradation of Mediterranean landscapes in Israel. *In* F. di Castri and H. A. Mooney, eds. Mediterranean type ecosystems: Origin and structure. New York: Springer-Verlag. 373–90.
Ohmart, C. P. 1984. Is insect defoliation in eucalypt forests greater than that in other temperate forests? Australian Journal of Ecology 9:413–18.
Ohmart, C. P., and S. Larsson. 1989. Evidence for absorption of eucalypt essential oils by *Paropsis atomaria* Olivier Coleoptera: Chrysomelidae. Journal of the Australian Entomological Society 28:201–6.
Old, K. M., G. A. Kile, and C. P. Ohmart, eds. 1981. Eucalypt dieback in forests and woodlands. Melbourne: CSIRO.
Painter, R. H. 1951. Insect resistance in crop plants. New York: Macmillan.
Parker, J. N. 1994. The plight of the planted eucalypt: Evening the odds. *In* Faces of farm forestry. Proceedings of the annual conference of Australian forest growers, Launceston, May. Canberra. 257–67.
Penfold, A. R., and J. L. Willis. 1955. The formation of essential oils in plants. Australian Journal of Pharmacy (Sept.).
Pook, E. W. 1981. Drought and dieback of eucalypts in dry sclerophyll forest and woodlands of the southern table lands, New South Wales. *In* Old et al., eds. Eucalypt dieback in forests and woodlands. 179–89.
Pryor, L. D. 1952. Variation in resistance to leaf-eating insects in some eucalypts. Proceedings of the Linnean Society, New South Wales. 27:364–69.
Readshaw, J. L. 1976. A theory of phasmatid outbreak release. Australian Journal of Zoology 13:475–90.
Reid, R., and R. D. Bird. 1990. Shade and shelter. *In* Cremer, ed. Trees for rural Australia. 319–35.
Roberts, B. R. 1981. Erosion in the pastoral areas. Australian Rangeland Society. Biennial Conference Papers 3:53–63.
Roberts, R. J., A. J. Campbell, M. R. Porter, and N. L. Sawtell. 1982a. The distribution and abundance of pasture scarabs in relation to eucalypt trees. *In* K. E. Lee, ed. Proceedings of the third Australian conference on grassland and invertebrate ecology, Adelaide, 30 November–4 December 1981. Adelaide, South Australia: South Australia Government Printer. 207–14.
Roberts, R. J., T. J. Ridsdell Smith, M. R. Porter, and N. L. Sawtell. 1982b. Fluctuations in abundance of pasture scarabs over an 18-year period of trapping. *In* K. E. Lee, ed. Proceedings of the third

Australian conference on grassland and invertebrate ecology, Adelaide, 30 November–4 December 1981. Adelaide, South Australia: South Australia Government Printer. 75–79.

Roffey, J. 1972. The effects of changing land use on locusts and grasshoppers. *In* C. F. Hemming and T. H. C. Taylor, eds. Proceedings of the international study conference on current and future problems of acridology. London: Center for Overseas Pest Research. 199–204.

Rosario, S., P. J. Gullen, and R. A. Farrow. 1992. Aspects of the biology of two species of eucalypt-feeding leafhoppers Hemiptera: Eurymelidae. Journal of Australian Entomological Society 31:317–25.

Schmedding, R. 1994. Saving Blakely's red gum. Trees and Natural Resources 36:20–23.

Schofield, N. J., and M. A. Bari. 1990. Valley reforestation to lower saline groundwater tables: Results from Stene's Farm, Western Australia. Australian Journal of Soil Research 29:635–50.

Schultz, J. C. 1983. Impact of variable plant defensive chemistry on susceptibility of insects to natural enemies. *In* P. A. Hedin, ed. Plant resistance to insects. Washington, D.C.: American Chemical Society. 37–54.

Siegler, D. S. 1977. Primary roles for secondary compounds. Biochemical Systems Ecology 5:195–99.

Tuomi, J., P. Niemela, F. S. Chapin, J. P. Bryant, and S. Siren. 1988. Defensive responses of trees in relation to their carbon/nutrient balance. *In* W. J. Mattson, J. Levieux, and C. Bernard-Dagan, eds. Mechanisms of woody plant defenses against insects: Search for patterns. New York: Springer-Verlag. 57–72.

Walker, P. H. 1986. The temperate Southeast. *In* J. S. Russell and R. F. Isbell, eds. Australian soils: The human impact. St. Lucia: University of Queensland Press. 36–62.

White, T. C. R. 1974. A hypothesis to explain outbreaks of looper caterpillars with special reference to populations of *Selidosema suavis* in a plantation of *Pinus radiata* in New Zealand. Oecologia 16:279–301.

———. 1978. The importance of a relative shortage of food in animal ecology. Oecologia 33:71–86.

———. 1986. Weather, *Eucalyptus* dieback in New England, and a general hypothesis of the cause of dieback. Pacific Science 40:69–89.

Whittaker, R. H., and P. P. Feeny. 1971. Allelochemicals: Chemical interactions between species. Science 171:757–70.

Williams, J., K. R. Helyar, R. S. B. Greene, and R. A. Hook. 1993. Soil characteristics and processes critical to the sustainable use of grasslands in arid, semi-arid, and seasonally dry environments. Proceedings of the seventeenth International Grasslands Congress. Wellington, New Zealand: SIR Publishing. 1335–50.

17 Savannization: An Ecologically Viable Management Approach to Desertified Regions

Menachem Sachs and Itshack Moshe

The Forest Department of the Land Development Authority (Jewish National Fund) initiated the Savannization Project to find ways to increase biological productivity and diversity in the desertified ecosystem in the Negev desert and to develop management practices for restoration and sustainability of arid lands.

Restoration included runoff harvesting in catchments, planting of trees and shrubs, and allowing for recovery of natural vegetation. This is based on an ecological understanding of the structure and function of the ecosystem. The restoration process stops desertification, increases the abundance and diversity of organisms, allows for multiple land use, and leads to environmental benefits.

The savannization approach establishes an ecologically based, human-made, artificial ecosystem in desertified environments, leading to partial restoration of environmental disturbances. This concept represents a continuous manipulation process controlled by natural succession and managed by humans.

Desertification is one of the most pressing problems in the world. Vast areas in semi-arid zones are going through a process of desertification due to overexploitation by humans, often exacerbated by climatic changes. Desertification processes are characterized by increased soil erosion and decreased productivity and diversity of plants and animals. When these processes occur in arid regions, they are almost irreversible.

Most of the land resources in the Negev desert, after being used continuously for more than 4,000 years, are in a state of environmental degradation. The Land Development Authority, through its Forest Department, began to develop and implement land management practices, attempting to achieve partial restoration of desertified areas beside and around settlements in the Negev and in agricultural landscapes.

Definitions

Savanna is grassland often dotted here and there with trees or patches of open forest. Some savannas are quite arid with only scattered thorny trees; others have an almost complete tree cover (Billings 1965).

Savannization is a human-managed process whereby the water, nutrients, and soil flows are controlled by increased productivity and biodiversity in desertified areas.

Desertification refers to the loss of the land's productive capacity—a change to desert-like conditions (Thames 1989b). The principal causes of desertification are external pressures that induce increasingly intensified use of the fragile resources, leading to their degradation and destruction.

The *Negev desert* lies in the southern region of Israel and covers about 11,000 km^2 (over 50% of the country) south to the west-east geographical line (lat. 31°37'N), forming a triangle bounded by the Mediterranean Sea and the Sinai desert (Egypt) border to the west, the Dead Sea and the Arava Valley to the east, and Elat on the Gulf of Aqaba at its most southern apex. Mean annual rainfall ranges from 400 mm in the northern Negev to as low as 30 mm in the southern Arava Valley. Precipitation is in the winter and falls mainly between December and March. Its spatial and monthly distribution, as well as the annual amount, varies markedly from year to year.

The term *arid region* is often used loosely to describe areas with relatively low rainfall and a prolonged dry season; thus, all of Israel can be referred to as arid. Yet, over large areas of the country, the annual rainfall may exceed 600 mm (or more), and the ecological systems are not deserts. In this chapter, we will refer to the widely accepted definition of the index of aridity (IA—annual precipitation/potential evapotranspiration). When the value of this index is below 0.2, we approach desert conditions (table 17.1).

Background

The Negev area in Israel includes several geographical regions characterized by their climatic and environmental conditions, both related to rainfall gradients:

1. The northern Negev includes the areas of the southern and western slopes of the Hebron Mountains and the southern regions of the coastal plain (250–400 mm rainfall/year).
2. The western Negev is characterized by light soil (250–300 mm rainfall/year).
3. The central Negev is characterized by loess soil and poor water penetration (150–200 mm rainfall/year).
4. The Negev Highlands, which are 400–900 m above sea level, are characterized by relatively extreme temperatures and the salinity of their soils (100–150 mm rainfall/year).
5. The Arava and Dead Sea Valleys are arid with saline soils or Hamada (30–80 mm rainfall/year).

There has been no significant change in the climate in Israel for the last 3,000 years (Horowitz 1968).

Despite the harsh conditions, the Negev region was settled in ancient times, and the land resources have been in continuous use for 4,000 years or more (Evenari et al. 1982). Limitations of water availability for personal and local agricultural uses were enhanced by developing efficient methods to collect runoff water into reservoirs (cisterns) and into open agricultural fields.

During the Hellenistic period, the Israelite Kingdom (332–37 B.C.), and later during the Roman-Byzantine period (63 B.C.–640 A.D.), urban and rural settlements and areas of runoff agriculture were established in the Negev.

In the seventh century, the urban and rural settlements in the region were abandoned. The permanent residents left, and migrating tribes invaded the area. The migrating tribes continued to use some of the water reservoirs and runoff-collecting installations but did not maintain or preserve them. The runoff agricultural systems that covered tens of thousands of hectares were gradually destroyed. Even in the last century, a small number of nomads still overexploited the opened areas by clear-cutting and overgrazing. Removal of the vegetation has caused accelerated erosion leading to desertification of the Negev (Lowdermilk 1944, Thirgood 1981).

Based on knowledge of the Negev desert ecology, desertification processes can be stopped. The trend of destruction and loss of land value can be reversed by ecological restoration and management. This will increase its value for different uses that will be determined by the people inhabiting the area. About 6% of the Israeli people live in the Negev. The goal of the Forest Department is to stop desertification and increase the value of the land for human use by ecological management of vegetation, soil, and water.

The Objectives of Savannization

The three main objectives of savannization are:

1. to develop management schemes for the rehabilitation of desertified areas in the degraded part of the Negev Desert;
2. to foster development of ecological and hydrological landscape management, increasing biotic productivity while conserving biodiversity in arid lands; and
3. to develop integrated models of rainfall, runoff, soil moisture, vegetation, and animal relationships that will enable efficient implementation of water harvesting techniques in arid regions.

From Desert Afforestation to Savannization

Since the 1950s, forestry development activities concentrated mainly in the northern Negev. Coniferous forests

Table 17.1. Precipitation (P), Evapotranspiration (ET), and Index of Aridity in Different Locations in the Negev

	Years Analyzed	Average Yearly Rainfall (mm)	Average Yearly Potential ET (mm; class A Pan)	Index of Aridity (P/ET)
Sede Moshe	21	385	2,060	0.19
Bet Qama	14	298	2,100	0.14
Gilat	36	242	2,250	0.11
Beer Sheva	68	202	2,340	0.09
Sede Boqer	38	96	2,555	0.04

(mainly) were planted in Shacharia, Amatzia, Lahav, and Yatir on the inclines of the Hebron Mountains. This created a chain of forests along the northern part of the Negev. In addition, planting (mainly eucalyptus) was carried out in eroded areas and badlands in the southern coastal and western plains of the Negev. Success was achieved by the use of conventional afforestation techniques with refined technologies for site preparation, sloping quality, planting, and management programs.

The planting and soil preservation activities expanded in the 1980s to the southern regions of the central Negev. Questions were raised about the feasibility of successful planting in areas where water availability for the establishment of forests and groves is marginal, at least when conventional methods and means of planting are used.

The developmental activities focused on reducing tree density, examining suitability of tree species, adding water through runoff water harvesting, improving propagation material, and intensifying the research activities.

Planting Density

Planting density in the northern Negev was in the range of 1,200–1,500 conifer trees/ha, while the range for broadleaf trees was 400–500 trees/ha.

In areas of 200 mm annual rainfall, planting density was reduced to 100–120 trees/ha. The objective was to establish a population of trees, enable them to use most of the rainwater, and minimize the competition on water resources between the newly planted trees. The open landscape between the planted trees is not altered, enabling development of herbaceous vegetation (mainly annuals) that utilize the water resources available after management.

As a result of the reduced planting density and the recovery of natural vegetation, the landscape developed into an area resembling a dry savanna.

The increase of soil moisture in patches in the human-made savanna, due to runoff harvesting, affects the dynamics of herbaceous vegetation and increases herbaceous layer productivity. This may be associated with lower soil temperature, intercepted rainfall, and greater soil fertility found under tree canopies (regardless of the tree species) in the natural savanna (Belsky et al. 1989).

Suitability of Trees

In the Negev, the variety of local trees that are suitable for planting under arid conditions is limited. Indigenous trees, such as *Tamarix aphylla, Acacia raddiana, A. negevensis,* and *Pistacia atlantica,* were planted together with other species (local and introduced) that are known for their relative resistance to drought. They include several species of *Eucalyptus, Acacia, Pinus, Cupressus sempervirens, Calitris verucosa, Ceratonia siliqua, Ziziphus spina-christi, Ficus sycomorus, Casuarina cunninghamiana, Schinus terebinthifolius, Parkinsonia aculeata,* several *Prosopis* sp., *Retama raetam, Pistacia palaestina,* and so on.

In habitats with better conditions, such as deep soil and low topography, species such as *Ficus carica, Washingtonia, Phoenix dactylifera, Olea europaea, Punica granatum,* and other trees were planted.

Water Harvesting

In semi-arid areas, one of the main factors limiting the development of natural vegetation and planted trees is the low amount of precipitation. The primary developmental activities are based on creating sites in drainage basins that are rich in soil moisture. This is achieved through manipulating water resources by collecting and storing runoff water in the soil. The planting is carried out in water catchments that are rich in soil moisture.

Runoff harvesting and plant introduction were initiated in the Middle East and in Israel (Thames 1989a) thousands of years ago, because our ancestors already knew how to harvest and store runoff water in cisterns for drinking or in terraces for irrigating agricultural fields (Evenari et al. 1982). These historic methods can be used to stop processes of desertification in open lands, increasing their use for the environmental benefit of the desert inhabitants.

The hydrological approach of the savannization project includes the management and harvesting of runoff water all along the watersheds. The watershed is divided into small areas in which water catchments were established along the contour lines. The distance between the water catchments varies according to the expected runoff, which depends on rock and soil types, natural vegetation, topography, slope aspect, and rainfall regime (fig. 17.1). This method of watershed management prevents soil erosion, headcuts, and gully formation that decrease habitat productivity and biodiversity. In addition, by deep infiltration of runoff into the soil, more water is available to natural and planted vegetation.

Quality of Propagation Material

Advanced nursery methods were developed by which the saplings are grown in minimal-volume containers on soilless growing medium and are planted under severe conditions in the desert when they are relatively young (6–8 months). Their high quality enables rapid development immediately after planting. The saplings, characterized by vigorous luxuriant growth, are naturally air pruned, have a dense root system, and are primed for rapid growth. The saplings are easily transplanted without root damage and have a fully active root system to ensure a maximum survival rate (85–90%) under desert conditions.

a.

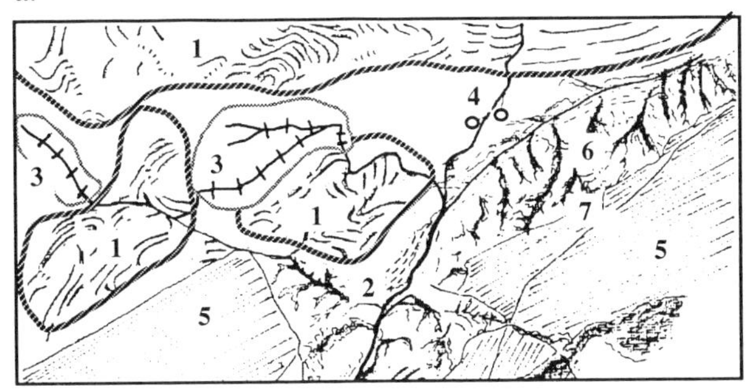

b.

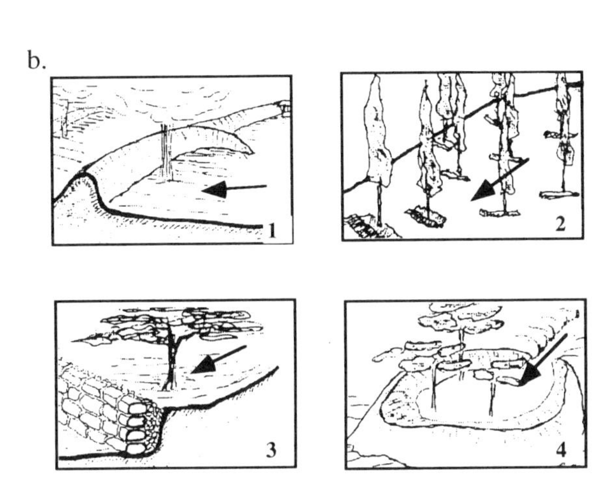

Figure 17.1. (a) A restored landscape: savannization in a desertified area in the northern Negev (after an aerial photograph 1:15,000). (b) Runoff after methods in restored landscape (1 = contour bench terrace; 2 = hillslope minicatchment; 3 = terraced wadi; 4 = wadi catchment basin; 5 = cultivated field; 6 = badlands; and 7 = gully headcut).

Research

When we began to develop degraded areas with 150–200 mm annual rainfall and modify the sensitive desertified ecological system, we acted with minimal theoretical and research background. To reduce the risks of mistakes concerning the environmental aspects and the sustainability of the system, a research team was established to develop the scientific basis for arid land ecological management. The combined activity of research and development was termed the *Savannization Project.*

The project brings together an interdisciplinary group of scientists and developers from various institutions in Israel and abroad. The range of scientific disciplines includes ecology, hydrology, geomorphology, soil conservation, range management, remote sensing, forestry, plant and animal physiology, microbiology, and resource management. The theme is to integrate the studies and the implementation of the different disciplines concerning the processes governing the biological resources and diversity into *one* ecosystem model.

Most of the research projects are carried out in three research stations that have been set up along a precipitation gradient: Lehavim (300 mm), Sayeret Shaked (200 mm), and Sede Boqer (100 mm average annual rainfall). Examples of the ongoing research and development projects include, for example:

1. the structure, function, and development of natural and human-made savanna landscapes and their effects on productivity and diversity of arid lands;
2. development of rainfall, runoff, soil moisture, and vegetation models for landscape units; models are based on rainfall, soil topography, vegetation, and animal characteristics;
3. soil crust (physical and biological) formation and its effects on water flow in arid lands using a portable rainfall simulator and in situ measurements of rainfall and runoff;
4. the dynamics of herbaceous production (a principal source of forage for grazing livestock) in stressed and water-enriched and/or nutrient-enriched habitats;
5. screening tree species to increase water-use efficiency and drought hardiness;
6. establishment of trees (local and introduced exotic species) and developing silviculture management practices for the human-made savanna; and
7. implementation of water harvesting techniques to determine optimum ratios between the catchment size and the target plot, depending on regional landscape characteristics.

Ecological Management

The ecological system of arid regions in Israel has been affected by the presence and interference of humans for thousands of years. The main influence of humans on open desert regions has been through developing settlements and building roads, establishing runoff agriculture, and intensive grazing activities. In view of the continuous activities of humans in the Negev, open lands can be defined according to several types of ecological systems:

1. natural ecological systems that function without the interference of humans;
2. human-made ecological systems that function only when humans interfere in land management (e.g., afforestation, agriculture);
3. desertified ecological systems that were destroyed by humans but still function without interference; and
4. mixed ecological systems—natural and human-made—that are based on limited development processes and watershed management (savannization).

The Israeli desert is a network of the four types of ecological systems. Each type is an open system with soil, organic matter, nutrients, seeds, and animal movement in and among them. External climatic factors and activities carried out by humans have a notable influence on the characteristics and reactions in these ecological systems. The ecological systems are exposed to frequent and various agents of change. The climatic factors are mostly unpredictable (e.g., floods, drought, heat) while the human factors are difficult to control (e.g., uncontrolled pollution from developed areas and/or uncontrolled intensive grazing by domestic animals). Since these are open systems by nature, a conversion from one type to another can occur due to management activities that are designed and regulated by humans. The above has determined in the past and still affects at present the dynamics of open lands (changes in ecological system types) in the Negev regions.

The interrelationships among the different ecological systems in the desert are characterized by three types of processes that could cause such a change:

1. *desertification processes:* degradation of ecological systems mainly due to plant cover reduction and soil erosion, resulting in a decrease in productivity and biodiversity;
2. *restoration processes:* specific management processes in desertified ecological systems (including developmental aspects) directed to halt the desertification processes and increase productivity and biological diversity; and
3. *conversion processes:* management and other processes in nondesertified systems that bring a direct change in ecological systems (changing the type of the ecological system) (fig. 17.2).

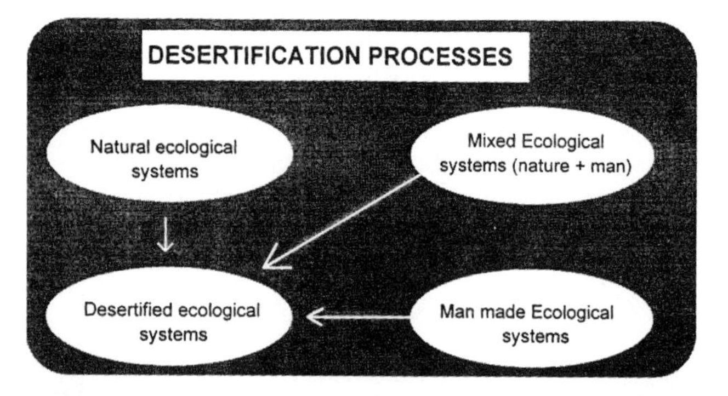

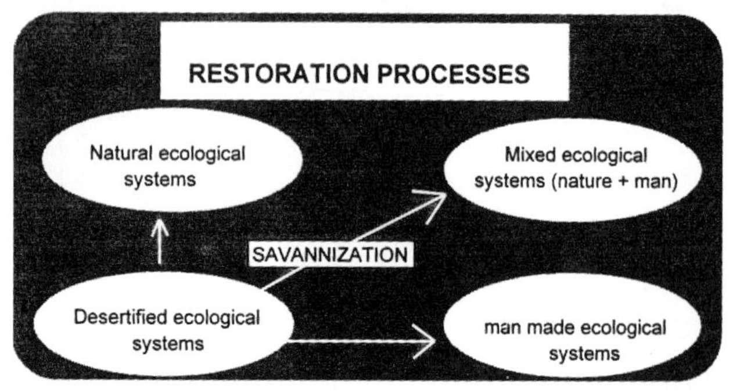

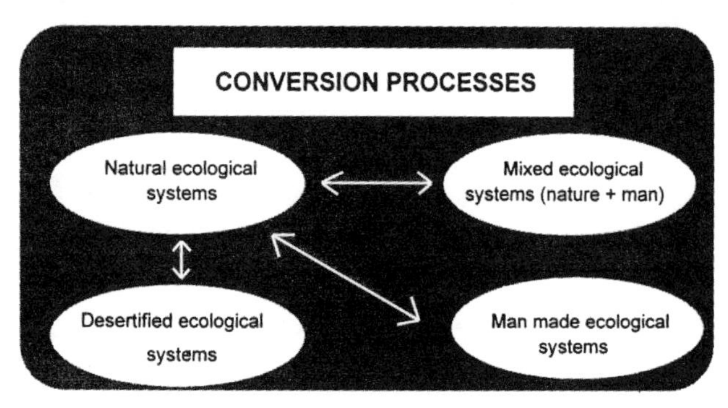

Figure 17.2. Desertification, restoration, and conversion processes and changes in ecological systems (savannization is a restoration process resulting in a mixed ecological system).

The management leading toward the creation of a human-made savanna is an example of a restoration process of the changing of a desertified ecological system into a mixed ecological system. This process differs from other developmental and management processes since it is guided by an ecological philosophy and by minimal human intervention. All the changes carried out during the restoration process are aimed at minimizing the impact on future land use. This is a watershed management approach that considers landscape units in trying to manage the diversity of several components of the ecosystem within a watershed.

The uniqueness of this ecological restoration of desertified regions results from the developmental activities as well as the human obligation for ecological administration of the system to assure its sustainability. One hopes that in this status the ecological system will be able to withstand environmental and climatic agitations with minimum effects.

The knowledge gained in the present management stage has encouraged us to continue the restoration of open lands even in regions with more extreme arid conditions (the Negev Highlands and even the Arava). Developmental processes in these regions should be carried out on a very small scale, and the management aspects must be implemented with even less intervention than in the current savannization project.†

Conclusion

Seven years of implementation experience yielded 2,300 ha of flourishing human-made savanna in the Negev desert, and we are still beginners and students.

The goal of the Savannization Project is to provide a restoration model for desertified areas in Israel as well for other arid countries. The use of the model should help to develop the desert landscape, restore its productivity, expand biodiversity, elevate and sustain the use of arid lands, and, thus, increase its value for humans.

Note

The authors wish to acknowledge the unique cooperation with the Savannization Research Team including Moshe Shachak, Rami Garti, Avi Perevolotsky, and Gabriel Schiller. Special thanks to Moshe Shachak for many fruitful discussions that brought theories to a successful field implementation and to Sol Brand for his useful comments, which were incorporated into the manuscript.

References

Belsky, A. J., R. G. Amundson, J. M. Duxbury, S. J. Riha, A. R. Ali, and S. M. Mwonga. 1989. The effect of trees and their physical, chemical, and biological environments in a semi-arid savanna in Kenya. Journal of Applied Ecology 26:1005–24.

Billings, W. D. 1965. Plants and the ecosystem. London: Macmillan.

Evenari, M., L. Shanan, and N. Tadmor. 1982. The Negev: The challenge of a desert. Cambridge, Mass.: Harvard University Press.

Horowitz, D. 1968. Upper pleistocene-halocene climate and vegetation of the upper Jordan Valley. Ph.D. diss. Hebrew University of Jerusalem. [Hebrew, with English summary]

Lowdermilk, W. C. 1944. Palestine: Land of promise. London: Victor Gollang.

Thames, J. 1989a. Water harvesting. *In* Role of forestry in combating desertification. FAO Conservation Guide 21. Roma. 234–252.

———. 1989b. Watershed management in arid-zones. *In* Role of forestry, in combating desertification. FAO Conservation Guide 21. Roma. 211–33.

Thirgood, J. V. 1981. Man and the Mediterranean forest: A history of resource depletion. London: Academic Press.

18 Managing Patchiness, Ecological Flows, Productivity, and Diversity in Drylands: Concepts and Applications in the Negev Desert

Moshe Shachak, Steward T. A. Pickett, Bertrand Boeken, and Eli Zaady

Dryland ecological systems can be studied and managed from various perspectives. Population, ecosystem, or landscape perspectives are perhaps the most familiar ones used by desert ecologists, although they are rarely unified in individual studies. It may be possible to integrate the various perspectives to bring the greatest weight of ecological knowledge to bear and to improve the success of management. Due to the limited amount of water, soil, and nutrients available to the organisms in drylands, clear patterns of spatial heterogeneity in the landscape have developed. Key features of dryland heterogeneity are the spatial distribution of rock and soil and the distribution of shrubs within the matrix of the soil crust formed by microphytes. We refer to such spatial heterogeneity as *patchiness* when an abiotic or biotic variable shows two discrete states at one level of resolution. In this chapter, we demonstrate the importance of dryland patchiness for understanding and management of ecological systems. We use ecological studies in the Negev desert as an illustrative case. Our objectives are (1) to show the role of patchiness in the functioning of dryland ecological systems and (2) to demonstrate the functional relationship between patchiness of ecological systems and management in the Negev.

In order to meet our objectives, this chapter addresses the following questions: (1) What are the patterns of patchiness in the Negev? (2) What are the relationships between patchiness and dynamics in the ecological systems of the Negev? (3) What are the relationships between patchiness and the productivity and diversity of the Negev? (4) What is the nature of human-made patches in the Negev, and what are their implications for the management of arid ecological systems in general?

Models of Ecological Systems for Drylands

In order to understand and manage dryland systems, tractable yet comprehensive models are required. Dividing ecological systems into population, community, biogeochemical, and landscape systems is a useful way to identify the key linkages that can reduce ecological complexity in addressing scientific and management problems while maintaining the richness of system components. Population ecological systems focus on the dynamics of individuals of the same species; community systems focus on between-species interactions; ecosystems deal with energy and materials flows; and landscape systems focus on the arrangement, origins, effects, and alterations of patches, that is, on patch dynamics. The dynamics of all these kinds of ecological systems can be cast in terms of flows because, in each case, there is a flux of the entities or quantities from one state to another. For example, if our main concern is to study and manage a population, our focus is on changes in the number of individuals through time, and the flow is of individuals among size, age, or reproductive classes and from birth to mortality. Concern with flows also focuses on the factors and agents that control the rate of the flow.

In essence, the study and management of an ecological system is a problem of selecting a flow of interest out of an ecological system comprising multiple flows and then identifying the interactions between the target flow and other important flows. For example, in the case of the population flow mentioned above, questions of interacting flows might focus on whether population dynamics were controlled by water availability and competition. Therefore, the interaction of the flow of organisms with the flows of water and other species must be assessed.

The strategy of constructing ecological systems by focusing on flows is a flexible and generally applicable approach. It can even be used in situations where trade-off or optimization is a concern. For example, the balance between one or more species that might have contrasting requirements in an area can be incorporated into a model of an ecological system by making that balance the state that is subject to a flow or transition. Alternatively, a composite variable, such as native species diversity, may be chosen to represent the status of all such entities in a system.

Patchiness in the Negev

The subdiscipline of ecology that emphasizes the role of spatial heterogeneity in the functioning of ecological systems is landscape ecology. Landscape ecology focuses on the development and dynamics of spatial heterogeneity and its effect on populations, communities, and ecosystems. In this chapter, landscape structure refers to the size, shape, number, type, and configuration of patches. Landscape functioning is defined as the interaction of species, communities, nutrients, and energy in the array of patches in a landscape and the changes in the status and configuration of patches. Reciprocal relationships may exist between landscape structure and function.

Types of Patches

Our classification of patch type in the Negev is based on (1) how patches affect the flow of water and (2) their origin. Patches can exist in two states relative to the flow of water: one state is a *source* and the other is a *sink* for water. We selected the control over water as the criterion for patch classification because water is the main limiting resource and its redistribution by landscape heterogeneity is a controller of landscape function, productivity, and diversity (Yair and Shachak 1987). Characterizing a patch as a source of water indicates that it is a source for runoff to some other patch downstream, not that it is the ultimate, original source of water in the entire system.

The other criterion by which patches can be characterized is by mode of origin. In the area of the Negev used to illustrate the principles of heterogeneity and management, patchiness originates in three ways: (1) physical, (2) biological, and (3) human construction. We explain these types below.

1. *Physical patchiness* is the result of two discrete states of the surface in the landscape: soil and rock. Rock is a source and the soil is a sink for runoff water. Physical patchiness is caused by two processes that act over different time and spatial scales but that interact to affect biodiversity and productivity. Geologically, patchiness results from the existence of different rock types, and geomorphologically, heterogeneity is caused by deposition, weathering, erosion, and redeposition of material. Both sources of physical heterogeneity affect the spatial configuration and size distribution of the physical patches of rock and soil.

2. *Biological patchiness* is determined by the distribution and performance of organisms and their by-products. Within the soil patches, shrub distribution determines biological patchiness. Shrub patches are characterized by an environment that differs in biological, physical, and chemical properties from that between shrubs. Under shrubs, the soil is often mounded, has higher nutrient and organic matter content, and contains a richer herbaceous understory relative to the surrounding area (Allen 1991, Halverson and Patten 1975, Muller 1953, Muller and Muller 1956, Weinstein 1975, Went 1942). The area between shrubs is covered by a soil crust community that may include algae, cyanobacteria, mosses, or lichens.

The two discrete states of biological patchiness are the level soil crust with its lower plant community ("microphytic patch"), and the soil mound with its community of higher plants ("macrophytic patch"). A downslope soil mound is a sink for runoff water and the upslope soil crust is the source of that runoff. Source and sink are defined in terms of the surface flow and absorption of water originating as rainfall.

3. *Human-made patchiness* is generated by constructing pits and mounds in the landscape. The constructed pits and mounds replace part of the natural environment and add new patches to the landscape that are used for cultivation. The two discrete states of human-made patchiness are the pit and the mound. The pits are constructed to collect runoff and the mound to prevent water leakage from the pits. Natural and introduced vegetation in the pits and on the mounds can be used for crops and firewood production and for livestock grazing. For example, several species of native and introduced trees have been successful in the enhanced moisture of the pits in spite of the annual rainfall totaling only 200 mm per annum. Between the sets of pits and mounds are more or less intact natural surfaces, although immediately upslope of the pits some initial plowing is often done.

Physical, biological, and human-made patches have an important effect on the distribution of resources, primarily water in the form of runoff (Shachak and Brand 1991, Shachak et al. 1991, Yair and Lavee 1982, Yair and Shachak 1987) but also nutrients and organic matter and seeds in the landscape (Price and Reichman 1987, Reichman 1984).

The three types of patchiness show a pattern on a regional scale along the rainfall gradient in the Negev (fig. 18.1). The percentage cover of rocky, microphytic, and natural physi-

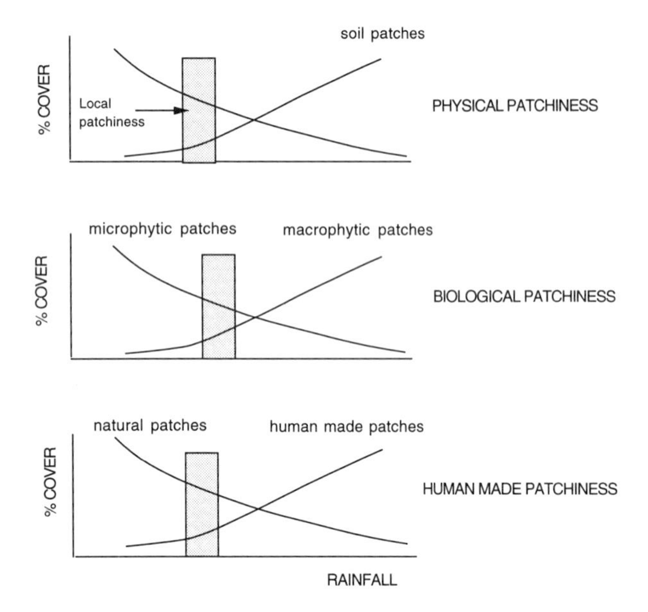

Figure 18.1. Changes in physical, biological, and human-made patches along a rainfall gradient in the Negev desert. The bars represent the location of a range of local patchiness under consideration as a part of the entire rainfall gradient.

cal and biological patches decreases as mean annual rainfall increases, while the percentage cover of soil, macrophytic, and human-made patches increases as mean annual rainfall increases. In terms of sources and sinks, the general trend is an increase in the cover of sink patches as mean annual rainfall increases. This trend implies that, as the input of water by rainfall to the landscape increases, proportionally more water will be stored in the area.

Patchiness and Level of Organization

We selected three case studies to demonstrate the relationships among landscape, biogeochemical, community, and population systems. The objective of the first case study is to show the relationship between physical patchiness and population dynamics. The second will show the effect of biological patchiness on the input-output relationship of dry and wet deposition. The third will demonstrate the effect of human-made pits and mounds on productivity and diversity.

Patchiness and Populations

Scales of Physical Patchiness Our study of the relationship between scale of patchiness and population dynamics occurred in two different landscape units: a loessial valley and a rocky slope under the same climatic regime (100 mm annual rainfall) about 4 km from each other. The loessial valley is covered by thick loess-derived serozem soil, and vegetation covers 10–15% of the surface. The loessial valley is part of a large basin that experiences flooding infrequently. However, during low frequency floods, the area experiences a high magnitude of runoff water input. The rocky hills are 40–80% massive limestone bedrock, and soil is found in strips located at the base of bedrock steps formed along major bedding planes. The vegetation covers from 5% to 10% of the surface and is concentrated along the soil strips and in soil-filled joints. Rocky hills exhibit runoff properties different than the loessial valleys. Soil downslope from rocky outcrops receives runoff water in addition to rain after most rain events (Yair and Lavee 1982). Soil closer to the rocky outcrops absorbs runoff water more frequently and in greater amounts than soil farther away from the rock outcrops. However, the quantity of runoff water generated on one slope does not reach the magnitude of runoff generated in a large basin, such as an entire loessial valley.

On a small scale, the rocky slope is patchy in rock and soil distribution, while the loessial valley is spatially homogeneous. However, if we consider the rocky hills and the loessial valley as one landscape unit on a larger spatial scale, then this unit is patchy. The hills represent the rocky patch, and the valley, a soil patch.

The Population Our long-term study on the population dynamics of the desert isopod, *Hemilepistus reaumuri,* on the loessial valley and the rocky hills can demonstrate the relationships among population abundance, physical patchiness, and scale of patchiness (Shachak and Brand 1991). *H. reaumuri* is an annual animal with a simple life cycle (Linsenmair 1972). It lives in monogamous family units in a single burrow dug by the family members. Each unit comprises parents that live together with the young until the parents' death, usually in September. During September and October, the siblings continue to live in the same family unit, spending most of their time in the burrow and only about 1–2 hours each day outside foraging on organic matter and the microphytic community of the soil crust. From November to February, the family is in a sedentary state and stays underground in its burrow, which is about 50 cm deep. Each burrow is occupied by about 30–50 siblings. In February, the isopods vacate the sites where they hatched and matured and search for new settling sites. In April, the females are gravid, and in May the young hatch. During the postsettling period, from April to February, the parents die, the offspring mature, and the isopods extend their burrow to a depth of about 50 cm in quest of relatively high soil moisture. The main factor in the isopods' survival is soil moisture. If the burrowing site at a depth of 40–60 cm does not maintain at least 6–10% soil moisture all summer, then its inhabitants perish (Coenen-Stass 1981, Shachak 1980).

In order to determine the abundance of *H. reaumuri* populations, we measured the densities of settlers and successful families in the loessial valley and rocky hills for fifteen generations (1973–87). Settlers are defined as the number of new families formed in February. Successful families are the number of families that survived after settling until the next settling season.

Population Dynamics We found that settlers and successful families have a pattern related to elevational contours on the rocky slope but no apparent spatial pattern in the loessial valley (fig. 18.2). On the rocky slope, there is a positive relationship between successful families and settlers along contour bands. On the rocky slope, the high variability in *H. reaumuri* abundance among contours is related to small-scale physical patchiness in terms of the rock-to-soil ratio.

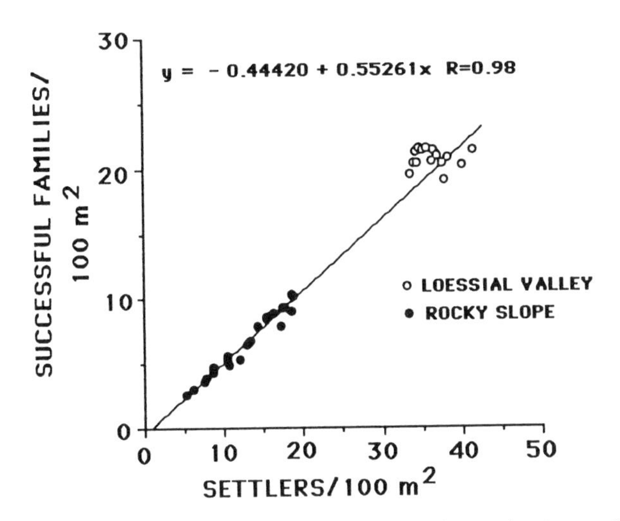

Figure 18.2. Relationship between physical patchiness (rock-to-soil ratio) and demographic variables (settling and establishment of successful families) on the desert isopod, *Hemilepistus reaumuri*, on heterogeneous rocky slope and on homogeneous loess-covered valley.

In the loessial valley, due to homogeneity in soil cover, we do not find an abundance-survivorship relationship.

Evenari and colleagues (1983) found that runoff frequency and magnitude is scale dependent. On small watersheds, runoff generation and infiltration is characterized by a relatively small magnitude but a high frequency in comparison to large watersheds. On the rocky slope, *H. reaumuri* abundance and variability is a response to small-scale processes of runoff generation and infiltration that create a typical spatial pattern of soil moisture distribution along the slope. The pattern is of relatively high water concentration in soil patches in areas with a high rock-to-soil ratio (Yair and Lavee 1982). Soil moisture distribution controls the distribution of sites available for the settling of isopods. Small-scale patchiness controls water flow that in turn controls the variability in abundance of settlers and successful families.

In the loessial valley, the high abundance of settlers and successful families is a response to large-scale drainage basin processes that collect runoff water from many small watersheds. Population dynamics in the loessial valley are controlled by water accumulation in the soil of the loessial valley from many rocky hillslopes. This is a population response to large-scale landscape patchiness.

Patchiness and Ecosystem Properties

Biological Patchiness We are studying the relationship between biological patchiness and ecosystem properties in Sayeret Shaked Park near Beer Sheva in the northern Negev desert (31°17'N, 34°37'E). Rainfall, which only occurs in winter between November and March, has a long-term annual average of 200 mm.

Spatial heterogeneity in the landscape is characterized by a *matrix* (Pickett and White 1985) of crusted soil with microphytic communities and distinct patches of perennials with a soil mound and herbaceous understory. The microphytic crust of the matrix consists of cyanobacteria, bacteria, algae, mosses, and lichens (Andrew and Lange 1986a, 1986b; Hacker 1984, 1987; McIlvanie 1942; West 1990). Soil covered with well-developed microphytic crust has a tightly structured surface (Fletcher and Martin 1948), primarily due to binding of soil particles by polysaccharides excreted by bacteria and cyanobacteria (Metting and Rayburn 1983). In contrast, the soil of shrub patches lacks a well-developed microphytic crust, and its surface is covered with loose soil particles (West 1989).

Patchiness and Flows We measured particulate deposition to the soil surface in the two classes of the biological patches, macrophytic and microphytic. This deposition, which includes both airborne dust and microparticulate litter fall, is higher in the macrophytic patches than in the microphytic patches (fig. 18.3a). We assume that the higher rates of particulate deposition in addition to sedimentation during runoff events are the two controllers over soil mound formation in the macrophytic patch. We also found that the deposition in two patch types differs in its chemical composition. Carbon-to-nitrogen ratio is significantly higher in the microphytic patch (fig. 18.3b). This implies that the two patch types control both the distribution and quality of particulate deposition in the landscape. A greater amount of high nitrogen materials is accumulated in the macrophytic patches.

To investigate the relationship between the biological patchiness and water flow, we have studied runoff generation in isolated enclosures containing either macrophytic patches or pairs of macrophytic patches and their adjacent upslope microphytic patches. We measured the runoff from each enclosure following each rainfall event. In addition, we analyzed the nitrogen content of runoff samples. A rainfall

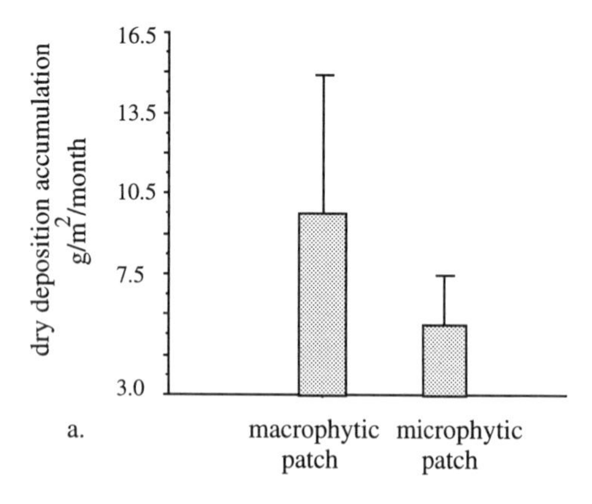

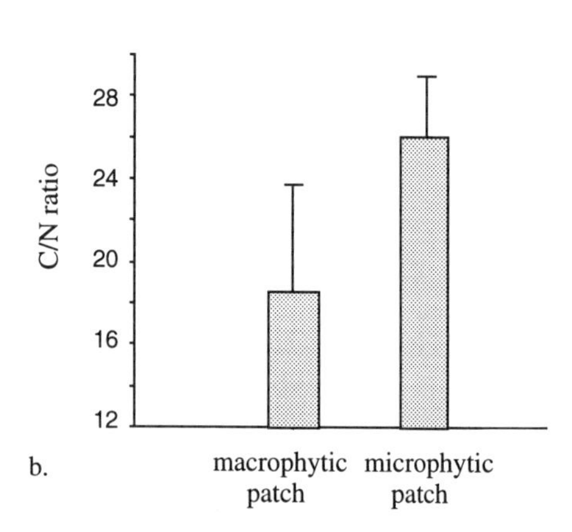

Figure 18.3. Relationship between biological patchiness (micro- and macrophytic) and dry deposition: (a) total dry deposition of nitrogen during the summer of 1993 and (b) the C:N ratio of dry fall during that same period.

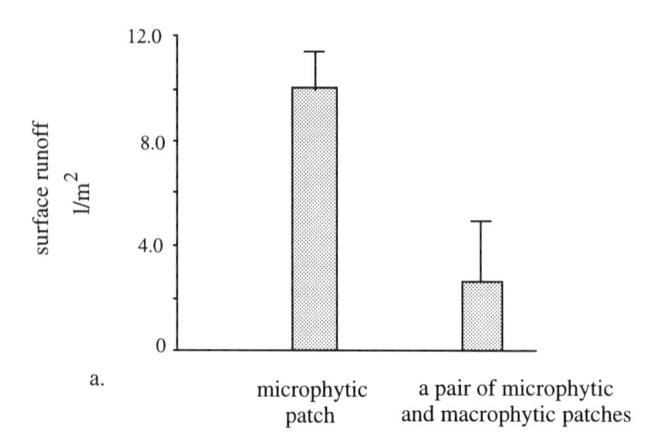

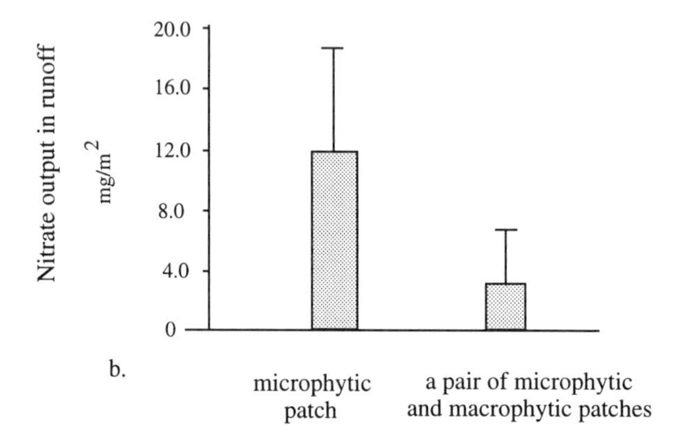

Figure 18.4. Relationship between biological patchiness and leakage: (a) runoff generated by one rainfall event in isolated enclosures of a microphytic patch paired with its downslope macrophytic patch in winter 1994 and (b) nitrate output in runoff from one rainfall event in the same exclosures used for data in (a).

event in winter 1994 illustrates the fact that runoff generation by the microphytic patches alone is significantly higher than runoff production by the pairs of the two patch types (fig. 18.4a). This occurs because the macrophytic patch acts as a site for runoff water from the microphytic patch. Nitrate output is also higher in the microphytic patches than in the pairs of the two patch types (fig. 18.4b). These data indicate that runoff and nitrate flows, which affect the spatial distribution of water and nutrients in the area, are controlled by the biological patchiness.

This case study demonstrates the complex relationships between biological patchiness and ecosystem properties such as soil and water flows. When the macrophytic patch functions as a trap for water or airborne materials, the result is an increase in its size. The consequences of a larger macrophytic patch are a smaller microphytic patch size in a given area, which implies changes in source-sink relationships in the landscape. More water and nutrients are stored and lesser amounts are leaked from the landscape.

Human-Made Patches and the Biotic Community

In the Negev, most human-cultivated patches were constructed in order to harvest runoff water (Bruins et al. 1986, Evanari et al. 1983, Kedar 1967). Thus, pits and their associated mounds are typical structures in human-made patches, both ancient and modern. A central question in relation to human-made patches is how plant communities respond to changes in the biological patchiness resulting from the digging of pits and placing the excavated soil in a mound covering the adjacent crusted soil.

Herbaceous Plant Communities in Pit and Mound Complexes The effect of small-scale human-made patches on herbaceous plant communities was studied at Sayert Shaked (Boeken and Shachak 1994). The area is covered with microphytic soil crust, consisting of bacteria, cyanobacteria, algae, mosses, and lichens, and with scattered patches of the shrubs *Noea mucronata* and *Atractylis serratuloides* (Feinbrun-Dothan and Danin 1991). The soil is loessial, with 14% clay, 27% silt, and 59% sand, and is at least 1.10 m thick (Teomim

1990). Salt content of the 0–25 cm soil layer is low with electrical conductivity of 0.4 mMho (Teomim 1990).

Pits (1 m x 30 cm) with a depth of 20 cm were dug in the microphytic soil crust parallel to the contour of the slope. Excess soil was deposited directly below the pits as 15 cm high mounds having a footprint of 1 m x 40–50 cm. During rainfall, the soil crust upslope of the pit functions as the contributing area for runoff that collects in the pits. In contrast to the pits, the mounds only receive direct rainfall. Soil moisture measurements at rooting depth showed that the pits were significantly wetter than the crusted soil while the mounds were drier.

Seventy-seven species were identified in the pits and mounds. Of these, 65 were annuals (84%) with an average frequency per species of 17 out of 60 samples. Total plant density and species richness differed significantly among the patch types. Density and species richness were higher in the moister pits relative to the undisturbed matrix. In the mounds, in spite of the decrease in soil moisture, density and species richness were also higher than the undisturbed matrix.

Biomass yield per sample also differed significantly among patch types, being greatest in pits, intermediate in mounds, and lowest in the matrix. The increase in total biomass per patch in mounds and pits relative to the matrix was primarily due to increased plant density.

The annual plant community composition changed in response to changes in (1) soil surface texture, from densely packed to loose soil in both pit and mound, (2) microtopographic structure, from flat surface to mound or pit, and (3) soil moisture availability, which was greater in the pits but lower in the mounds. Successional changes in macrophytic patches are expected to take roughly one hundred years, while development of microphytic patches occurs on the time scale of decades.

The changes in species richness, plant density, and biomass are the result of the disturbance of the soil surface. The differences in height of the pits and mounds relative to the surrounding matrix function as traps for seeds and as a filter that affects seed size distribution. The absence of a microphytic crust causes higher seed densities, greater availability of protected sites, and a shift in seed size distribution. In addition, soil crust disturbance in the form of a pit increases soil moisture while a mound decreases soil moisture relative to undisturbed soil crust. The pits illustrate positive changes in relation to seed and water flows, as they function as a sink for runoff water and seeds. The mounds slow the wind and, therefore, function as a seed filter, but their combined net effect on plants is negative due to a decrease in water availability.

Patchiness and Human Activities

Throughout history, humans have controlled the ecological systems of the Negev (Evanari et al. 1983). The impacts are diverse and span the ancient to industrial eras. Here we describe the variety of impacts in order to lay the groundwork for showing how patchiness can be exploited to enhance ecological management in both the Negev and other arid lands subject to patchiness. In essence, throughout the long history of dryland management in the Negev, people have managed the patchiness of the landscape.

Grazing, Runoff Farming, and Patchiness

Grazing and runoff farming were the main methods of using the limited resources of the Negev in order to support human populations throughout the preindustrial era. Both herding and runoff farming changed the patchiness, flow of resources, and the productivity and diversity of the Negev (fig. 18.5).

Domestic animals feed on higher plants and disturb the soil crust and mounds by trampling. Therefore, we envision changes in this landscape as a result of millennia of intense human use. Grazing animals must have modified the size, shape, number, and configuration of the macrophytic and microphytic patches (i.e., they altered the biological patchiness). When the rate of feeding on macrophytic patches was higher than their recovery, growth and coverage of macrophytic patches would have been reduced. As a consequence of the reduction in the macrophytic patches, the area covered by microphytes would have increased. The changes in landscape structure due to modification of the biological

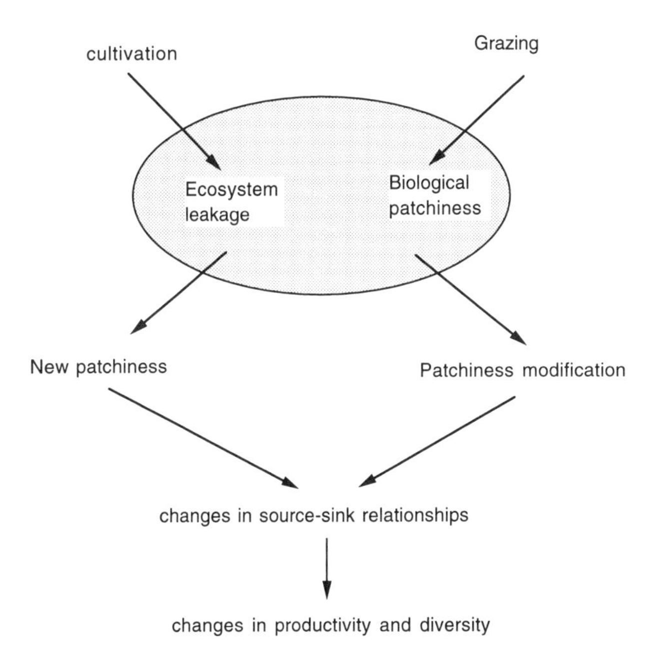

Figure 18.5. Human activities and their effects on patchiness, flows, productivity, and diversity in the Negev desert. Grazing modifies the existing biological patchiness, and cultivation creates new patches in the landscape. Both change the relationships between sources and sinks and result in altered productivity and diversity.

patchiness by grazing would have changed the flow of water in the ecosystem. Runoff generation by the microphytes would have increased while runoff absorption by the macrophytes would have decreased. The net effect of grazing in the long run would have been to increase the microphytic-to-macrophytic patch ratio. This would have resulted in decreasing productivity and increasing the leakage of water and nutrients from the ecosystem. Note that while these net effects were in progress, other relationships between microphytic and macrophytic patches may have existed.

Runoff farming was the first attempt to utilize ecosystem leakage by adding human-made cultivated patches into the Negev desert (fig. 18.5). This technology was based on the idea that ecosystem leakage can support only small patches of human-made agricultural systems richer in resources relative to the natural matrix. For example, in the Negev Highlands, with about 100 mm rain/year during the peak period of rainwater harvesting agriculture during the Nabatean period, the ratio of cultivated to noncultivated areas was 1:18 (Evenari et al. 1983, Kedar 1967). The cultivated patches were constructed to utilize the leakage of water and soil from the noncultivated landscape. The input of water, soil, and nutrients into the fields was a function of landscape structure and interaction among the patches.

Human-Made Cultivated Patches Seven types of human-made cultivated patches have been identified in the Negev. They differ in their location in space, methods of runoff water distribution, spatial scale of runoff water contributing area, and heterogeneity of resources distribution (Bruins et al. 1986, Evenari et al. 1983, Kedar 1967, Yair and Shachak 1987). Some of these are ancient while some represent both ancient and modern runoff collection techniques.

1. *Terraced wadi:* In this type, the fields are located in the wadis, or streambeds, of small watersheds. The fields are among a series of low check dams constructed across the wadi. The fields are small (102–3 m^2) and vary in size. The scale of the runoff contributing area is smaller than a watershed. Runoff water distribution among the fields is nonuniform. The frequency and the magnitude of runoff collected in the fields at the head of the valley are higher than water collected in the downstream fields. Note that this pattern is the opposite of the relationships expected in mesic environments.

2. *Diversion:* The area contributing runoff to the diversion systems is a large drainage basin that collects the water flow from many small watersheds. In diversion fields, the runoff water from a point in the wadi is diverted to a large arable area at an elevated position with respect to the downstream wadi bed. This method of cultivation allows for a higher degree of homogeneity in water distribution in comparison to the terraced wadi method.

3. *Hillslope minicatchment:* These cultivated patches are located on the soil patches of the hillslope to collect runoff generated by the rocky patches. The patches are made of small furrows (about 4 x 2 m) dug across the slope for water collection. The runoff contributing area is small, consisting of the rocky patches on the slope. Due to the variability of the physical patchiness, there is high variability in the soil moisture of the minicatchment.

4. *Mini-liman:* The fields are located either at the transition between rocky and alluvial parts of very small tributaries that join the main wadi or at the mouth of a small watershed. One check dam is built across the tributary to collect runoff generated in that part of the watershed flowing into the tributary. A spillway regulates the water level in the mini-liman. The mini-limans vary in size and water resource base. Therefore, there is great variation in the resource availability of this human-made patch type. Both the hillslope minicatchment and the hillslope mini-liman are constructed to utilize small-scale ecosystem leakage based on physical patchiness of high rock-to-soil ratio.

5. *Liman:* These cultivated patches are similar in structure to the mini-limans. However, they differ in their spatial location and the scale of runoff contributing area. Limans are built across the wadi in large plains. Runoff water is collected from a network of watersheds or from a large drainage basin. Due to their location and water resource base, limans represent a higher degree of homogeneity in resource distribution than mini-limans.

6. *Contour field* (*shikim*): In this type of field, which is dyked along the contours at roughly 20-m intervals, runoff water is collected from long catchments above each dyke. Shikim may be viewed as a long minicatchment. The runoff contributing area is small; it catches the leakage from the biological or physical patchiness on the slope below the next highest dike. Due to the variability of the biological and physical patchiness on slopes, there is high variability in water yield, soil, and nutrients in a contour field.

7. *Microcatchment:* These fields differ from the other types because the runoff-producing areas as well as the runoff-receiving areas are manipulated. The microcatchments are located at the lower gentle slopes of the hillsides. A microcatchment field is composed of many microcatchments divided into two parts: a relatively large area with a gentle slope that produces runoff and a small pit in the lowest section used as a runoff collector and infiltration basin. As all the microcatchment fields are designed, the degree of heterogeneity in resources is low. However, due to their small size, the amount of resources in the microcatchment is relatively small.

Patterns in Cultivated Patches

Human-made patches are a major management tool for afforestation and restoration of desertified areas in the Negev (Sachs and Moshe chap. 17, this volume). In order to ensure sustainability of both productivity and natural diversity, the effect of human-made patches in the landscape on the functioning of the larger ecological systems of which they are a part must be taken into consideration. The properties of patches that must be accounted for in assessing their role in the system of both natural and artificial patches are the following: patch productivity, patch diversity, patch integration, and patch management (fig. 18.6). *Patch production* is defined as biomass production per unit area of natural or introduced vegetation. *Patch diversity* is the degree of heterogeneity in the distribution and abundance of resources over space and time in the patches. *Patch integration* is the net resource yield (or demand) of the human-made macrophytic patches in relation to the physical and biological patches that they interact with. The level of *patch management* is the amount of human resources required to construct and maintain the patch and the effect of those management practices on the natural patches in the landscape.

There are trade-offs between the diversity of resources in a patch type and its productivity. For example, hillslope minicatchments, because they are small, heterogeneous in location in relation to physical and biological patches on the slope, and act as a sink of resources from a small contributing area, collect small but variable amounts of resources. In contrast, limans, which are located in the wadi and, therefore, are a sink for resources from a large contributing area, are relatively more homogeneous and high in resources. Note that limans, with their high soil water content, act as mesic environments—thus, the relationships between productivity and diversity are different than in the unmodified desert. Therefore, if higher variability in resources can support higher species diversity, then in selecting among human-made patches, we have to trade between biological diversity and productivity. An additional trade-off is between scale of integration and production. The functioning of a human-made patch, such as a diversion system, is based on a large runoff contributing area. Its size in relation to the area covered by the physical and biological patchiness is relatively small in comparison to contour fields that cover up to 20% of the landscape. However, the biological production per unit area is much higher in the diversion system than in the contour fields.

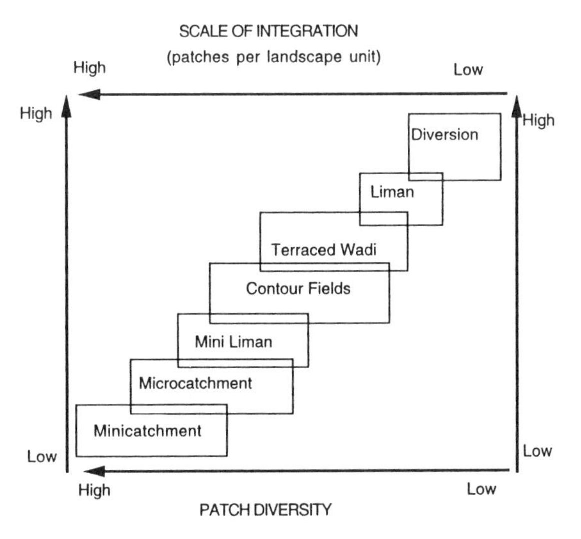

Figure 18.6. Trade-off properties of seven human-made cultivated patch types in the Negev desert. The patch types are defined in the text.

To summarize these examples, the benefits of adding minicatchments are high coverage, high diversity, and low management level at the cost of low production. The benefit of constructing diversion systems is in high production at the costs of high management and low coverage. The essence of adding human-made patchiness as a management tool is, therefore, an optimization problem with different solutions in relation to productivity and diversity.

Implications for Management

We have demonstrated that patchiness controls population abundance, species richness, biomass, and flows of water and nutrients in the Negev. This implies that the management of patchiness, in order to control abiotic and biotic flows, is an important tool for the management of ecological systems in the Negev. Adding human-made patches to the landscape can be used for decreasing ecosystem leakage and increasing local biological diversity and productivity.

Adding human-made cultivated patches can restore key ecological features of degraded arid and semi-arid landscapes. *Desertification,* defined as a decline of landscape productivity and biodiversity (Graetz 1991), is usually associated with changes in the biological patchiness, such as a loss of higher plants and an increase of the area covered with microphytic soil crust as a result of heavy grazing and woody plant harvesting (Aschman 1991, Fuentes 1991, Graetz 1991). Constructing human-made patches to replace part of the microphytic crust will add patches that are rich in herbaceous plants, resulting in higher biological productivity and diversity of the landscape. In addition, human-made patches can be utilized for introduced plants, such as trees in the pits, in combination with denser natural vegetation in the pits and on the mounds. The trees can be used for firewood production as well as for shade for humans and livestock. The herbaceous productivity in the human-made patches can be utilized for livestock grazing.

Creating pits and mounds may also be a powerful tool for species conservation. If global climatic change results in xerification of arid regions (Schlesinger et al. 1990), extinction of species may ensue. Based on their positive effect on the richness of species, human-made patches may provide suitable refuges for species that require relatively high soil moisture, conditions common in the landscape prior to climate change. Care must be taken in designing and managing human-made patches to minimize the negative impact of nonnative and invasive exotic species on the landscape as a whole.

In conclusion, the ultimate objective of the management of ecological systems is to control the production of a species or community and the diversity of the community. This can be done by controlling the flows of water, soil, nutrients, and organisms. In the case study of pits and mounds, we demonstrated that physical, biological, and human-made patchiness controls the flows of water, soil, nutrients, and organisms in the Negev. Thus, managers, by controlling patchiness, can control the flows that in turn impact productivity and diversity of drylands.†

To increase our knowledge of patch management, future research should emphasize the theoretical and applied aspects of the relationships among patchiness, abiotic and biotic flows, and productivity and diversity of ecological systems. Theoretical studies should emphasize the above interactions in relation to substrate properties (e.g., sand vs. loess), scale, and environmental gradients. The practice of landscape management by adding human-made patches can further be improved if we better understand the dynamics of the physical and biological patches and the effects of their spatial configuration on the leakage of resources on various scales (Pickett and White 1985). Soil deposition and erosion and the reciprocal dynamics of macrophytic and microphytic patches are of particular interest. In addition, applied aspects of research in the field of patch management should emphasize the following kinds of studies:

- the effects of dryland settlement on patchiness, flows, productivity, and diversity of nonsettled landscapes;
- the management of patchiness in landscapes and its effects on flows in productivity and diversity;
- the effects of human-made cultivated patches on human and natural populations, communities, and ecosystems; and
- the effects of natural populations, communities, and ecosystems on human-made cultivated patches (e.g., through pest and weed dynamics).

Future studies on the relationships among patchiness, flows, productivity, and diversity could potentially generate a comprehensive conceptual scheme, along with models, methodology, and technologies for sustainable ecological management of drylands. The understanding of the generation, maintenance, and effects of patchiness in drylands is a powerful tool for desert conservation, restoration, and sustainable resource use.††

Note

We are grateful to Gary M. Lovett for helpful criticism of the manuscript.

References

Allen, E. B. 1991. Temporal and spatial organization of desert plant communities. *In* J. Skujins, ed. Semiarid lands and deserts: Soil resource and reclamation. New York: Marcel Dekker. 193–208.

Andrew, M. H., and R. T. Lange. 1986a. Development of a new biosphere in arid chenopod shrubland grazed by sheep. Pt. 1: Changes to the soil surface. Australian Journal of Ecology 11:395–409.

———. 1986b. Development of a new biosphere in arid chenopod shrubland grazed by sheep. Pt. 2: Changes in the vegetation. Australian Journal of Ecology 11:411–24.

Aschman, H. 1991. Human impact on the biota of Mediterranean-climate regions of Chile and California. *In* R. H. Groves and F. Di Castri, eds. Biogeography of Mediterranean invasions. Cambridge: Cambridge University Press. 33–41.

Boeken, B., and M. Shachak. 1994. Desert plant communities in human-made patches: Implications for management. Ecological Applications 4:702–16.

Bruins, H. J., M. Evenari, and U. Nessler. 1986. Rainwater harvesting agriculture for food production in arid zones: The challenge of the African famine. Applied Geography 6:13–32.

Coenen-Stass, D. 1981. Some aspects of the water balance of two desert woodlice, *Hemilepistus aphganicus* and *Hemilepistus reaumuri* (Crustacea, Isopoda, Oniscoidea). Comparative Biochemical Physiology 70A:405–19.

Evenari, M., L. Shanan, and N. Tadmor. 1983. The Negev: The challenge of a desert. London: Oxford University Press.

Feinbrun-Dothan, N., and A. Danin. 1991. Analytical flora of Eretz-Israel. Jerusalem: Cana Publishing.

Fletcher, J. E., and W. P. Martin. 1948. Some effects of algae and molds in the rain-crust of desert soils. Ecology 29:95–100.

Fuentes, E. R. 1991. Central Chile: How do introduced plants and animals fit into the landscape? *In* R. H. Groves and F. Di Castri, eds. Biogeography of Mediterranean invasions. Cambridge: Cambridge University Press. 43–49.

Graetz, R. D. 1991. Desertification: A tale of two feedbacks. *In* E. Medina, D. W. Schindler, E.-D. Schulze, B. H. Walker, and H. A. Mooney, eds. Ecosystem experiments SCOPE 45. New York: John Wiley and Sons. 59–88.

Hacker, R. B. 1984. Vegetation dynamics in a grazed mulga shrubland community. Pt. 1: The mid-storey shrubs. Australian Journal of Botany 32:239–29.

———. 1987. Species responses to grazing and environmental factors in an arid halophytic shrubland community. Australian Journal of Botany 35:135–50.

Halverson, W. L., and D. T. Patten. 1975. Productivity and flowering of winter ephemerals in relation to Sonoran desert shrubs. American Midland Naturalist 93:311–19.

Kedar, Y. 1967. The ancient agriculture in the Negev mountains. Jerusalem: Bialik Institute.

Linsenmair, K. E. 1972. Die Bedeutung familienspezifischer "Abzeichen" für den Familienzusammenhalt bei der sozalien Wustenassel Hemilepistus reaumuri (Crustacea, Isopoda, Oniscoidea) Z. Tierpsychology 31:131–62.

McIlvanie, S. K. 1942. Grass seedling establishment and productivity: Overgrazed and protected range soils. Ecology 2:228–31.

Metting, B., and W. R. Rayburn. 1983. The influence of microalgal conditions on selected Washington soils: An empirical study. Soil Science Society of America Journal 42:682–75.

Muller, C. H. 1953. The association of desert annuals with shrubs. American Journal of Botany 40:53–60.

Muller, W. H., and C. H. Muller. 1956. Association patterns involving desert plants that contain toxic products. American Journal of Botany 43:354–61.

Pickett, S. T. A., and P. S. White. 1985. Patch dynamics: A synthesis. *In* S. T. A. Pickett and P. S. White, eds. The ecology of natural disturbance and patch dynamics. Orlando, Fla.: Academic Press.

Price, M. V., and O. J. Reichman. 1987. Distribution of seeds in Sonoran desert soils: Implications for heteromyid rodent foraging. Ecology 68:1797–1811.

Reichman, O. J. 1984. Spatial and temporal variation of seed distributions in Sonoran desert soils. Journal of Biogeography 11:26.

Schlesinger, W. H., J. F. Reynolds, G. L. Cunningham, L. F. Huenneke, W. M. Jarrell, R. A. Virginia, and W. G. Whitford. 1990. Biological feedbacks in global desertification. Science 247:1043–48.

Shachak, M. 1980. Feeding, energy flow, and soil turnover in the desert isopod, *Hemilepistus reaumuri.* Oecologia 24:57–69.

Shachak, M., and S. Brand. 1991. Relations among spatiotemporal heterogeneity, population abundance, and variability in a desert. *In* J. Kolasa and S. T. A. Pickett, eds. Ecological heterogeneity. New York: Springer-Verlag. 202–23.

Shachak, M., S. Brand, and Y. Gutterman. 1991. Porcupine disturbances and vegetation pattern along a resource gradient in a desert. Oecologia 88:141–47.

Shachak, M., and A. Yair. 1984. Population dynamics and the role of *Hemilepistus reaumuri* in a desert ecosystem. *In* S. L. Sutton and D. M. Holdich, eds. The biology of terrestrial isopods. Symposium of the Zoological Society of London 53:295–314.

Teomim, N. 1990. Soil survey in the Sayeret Shaked Park. Gilat, Israel: Jewish National Fund.

Weinstein, N. 1975. The effects of a desert shrub on its micro-environment and the herbaceous plants. Master's thesis. Jerusalem: Hebrew University.

Went, F. W. 1942. The dependence of certain annual plants on shrubs in a southern California desert. Bulletin of the Torrey Botanical Club 69:100–114.

West, N. E. 1989. Spatial pattern: Functional interactions in shrub dominated plant communities. *In* C. M. McKell, ed. The biology and utilization of shrubs. London: Academic Press. 283–305.

———. 1990. Structure and function of microphytic soil crusts in wildland ecosystems of arid to semi-arid regions. Advances in Ecological Research 20:179–223.

Yair, A., and H. Lavee. 1982. Factors affecting the spatial variability of runoff generation over arid hillsides, southern Israel. Israel Journal of Earth Sciences 31:133–43.

Yair, A., and M. Shachak. 1987. Studies in watershed ecology of an arid area. *In* L. Berkofsky and M. G. Wurtele, ed. Progress in desert research. Totowa, N.J.: Rowman and Littlefield. 145–93.

Summary and Conclusion: Synthesis of Research and Management Implications

Thomas W. Hoekstra and Moshe Shachak

The purpose of the workshop on Arid Lands Management: Toward Ecological Sustainability that culminated in this volume was to establish the state of knowledge on arid lands ecological systems and their management for two uses. The first use of the information is for the International Arid Lands Consortium and similar granting institutions where it is necessary to comprehend the state of knowledge. That informaton is used to determine where to best expend limited resources for additional research in arid lands ecology. The second use is by arid lands managers who wish to implement the state of knowledge in their management programs.

A Framework for Arid Lands Ecological Systems

The subject of arid lands ecology is complex. Here we recommend a framework to assist in making the subject matter more tractable. We provide this concept for use as (1) a guide for understanding and changing scales of different ecological systems and among ecological systems and (2) to disaggregate the complex state of knowledge into useful segments. The framework uses several criteria for systematically considering management and research of ecological systems. The first criterion uses the distinction between predominant influences on the ecological system, namely, natural biophysical phenomenon, and human-engineered phenomenon. The second criterion defines what we refer to as a simple or singularly defined system versus a complex or multiple system—the distinction constitutes a change in scale of complexity. The third criterion for classification that is not scale-related explicitly identifies the kind of ecological system, that is, population, ecosystem, landscape, community, and so on. The fourth criterion is the spatial and temporal dimension of one or more kinds of ecological systems, obviously a scaling criteria. It is our expectation that the use of these criteria will reduce confusion regarding what is being managed or studied.

With the above framework as a guide, the authors have reviewed the chapters in this volume and have documented the following examples for management and research implications of arid lands ecological systems.

Management Implications

Arid and semi-arid ecological systems are more generally recognized today as ecologically sensitive systems rather than wastelands, as was the case in the past. Managers are often restoring systems from former uses (i.e., grazing, military, etc.) or manipulating systems where small incremental changes in the structure or process of a system can have the desired effects. In most ecological systems in Israel, many in the United States and Mexico, and some in Australia, active management is required to maintain the arid and semi-arid systems because of the important impact that humans have had and continue to have on these systems. The option of not managing is largely out of the question in all of these countries. Every chapter in this volume is a repository of information for managers of arid and semi-arid lands. The purpose of this section, therefore, is to lift out some of the key management considerations in the context of the above framework.

The knowledge to manage arid and semi-arid ecological systems is imperfect and inadequate—always has been and always will be. The consequence of that fact is explicitly acknowledged in the adaptive management approach to implementing what we know and continuing to improve our management with the addition of new knowledge. Adaptive management is further described in Hoekstra and Joyce, and managers should seriously consider the implications of implementing this concept. Risser suggests that managers need to develop indicators of sustainability, whether estimated or direct measures. The determination of when an ecological system is sustainable is a key to management success or failure. Ecological systems cannot be "almost" sustainable; they are either sustainable or not, and if they are not sustainable, they will change to some other kind of system and generally become less productive in meeting humans' needs. Sustainable management and sustainable development are only possible when ecological systems are sustainable.

Ecological systems generally have a surplus of some components in some places and times that are available for human use as natural resources. The kinds of natural resources available vary for different kinds of ecological systems (i.e., ecosystems monitor natural resource use in energy and matter metrics, communities score natural resource use in terms of species composition and competition, for example, and populations in terms of numbers of individuals and reproduction and mortality). Therefore, managers must measure the effects of their actions in terms of the metric of the ecological systems affected by their actions. These ideas are found in chapters in this volume (e.g., ecological systems can be managed for a variety of uses as described by Pickett et al., and management options vary with the characteristics of the ecological system, specifically communities, as discussed in Archer et al.). Population management as described by Saltz and colleagues requires understanding the contextual factors affecting the distribution and abundance of plants and animals. Joyce and colleagues further develop the idea that natural resources production interacts with ecosystem processes and states with implications for management of that ecological system.

More specific management implications are found in the chapters we have incorporated under the heading of "Progress toward Ecological Sustainability." Yair found that collecting water in arid systems using limans generally provides adequate water for use of trees planted there; however, annual variation in runoff means that sustainable afforestation must be based on long-term records. Sachs and Moshe describe the Israeli Savannization Project that has demonstrated the importance of a close relationship between research and management in the application of adaptive management. Shachak and colleagues develop the importance of managers' understanding of patch dynamics in controlling the flow of water, soil, nutrients, and organisms. The chapters by Farrow and Perevolotsky identify the difference between a long and short history of ecological manipulation related to pastoral systems. In Australia, with the shorter history, ecological systems are identifiably changing in response to grazing and alteration of forests to grasslands, while in Israel the grazing disturbance has been incorporated into the ecological systems and the disturbance would now be the removal of grazing. The comparison of these two studies indicates the effect that management can have on ecological systems and how important it is to know the current sustainable attributes of a system to be managed.

Four chapters provide the essentials of management planning for ecological systems in terms of natural resource production. The chapter by Fisher and colleagues identifies differing objectives and approaches to arid and semi-arid lands management in Israel, Australia, the United States, and Mexico. For example, the current approach in the United States is referred to as "ecosystem management" that explicitly attempts to consider the structure and function of different kinds of ecological systems and includes humans as integral components of systems. Israel's approach emphasizes interagency cooperation and closer ties between research and management. Land uses and associated management problems in Australia focus on grazing, tourism, mining and exploration, Aboriginal lands, and conservation considerations. Ffolliott and colleagues describe a planning process as a general context for consideration of when managers should employ the analytical approaches discussed by Hof (operations research) and Pickup and Stafford Smith (simulation) in their respective chapters.

Research Implications

Earlier in this summary, we indicated that managers should implement adaptive management in their planning and conduct of natural resources management. The rational for that recommendation was that we do not have all the information needed to make our natural resources management decisions. The direct implication for managers in adopting adaptive management is that the research and management communities must work together to determine what kind of information is missing and what the priority is for obtaining that information. This is not to imply that all research should be devoted to solving management needs, however. Hoekstra and Joyce identify the need for research to further elucidate the relationships among management actions, kinds of ecological systems and their interactions, and the scale of those actions and systems. They recognize that managers will influence a number of different ecological systems, potentially at several different scales, each time they manipulate the land; however, research has not demonstrated how to determine which systems will be influenced and at what scales. Managers need a straightforward process in that determination.

Risser identifies subject areas of research that have also been highlighted by the research community, including the following: paleoecology, past biogeographic patterns, multi-scales associated with ecological systems, global changes, biological diversity, and sustainability. Any improvement in our understanding of these subjects will be valuable to managers. From the management perspective, Fisher and colleagues outline the approach to research that considers management needs and identify the following research issues to be addressed: land stratification, water use, inventory, mixed land uses, and resource maintenance in contrast to resource enhancement. It is common for managers and researchers to develop different lists of research subjects; the important implication for research is how to fulfill the needs of managers while advancing the state of knowledge.

More specific research implications are identified by several authors. Arid and semi-arid ecological systems are sensitive to minor changes in atmospheric conditions, and certainly water is one of the most significant limiting factors in these systems. Nicholson identifies the priority atmospheric research as seeking improved understanding of atmosphere-land interactions. More specifically, Morin recommends additional research to enhance the determination and prediction of rainfall events and their interaction, using hydrologic models for improved estimation of timing and quantities of water. These atmospheric research needs are closely associated with the research implications that Joyce and colleagues identify for ecosystems. From their perspective of ecosystems, they believe that the dynamics of ecological system interactions and of ecosystem structure and function are insuffiently understood. Three additional research questions are identified: What is the optimal biomass removal? What is the functional significance of species? And what is the impact of management on the distribution of landscape components? Archer and colleagues, looking at ecological systems from the community perspective, suggest that we know little about the ecological attributes of communities that predispose them to abrupt change. They cite the example of grass-shrub interactions, about which research has been able to establish an adequate base of knowledge. Following on that theme but from an ecosystem perspective, Shachak and colleagues indicate that research should emphasize the theoretical and applied aspects of the relationships among patches, abiotic and biotic flows, and productivity and diversity of ecological systems.

Conclusion

The workshop that brought together an international group of managers and research scientists facilitated an excellent body of knowledge for use in management and for sparking the imaginations of research scientists interested in arid lands ecology. But this volume is not the complete story, for additional areas of ecology and especially social sciences are inadequately represented or absent. However, for the topics selected for the workshop and this volume, the various authors have accomplished a great deal. Several research scientists and arid lands managers left the workshop with plans to work collaboratively on subjects of mutual interest. In that regard, the workshop will continue to pay dividends.

Afterword: Reflections and Needs

Gene E. Likens

The following reflections and needs were derived largely on the basis of my reactions to presentations made at the workshop on 20–21 June 1994 in Jerusalem. These comments are not intended to be a detailed summary of the workshop but rather an attempt to provide some integration and elaboration of the general concepts presented and considered.

Reflections

Semi-arid and arid (drylands, according to the UNESCO definition) ecosystems are characterized by physical factors that lead to severe desiccation, such as (1) intense solar radiation, (2) small amounts and variable intensities and timing of precipitation, (3) variable amounts, rate, and depth of infiltration of water into the soil, and (4) wind. However, the structure, function, and change over time of dryland ecosystems are also regulated and modified in many important ways by the biota. Extreme variability in the abiotic and biotic components, that is, heterogeneity in space and time, is another major characteristic of these dryland ecosystems.

A major breakthrough in the ecological understanding of this variability came from Negev desert studies that discovered the functional role of microphytic and macrophytic patches of vegetation (e.g., Boeken and Shachak 1994, Shachak and Brand 1991, Shachak et al. 1991). The identification of this mosaic of patches provided a key concept toward understanding how the landscape functions and changes with time (that is, for example, how human management can convert deserts into new savannas, if indeed this change were desired). The Israeli Savannization Project is based on this developing understanding of the ecological role of patches in the desert landscape. There now appears to be a major need to quantify the structure, function, and change of these patches over time and in different environmental conditions. In addition, there should be quantitative studies of the effect of patchiness on the flow of water, nutrients, and soil. What are the effects of human-induced disturbances, such as grazing, on the structure and function of these patches?

Currently, the earth's surface is undergoing massive land-use changes that are now a major component of what I have called "human-accelerated environmental change" (Likens 1991). These changes are widespread and increasing because of human activities (fig. A.1).

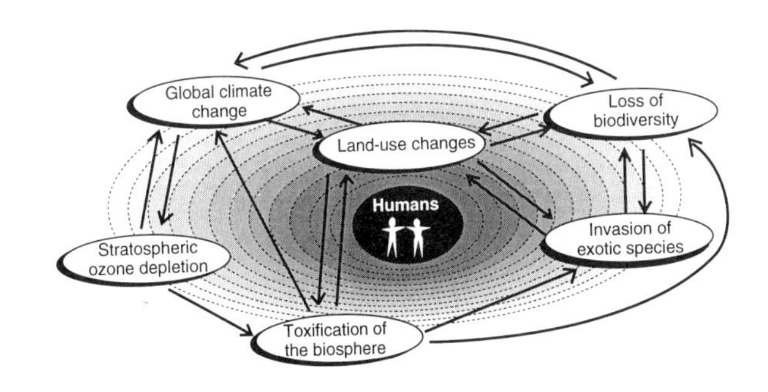

Figure A.1. Human-accelerated environmental change.

Two forms of land-use change, desertification and salinization, represent major environmental problems because they lead to deterioration of vital resources with regard to sustaining an ever-increasing number of humans and their insatiable appetite for energy, food, and water resources. Critical questions here are: What is natural and what is human-caused desertification and salinization? And what are the specific causes and rates of change?

It is important to define clearly our terms relative to these environmental issues. For example, because of their increased popularity and use, there often is much confusion about the terms *preservation, conservation, restoration, mitigation,* and *sustainability*. These terms have become "buzzwords" for politicians and the news media. For example, reference now is made to sustainable environment, sustainable development, sustainable forestry, sustainable yield, and so on. Even this workshop's title includes ". . . toward Ecological Sustainability." What exactly is meant by these terms? Simple statements of "ecological sustainability" or "environmental sustainability" are especially ambiguous. Minimally, clarification is required as to what stage, condition, or function of the ecology or the environment is to be sustained.

The concept of sustainability combines the simultaneous components of *use* of resources by humans and *protection* of those resources for future populations of humans. As such, as most commonly used, this concept is desirable only from an anthropocentric point of view. What about the needs of other organisms for these resources? There are clearly important ethical considerations regarding these issues as well. Thus, there is a need to manage for use and protection—but how is this—or can it be—done?

Because of aesthetics and other reasons, there has been a growing approach during the past twenty years or so of *protectionism* in the conservation movement. The idea simply expressed is that all that is needed is to acquire ownership of a piece of the earth's surface, put a fence around it, and it will be protected from deterioration into the future. Nonsense! This is simply an inaccurate view. Areas are not protected or "sustained" by a fence for two reasons: (1) ecosystems are living entities and, as such, they change naturally with time (e.g., Bormann and Likens 1979), and (2) pollutants moving in the biosphere, such as air or groundwater pollutants, do not respect boundaries of ownership or fences.

Ecologically acceptable (or sustainable) management has as a goal to mimic natural processes and natural disturbance of the structure and function of ecosystems. Obviously, to do this we absolutely must understand the sources and effects of natural disturbance in ecosystems.

In 1987, Cary Conference participants proposed and unanimously adopted a statement that incorporated a call for a new partnership between scientists and managers. This statement would seem to be applicable here as well. It reads in part:

> Because they have common long-term goals, we propose a new partnership between scientists and resource managers. Elements of this partnership include:
>
> (1) Agreement by scientists to answer the questions put to them by managers while making clear the level of uncertainty that exists and what additional research needs to be done.
> (2) Agreement by managers to give serious consideration to these answers and to support the continuing research toward better answers. (Likens 1989)

This interaction partnership for adaptive management (sensu Holling 1978) holds great promise for management. Again, ecosystems represent living entities and change with time. Understanding that these systems change with time means that management options and procedures may need to change as well. Both may change but, within the context of this partnership, one hopes these changes can be more efficient and less contentious. A common criticism from managers is, "Just give me a target toward which I can manage—the target keeps changing!" If the partnership is started *very* early in the process, scientists and managers can work interactively to set targets and to develop new information. Because of the enormous complexity of ecological systems and environmental problems, the search for reasonable solutions begs for such cooperation.

The idea of "balance of nature," of how ecosystems are structured, function, and change with time, has been replaced by another informal metaphor, "the flux of nature," as the current paradigm in ecology (Pickett and Ostfeld 1994). This paradigm is characterized by multiple equilibrium points and change and explicitly includes humans as an integral part of ecosystems. However, we ecologists usually search out the most remote lake, forest, grassland, or desert for study, while, in fact, humans now have an important role in ecosystems around the globe, either directly or indirectly (e.g., McDonnell and Pickett 1993, Turner et al. 1990).

The modern paradigm of ecology is that ecosystems (1) are open, (2) are regulated by processes arising outside their individual boundaries, (3) exhibit multiple equilibria or end points, (4) have multiple and probabilistic successions, (5) are subject to natural disturbances, and (6) incorporate humans and their effects (Picket et al. 1992, Pickett and Ostfeld 1994). The specific dynamics of any one system are contingent upon (1) the history, (2) the accidents of arrival and survival of a species at a site, and (3) the diversity and nature of connections of the ecosystem to the surrounding landscape (Pickett and Parker 1994). Therefore, ecological systems' structure, function, and change over time depend importantly upon their unique parts and their specific spatial and temporal influences, which constitutes their patch dynamics. Ecosystems, such as a forest, a lake, or a desert patch, are embedded in heterogeneous, changing, and influential landscapes and airsheds (Forman 1987, Likens and Bormann 1974). Thus, these paradigm contingencies may serve as a guide to long-term monitoring as they identify the important dynamics of ecosystems. For example, three of these critical dynamics are (1) the flux of water and chemicals that provide important connections with the biosphere, (2) the dynamics of patches, and (3) bacterial processes. These are critical because they help to indicate the structure, functions, and change in ecosystems.

There is great value in long-term studies, particularly at the watershed/landscape scale, for management. From such long-term information, it is possible to pose realistic and important questions whose answers can lead to management decisions. Based on a long-term perspective, it is possible to determine when an event is unusual. What are the effects of five years of drought in a row? What are the quantitative interactions among drought, salinization, and rate of irrigation? It is extremely difficult to achieve an overall *integrated* understanding for an ecosystem or landscape, but this still remains the *goal* (and is important to minimize the occurrence of some nagging, unanswered, yet critical questions) of ecology. Large-scale (e.g., watershed-scale) experiments are valuable for more realistic tests of management options and for validation of simulation models.

Recently, Mele (1993) calculated in *Polluting for Pleasure* the amount of oil that is added to surface waters in the United States by emissions from two-stroke outboard motors. He calculates that fifteen times more oil and hydrocarbons are added to these surface waters each year than were released in the Exxon Valdez oil spill in Alaska. This pollution is an excellent example of the "invisible" or subtle effect of humans in degrading natural ecosystems. An Exxon Valdez spill gets widespread publicity by the news media, but these "secret spills" (Mele 1993) do not. Indeed, by-products of human activity are found in and have degraded even the most remote parts of the earth.

Needs (Regarding Management of Drylands)

A list of needs regarding management of drylands includes:

1. Utilization of the ecosystem/watershed/landscape approach, incorporating components of physical, chemical, hydrological, geological, population, and community approaches into an *integrated* understanding of ecological systems.
   - Synthesis and integration of information is the hallmark of ecology (Likens 1992).
2. Better understanding of nutrient flux and cycling.
   - Is there a limitation of biological productivity by nutrients in drylands and/or colimitation with water?
   - What is the nutrient input from dew, fog water, and dust to dryland ecosystems?
   - Lowdermilk's (1944) "Eleventh Commandment" focuses on the quantity of water in drylands. We also need to focus on the quality of this water in terms of both nutrients and pollutants.
3. Quantitative information on the interaction among watersheds and airsheds.
   - What are these quantitative connections with the remainder of the biosphere?
   - Can we sustain the source/sink linkages?
4. Consider humans as integral components of ecosystems.
   - The increasing number of people and particularly the increase in per capita use of resources can no longer be ignored.
5. Reduce piecemeal approaches to problem solving and decision making.
6. Quantify ecological and biogeochemical dynamics of patches.
7. More effective utilization of data from large-scale experiments and models.
   - We need to try different combinations of variables to seek simplifying approaches to developing an understanding and better management of landscapes and regions.
8. Intelligent monitoring and sustained ecological studies that interact meaningfully with management programs.
9. Development of a long-term ecological research program along the rainfall gradient of < 100 mm/yr to about 300 mm/yr.

Scientific interaction, cooperation, and information can be important ingredients for sustaining communication and peace. In the Middle East, where water is the major limiting resource for humans, conflicts are commonplace. Such conflict will become a more common problem in other drylands in the near future.

At the beginning of the workshop, Moshe Rivlin stated that his dream fifty years ago was that the only "war" for Israel would be with the desert. I would suggest that, instead of war with the desert, ways must be found for more humans to harmonize with the desert.

References

Boeken, B., and M. Shachak. 1994. Desert plant communities in human-made patches: Implications for management. Ecological Applications 4:702–16.

Bormann, F. H., and G. E. Likens. 1979. Pattern and process in a forested ecosystem. New York: Springer-Verlag.

Forman, R. T. T. 1987. The ethics of isolation, the spread of disturbance, and landscape heterogeneity. *In* M. G. Turner, ed. Landscape heterogeneity and disturbance. New York: Springer-Verlag. 213–29.

Holling, C. W., ed. 1978. Adaptive environmental assessment and management. Chichester: John Wiley and Sons.

Likens, G. E., ed. 1989. Long-term studies in ecology: Approaches and alternatives. New York: Springer-Verlag.
Likens, G. E. 1991. Human-accelerated environmental change. BioScience 41(3):130.
———. 1992. The ecosystem approach: Its use and abuse. Excellence in ecology, vol. 3. Oldendorf/Luhe, Germany: Ecology Institute.
Likens, G. E., and F. H. Bormann. 1974. Linkages between terrestrial and aquatic ecosystems. BioScience 24(3):447–56.
Lowdermilk, W. C. 1944. Palestine: Land of promise. London: Victor Gollang.
McDonnell, M. J., and S. T. A. Pickett, eds. 1993. Humans as components of ecosystems: The ecology of subtle human effects and populated areas. New York: Springer-Verlag.
Mele, Andre. 1993. Polluting for pleasure. New York: W. W. Norton.
Pickett, S. T. A., and R. S. Ostfeld. 1994. The shifting paradigm in ecology. Ecological Environments 1:151–59. [in Hebrew]
Pickett, S. T. A., and V. T. Parker. 1994. Avoiding the old pitfalls: Opportunities in a new discipline. Restoration Ecology 2:75–79.
Pickett, S. T. A., V. T. Parker, and P. Fiedler. 1992. The new paradigm in ecology: Implications for conservation biology above the species level. *In* P. Fiedler and S. Fain, eds. Conservation biology: The theory and practice of nature conservation, preservation, and management. New York: Chapman and Hall. 65–88.
Shachak, M., and S. Brand. 1991. Relationships among spatiotemporal heterogeneity, population abundance, and variability in a desert. *In* J. Kolasa and S. T. A. Pickett, eds. Ecological studies, series 86. New York: Springer-Verlag. 202–22.
Shachak, M., S. Brand, and Y. Gutterman. 1991. Patch dynamics along a resource gradient: Porcupine disturbance and vegetation pattern in a desert. Oecologia 88:141–47.
Turner, B. L. II, W. C. Clark, R. W. Kates, J. F. Richards, J. T. Matthews, and W. B. Meyer, eds. 1990. The earth as transformed by human action. Cambridge: Cambridge University Press.

Contributors

Technical Editors

THOMAS W. HOEKSTRA is the director of the Inventory and Monitoring Institute, U.S. Department of Agriculture Forest Service, Fort Collins, Colorado.

MOSHE SHACHAK is a professor at the Jacob Blaustein Institute for Desert Research, Ben Gurion University of the Negev.

Senior Authors

STEPHEN ARCHER is a professor in the Department of Rangeland Ecology and Management at Texas A&M University, College Station.

ROGER FARROW retired in 1996 from the Division of Entomology, Commonwealth Scientific Industrial and Research Organization, Canberra, Australia, and now does private consulting in insect ecology.

PETER F. FFOLLIOTT is a professor in the School of Renewable Natural Resources at the University of Arizona, Tucson.

JAMES T. FISHER is a professor in the Department of Agronomy and Horticulture at New Mexico State University, Las Cruces.

KENNETH E. FOSTER is the director of the Arid Lands Studies Program at the University of Arizona, Tucson.

JOHN HOF is a research project leader at the USDA Forest Service Rocky Mountain Research Station, Fort Collins, Colorado.

LINDA A. JOYCE is a research project leader at the USDA Forest Service Rocky Mountain Research Station, Fort Collins, Colorado.

GENE E. LIKENS is the director and president of the Institute of Ecosystem Studies in Millbrook, New York.

JOSEPH MORIN served as a senior soil scientist at the Soil Conservation Research Center in Israel and as chief research advisor for the Soil and Water Conservation Authority of Israel.

DAVID NAHMIAS is the director of the Land Development Authority of Israel.

SHARON E. NICHOLSON is a professor of meteorology at Florida State University, Tallahassee.

AVI PEREVOLOTSKY is an ecologist and researcher at the Agricultural Research Organization of the Ministry of Agriculture in Israel.

STEWARD T. A. PICKETT is a researcher at the Institute of Ecosystem Studies in Millbrook, New York.

GEOFFREY PICKUP is a researcher at the Centre for Arid Zone Research, Commonwealth Scientific Industrial and Research Organization, Canberra, Australia.

PAUL G. RISSER is president of Oregon State University, Corvallis.

MENACHEM SACHS is the forest department director of the Land Development Authority of Israel.

URIEL N. SAFRIEL is director of the Jacob Blaustein Institute for Desert Research, Sede Boqer Campus of Ben Gurion University of the Negev.

DAVID SALTZ is an assistant professor of conservation biology at the Mitrani Center for Desert Ecology, Jacob Blaustein Institute for Desert Research, Ben Gurion University of the Negev.

A. YAIR is a professor in the Department of Geography at the Hebrew University, Mount Scopus Campus, Jerusalem.

Index

Typeset in 9/12 Stone Serif with Stone Serif display
Book design by Paula Newcomb
Composed by Jim Proefrock
at the University of Illinois Press
Manufactured by Cushing-Malloy, Inc.